The Science and Technology
of Polymer Films

VOLUME I

POLYMER ENGINEERING AND TECHNOLOGY

Executive Editor: D. V. Rosato
Editors: R. B. Akin, H.F.Mark, J. J. Scavuzzo,
S. S. Stivala, L. J. Zukor

The Science and Technology of Polymer Films

Edited by
ORVILLE J. SWEETING

Yale University

VOLUME I

Interscience Publishers

a division of John Wiley & Sons
New York · London · Sydney · Toronto

To my parents who
put up with me then

Rose E. Bailey, 1876–1935
Burton C. Sweeting, 1876–1939

and to Mary, Janet, Richard, and Barbara
who put up with me now

PREFACE

It is the hope of the authors that this book will be of value to a wide audience. An attempt has been made to correlate the recent past in polymer chemistry with current theories in such a way as to be comprehensible to a newcomer in the field of polymer films, yet not insult the more sophisticated. The subject is a growing and rapidly changing field of chemistry, and the kaleidoscope which chemists amuse themselves by calling "the literature" makes it difficult to present a basis for understanding the structure and properties of thin self-supported polymeric films. This is a field in which the crystal ball gets cloudy fast. To paraphrase a remark made long ago by Mr. Justice Pound, with reference to the law, polymer chemistry should strive to be stable, yet not stand still.

This first volume presents the scientific basis for polymer films, with special, though not exclusive, emphasis upon packaging films. The chemistry and physics involved are complicated and in many respects are in need of further study and refinement. In some instances, basic explanations are lacking. Nonetheless, we have tried to present as much theoretical material as is consistent with an understanding of packaging film principles and applications, without the inclusion of highly abstruse material or speculative theories which must in future rest or fall on experiments not yet done.

It had been originally planned that one additional basic chapter entitled "Barrier Properties of Polymer Films" would appear in Volume I, but because of extraordinary delays in putting this chapter into final form, it will appear in Volume II.

Volume II, now in active preparation, will bring together the packaging-film facts and raison d'etre for many of the important films in use today. Though much of this material may be found scattered in commercial brochures, books on polymer chemistry, and monographs on certain individual films, a modern comprehensive treatment of salient facts may be of value both to students and to specialists in the field.

I wish to express my thanks to all of the authors who have contributed, many of them former colleagues in the Film Division at Olin.   Thanks are due also to the late Edward L. Lynn, then General Manager of Olin's Film Division, who encouraged the initial effort, and to Philip C. Brownell and Dr. Linton E. Simerl who approved the project.   To all of these three who enthusiastically gave the idea life, I am grateful.

The editor welcomes comments and criticism.   In a book as complex as this, in a field as rapidly changing as this, errors must certainly have infiltrated.   I take full responsibility for them—and hope that they are small ones.

ORVILLE J. SWEETING

New Haven, Connecticut
May, 1968

AUTHORS

HARRIS J. BIXLER
   The Amicon Corporation, Cambridge, Massachusetts

GEORGE L. BOOTH
   Dilts Division, The Black-Clawson Company, Inc., Fulton, New York

JOHN G. COBLER
   The Dow Chemical Company, Midland, Michigan

VERNON C. HASKELL
   Film Department, Spruance Film Research and Development Laboratory, E. I. du Pont de Nemours & Co., Richmond, Virginia

PETER M. HAY
   Sandoz, Inc., Hanover, New Jersey

JAMES L. HECHT
   Research Supervisor, Film Department, Research and Development,
   E. I. du Pont de Nemours & Co., Buffalo, New York

RICHARD HENDERSON
   Olin Mathieson Chemical Corp., New Haven, Connecticut

HENRY J. KARAM
   The Dow Chemical Company, Midland, Michigan

JOSEPH J. LEVITZKY
   Packaging Division, Olin Mathieson Chemical Corp., New Haven,
   Connecticut

RICHARD N. LEWIS
   Stauffer Chemical Company, Adrian, Michigan

MERTON W. LONG, JR.
   The Dow Chemical Company, Midland, Michigan

JEAN B. MAURO
   U.S. Industrial Chemicals Co., Bridgeport, Connecticut

KARL W. NINNEMANN
   Fabricated Products Division, Allied Chemical Corp., Morristown,
   New Jersey

D. K. OWENS
   Film Department, Spruance Film Research and Development Laboratory, E. I. du Pont de Nemours & Co., Richmond, Virginia

E. Guy Owens II
The Dow Chemical Company, Midland, Michigan

F. J. Reidinger
Film Division, Olin Mathieson Chemical Corp., Pisgah Forest, North Carolina

Robert J. Samuels
Research Center, Hercules Incorporated, Wilmington, Delaware

Orville J. Sweeting
Department of Chemistry, Yale University, New Haven, Connecticut

Donald R. Walton
Rayonier, Inc., New York, New York

Eric Wellisch
Film Division, Olin Mathieson Chemical Corp., Pisgah Forest, North Carolina

CONTENTS

CHAPTER 1

INTRODUCTION

ORVILLE J. SWEETING*

Yale University, New Haven, Connecticut

I. General Aspects

The journal articles of 20 or 30 years ago occasionally reported the failure of an organic synthesis with a discouraging comment that failure was signified by formation of an intractable tar (usually black) from which no useful compound could be extracted. Who knows what gold was discarded by graduate students of yesteryear,

* Formerly Associate Director of Research and Development, Olin Film Division, Olin Mathieson Chemical Corp., New Haven, Conn.

1

always in search of simple, crystalline, easily characterized solids, melting preferably in the range 75–200°C?

In my day, the word *polymer* was a polite synonym for the gunks toward which many highly reactive compounds seemed to trend when heated at reduced pressure; the accursed polymers reduced the yields of useful compounds and resulted in the irretrievable loss of time and money.   Today chemists in industrial laboratories (and even a few in academic posts) are attracted by reports in the journals that a "worthless tar insoluble in all common solvents was formed in large amount."   Such a comment is a challenge to the chemist working with synthetic polymers, for here may be the clue to a stable, insoluble, nonhydrolyzable, high-melting polymer of value as a structural material.   It needs only to be made in quantitative yield and freed of the impurities that discolor it!

Today it is not necessary to explain even to most school children what a polymer is (they have a feel for the word), but large books are required to explain what polymers are, how they behave, and what they can do.   In a short historical introduction, Rowland Hill, writing in 1953 (1), commented that "high polymer chemistry, or macromolecular chemistry as it is sometimes called, can now very properly be accepted as a science in its own right."   Yet I can recall that in my own graduate school days, 10 years before, a faculty member of the department of chemistry where I was studying who wished to offer a graduate course in the chemistry of high polymers was, after due deliberation of the faculty, denied permission, on the grounds that polymer chemistry could be quite adequately taught within the subdivisions of chemistry then in vogue.

In this country, Herman Mark, formerly at Staudinger's laboratory at Freiburg Breisgou where they had collaborated on fundamental research into the structure of cellulose, and afterwards a director of research at I. G. Farben and professor at the University of Vienna, helped to publicize the field of high polymers, as these intractable organic compounds began to be termed, and put the study of them on a scientific basis.   He was audacious enough to entitle Volume I of a new series of abstracts *The Science of Plastics. A Comprehensive Source Book Based on the Original Literature for 1942–1946* (2).   Thus was the subject dignified: a "science" of plastics drawn from the original "literature"!   Now textbooks exist on the subject, eloquent testimony to the commercial worth of writing for those who learn of polymers in schools, (3,3a).   Although it is said (3) that the first chemists to define certain natural products as polymeric in the modern sense were H. Hlasiwetz and J. Habermann in 1871, Staudinger first introduced the term *macromolecule* (4) for these substances of high molecular weight, e.g., proteins, cellulose, and rubber.   In the 20 years which followed, many chemists developed new methods such as the use of x-rays for investigation of the fundamental properties of high polymers.   In the late 1920's, Wallace H. Carothers at the E. I. du Pont de Nemours & Co. Experimental Station in Wilmington, Del. began a series of brilliant researches on the synthesis of molecules of very high molecular weight.   The results appeared in a steady flow of articles in the *Journal of the American Chemical Society,* the work duly protected by United States and foreign patents (5).   One of the first patents in this field granted to Carothers (6) has rarely been equalled in its clarity of description of crucial experiments and the massive bulwark presented in basic coverage of an important scientific and commercial development.   For the first time, synthetic condensation polymers of nearly all imaginable sorts had been made and shown to have a useful degree of pliability, strength, and elasticity. Carothers had reexamined the pioneering work of Staudinger and by a close look at fundamentals, he made for the first time materials with useful (not merely interesting) properties.

Carothers writes (6):

. . . The synthetic linear condensation superpolymers produced in accordance with the present invention are suitable for the production of artificial fibers which are pliable, strong, and elastic and which show a high degree of orientation along the fibre axis.   In this respect they resemble cellulose and silk which, as recent researches have proved, are also linear superpolymers. So far as I am aware, no synthetic material has hitherto been prepared which is capable of being formed into fibres showing appreciable strength and pliability, definite orientation along the fibre axis, and high elastic recovery in the manner characteristic of the present invention.   It is true that Staudinger has frequently emphasized the probable structural analogy between polyoxymethylene and cellulose, and he has shown that it is possible to obtain polyoxymethylene in the form of oriented fibres, but these fibres are only a few millimetres in length and they are very fragile.   It is true also that threads or filaments can be drawn from any tough thermoplastic resin, and British Patent 303,867 discloses a process for making artificial silk in which a condensation product of a polyhydric alcohol and a polybasic acid or its anhydride is employed as a raw material.   British Patent 305,468 discloses a process for making synthetic fibres from a urea-formaldehyde resin. But there is nothing in the disclosures of these references to indicate that the filaments or fibres are sufficiently strong or pliable to have any utility, and insofar as I am able to ascertain, filaments or fibres produced in accordance with the disclosures of these patents do not have any useful degree of pliability, strength, or elasticity.

After defining the terms, condensation ["new bonds between atoms not already joined . . . with the elimination of elements (H_2, N_2, etc.) or of simple molecules (H_2O, C_2H_5OH, HCl, etc.)"], linear polymer, unit of a polymer ("—A—"), superpolymer, Carothers showed how polyesters, polyethers, polyamides, and polyanhydrides, as examples, could be made, theoretically with infinite chain lengths, but was careful to limit and discuss the polymerization reaction (6). He says:

> . . . It may be observed that each of these starting compounds contains two (and only two) functional groups capable of participating in the condensation. I call such compounds bifunctional compounds, and their reactions bifunctional reactions. My invention is concerned not with polyfunctional condensations generally, but only with bifunctional condensations. This restriction is necessary since the presence of more than two functional groups in any of the reacting compounds introduces the possibility of developing a three-dimensional polymeric structure; and this involves a complication with which my invention is not concerned.
>
> It is a characteristic feature of bifunctional condensations . . . that they present the formal possibility of producing molecules of infinite length. Thus the self-esterification of ten molecules of hydroxy acid,

$$HO—R—CO—OH$$

> would lead to the formation of the polyester,

$$HO—R—CO—(O—R—CO—)_8—O—R—CO—OH$$

> and this, since its molecule still bears the terminal groups which were responsible for the initial reaction, is potentially capable of reacting with itself to produce a new molecule twice as long. A continuation of this progressive coupling would finally yield a molecule of infinite length.
>
> In practice there are several factors that may be expected to interrupt this progressive coupling before the molecules have grown to any very great length. The following may be mentioned:
>
> (1) Reaction may be intramolecular at some stage (i.e., it may lead to the formation of a ring).
>
> (2) The terminal functional groups responsible for the progressive coupling may be lost or mutilated through side reactions.
>
> (3) Mechanical factors such as solubility and kinetic effects may come into play.

I have devoted considerable study to reactions of the type defined above as bifunctional condensations and have discovered the following facts: Such reactions are usually exclusively intermolecular at every stage, and the terminal groups responsible for the initial reaction are still present at the ends of the product molecule. Interruption of the progressive coupling through intramolecular reaction (ring formation) occurs generally only through the formation of 5-atom rings, less frequently through the formation of rings of 6 atoms, and rarely through the formation of rings of 7 or 3 atoms. The loss or mutilation of terminal groups through side reaction occurs only when patently inappropriate experimental conditions are adopted. Thus in most cases there is nothing theoretically to preclude the possibility of producing exceedingly long molecules in bifunctional condensations.

It is well known that reactions such as esterification, ester interchange, amide formation, etherification, anhydride formation or acetal formation are reversible reactions, and that such reactions can be forced to completion by the separation of the reaction products as they are formed. But from the facts outlined above it follows that if the reversible reaction is a bifunctional condensation, the degree of completeness of the reaction will regulate the size of the molecule in the polymeric product; the more nearly complete the reaction, the longer the molecule; and the reaction can be absolutely complete only when the product molecule is infinitely long. According to the present invention reactions of this type are brought to a degree of completeness heretofore unknown.

This invention has as an object the preparation of linear condensation superpolymers, and by this I mean linear condensation polymers capable of being formed into useful fibres. A second object is the manufacture of synthetic fibres. A third object resides in a method of propagating reversible chemical reactions involving the simultaneous formation of volatile and non-volatile products.

The first of these objects, briefly expressed, is accomplished by subjecting the linear polymers resulting from reversible bifunctional condensations to the action of heat under conditions which particularly facilitate the removal of any possible volatile reaction products. The second of these objects is accomplished by spinning or drawing filaments from a synthetic linear condensation superpolymer. The third object is accomplished by utilizing a molecular still for removing the volatile product or products of a reaction where their effective vapor pressure is too low to be removed by the usual distillation technique and where the non-volatile product is substantially completely non-volatile.

It may be observed that the results described herein furnish a very satisfactory verification of my theory. It is to be understood, however, that I do not desire the claims to be limited thereby inasmuch as the theory has been presented in detail merely to make clear the nature of the invention and especially to explain the terminology used in describing it.

Carothers did much more than prepare and describe a few "superpolymers." He showed in summary that a polymer capable of forming strong fibers must consist of very large molecules, of molecular weight 10,000–12,000 and a chain length of 1000 Å or more—and furthermore that it must be capable of crystallizing, this being a property possessed by linear polymers which have no bulky side groups to destroy symmetry. He explained why three-dimensional polymers such as the glyptals then in common use could never be used as fiber-forming materials. And to complete the coverage, Carothers described a molecular still which he used for propagating the chemical reaction, the first use of this device for such a purpose.

Yet one may find it stated in a reputable book dealing with polymerization processes published as late as 1948 that polyesters are of little value as fibers (7). True, Carothers had prepared many superpolyesters from glycols and dicarboxylic acids during the early thirties and had concluded that the linear polyesters may be too low melting to be good fiber-forming materials. But these authors went so far as to explain why this should be so on theoretical grounds (7).

> The fact that after years of research it was finally possible to produce wholly synthetic fibers was highly gratifying; but, for all that, the results were disappointing. Polyester filaments are of little practical value. Their softening points are so low that they melt at ironing temperatures. They are not strong enough to compete with cotton and silk, and they are not sufficiently resistant to solvents to withstand dry cleaning and laundering.
>
> In retrospect, it is easy to explain what required years of research to discover. Carothers finally abandoned the attempt to produce commercial polyester textile fibers. Their shortcomings, enumerated above, are all accounted for on the simple assumption that the inter-molecular forces of attraction are not strong enough. The problem, then, is to make linear polymers for which these forces are greater. As discussed earlier, four factors may be exploited toward this end: (1) increase in degree of polymerization, (2) increase in molecular symmetry, (3) increase in degree of orientation, (4) increase in polarity of the mer. In his experiments with polyesters Carothers had about exhausted these potentialities. He had produced streamlined polyesters of high molecular weight and oriented them by elongation under tension, but they were still inadequate.

All of which simply proves that there are times when research should be reexamined, along with the utter foolishness of making categorical scientific statements. But beyond that, this quotation ignores the interesting set of facts discovered in the laboratories of the Calico Printers Association by Whinfield and Dickson in 1941, who found that the tidy generalizations about polyesters did not apply to certain polyesters of terephthalic acid (8).

Carothers and other early chemists who worked with polymerizable molecules stressed their fiber-forming properties, since much effort in those days was directed toward the discovery of synthetic substitutes for silk. And it soon became apparent that in many instances a satisfactory fiber former was also capable of being formed into a transparent film under appropriate conditions.

For about 20 years, regenerated cellulose under various trade names (Cellophane, Transparit, Diophane) had been almost unchallenged for transparent film applications, particularly for packaging. Competitors such as cellulose acetate were too expensive to be used extensively. The discovery of a strong polyester (polyethylene terephthalate) by Whinfield and Dickson (8), termed Terylene in Britain, and soon to become Mylar and Cronar in the United States under development by the du Pont Company, posed the first serious threat to regenerated cellulose film in many markets (some of which cellophane met badly).

Yet it was the advent of polyethylene, first in a trickle in the late 1940's and early 1950's, soon to become a flood of many varieties of polyethylenes in huge quantities in 1955–1960 that provided a major threat to the commercial use of cellophane.

The first commercial use of polyethylene had been based on its flow characteristics and dielectric properties which made it suitable for extrusion coating of wire and cable. The entire available supply was preempted during the war for radar equipment.

In the second phase of development, the growth of polyethylene as a film was based on flow properties and also on its resistance to failure when stressed. We pass quickly over 15 years of development, a period in which special resins and special processing methods were able to eliminate fisheyes, haze, color, odor and to improve transparency. Gradually polymers and films with acceptable toughness were available (toughness usually taken to include tear strength and impact strength) for a burgeoning transparent film packaging market.

Polyethylene competed successfully with cellophane in properties as well as in cost, and at the present time it appears that it will eventually take most, if not all, of the transparent film market in many large-scale applications. In recent years, other competitors of cellophane have appeared, but none has posed the same serious competition as polyethylene.

The rise and fall of the cellophane market is shown in Tables 1–4. Except for the war years, a steady gain of about 3–5% a year was registered until the market faltered in 1957, recovered briefly, and then in 1961 nosed sharply down.

Food application (Table 2) dropped 24 million pounds in 1961 and 24 in 1962 for a total decrease of 48 million pounds from the all-time high of 339 million pounds in 1960. In 1962, six leading food categories consumed 59.4% of all cellophane and accounted for 80% of all food applications (Table 3). The loss of any one of the large categories was sure to have a profound effect on cellophane sales. The overall drop was the result of steady loss of the bread, cake, and sweet dough wrapping market, most of it to polyethylene.

In 1962, tobacco, paper products, textiles, hosiery, and laundry accounted for more than half of all non-food consumption of cellophane (Table 4). Except for tobacco, about which a question still

TABLE 1
Cellophane Produced

| Year | Millions of pounds | Increase over prior year | |
		Millions of pounds	% per year
1940	111		
1945	127	16	2.9
1950	265	38	2.7
1954	360	95	7.2
1955	375	15	4.2
1956	398	23	6.1
1957	388	−10	−2.5
1958	403	15	3.9
1959	436	33	8.2
1960	439	3	0.7
1961	423	−16	−3.7
1962	404	−19	−4.5
1963	405	1	0.2
1964	410	5	1.2
1965	405	−5	−1.2
1966	395	−10	−2.5

TABLE 2
Cellophane Uses

Year	Foods		Nonfoods		Total weight
	Weight[a]	Per cent	Weight[a]	Per cent	
1955	277	73.8	98	26.2	375
1956	299	75.0	99	25.0	398
1957	287	73.9	101	26.1	388
1958	306	75.9	97	24.1	403
1959	329	75.5	107	24.5	436
1960	339	77.1	100	22.9	439
1961	315	74.5	108	25.5	423
1962	291	72.0	113	28.0	404
1963	324	80.0	81	20.0	405
1964	324	79.0	86	21.0	410
1965	304	75.0	101	25.0	405
1966	296	75.0	99	25.0	395

[a] In millions of pounds.

exists, cellophane never did meet these markets very satisfactorily, and they were vulnerable to the inroads of polyethylene. Tobacco application in 1962 was about equal to its 1961 peak and textiles, etc. use was about equal to the 1959 peak. Paper products had their first even year in 1962 since poundage started to fall off seriously in 1957.

TABLE 3
Cellophane Applications in Certain Food Packaging

Applications	Per cent					
	1957	1958	1959	1960	1961	1962
Breads	15.8	17.5	17.8	18.5	15.4	12.6
Cakes and sweet doughs	10.5	10.4	9.9	9.6	8.4	7.6
Cookies, crackers, biscuits	8.0	7.3	7.2	7.3	7.2	7.5
Confectionery and gum	6.7	6.7	6.4	6.4	6.5	6.7
Meat, poultry, fresh fish	15.9	16.2	16.8	17.1	16.7	16.7
Popcorn, chips, other snack items	5.5	5.4	5.4	5.7	6.9	8.3
Totals	61.9	63.5	63.5	64.6	61.1	59.4

TABLE 4

Cellophane Applications in Certain Nonfood Packaging

Applications	1962[a]	Per cent of industry total	1962 Poundage loss[a]
Tobacco	45.1	11.2	0.1
Textiles, hoisiery, laundry	7.4	1.8	0.1
Packing of paper products	4.2	1.1	0.2
Totals	56.7	14.1	0.4

[a] In millions of pounds.

Nonpackaging uses of cellophane (mostly in pressure-sensitive tape) have for many years consumed 9–12 million pounds annually, or about 2.6% of the total produced for all purposes.

In contrast to what was happening to cellophane, polyethylene film markets grew enormously. After a beginning of 4 million pounds in 1948, consumption grew slowly. In 1953, only 35 million pounds of film for all purposes was sold, and it was estimated that the markets were largely unsatisfied, the potential being above 400 million pounds a year. By 1950, sales had reached 255 million pounds; by 1963, 455 million pounds; and by 1966 an estimated 730 million pounds (Table 5). Much of this market can be accounted for by the population increase, and by changing marketing and packaging habits, but a substantial chunk came out of the markets of competitive films.

The rest of the film market was divided in 1950 as follows: about 10 million pounds each to cellulose acetate, Pliofilm, and vinyls; in 1960: 35 million to sarans and vinyls, 8 million pounds to polyesters, and 6 million pounds to all others. By 1965, these poundages had increased very little for acetate and Pliofilm, but vinyls had about doubled, sarans had increased to about 30 million pounds, polyesters to about 20 million pounds, and all others to about 15 million pounds annually. In addition, a newcomer, polypropylene, sold about 40 million pounds for packaging. Table 6 summarizes the best estimates available of film poundage consumed by packagers from 1960 to 1967.

Thus, at the present time polyethylene is the transparent film

TABLE 5
Consumption of Polyethylene Film in Packaging in 1965 and 1966[a,b]
(Millions of pounds)

	1965	1966
Food packaging		
Candy	16	20
Bread, cake	80	85
Crackers, biscuits	5	8
Meat, poultry	16	20
Fresh produce	145	160
Snacks	2	2
Noodles, macaroni	6	7
Cereals	3	4
Dried vegetables	8	10
Frozen foods	14	19
Dairy products	8	12
Other foods	7	13
Total food uses	310	360
Nonfood packaging		
Shipping bags, liners	80	95
Rack and counter	40	50
Textiles	60	90
Paper	25	30
Laundry, dry cleaning	60	75
Miscellaneous	40	30
Total nonfood uses	305	370
Total packaging uses	615	730

[a] Based on industry estimates. One source, based on a private survey of film extruders, reports 1965 production of polyethylene film (millions of pounds) as follows: bakery, 120; produce and meat, 160; frozen food, 10; other food, 17; shipping bags, 12; drum and box liners, 80; rack, counter, laundry, dry cleaning, 100; textiles, 50; other nonfood, 100; for a total of 649 million lb. This source points out that partial-year tariff reports indicated a plus 800 million lb production for 1965. Three-fourths of the polyethylene film production is customarily considered packaging. The one sure conclusion: polyethylene film is big, growing, and size at present is hard to pinpoint.

[b] From *Modern Packaging Encyclopedia*, 1967, McGraw-Hill, New York, p. 27.

TABLE 6
Consumption of Films in Packaging, 1960–1966[a]

	1960	1961	1962	1963	1964	1965	1966
Cellophanes	439	423	404	405	410	405	395
Polyethylenes	275	370	400	455	500	615	730
Polypropylenes		3	15	25	28	40	45
Polystyrenes	3	5	7	11	15	15	15
Pliofilm	12	13	14	15	15	15	15
Vinyls	18	20	22	24	30	40	40
Sarans	17	19	21	23	26	30	30
Polyesters	8	8	10	10	10	20	20
Cellulose acetates	10	10	11	12	13	15	15
Total packaging uses	782	871	904	980	1047	1195	1305

[a] Millions of pounds. Compiled from various sources and normalized to agree with certain reliable figures.

sold in largest poundage, and cellophane is a strong second. Other films which are major factors in the market include cellulose acetate, Pliofilm (rubber hydrochloride), vinyl chloride and copolymers, sarans (vinylidene chloride copolymers), Mylar (polyethylene terephthalate), ethylcellulose, polypropylene, and polystyrene.

Cost to the packager, based on coverage (cents per 1000 sq in.) is shown in Table 7.

II. General Film Specifications

Fifteen years ago a great deal of discussion took place concerning the properties of an ideal film. This terminology signified that the chemical and physical properties of a film should be such as to require no modification (as by coating or laminating) to secure complete protection for the item packaged. But as packaging became diversified and the range of materials to be packaged broadened, the number of specific requirements grew enormously. One hopes for as many of the desirable properties in a single film as possible, but it is probably safe to say that no packager today believes that an unmodified film is likely to meet all requirements. The ideal film, like the ideal gas, probably will never be found in practice, but nonetheless the concept is a useful one.

The general specifications for a packaging film can be outlined in

TABLE 7
Comparative Costs: Films, Foils, Laminates[a]

Material	Cost,[b] $/lb	Coverage, in.²/lb	Cost, cents/10³ in.²
Cellophane			
MS, 195[c]	0.64	19,500	3.3
MS, 220	0.64	22,000	2.9
MSA or MSB, 195	0.71	19,500	3.6
Saran-coated, 140	0.79	14,000	5.6
Saran-coated, 195	0.79	19,500	4.1
Saran-coated, 250	0.73	25,000	2.9
Polyethylene-coated, 182	0.75	18,250	4.1
Vinyl-coated, 210	0.69	21,000	3.3
Cellulose acetate			
Cast, 1 mil	0.97	22,000	4.4
Extruded, 1 mil	0.74	22,000	3.4
Cryovac S, $\frac{6}{10}$ mil	1.09	27,300	4.0
Fluorohalocarbon, 1 mil	6.60	13,000	50.8
Nylon, 1 mil	2.15	24,500	8.8
Pliofilm, $\frac{3}{4}$ mil	0.95	33,000	2.9
Polycarbonate, 1 mil	2.07	23,100	8.9
Polypropylene			
Cast, 1 mil	0.59	31,000	1.9
Bioriented, $\frac{3}{4}$ mil	1.14	40,700	2.8
Polystyrene, oriented, 1 mil	0.63	26,300	2.4
Polyester			
Nonheat-sealing, 0.92	1.80	21,500	8.4
Nonheat-sealing, $\frac{1}{2}$ mil	2.09	40,000	5.2
Heat-sealing,[d] $\frac{1}{2}$ mil base	2.00	27,600	7.3
Polyethylene			
Low density, 1 mil	0.34	30,000	1.1
Heat shrinkable,[e] 1 mil	1.00	30,000	3.3
Medium density, 1 mil	0.38	30,000	1.3
High density, 1 mil	0.50	29,000	1.7

(continued)

[a] After *Modern Packaging Encyclopedia*, 1967, p. 26.

[b] Based on reported 1966 prices for representative standard or basic materials and substantial orders. Many other grades, types, gauges, combinations, and resulting price variations exist.

[c] MS indicates moistureproof, heat sealing.

[d] Polymer-coated, bag-making grade.

[e] Special resins and processing are required to develop shrink properties. Cheaper films are available.

TABLE 7 (*continued*)

Material	Cost,[b] $/lb	Coverage, in.²/lb	Cost, cents/10^3 in.²
Polyethylene–cellophane			
1 mil/195 MS	1.05	11,800	8.9
Polyethylene–polyester			
1½ mil/½ mil	2.00	13,200	15.2
Saran, 1 mil	1.08	16,300	6.6
Vinyl			
Cast, 1 mil	0.83	21,600	3.8
Extruded, 1 mil	0.55	21,500	2.6
Water-soluble film,[f] 1½ mil	1.45	15,000	9.7
Aluminum foil			
0.00035 in.	0.65	29,300	2.2
0.001 in.	0.56	10,250	5.5
Foil-acetate, 1 mil/1 mil	1.49	6,750	22.1

[f] Polyvinyl alcohol and polyethylene oxide films. Calculation is based on alcohol.

terms of physical and chemical properties, grouped functionally in relation to uses. Each application represents a compromise of requirements, with economy often one of the most important determining factors. One should note, however, that cost *alone* should never decide, since a package that does not protect is simply not an acceptable package, regardless of how low the cost may be. Function and performance, therefore, must control any decision about the optimum packaging material for a specific application.

It is also easy to leap to erroneous conclusions, for an approach to the ideal in one set of properties may not ignore the existence of other seemingly unrelated properties. As the underlying physical properties that control behavior of polymer films become better understood, it becomes clear that gross behavior characteristics which the practical man may regard as unrelated are indeed mutually exclusive on basic chemical or physical principles. An example which may be cited is the specification that a packaging film shall be chemically inert, and as corollaries of this, the additional specifications that the film shall be tasteless, odorless, nontoxic, resistant to solvents, fats and oils, and impervious to gases. Less obvious, however, are the facts that a film which is inert by this standard cannot readily be printed or sealed, and that it cannot be readily plasticized to modify its properties. Further, the ease with which a film develops troublesome static charges appears to be largely a function of the ease of removal of electrons from its surface and the inability of the electrons to traverse the film surface. The same inertness minimizes any tendency of the film surface to absorb moisture which would act to permit dissipation of static electricity.

In the selection and design of packaging films, approximately thirty criteria can be applied. These deal with appearance, physical properties, chemical properties, and economics. The precise manner in which properties of materials in use are considered will depend upon the specific application, but we should consider here the broad aspects of film specifications. Specific considerations will be apparent in the reading of subsequent chapters.

III. Appearance

A. Transparency and Gloss

For most applications a polymer film should be transparent, preferably optically clear and free from streaks, haze, blush, or an off-white color (colored films must have full color values, lest they look merely old and soiled). This specification requires normally a high gloss which will result in live printing.

Commercial films which meet this requirement include polystyrene, ethylcellulose, sarans, vinyls, nylons, cellophane, Kel-F, polyvinyl alcohol, butacite, Pliofilm, cellulose nitrate and acetate, polyethylene, polypropylene, polycarbonates, and polyesters. At one time it was believed that a high degree of crystallinity (as revealed by x-rays, cf. Chapter 2) was inimical to optical clarity, but methods are now known to control both crystallinity and clarity, notably by orientation of the ordered regions or by controlling the size of crystallites.

B. Printability

A film should be printable by standard techniques on current equipment, at modern press speeds, and with little or no press modification. The film surface should yield high resolution with finest screens and gradation, be free from bleeding, smudging, offsetting, flaking, and pinholing, and be resistant to stripping tests (such as the Scotch-tape test commonly used in pressrooms).

This requirement implies that the film surface not be inert to wetting, to solvents, or to temperatures above room temperatures, yet requires that the film not be wet well enough by inks as to flow and spoil fine screen printing. Ink formulation and surface treatments of films have reached such a state of refinement today that polymers once considered unprintable (e.g., polyethylene and polypropylene) can be treated satisfactorily.

IV. Physical Properties

A. Stability

In general, useful films must possess dimensional and compositional stability. Or if they are to be used as shrink-wrapping materials, the films must be stable dimensionally at one condition, be predictably unstable at another, and finally be stable when on the package.

Film should be stable in storage for at least 6 months at temperatures up to 120°F in stacked packages. Color, heat sealability, blocking, transparency, and flexibility should not suffer adversely. The film should not lose plasticizers, or bleed inks or other constituents under normal warehousing conditions.

Polymers used as films should not depolymerize appreciably below approximately 400°F, nor should they be degraded by light and other forms of ionizing radiation, nor should they be readily hydrolyzed or oxidized by oxygen, ozone, water vapor, or dilute acids and bases. Formation of monomers or of a polymer of low-molecular weight normally results in objectionable odor, color, or taste and loss of strength and utility. Serious failure may cause the film to run afoul of regulations issued under authority of the federal Food, Drug, and Cosmetics Act.

In general, the film should be so constructed and processed that environmental conditions have no adverse effect. Dimensional stability will depend to a large extent upon how well equalized in machine and transverse directions are the physical properties of the sheet such as tensile strength and elongation, as well as the manufacturing parameters such as thickness.

For use in food packaging, it is especially desirable that films contain only additives that are not extractable by water, hexane, dilute acids, or fats, or that any material so extracted be unexceptionable from a health standpoint (whether the matter extracted be monomer, polymer of low molecular weight, plasticizer, dye, antioxidant, or other material). If heating or exposure to light, for example, reduces stability, the extractable or migrating components must be studied.

B. Gas Transmission

The permeability of films to gases can be varied widely by varying polymer composition and by processing, including coating. Thus, the industry has available packaging films whose gas transmission rates vary from near-zero to high rates. Furthermore, it is possible to provide for low transmission of water vapor with concomitant high transmission of oxygen, or for near-zero transmission of oxygen accompanied in the same film by very high rates of transfer of carbon dioxide. Chapter 12 covers this subject in detail.

C. Durability

An acceptable scientific definition of durability is still wanting, although all agree that standards of durability are desirable for polymer films.

Durability for any specific application appears to rest on a complex interplay of film tenacity, capacity for deformation, impact strength, resistance to abrasion, capacity to endure folding and flexing, and possibly other more subtle factors. In addition, the mode of packaging, environmental conditions, cyclical changes and the severity, character, and frequency of physical stresses on the package are also important variables.

D. Tenacity

To be useful as a packaging film, it is desirable that the sheet have a minimum tensile strength of approximately 15,000 psi, i.e., 15 lb per linear inch for 1-mil film. But some useful films have much lower tensile strengths, even as low as 2000 psi, which limits their use in certain applications. Yet, some other desirable property or properties (as, for example, a high oxygen permeability or exceptionally low oil transmission rate) may promote their use in special ways. The tensile strength should undergo little decrease with increase in temperature within a useful range (e.g., from 20 to 100°C).

Tensile strength in polymers is doubtless related to the orientability and ordering of the individual chainlike molecules (see Chapter 2).

To get orientability, the chains must display a high order of flexibility but this is not the only factor which is necessary. Natural rubber at break has a tensile strength as high as 50,000 psi, but this orientation is very sensitive to temperature. Thus one must think of tensile properties as if tenacity has a rate of exchange for elongation and, moreover, presume that such developed tenacities may be easily exchanged for retractive forces.

Usually, to realize highest polymer strengths stable at temperatures up to 200°C, bonding between chains to stabilize crystal structure is sought. Likewise a high and uniform frequency of interacting functional groups along the chain is desirable. Thus, 6,6-nylon and polyethylene terephthalate fibers are found to give strengths of 80,000–200,000 psi, whereas the strongest polyethylene ever made had about 65,000 psi. Certainly it is not possible to say how much regularity or with what frequency functional groups should appear along the chain to give any stated tenacity. Let it suffice to say that such strength seems to require orientation to develop it and in most polymers this does not necessitate very much orientation.

E. Tear Strength

In most applications, film must have a moderately high tear resistance (2–16 oz for 1-mil film) but no arbitrary values can be set. As tear strength has increased in recent years, packages have been criticized for being troublesome to open, but by proper package design most of these objections can be overcome.

Tear strength appears to be related to chain flexibility, tensile strength, and elongation. Tear appears to bear a relation to tenacity in much the same way that brittleness is related to hardness. Such qualitative reasoning implies that polymer films of only moderate tenacities should be sought, in order to avoid brittleness, but even such tenacities should be accompanied by moderate elongation (30–70%).

F. Impact Strength

The durability of a film must be sufficient to withstand shipping vibration and sudden shock. This requirement implies a material which is capable of distributing stresses within the polymer without rupture. Measurement of impact resistance, especially resistance to stresses suddenly received, is discussed in Chapter 14; much work is still necessary on the physics of stress–strain distribution at high speeds. A film that is sufficiently extensible and flexible under low rates of loading may fail wretchedly under the same load at a faster rate or a lower temperature.

G. Sealability

Most packaging film must be amenable to sealing by heat, solvents, or adhesives without appreciable loss in seal strength with wide changes in humidity. This specification requires that the film show no significant softening below approximately 100°C, and that the melting or softening range within which good heat seals result lies between 100 and 200°C in the body of the film. The melting point of the film or its heat-sealable coating should not be too sharp, since a broad temperature range permits simpler operation. Further, the film surface should not lift or decompose under the heat-sealing implement, nor leave any deposits on the heated parts. There should be no thinning, weakening, or depolymerization of the film in or near the heated areas.

It should be noted that sealing with solvents or adhesives is contrary to other requirements that the film shall be resistant to absorption of or modification by solvents, fats, or organic vapors. This idea, however, must in practice be intentionally limited to render the film capable of being sealed, printed upon, or laminated to other materials.

H. Lubricity (Slip)

In general, a film should have good slip from the standpoint of packaging machine operability and easy separation of sheets and fabricated bags, yet the finished product should be capable of stacking without undue slippage. Satisfactory coefficients of friction fall often between 0.30 and 0.35. A nice balance of surface properties is required, but they are elusive to define. Chapter 9 discusses these problems in detail.

I. Stiffness

A wide range of stiffness is desirable and can be obtained either by polymer modification or by additives (plasticizers). Control is essential to permit the film to operate satisfactorily on automatic packaging machines as well as to satisfy subjective judgments of the final package. For instance, a tight hard wrap is desired in the packaging of hardware and cookies, but a supple, soft wrap is desired for soft goods such as sweaters. Packaging machines have been modified in recent years and new machines developed to such an extent that the limitations of modulus and slip which governed operations 10 years ago are no longer controlling. The use of films of higher coverage yield in recent years will undoubtedly continue to expand, thus prompting further machine adaptations.

J. Abrasion Resistance

Films should have abrasion resistance such that packages that fail in normal handling always fail in tensile, tear, or impact, but never from abrasion of the surface. The film should be resistant to marring, crazing, scratching, and creasing in normal handling. There should be an absence of pinholes or a tendency to form pinholes in handling.

It is possible that abrasion resistance is not a separate durability characteristic but is a combination of lubricity, hardness, impact strength, modulus, and perhaps other factors measurable in a fundamental way. Further study should be given to this factor.

K. Dielectric Properties

For certain obvious applications, dielectric properties of a film will be controlling. These properties seem to be dependent on polymer structure, catalysts, plasticizers or other additives primarily and will be discussed in later chapters.

L. Miscellaneous Film Properties

Other properties that must sometimes be taken into account are resistance to insects (probably no polymer film is truly insect-proof), resistance to burning, and resistance to biolytic degradation by microorganisms.

M. Economic Considerations

The practical manufacture of film must rest upon a plentiful low-cost supply of raw materials which are available or can be developed, a low-cost manufacturing operation, and a product capable of high packaging yield (square inches of coverage per pound of film), and high converting speeds on automatic machinery.

The technique of film manufacture should be such as to facilitate close control of all manufacturing variables throughout the process and to yield a film having a minimum of differences in physical properties in machine and transverse directions. The physical nature of the film should be such as to permit manufacture without significant curl, in widths of at least 50 in. The film should be capable of being produced as satisfactory film in gauges varying between 0.25 and 4 mils.

From the preceding discussion, it is clear that it would be very difficult to generalize upon the critical characteristics of a film, since they depend upon the product involved and the length of time and the conditions under which protection is desired.

In general, however, the properties most highly sought include nontoxicity, resistance to water vapor or liquid water, greaseproofness, a controllable permeability to specific gases, clarity, sealability by some simple means, strength and durability, ease and quality of printing, convenient handling on automatic high-speed machines, and low cost. To varying degrees, all polymeric films in use possess several of these properties to a desirable extent. It is clear that no single ideal film can be expected to perform all film functions in a completely acceptable manner.

V. Trends

On chemical grounds, polymeric film materials can be broadly classified into two groups: (1) the cellulosics (cellophanes, hydroxyethylcellulose) and (2) those derived from petroleum, natural gas, or coal (polyethylene, polypropylene, polystyrene, vinyls, sarans). Following its introduction in the United States in 1924, cellophane displaced paper in many applications, primarily because of transparency. In turn, the polymers not based on cellulose had taken about 40% of the film market by 1960 and as the surface and strength properties of these newer films continue to be improved, the percentage of the market held by cellophanes will continue to decrease. Laminations now being used to a very large extent will continue to consume huge quantities of cellophane, primarily because of its high modulus, its dimensional stability, its nonmelting even at high temperatures, and its transparency and gloss which allow it to be beautifully printed, even in reverse printing. One fact must not be overlooked, however: laminations are an expensive solution to a specific problem which may ultimately be met by an integral sheet more cheaply.

It has been pointed out that we must assume further major changes in the use of polymer films (particularly in packaging) in the next 40 years (9). New polymer films, not yet synthesized, will secure a part of the market. It is also assumed that petroleum-based polymers will rise from 40% of the packaging film market to about 70% by the year 2000 (this percentage rose from 12 to 38% in the decade 1950–1960). These authors (9) have tabulated the ranges for consumption of polymers in packaging films, after due allowance for different allocations of the plastic container and packaging market (for example, an increase in market use of rigid molded and blown containers) (see Table 8).

TABLE 8

Projections of Film Consumption to A.D. 2000

	1960		Estimate[a]		
			1980	1990	2000
Transparent films based on	270	Low	920	1,180	1,460
petrochemicals		Medium	1,410	2,320	3,360
		High	3,310	6,640	11,990
Cellophane	440	Low	370	290	290
		Medium	940	1,110	1,440
		High	2,070	3,800	6,660

[a] Millions of pounds.

References

1. R. Hill, *Fibres from Synthetic Polymers*, Elsevier, New York, 1953, p. 1.
2. H. Mark and E. S. Proskauer, Eds., *The Science of Plastics*, Interscience, New York, 1948.
3. J. K. Stille, *Introduction to Polymer Chemistry*, Wiley, New York, 1962.
3a. F. W. Billmeyer, Jr., *Textbook of Polymer Science*, Interscience, New York, 1962.
4. H. Staudinger, *Helv. Chim. Acta*, **5**, 785 (1922).

5. H. Mark and G. S. Whitby, Eds., *Collected Papers of Wallace Hume Carothers on High Polymeric Substances*, Interscience, New York, 1940.
6. W. H. Carothers, "Linear condensation polymers," U.S. Pat. 2,071,250 (Feb. 16, 1937). The application was filed July 3, 1931, and the Patent Office took 6 years to decide whether to issue a patent.
7. A. Schmidt and C. A. Marlies, *Principles of High Polymer Theory and Practice*, McGraw-Hill, New York, 1948, p. 58.
8. J. R. Whinfield and J. T. Dickson, Brit. Pat. 578,079 (June 14, 1946); U.S. Pat. 2,465,319 (Mar. 22, 1949).
9. H. H. Landsberg, L. L. Fischman, and J. L. Fisher, *Resources in America's Future—Patterns of Requirements and Availabilities 1960–2000*, Johns Hopkins Press, Baltimore, 1963, pp. 168, 701.

CHAPTER 2

MOLECULAR CONSTITUTION OF FILM-FORMING POLYMERS

ORVILLE J. SWEETING*

Yale University, New Haven, Connecticut

and

RICHARD N. LEWIS

Stauffer Chemical Company, Adrian, Michigan

I. Definitions

The term *polymer* was originally reserved for those molecules which could be thought of as consisting of several simple units of the same composition ($\pi o \lambda v \sigma$ many, and $\mu \epsilon \rho o \sigma$ part). Thus, sulfur (S_8) is normally a polymeric ring consisting of eight sulfur atoms, benzene (C_6H_6) is a polymer of acetylene (C_2H_2), and trioxane

* Formerly Associate Director of Research and Development, Olin Film Division, Olin Mathieson Chemical Corp., New Haven, Conn.

($C_3H_6O_3$) is a polymer of formaldehyde (CH_2O).   The terms *dimer*, *trimer*, and *tetramer* are commonly used to designate double, tripled, and quadrupled molecules, respectively.   Thus, nitrogen tetroxide (N_2O_4) is a dimer of nitrogen dioxide (NO_2), diisobutylene (C_8H_{16}) is a dimer of isobutylene (C_4H_8), and benzene may be called a trimer of acetylene.   All of these simple built-up molecules have properties in solution and vapor phase similar to the familiar properties of simple compounds of similar molecular weight and constitution, although many of them decompose under appropriate conditions into monomer.

But in the early 1900's, the study of natural products and the researches of Staudinger, Carothers, and many others demonstrated clearly the existence of a class of materials which were related to the simple structural units composing them only in a remote way.   The gross physical properties of these polymeric materials in no way resembled the physical properties of the component simple molecules, and in solution they behaved in a hitherto unknown manner.   These substances were termed *macromolecules*, and later, *high polymers*, but with recent years has come familiarity with the substances, and it is these materials of very high molecular weight (thousands or millions of atomic weight units), which are now usually simply called *polymers.*

Furthermore, the polymer of today may differ in composition from its monomer or monomers.   For instance, polyethylene glycol, $H(—OCH_2CH_2—)_xOH$, is considered a polymer of ethylene oxide, C_2H_4O; 6,6-nylon, $H[HN—(CH_2)_6—NH—CO—(CH_2)_4—CO—]_xOH$, is a polymer of hexamethylenediamine and adipic acid, although water is lost in the polymerization process.   Copolymers may have no definite composition.

As used here, the term *polymer* will refer to macromolecular compounds consisting of a hundred or more monomer units.

Copolymers are polymers which are made up of two or more kinds of monomer units.   These have been classified as random, alternating, block, and graft (Fig. 1).

II. Classification Based on Methods of Polymerization

The most convenient classification of polymers is based on the method of preparation (Chapter 3).   Carothers grouped them broadly as *addition polymers*, in which the structural unit has the

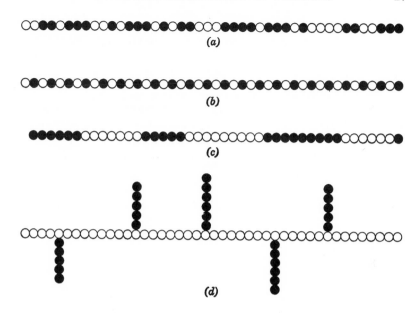

Fig. 1. Copolymers consist of polymer chains in which more than one kind of monomer unit occurs. Here one kind of monomer is represented by an open circle; the other, by a filled circle. The structure of copolymers may be (a) random, (b) alternating, (c) block, or (d) graft.

same empirical formula as the monomer, and *condensation polymers*, in which a part of the monomer or monomers is lost in the polymerization process (1). Polymers of ethylenically unsaturated monomers, as well as the polyethylene glycols made from ethylene oxide, are of the addition variety. Polyesters made from bifunctional acids and alcohols, polyamides from dibasic acids and diamines, and phenol–formaldehyde resins are typical condensation polymers.

As for polymer properties, the distinction between the two types is not a fundamental one. Indeed, there are many examples in which the same polymer results from both types of polymerization. For instance, linear polyethylene is usually made by the addition polymerization of ethylene, but it can also be made from diazomethane, CH_2N_2, by condensation polymerization, with loss of nitrogen. Low polymers (i.e., polymers of low molecular weight) have also been made by the reaction between decamethylene dibromide, $Br(CH_2)_{10}Br$, and sodium. 6-Nylon may be made equally well from aminocaproic acid, $H_2N(CH_2)_5CO_2H$, by condensation, or from caprolactam $NH(CH_2)_5CO$, by addition. Polyethylene glycol was originally made by condensation of ethylene glycol and ethylene dichloride but is now made by addition of ethylene oxide. Polydimethylsiloxane, $[-OSi(CH_3)_2-]_x$, can be made by condensation from dimethylsilanediol, $HOSi(CH_3)_2OH$, or by the catalytic opening of small-ring siloxanes. In such ambiguous cases, it appears to be satisfactory to classify the polymers according to the most characteristic method of preparation.

Addition polymers from nonethylenic compounds are often ignored; in order to emphasize the wide variety that is possible, a separate category of such polymers is included here. The best-known examples are the polyethers, polymerized aldehydes, epoxy resins, and isocyanate polymers.

Separate categories are also set apart for polymers of biological (Chapter 4) or geological origin, as well as synthetic inorganic polymers (Chapter 5). It is often not clear whether they originated by addition or condensation methods. Furthermore, a separate discussion is warranted by their unusual properties.

III. Factors which Influence Polymer Properties

The properties of a polymer are determined by its molecular structure, molecular weight, degree of crystallinity, and chemical composition. Other factors may exist, but they are not yet well known.

A. MOLECULAR STRUCTURE

The shapes and sizes of polymer molecules are perhaps the most important factors which influence the properties of a polymer.

1. Linear Polymers

Difunctional monomers, i.e., those which are capable of joining with two other monomer units, form linear polymer molecules. Because of flexibility in these molecules they will not normally be extended, but exist randomly bent and coiled. They are capable of being extended, however, at least in theory. Even amorphous polymers like polystyrene show some degree of orientation when stretched or extruded. Polymers suitable for film and fiber are found chiefly among these linear polymers, which are the most likely to give a flexible product.

The various structural types may all be illustrated among the polyesters. An equimolar mixture of ethylene glycol and phthalic acid or anhydride can polymerize indefinitely to products of extremely high molecular weight. If there is a slight excess of either component, say glycol, a point is reached where no carboxyl groups remain to be esterified and no further growth takes place. The same is true if part of the phthalic acid is replaced by a mono-functional acid. In this particular series, a range of polymers may be obtained, from viscous liquids to amorphous thermoplastic solids, depending on the molecular weight.

2. Branched Polymers

If the glycol in a glycol phthalate is replaced by glycerol and the mole ratio of difunctional acid is not increased, the chief effect is to introduce hydroxyl groups along the polymer chain. Each glycerol unit will, on the average, be attached to two phthalic acid groups, though some will be attached to only one and others to three. The overall effect is a branched structure (Fig. 2a).

Slightly branched polymers are not much different from linear polymers of the same molecular weight, but highly branched molecules cannot be extended and are not capable of orientation. There is one example where branched polymers have an advantage: the branched silicone oils, which have lower solidification temperatures (pour points) than their linear analogs. Here branching is produced by incorporating a small fraction of trifunctional units in the polymer and preventing crosslinking by adding an equivalent amount of monofunctional units.

3. Crosslinked Polymers

If, in the glycerol phthalate cited above, any excess of phthalic acid is used over the 1:1 ratio, the chains will be linked together by further esterification reactions to form much larger molecules. Only a small excess of phthalic acid is necessary to link together all the molecules in one container, thus producing a single giant visible molecule. Such a structure is said to be crosslinked (Fig. 2b). It is normally insoluble in any solvent and infusible, although solvents may diffuse into it and cause it to swell. If additional

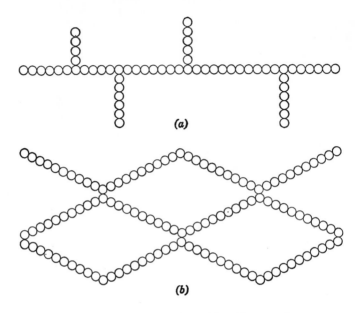

Fig. 2. Polymer chains can grow by branching (*a*), or can become connected by crosslinking (*b*). The circles represent not single atoms, but monomer units.

crosslinks are introduced, by raising the phthalate–glycerol ratio to a maximum of 2:2, the structure becomes more rigid until it is very hard. In this state it is practically unaffected by solvents. The theoretical ultimate in rigidity, however, may never be reached, because some of the potential crosslinking groups become immobilized and unavailable for reaction.

Crosslinking is not restricted to condensation polymers, nor is it necessary that the crosslinks be formed in the same way as the links in the basic polymer chains. Modern alkyd resins, which were originally glycerol phthalates only, now incorporate maleic anhydride and unsaturated fatty acids, which produce crosslinks through addition polymerization after the condensation polymerization has been completed.

The rigidity of most crosslinked polymers makes then unsuitable for film and fiber applications, but they are much used in molding compounds. Before molding, they are thermoplastic solids, incompletely polymerized; the heat of the mold completes polymerization and crosslinking, and they become *thermoset*.

A slight degree of crosslinking is sometimes incorporated in essentially linear polymers to improve stiffness and dimensional stability. Examples are vulcanized rubber and irradiated polyethylene.

It must not be thought that all reactions which produce branching will ultimately produce crosslinks. An example of a naturally occurring branched polymer is amylopectin, the insoluble fraction of starch. Here it would appear that the growing chain has several growing ends, capable of adding more glucose units, but not capable of combining with a polyglucose. The situation must be very similar to that in the radical polymerization of ethylene, where branching occurs because of chain transfer; a growing chain plucks a hydrogen atom from a polymer molecule, which in turn can add ethylene and start growing again at that point. The growing branch cannot combine with a saturated polymer molecule, however.

Another situation is illustrated by the polymerization of ε-caprolactam in the presence of a polyfunctional acid. There is no way to get more than one branch site in each polymer molecule, which therefore has as many branches as the acid has carboxyl groups.

4. Cyclic Polymers

In condensation polymerization there is always the possibility that the reactive ends of a growing chain will unite to form a ring. Five- and six-membered rings form very readily, but larger ones form with great difficulty. The explanation does not lie primarily in any kind of strain in the larger rings, but in probability. An extended chain has so many possible configurations that at any moment it has only a slight probability of being in a position to cyclize, and the chance of finding a reactive group on another molecule is much greater. Ruzicka obtained large rings by the technique of high dilution, but the yields were low.

Often in the process of polymerization, conditions of equilibrium are established, and it is possible to detect ring compounds. In the catalytic bulk polymerization of a dimethylsiloxane at room temperature about 2% of volatile cyclic compounds are always found. These are found to consist mainly of trimer, tetramer, and pentamer, with decreasing amounts of higher polymers through the cyclic enneamer (an 18-membered ring).   Although small rings are favored somewhat over chains at elevated temperatures, or in solution, it is doubtful that significant amounts of very large cyclic molecules will ever be obtained.   The largest rings ever isolated in any reaction consist of only about 30 atoms.   In the depolymerization of 6-nylon, for instance, the largest molecules that have been been detected are the 28-membered rings of the cyclic tetramer.

In a crosslinking system, opportunity for ring formation is much greater.   To take another example from the field of silicones, the equilibration of difunctional dimethylsiloxane, $[(CH_3)_2SiO-]_n$, and trifunctional methylsiloxane, $(CH_3SiO-)_n$, in a solvent results in a stable solution that obviously contains polycyclic compounds. Among those that have been identified are the following (symbols D and T indicate the two kinds of units):

Upon evaporation of the solvent, and baking, these molecules rearrange and polymerize to form a crosslinked resin.

5. Microgels

Large polycyclic molecules of the type just mentioned are crosslinked resins on a miniature scale.   If the molecules become large enough, they are better thought of as colloidal particles.   The phenomenon is encountered in the emulsion polymerization of dienes.   Ordinarily, a suspended polymer particle consists of many discrete molecules, but occasionally a crosslinking occurs and the entire particle is converted to one more-or-less rigid molecule.   Such a particle is called a *microgel*.

B. MOLECULAR WEIGHT

In linear polymers of a given chemical type, one kind of molecular weight is obtained simply by dividing the weight of a sample by the number of molecules in it, which can be determined by a counting method such as end-group analysis, freezing-point depression, or osmotic pressure. Since not all of the molecules in the sample are of the same size, the value obtained is an average value called the *number-average molecular weight*, $\bar{M}_n$ (2). This particular molecular weight is defined as

$$\bar{M}_n = \frac{\sum_i n_i m_i}{\sum_i n_i}$$

where n_i is the number and m_i is the molecular weight of the ith species, summed over all species.

Many properties depend as much on the size of the molecules as on their number. One such property is light scattering, which is proportional to the square of the size. A few large molecules will therefore scatter more light than many small ones. The amount of scattering is found to be proportional to another function called *weight-average molecular weight*, $\bar{M}_w$, defined as follows:

$$\bar{M}_w = \frac{\sum_i n_i m_i^2}{\sum_i n_i m_i}$$

$\bar{M}_w$ can also be written in the form $\Sigma\, w_i m_i$, where w_i is the weight fraction of molecules of molecular weight m_i.

Measurements of viscosity lead to another molecular weight, known as the *viscosity-average molecular* weight, $\bar{M}_v$, which in some cases may be identical with the weight-average molecular weight but more often is a complex average which has been defined by Flory as

$$\bar{M}_v = \left[\frac{\sum_i n_i m_i^{\beta+1}}{\sum_i n_i m_i}\right]^{1/\beta}$$

where β is a constant (3). When $\beta = 1$, it is evident that $\bar{M}_v = \bar{M}_w$.

The Flory definition is a more sophisticated version of the old Staudinger relation among specific viscosity η_{sp}, concentration c, and molecular weight $\bar{M}$, viz.,

$$\eta_{sp}/c = K\bar{M}$$

where K is a constant characteristic of the system, polymer plus solvent. The Staudinger relation holds amazingly well for a wide variety of polymeric substances, despite the simplicity of the assumptions made in deriving it. One simple modification of the Staudinger equation is $[\eta] = KM^\beta$, where $[\eta]$ is the intrinsic viscosity (obtained from a plot of specific viscosity as a function of concentration, extrapolated to zero concentration, the so-called viscosity of the polymer at infinite dilution) and β is a characteristic constant. Numerous studies have shown that β is not constant for all polymers as assumed by Staudinger, but varies from polymer to polymer. Tables relating K and β for a great many polymer systems are given in the literature. The viscosity method is a very simple method of getting relative molecular weights for comparison (see Chapter 16).

Other types of averages can be obtained by sedimentation equilibrium and sedimentation rate experiments. For these less frequently used averages, the reader is referred to the standard works (3–8).

Condensation polymers prepared under equilibrium conditions have a distribution of molecular weights which depends only on the degree of polymerization (DP). At high DP the ratio of $\bar{M}_w$ to $\bar{M}_n$ reaches a constant value of 2. Polymers prepared under nonequilibrium conditions tend to have broader distribution curves and $\bar{M}_w/\bar{M}_n$ increases to much higher values. A not entirely ridiculous example of such a polymer might be one comprising 10 molecules with a molecular weight of 10,000 and one with a molecular weight of 1,000,000. In this case $\bar{M}_n$ is 100,000, $\bar{M}_w$ is 900,000, and $\bar{M}_w/\bar{M}_n$ is 9.

In a homologous series of linear polymers, the lower members are likely to be viscous oils. With larger molecules the viscosity increases until a rigid state is reached. At that point the polymer becomes crystalline, glassy, or rubbery, at least at room temperature. Almost all linear polymers are thermoplastic, which means that they become fluid on heating, and soluble in suitable solvents, usually at room temperature, although a few remain insoluble until near their softening or melting points. One exception is polytetrafluoroethylene, $(-CF_2CF_2-)_n$, which neither melts nor dissolves below its decomposition temperature of about 400°C. Another is cellulose, which chars without melting.

The softening point of amorphous polymers generally increases with molecular weight. Crystalline polymers, however, have characteristic sharp melting points which are constant above a certain minimum molecular weight. Toughness and mechanical strength in general increase with molecular weight. For use as fiber or film, desirable properties do not appear in any polymer until the molecular weight, $\bar{M}_w$, reaches about 10,000. The figure is about 20,000 for polyethylene. Some polymers lack the required strength at all molecular weights. For instance, polydimethyl-siloxane remains fluid up to a molecular weight of 75,000, beyond which it is a rather weak rubbery material up to molecular weights of over a million. Crosslinking improves dimensional stability, but does not improve mechanical strength appreciably in most instances.

C. Crystallinity

The degree of crystallinity is an important factor affecting polymer properties. Certain polymers are composed of chains which are so constructed that they can pack well in ordered arrangements. The order may extend over a finite volume, designated by the term *crystallite*. Unlike crystals of small molecules, crystallites are not composed of whole molecules, or molecules of uniform size. They do possess certain properties of crystals, however. For instance, they exhibit typical x-ray diffraction patterns. They are also denser than surrounding disordered (amorphous) regions, and have a higher refractive index. If the crystallites are large enough to scatter light, say 500 Å or more in diameter, or if they form visible spherulites (9), the polymer will be hazy.

An old, but still useful concept of crystallinity in synthetic polymers is that due to Bryant (10) who modified suggestions made many years before concerning the structure of gelatin (11). He pictured semicrystalline polymers as consisting of groups of aligned molecules in ordered, or crystalline, regions which pass through a less well-ordered region into a region of random conglomerations of molecules without regular orientation, the so-called amorphous region. This "fringed micelle" picture (Fig. 3), or "fringed crystallite" model ("fringed" because the crystallite domain changes to an amorphous domain through a "fringe" of ill-defined structure), has limitations in explaining the fine structure of polymers, and

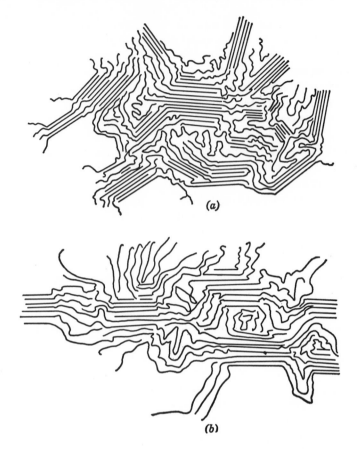

Fig. 3. Fringed micelle picture of a crystalline polymer (*a*) unoriented and (*b*) oriented. After Bryant, ref. 10.

does not account well for larger structures such as spherulites, but it does allow a simple interpretation of the effects observed when semicrystalline polymeric films undergo stretching. Stretching a film or fiber orients the crystallites and realigns other molecules, or segments of molecules (Fig. 3*b*), thereby causing the total crystallinity of the specimen to increase. Such oriented films and fibers are said to be *tensilized;* they are in general tougher than either amorphous or unoriented crystalline materials (see Chapter 10).

Recent views of crystalline structure in polymers are reviewed in Chapter 7.

A single polymer molecule may run through several crystalline regions, where it is bound more or less firmly to other molecules. Crystalline polymers tend to be stiffer and tougher than amorphous polymers. The crystallites disappear when the polymer is heated, and the temperature at which crystalline regions vanish is called the crystalline melting point.

Polyethylene terephthalate (Mylar) is a good example of an orientable, crystallizable polymer. If a film of Mylar is prepared above its melting point (265°C) and quenched at room temperature, it remains clear and amorphous. It is flexible, but has a low impact strength; it appears to be, and actually is, in the glassy state. The same film heated to temperatures above the second-order transition temperature (12), 110°C, becomes opaque, since the molecules have become mobile enough to form rather large crystallites. The crystallized film is, however, not appreciably stronger. On the other hand, if the film is heated and stretched in both directions, it remains clear and is now tensilized. It is not only flexible, but very tough. X-ray diffraction shows a crystalline pattern, with definite evidence of orientation.

Orientation and crystallinity are related in that only polymers which are capable of crystallization can be oriented. During orientation, crystallites become aligned and the total amount of crystalline material increases. It appears, therefore, that the toughest polymers are to be found among those which are crystalline.

The factors which favor crystallinity in a polymer are rigidity, linearity, regularity and symmetry, and polarity (13).

1. Rigidity

Rigid molecules are known to have generally higher melting points than flexible ones. Thus, cyclic molecules—even unsymmetrical ones—are higher melting than acyclic ones, and bicyclic ones, much higher. To take a single example, camphor, an unsymmetrical bicyclic ketone, melts at 178°C, while most monocyclic and acyclic ketones of similar molecular weight are liquids at room temperature.

Similar effects are observed in polymers. The high melting point of cellulose may result in part from the rigid ring structure in each glucose unit. Polyethylene terephthalate has a higher crystalline melting point than that of any aliphatic polyester. Even among aliphatic polymers there are obvious differences in rigidity, e.g., where crowding may be a factor. The stiffness and high melting point of polytetrafluoroethylene, compared with polyethylene, may be ascribed in part to restricted rotation in the fluorine compound.

Oxygen atoms in a chain may perhaps decrease rigidity, as the oxygen bond angles are less firm than carbon bond angles (they may vary from 90 to almost 180° in different compounds, while the tetrahedral carbon angle is always close to 110°). Perhaps this is a reason for the non-fiber-forming properties of polyethylene oxide. Oxygen bond angles are particularly indefinite in the silicones, which seem to have a good deal of internal freedom of motion. Polydimethylsiloxane is vastly different from polyisobutylene, which is a tough rubbery material, although their structures are rather similar.

Linear polymers obviously have a much greater chance of having their molecules aligned than do branched or crosslinked polymers. In fact, only a small degree of branching is required to erase evidence of crystallinity altogether.

2. Symmetry

The question of symmetry arises concerning polymers of the type $(—CH_2—CHX—)_n$, where X is an atom or group larger than hydrogen. Until recently, examples of this unsymmetrical class were all amorphous. Within the last few years, however, crystalline polymers of propylene, styrene, and other olefins have been obtained (Chapter 7); this type of asymmetry in the monomer, provided X is not a long chain, obviously does not of itself preclude crystallinity. The explanation lies in the regularity of the substituents X which are identical *and occur at every second carbon atom along the chain*, because of the nature of the polymerization. An added feature in surface-catalyzed polymerization, which promotes regularity, is that the configuration of each branch-bearing carbon atom is the same. This means that, if the polymer molecule is laid out with its backbone flat, all of the branches will lie on one side (Fig. 4) or the other, instead of randomly on both sides (Fig. 5). The former structure is called *isotactic* (Fig. 4). Another regular structure, called *syndiotactic*, is found wherein the branches lie

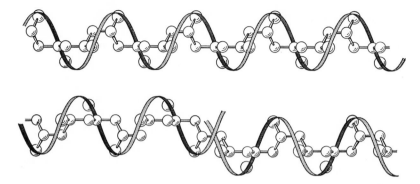

Fig. 4.   Degrees of order in molecules can be varied.   This drawing represents polypropylene.   The isotactic molecule at the top has methyl side chains (the balls through which the imaginary ribbon passes) in a regular helical sequence around the central chain of carbon atoms.   It forms crystalline films.   The lower drawing represents a section of a block polymer with two isotactic parts.   It forms a less crystalline material.

alternately on opposite sides of the chain.   There is less interference between adjacent side chains, and it may be possible to construct a polymer with such bulky side chains that only the syndiotactic structure is possible.   Perhaps this is the explanation of the crystallinity of polyvinyl isobutyl ether when prepared at low temperature with a cationic initiator but no solid catalyst (14).

A growing polyolefin chain (except polyethylene) has an asymmetric carbon atom in the terminal unit.   Thus, according to the Ziegler mechanism, the structure is as follows, the β carbon atom

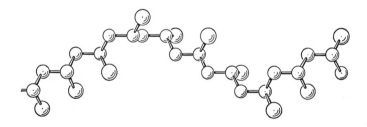

Fig. 5.   A random molecule of polypropylene (cf. Fig. 4).   This material is a rubbery, amorphous substance.

(C*) being asymmetric:

$$\sim\overset{\displaystyle R}{\underset{\displaystyle H}{\overset{|}{\underset{|}{C^*}}}}-CH_2-\overset{|}{\underset{|}{Al}}-$$

The polymer as a whole is symmetrical, with the possible exception of the end groups, which can be neglected. The body of the chain has alternate —CH_2— and —CHR— groups, all the H and R groups being above or below the plane of the chain. If all R's are on one side of the chain there is a plane of symmetry at every carbon atom.

$$\diagdown_{CH_2}\diagup^{CHR}\diagdown_{CH_2}\diagup^{CHR}\diagdown_{CH_2}\diagup$$

The configuration in which all R's are up is identical with that in which all R's are down, as may be seen by turning the model end for end. Therefore, it is not correct to speak of an all D or all L structure; it is actually all *meso*.

In polypropylene oxide, on the other hand,

$$(-CH_2-\overset{\displaystyle CH_3}{\underset{\displaystyle H}{\overset{|}{\underset{|}{C}}}}-O-)_n$$

the monomer itself is asymmetric and it is possible to prepare isotactic polymers which are optically active.

In order to achieve crystallinity, one further requirement has to be met. The units in the chain must be directly superimposable, without large rotation, on preceding units, either immediately adjacent or separated by a small whole number of units. The simplest type is illustrated by crystalline polyethylene, in which all the carbon atoms of a chain lie in one plane, and each unit is directly superimposable on adjacent units by translation in the plane, or by rotating around the long dimension (c axis) of the molecule one-half revolution and moving simultaneously a short distance along the c axis (15).

Polyisobutylene might be expected to have a similar structure, but because of steric interaction between methyl groups in adjacent units, successive units are twisted out of the plane. As a result, the polymer is amorphous—rubbery, in fact. If it is stretched, however, the molecules assume the extended configuration and crystalline x-ray patterns appear.

Polyvinylidene chloride is somewhat similar, adjacent units being twisted through an angle of about 60°. The direction of twist alternates, however, so that every second unit is superimposable, and the polymer is crystalline.

Polyvinyl chloride is only slightly crystalline, as would be expected, since the bulky chlorine atoms do not fit snugly into the polyethylene structure. Polyvinyl alcohol, on the other hand, is more crystalline because the hydroxyl groups pack more readily and the electronegativity of the oxygen atoms (and therefore their repulsion one from another) is countered to a large extent by the hydrogen atom embedded in each one. The repeat distance in polyvinyl alcohol is the same as that found for polyethylene, but the repeat distance for polyvinyl chloride is twice as great. Thus, the position of the chlorine atoms alternates, for reasons which have never become very clear (16). In vinyl esters and acrylics, the side groups are too large to fit and the polymers are all amorphous.

Polyvinyl fluoride is highly crystalline. The structure is not known, but it probably fits very well into the polyethylene arrangement.

Polypropylene and polystyrene are normally amorphous, because of the random position of the side groups. The isotactic polymers are highly crystalline, however, and have been shown to have a helical structure, with three units per turn (9,17).

Polymers in which the chain backbone contains an odd number of atoms per unit must have at least two units before superimposability occurs. Thus, cellulose has two glucose units per repeat unit, and 6-nylon has two 7-atom caprolactam units per repeat unit. 7-Nylon, with a polymer unit and repeat unit of eight atoms is more crystalline and has a higher melting point than 6-nylon.

Starch is believed to have a wide helical structure with about six glucose units per turn. This leaves a cylindrical hole in the center which absorbs water or iodine molecules very strongly (hence the peculiar starch–iodine complex).

Copolymers generally have a random sequence of monomers and if a significant amount of each is present, regularity is not possible unless the monomer units are so nearly of the same shape that they are interchangeable in a crystalline geometry. The degree of crystallinity of linear polyethylene is reduced from about 90 to about 20% by the incorporation of only 6% propylene.

The occurrence of rubbery properties in polymers is surprisingly common. A typical rubber, such as polyisoprene, is an amorphous polymer. Unless it is rather highly crosslinked, it becomes a viscous fluid at elevated temperatures. At low temperatures it stretches, but does not snap back when released; it crawls back. It is leathery rather than rubbery. At still lower temperatures it becomes glassy and hard. Silicone rubber goes through exactly the same transformations, but the transition temperatures are lower. Many polymers are in the glassy state at room temperature, and when heated, pass through a rubbery state before melting. Indeed, it is characteristic of amorphous polymers that they have some temperature range in which they exhibit rubbery properties. The phenomenon is rarely encountered in polymers that are capable of crystallizing, because they usually do so upon being heated to the second-order transition temperature.

3. Polarity

The final basis for the classification of polymers is their chemical nature, and in particular the degree of polarity of their functional groups. The tendency of molecules to associate, which is a very important factor in polymer properties, is a measure of their polarity.

The high strength of polyamides undoubtedly results from the strong intermolecular forces between the polar amide groups (18). These forces no doubt assist in the alignment of the molecules in crystalline order. At the same time, the ordering effect of crystallization tends to create the maximum number of intermolecular hydrogen bonds.

Crystallization is not absolutely necessary to the formation of strong intermolecular forces. A copolymer of 6- and 11-nylon is believed to be completely amorphous, yet it is as tough as the more crystalline homopolymers, though perhaps less stiff.

On the other hand, polarity is not absolutely necessary either. The highly crystalline polymers made from linear polyethylene, isotactic polypropylene, and polyethylene terephthalate are also very tough, despite their low polarity.

Polymers which are neither polar nor crystalline will probably not make good films or fibers. Perhaps the best polymers will turn out to be both polar and crystalline.

References

1. H. Mark and G. S. Whitby, Eds., *Collected Papers of Wallace Hume Carothers on High Polymeric Substances*, Interscience, New York, 1940.
2. R. U. Bonnar, M. Dimbat, and F. H. Stross, *Number-Average Molecular Weights*, Interscience, New York, 1958.
3. P. J. Flory, *Principles of Polymer Chemistry*, Cornell Univ. Press, Ithaca, 1953, p. 311.
4. C. E. H. Bawn, *The Chemistry of High Polymers*, Interscience, New York, 1948.
5. G. V. Schulz, H.-J. Cantow, and G. Meyerhoff, "Molekulargewichts-bestimmung an makromolekularen Stoffen," in Houben-Weyl *Methoden der Organischen Chemie*, Eugen Müller, Ed., Vol. 3, Part 1, Georg Thieme Verlag, Stuttgart, 1955, p. 371 ff.
6. H. Tompa, *Polymer Solutions*, Butterworths, London, 1956.
7. M. L. Huggins, *Physical Chemistry of High Polymers*, Wiley, New York, 1958.
8. F. W. Billmeyer, Jr., *Textbook of Polymer Science*, Interscience, New York, 1962.
9. P. H. Geil, *Polymer Single Crystals* (*Polymer Reviews*, Vol. 5, H. F. Mark and E. H. Immergut, Eds.), Interscience, New York, 1963, p. 223 ff.
10. W. M. D. Bryant, *J. Polymer Sci.*, **2**, 547 (1947).
11. K. Hermann, O. Gerngross, and W. Abitz, *Z. Physik. Chem.*, **B10**, 371 (1930).
12. A. V. Tobolsky, *Properties and Structure of Polymers*, Wiley, New York, 1960, p. 61 ff.
13. R. J. W. Reynolds, in *Fibres from Synthetic Polymers*, R. Hill, Ed., Elsevier, Houston, 1953, pp. 115–197.
14. C. E. Schildknecht, *Vinyl and Related Polymers*, Wiley, New York, 1952, pp. 609–614.
15. S. L. Aggarwal and O. J. Sweeting, *Chem. Rev.*, **57**, 679 (1957).
16. Reference 4, p. 192.
17. N. G. Gaylord and H. F. Mark, *Linear and Stereoregular Addition Polymers: Polymerization with Controlled Propagation* (*Polymer Reviews*, Vol. 2), Interscience, New York, 1959, p. 45.
18. D. R. Holmes, C. W. Bunn, and D. J. Smith, *J. Polymer Sci.*, **17**, 159 (1955); see also reference 4, p. 198.

CHAPTER 3

SYNTHETIC ORGANIC POLYMERS
AS FILM FORMERS

ORVILLE J. SWEETING*

Yale University, New Haven, Connecticut

and

RICHARD N. LEWIS

Stauffer Chemical Company, Adrian, Michigan

* Formerly Associate Director of Research and Development, Olin Film Division, Olin Mathieson Chemical Corporation, New Haven.

As we have pointed out in Chapter 2, classification of polymers can be done in more than one way, but the most useful classification for our purpose will be that based upon method of preparation, namely, condensation polymers and addition polymers.

In the following overview of the more important types of condensation polymers that can be converted to thin films we shall place special emphasis upon those that appear particularly attractive on the basis of known properties or of possible ease and economy of preparation. Our intent in this chapter, and also in Chapters 4 and 5, is to survey the entire domain, calling attention at a few points to areas of research where possibly some potentially useful polymers may have been overlooked.

I. Condensation Polymers

Condensation reactions in polymer chemistry are in a formal way a metathetical reaction, that is, one in which polar compounds exchange partners.

$$AX + BY \rightarrow AB + XY$$

XY is usually a small molecule, e.g., H_2O, C_2H_5OH, N_2. A great many polymer syntheses fall into this category. If both A and B are bifunctional, linear polymers result. Linear polymers are also formed when X and Y groups are found in the same molecule, as in amino acids or hydroxy acids.

To obtain a polymer of high molecular weight, the condensation reaction must proceed to give a high yield. There must be no side reactions, and there must be no strong tendency to form ring compounds (Chapter 2). Of the many possible types of condensation polymers, only a few are produced commercially; they include poly-amides, polyesters, polyethers, and polysulfides.

A. POLYAMIDES, INCLUDING POLYUREAS

Polyamides are made most readily by the dehydration of (a) a diamine salt of a dibasic acid or (b) an amino acid. Dehydration takes place at temperatures above 180°C. The three most impor-tant polyamides now made for fiber and molding as well as for film are 6,6-nylon (from hexamethylenediamine and adipic acid), 6,10-nylon (from hexamethylenediamine and sebacic acid), and 6-nylon (from caprolactam, the cyclic anhydride of ε-aminocaproic acid). Several binary and ternary copolymers of these monomers are also produced. The various dibasic acids, diamines, and amino acids that might be considered for other polyamides are discussed below, always in relation to their uses in film-forming polymers.

1. Difunctional Acids

Carbonic Acid, H_2CO_3. This is by far the cheapest dibasic acid. It forms strong linear polyamides that contain the group —NH—CO—NH— and are therefore called polyureas. In the past, such polymers have been made from the rather expensive diisocyanates, and have therefore not received very serious attention. Because of recent commercial production of diisocyanates they are worthy of further consideration. They can also be made directly from carbon dioxide (1).

Oxalic Acid, $(CO_2H)_2$. This acid is also cheap. It is rather unstable thermally, but polyamides have been made from oxalyl chloride and hexamethylenediamine (2). Film-forming aromatic polyoxamides melting above 400°C have been made, e.g., from 2,2-bis(4-aminophenyl)propane (2a).

Malonic Acid, $HO_2CCH_2CO_2H$. This acid is very unstable ther-mally and is available only as the ethyl ester.

Succinic Acid, $HO_2C(CH_2)_2CO_2H$. The acid is available at mod-erate price. Condensation polymers have desirable properties, but are not superior to related polymers.

Glutaric Acid, $HO_2C(CH_2)_3CO_2H$. This dibasic acid is not read-ily obtainable.

Adipic acid, $HO_2C(CH_2)_4CO_2H$, is obtainable by the oxidation of cyclohexane and cyclohexanone at a low cost.

Pimelic acid and *suberic acid*, the C_7 and C_8 difunctional acids, are unobtainable in quantity in any practical way at the present time.

Azelaic acid, $HO_2C(CH_2)_7CO_2H$, has become available within the past 10 years by the ozonolysis of oleic acid, $CH_3(CH_2)_7CH=CH-(CH_2)_7CO_2H$.

Sebacic acid, $HO_2C(CH_2)_8CO_2H$, from castor oil, may be almost competitive with adipic acid, because it requires a smaller amount of diamine per pound of polyamide.

Higher paraffinic dibasic acids are not available. Very long-chain dibasic acids may some day be made by the oxidation of ethylene polymers and copolymers, however, and these should produce useful polymers.

Maleic acid, $HO_2CCH=CHCO_2H$, forms cyclic imides too readily to be considered for polyamides unless new techniques can be developed.

γ-Ketopimelic acid, $HO_2C(CH_2)_2CO(CH_2)_2CO_2H$, is not available commercially, but is easily made from furfural, and has been used for polyamides in Germany. A so-called *isosebacic acid*, a mixture of isomers, mainly α-ethylsuberic acid, is available from the condensation of butadiene with carbon dioxide and sodium. The lack of either symmetry or purity makes it an unlikely candidate for film.

Dimerized linoleic acid is a cheap compound of indefinite structure. Condensed with ethylenediamine it forms a low-melting polyamide suitable for adhesive and coating purposes, but it seems unsuitable for film.

Several dibasic acids are potentially obtainable by the hydrolysis of the cyanoethylation products of difunctional compounds such as water, methylamine, hydrogen sulfide, and glycols. These are all cheaper than acrylonitrile, $CH_2=CHCN$, which is itself not very costly.

Diglycolic acid, $O(CH_2COOH)_2$, is now commercially available, if required, as a by-product from the manufacture of glycolic acid.

Phthalic acid cannot be used, for it forms cyclic imides rather than polymers when condensed with amines.

Isophthalic acid, $m\text{-}C_6H_4(CO_2H)_2$, and *terephthalic acid*, $p\text{-}C_6H_4(CO_2H)_2$, are of considerable interest. Isophthalic acid and diamines produce amorphous polymers that do not appear to be film-forming. Terephthalic acid and hexamethylenediamine produce a polymer melting above 300°C, which is too high for practical extrusion. (There are few organic compounds that do not begin to decompose at such temperatures. In general, the practical upper limit of melting point is about 275°C.) Copolymers with various amounts of isophthalic acid are lower melting, and of course other diamines may be found to give suitable polymers with terephthalic acid alone.

2. Diamines

In contrast to the large number of available diacids, only two diamines are articles of commerce: ethylenediamine and piperazine. The others are not easy to synthesize.

Ethylenediamine, $NH_2CH_2CH_2NH_2$, in polyamide formation suffers from the disadvantage that it tends to form cyclic amidines.

Piperazine, $NHCH_2CH_2NHCH_2CH_2$, forms atypical polyamides that have no active hydrogens. The polymers are very high melting, however, because of their symmetrical, cyclic chain units. Those made from azelaic acid or sebacic acid are fiber-forming (3).

2-Methylpiperazine has been described as a major product in the reductive ammonolysis of sugar (4). Polymers obtained from it had unexpectedly high melting points—in most cases, over 300°C (4a). Copolymers may be more suitable.

Tetramethylenediamine, $NH_2(CH_2)_4NH_2$, is possibly available by way of succinonitrile or tetrahydrofuran. Its polymers are higher melting and possibly tougher than those of hexamethylenediamine.

Hexamethylenediamine, $NH_2(CH_2)_6NH_2$, is made in large quantities by du Pont and Chemstrand, but only for internal consumption. It is probably more expensive than adipic acid. Possible raw materials for commercial production are butadiene, chlorine, and sodium cyanide. Both 6,6-nylon and 6,10-nylon are attractive as film-forming polymers for special applications.

m-Xylylenediamine, m-$C_6H_4(CH_2NH_2)_2$, and *p-xylylenediamine*, could be made commercially from the corresponding phthalic acids, if a demand for these amines existed. They make satisfactory film- and fiber-forming polymers with adipic and sebacic acids (5).

Aromatic diamines such as *p*-phenylenediamine, $1,4$-$(NH_2)_2C_6H_4$, have never been found suitable for polymers.

4,4'-Diaminodicyclohexylmethane, $NH_2C_6H_{10}CH_2C_6H_{10}NH_2$, is obtainable from 4,4′-diaminodiphenylmethane, and produces crystalline polymers with adipic acid (6). A mixture of *cis–trans* isomers is obtained on reduction, but the mixture produces film-forming polymers with softening temperatures of 250–275°C.

As with the dibasic acids, dinitriles are suitable intermediates for diamines. It is of particular interest that a biscyanoethyl compound could be converted half to a diacid by hydrolysis, and half to a diamine by reduction, and both of the monomers required could thus be obtained from a single intermediate. *Oxy(bispropionitrile)*, $O(CH_2CH_2CN)_2$, and *ethylenedioxy(bispropionitrile)*, $NCCH_2CH_2OCH_2CH_2OCH_2CH_2CN$, from water and ethylene glycol, respectively, would produce diamines containing ether links, which might impart unusual properties. The ether group is flexible and might counteract the rigidity found in cyclic molecules such as terephthalic acid, thereby producing a polymer with a reasonably low melting point. Those compounds have not found application as yet, however, perhaps because they are hygroscopic, if not water-soluble.

3. Amino Acids

Polyamides may be obtained by the dehydration of certain types of amino acids, or by a rearrangement of their lactams (cyclic anhydrides). The latter method is actually an addition polymerization, but is included here for convenience. There is an advantage in this type of polymer in that only one monomer is required.

α-Amino Acids, NH_2CHRCO_2H. These are expensive compounds. Whereas living matter can convert them enzymatically to polypeptides, attempts to polymerize them by heating results in decarboxylation, not polymerization. Synthetic polypeptides have been made from the still more expensive N-carboanhydride derivatives.

β-Amino acids lose ammonia on heating, and are not promising as monomers. Poly-β-alanine has been prepared, however, by an unconventional method which comprises the treatment of *acrylamide* with a strong base (7).

$$n CH_2 {=} CH{-}CO{-}NH_2 \rightarrow (-CH_2{-}CH_2{-}CO{-}NH{-})_n$$

This polymer might be film-forming.

2,2-Dimethyl-β-alanine, $NH_2CH_2C(CH_3)_2CO_2H$, can be heated without loss of ammonia and produces a high-melting polymer (270°C) (8).

2-Pyrrolidone, $NHCH_2CH_2CH_2CO$, has been polymerized (9), but the product has a tendency to depolymerize above its melting point, 245°C. *2-Piperidone*, $NH(CH_2)_4CO$, has been polymerized (10) but properties of the polymer have not been published in detail.

Caprolactam, $NH(CH_2)_5CO$, is now polymerized on a commercial scale. It is readily polymerized (11,12), and makes a strong film with some unique and useful properties.

The C_7 and C_8 ω-amino acids are practically unobtainable.

9-Aminononanoic acid is not now available, but it is potentially obtainable from oleic or linoleic acid (a mixture of the two, probably from tall oil, is available commercially). Possible intermediates are oleonitrile and azelaic acid half-nitrile (by ozonolysis). The polymer, 9-nylon, melts at 198°C, and has certain advantages over 6-nylon, especially as regards a tendency toward spontaneous depolymerization.

11-Aminoundecanoic acid, $NH_2(CH_2)_{10}COOH$, is obtained by a six-step synthesis from castor oil (13).

Glyceryl triricinoleate $+ 3CH_3OH \rightarrow$

$$3CH_3(CH_2)_5CHOHCH_2CH\!=\!CH(CH_2)_7CO_2CH_3 + CHOH(CH_2OH)_2 \quad (1)$$
$$\mathbf{I}$$

$$\mathbf{I} \xrightarrow{\text{heat}} CH_2\!=\!CH(CH_2)_8CO_2CH_3 + CH_3(CH_2)_5CHO \quad (2)$$
$$\mathbf{II}$$

$$\mathbf{II} \xrightarrow[\substack{1.\ NaOH/H^+ \\ 2.\ +HBr(O_2)}]{} BrCH_2(CH_2)_9CO_2H \quad (3)$$
$$\mathbf{III}$$

$$\mathbf{III} + NH_3 \rightarrow H_2NCH_2(CH_2)_9CO_2H \quad (4)$$
$$\mathbf{IV}$$

$$\mathbf{IV} \rightarrow [\!-\!NH(CH_2)_{10}CO\!-\!]_n + nH_2O \quad (5)$$
$$\text{11-Nylon}$$

The polymer is manufactured in France and sold under the trade name Rilsan (14). It forms a beautiful, durable film when melt-extruded.

Of the possible aromatic amino acids one might be considered: *p-aminobenzoic acid*, $p\text{-}NH_2C_6H_4COOH$. Polymers of this amino acid have not been reported. The hydrogenated derivative, *4-*

aminocyclohexanecarboxylic acid, might possibly make a satisfactory film-forming polymer, even though it exists in two isomeric forms.

Acids with the amino group not in the omega position are generally to be avoided, since they give amorphous polymers. One, however, 9- (or 10-) aminostearic acid, $CH_3(CH_2)_8CHNH_2(CH_2)_7$-$CO_2H$, obtainable from oleic acid by the Ritter reaction, has potential and should be given some consideration. 12-Aminostearic acid (from castor oil) makes a fiber-forming polymer melting at 109°C (15).

Polyfunctional amino acids generally give branched or crosslinked polymers, but a polyhexamethylenediketopiperazine has been made from α,α-diaminosebacic acid, in which the latter's tetrafunctionality is exactly counteracted by its ring-forming tendency (16). The polymer has the structure indicated below.

$$\left[-(CH_2)_6-CH \underset{\underset{\displaystyle CO-NH}{\diagdown}}{\overset{\overset{\displaystyle NH-CO}{\diagup}}{}} CH- \right]_n$$

B. Polyesters

The term *polyester* covers several different classes of compounds, including (*1*) linear condensation products of glycol and dibasic acids (or of monohydroxy acids), (*2*) crosslinking alkyd resins based on glycerol and other polyhydroxy compounds, and (*3*) crosslinking addition polymers from unsaturated esters such as diallyl phthalate. Other polymers containing ester side chains, but not called polyesters, include cellulose acetate, polyvinyl acetate, and polymethyl methacrylate. We are concerned here only with linear polymers with the ester group in the chain.

1. Difunctional Acids

In general, the same acids may be used for polyesters as for polyamides.

Carbonic Acid. Ethylene carbonate, $O-CH_2CH_2OCO$, does not polymerize (17). Polycarbonates from higher aliphatic glycols have very low melting points. Aromatic polycarbonates are discussed further under dihydroxy compounds.

Oxalic Acid. Carothers (17) found that polyethylene oxalate has too low a melting point (172°C) to be suitable for films or fibers; moreover, it is readily depolymerized, and the higher homologs have progressively lower melting points (tetramethylene oxalate has, however, not been prepared, and therefore should be studied before one excludes the polyene oxalates entirely from futher consideration as film-forming polymers).

Succinic Acid. Polyethylene succinate melts at 108°C and the higher succinates melt at even lower temperatures (17).   Higher dibasic acids also give only low-melting polymers.   Thus, the only real possibility for using an aliphatic dibasic acid in making a polymer film is to combine it with an aromatic glycol. The recent trend, however, has been toward the opposite combination—aromatic acids with aliphatic diols.

Terephthalic acid is the basic ingredient of polyethylene terephthalate (Terylene, Dacron, Mylar, Cronar).

Isophthalic acid and *phthalic acid* produce amorphous polymers that are too brittle for film use.   It has been found, however, that up to 20% isophthalic acid may be substituted for terephthalic acid in a polyester without sacrificing the orientability necessary to produce a strong film (18).   There is not much advantage in cost with such a copolymer, but there may be possible processing advantages arising from a lower melting point (about 240°C as compared with 265°C for the pure terephthalate).

Other aromatic dibasic acids have been rather thoroughly investigated, but do not appear to offer any special advantages.

2. Dihydroxy Compounds

Ethylene glycol, $HOCH_2CH_2OH$, is the best ingredient in film-forming polymers, except when combined with carbonic or other aliphatic acids.   It is, and will probably remain, the cheapest glycol.

Methylene glycol (hydrated formaldehyde), $CH_2(OH)_2$, forms esters with simple acids, properly called acylals.   They are prepared by an addition reaction between formaldehyde and acid anhydrides. They are more susceptible to hydrolysis than are simple esters, but their potential cheapness makes them worth consideration.   If methylene carbonate could be made from carbon dioxide and formaldehyde, it would be extremely cheap. Perhaps a methylene maleate, succinate, or phthalate could be made more readily than the carbonate.

Propylene glycol, $CH_3CHOHCH_2OH$, gives low-melting polymers in general. This is probably because of its twofold asymmetry: the monomer itself is a mixture of D- and L-isomers, and in a polymer the methyl group may be found randomly at either end of the unit.

Trimethylene glycol, $HO(CH_2)_3OH$, forms a crystalline polymer with terephthalic acid, mp 221°C, but at present the glycol is not readily obtainable. The analogous neopentylene glycol, 2,2-dimethylpropanediol, forms a somewhat lower-melting polymer, 140°C (19), and is commercially available.

Tetramethylene glycol, $HO(CH_2)_4OH$, forms polymers with only slightly lower melting points than those from ethylene glycol, and is more expensive.

Hydroquinone, p-HOC_6H_4OH, is of academic interest. It probably does not form polyesters, though this is not certain. It does form derivatives of possible interest, especially the biscarboxymethyl and the bishydroxyethyl compounds. The latter forms a polycarbonate (20) and a terephthalate (21).

p-Xylyleneglycol, p-$C_6H_4(CH_2OH)_2$, can be converted to a high-melting polycarbonate, mp 239°C (20). This glycol is not commercially available, but could be readily synthesized if its polymers showed any desirable advantages.

3. Hydroxy Acids

Aliphatic ω-hydroxy acids generally produce low-melting polymers, although Carothers found that ω-hydroxydecanoic acid, HO-$(CH_2)_9COOH$, forms strong, orientable fibers. β-Propiolactone has become available, and polymerizes readily with acidic or basic catalysts; the polymer melts at 86°C and is readily decomposed to acrylic acid (22). A single methyl branch usually lowers the melting point, but a double branch at the same carbon atom restores symmetry.

Hydroxypivalic acid (2,2-dimethyl-3-hydroxypropanoic acid) produces a tough, orientable, water-resistant polymer with a melting point of 230°C (23). It could probably be made from neopentylene glycol.

Glycolic acid, $HOCH_2CO_2H$, is now available by the reaction of formaldehyde, carbon monoxide, and water. On heating, it is converted partly to a polymer and partly to a cyclic dimer, glycolide. The polymer melts at 223°C and is reversibly depolymerized on heating (24). If the polymer could be stabilized, glycolic acid would be a very promising monomer.

Of the large number of possible cyclic hydroxy acids, a few may be worthy of additional study, viz., p-hydroxybenzoic acid, p-(hydroxymethyl)benzoic acid, p-(hydroxyethoxy)benzoic acid (21), and trans-p-hydroxyhexahydrobenzoic acid, $HOC_6H_{10}CO_2H$.

C. Polycarbonates

A series of patents has been issued on various aromatic carbonates (20,25). These include bishydroxymethyl, bishydroxyethyl, and bishydroxyethoxy compounds containing one benzene ring or two, attached directly in the *para* positions (as in 4,4'-bishydroxyethyl-biphenyl) or through oxygen, sulfur, sulfone, methylene, amide, sulfonamide, ketone, and other groups. The general procedure is to convert to the bis(ethyl carbonate) by treatment with ethyl chlorocarbonate (made from phosgene), and thence to a polymer with the aid of a titanium-containing catalyst. Schnell (26) has described polycarbonates made from bisphenol-A [4,4'-bishydroxy-phenylpropane, $(CH_3)_2C(C_6H_4OH)_2$], which is made from phenol and acetone, as well as other bisphenols. They are made at room temperature by dissolving the bisphenol in alkali and adding phosgene. These polycarbonates are the only polyesters based on phenolic compounds. They are thermally stable at temperatures of up to 300°C, and films of the polymers have very low water-vapor permeability. These materials are described in detail in Volume II.

D. Other Polymers from Difunctional Acids

Polyester amides. Carothers copolymerized ε-aminocaproic acid with varying proportions of hexadecamethylene dicarboxylic acid and trimethylene glycol in order to reduce the melting point of the polymer (17). At present, 6-nylon is not considered to have too high a melting point, however, and there is little interest in this kind of approach.

Polyanhydrides. Dibasic acids containing six or more carbon atoms are susceptible to dehydration with acetyl chloride to give polymeric products. Polyanhydrides are film- and fiber-forming, but are too sensitive to water to have any present practical use.

Mixed anhydrides of dicarboxylic acids and other weak dibasic acids, such as hydroquinone, benzeneboronic acid, and benzenephosphonic acid, are theoretically possible but have not been studied.

Polyimides. Cyclic imides such as succinimide are quite inert, but acyclic imides are susceptible to hydrolysis. *Tricarballylic acid*, $HO_2CCH_2CH(CO_2H)CH_2CO_2H$, a derivative of citric acid, and hexamethylenediamine form a linear polyimideamide containing a cyclic imide group.

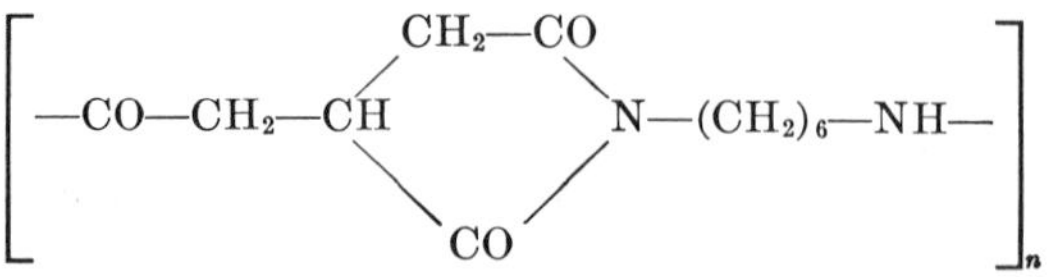

$$\left[-CO-CH_2-CH \underset{CO}{\overset{CH_2-CO}{<}} N-(CH_2)_6-NH- \right]_n$$

Aromatic polybasic acids give quite stable polyimides. In particular, *pyromellitic acid* (1,2,4,5-benzenetetracarboxylic acid) gives linear polyimides with extremely high thermal stability, especially with aromatic diamines (26a).

Polyimidazoles. Aromatic tetramines with two *ortho*-diamine groups produce linear benzimidazoles with dibasic acids. For instance, 3,3'-diaminobenzidine and isophthalic acid give a fully aromatic polymer that has extraordinary thermal stability (26b). It has the following structure:

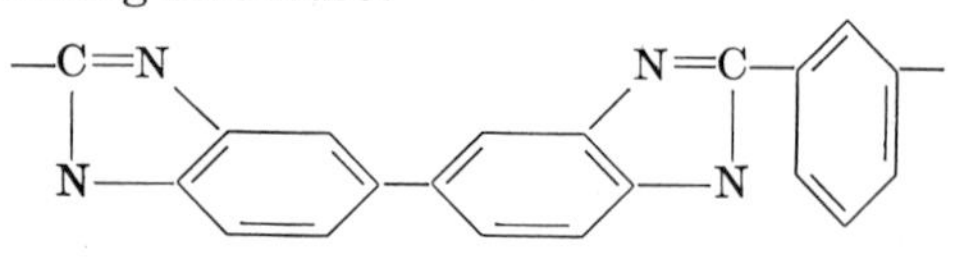

Polythioamides have lower melting points than do the corresponding oxygen compounds, and tend not to be crystalline. Polythiocaprolactam melts at 120°C (19). Even a slight tendency to hydrolyze would make these compounds objectionable because of the liberation of odoriferous sulfur compounds.

Polyhydrazides. Hydrazine, H_2NNH_2, is the simplest diamine. With dibasic acids it gives polymers containing the group —CO—NH—NH—CO—, which is transformed on heating to the oxadiazole group

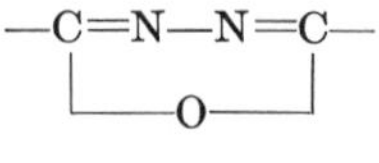

$$-C=N-N=C-$$
$$\underset{O}{\llcorner \quad \lrcorner}$$

Both groups are susceptible to hydrolysis, but on heating in the

presence of ammonia or excess hydrazine, the very stable *triazole* or *aminotriazole* group is formed.

$$—C=N—N=C—$$
$$\qquad\ \ \underset{\displaystyle NH}{\big|\underline{\qquad}\big|}$$

$$—C=N—N=C—$$
$$\qquad\ \ \underset{\displaystyle N(NH_2)}{\big|\underline{\qquad}\big|}$$

These polymers, as well as copolymers with polyamides form strong fibers and films, according to a series of Celanese patents issued in 1950 (27). The polymers are resistant to boiling hydrochloric acid, and if the cost of manufacture were lower, important commercial applications might be found for them.

Salts. Such compounds as 6,6-salt (adipic acid and hexamethylenediamine) are crystalline, water-soluble materials with no polymeric characteristics. Salts of the bivalent metals may be similar, yet it appears possible that the salts of the less electropositive metals will prove to be sufficiently covalent to exhibit polymeric character, especially if the acid has a fairly long chain, as in azelaic or sebacic acid.

Du Pont's ionomer film, Surlyn A, is described as being linked with metallic ions (28). Though it is probably not a salt in the ordinary sense, this polymer is proof that new consideration of ionic crosslinking as a method of securing useful properties is needed.

Polyketones. Carothers prepared a polyketone by the high-temperature decomposition of the thorium salt of octadecanedioic acid, but these polymers probably will not be useful substances (17).

E. Other Derivatives of Diamines

Amides of inorganic acids. Sulfamide and *sulfonamide* groups are little known in polymers. These groups are rather stable and not very susceptible to hydrolysis. Sulfuryl chloride is very cheap and should react with diamines to form polysulfamides and hydrogen chloride. With simple amines the reaction is best carried out in the presence of acetic acid and sodium acetate. Sulfur trioxide and sulfuric acid might be suitable starting materials.

Disulfonyl chlorides would also produce polymers. The naphthalenedisulfonyl chlorides are obtained readily, as are the disulfonyl chlorides of biphenyl and of diphenyl ether. All should produce polysulfonamides with diamines.

Sulfanilic acid, p-$NH_2C_6H_4SO_3H$, may possibly be convertible to a polymer on heating, but the product, $(—NH—C_6H_4—SO_2—)_n$, probably would have a very high melting point. It might make useful copolymers.

Sulfamide itself, $SO_2(NH_2)_2$, may be used in the formation of mixed amides with dicarboxylic acids, e.g., ($-CO-(CH_2)_4-CO-NH-SO_2-NH-)_n$ (29).

Sulfonimides. Polymers containing the grouping $-SO_2-NH-SO_2-$ might arise from disulfonyl compounds and ammonia, but they are probably easily hydrolyzed.

Thionamides. The reaction of *thionyl chloride*, $SOCl_2$, with amines has not been studied enough to formulate a possible polymerization reaction. One possibility is the formation of a thionimide, $R-N{=}S{=}O$, which is analogous to an isocyanate and would react with additional amino compound to produce a polymer.

Phosphorus chlorides. *Phosphorus trichloride* can be converted to *phenyldichlorophosphine*, $C_6H_5PCl_2$, which might form useful polymers with diamines, and possibly with monoamines or ammonia, e.g., $[-NH-(CH_2)_6-NH-PR-]_n$ or $(-NH-PR-)_n$. Benzenephosphonyl chloride, $C_6H_5P(O)Cl_2$, should react similarly. None of these polymers has been described.

Silicon chlorides. Polysiloxanes and polysilazanes are discussed in Chapter 5 as inorganic compounds. Analogous organic compounds of diamines are unknown, although they should be of considerable interest. Derivatives of dialkoxydichlorosilanes have been prepared (30).

Aldehyde derivatives. Primary amines form polymers with aldehydes and ketones, but they are usually cyclic trimers. Aniline tends to behave as a monofunctional compound, forming imines (Schiff's bases), $C_6H_5N{=}CHR$, as does hydroxylamine. A disecondary amine, such as piperazine, quite likely would form a linear polymer, but none has been reported yet. An amine or amide containing more then two active hydrogens, as in urea or melamine, gives crosslinked polymers with formaldehyde, furfural, and other aldehydes. These polymers are well known, but they are not useful as self-supporting films.

F. OTHER DERIVATIVES OF DIOLS

Esters of sulfur oxyacids. Esters of sulfuric acid are extremely susceptible to hydrolysis and therefore are of no practical interest in polymer formation.   Esters of sulfonic acids, RSO_3H, are also very reactive; perhaps polyesters of disulfonic acids can be made, however.   Little is known of the properties of alkyl sulfites.   A polymer from a glycol and sulfur dioxide, e.g., tetramethylene sulfite, $O(CH_2)_4OSO$, would probably be quite inexpensive to manufacture.

Phosphorus esters hydrolyze only in strong acid or alkali.   The most likely candidates are those from difunctional compounds like phenyldichlorophosphine and benzenephosphonyl chloride.   They give film-forming polymers with hydroquinone (31).

Silicon derivatives.   Almost all compounds having an Si—O—C linkage are hydrolyzed by atmospheric moisture, including ethyl silicate and polymers such as $(-O-CH_2-CH_2-O-SiR_2-)_n$; with respect to film formation these compounds do not seem promising (32).

Polyacetals.   Unlike most diamines, diols readily form linear polymers with aldehydes.   The aliphatic polyformals have low melting points and are not suitable for film.   It seems probable that those made from cyclic or aromatic diols will be better.   The hydrolytic stability increases in the acetals of the series: formaldehyde (CH_2O), acetaldehyde (CH_3CHO), acetone (CH_3COCH_3).   Naturally occurring polyacetals such as cellulose are discussed in Chapter 4.

G. DITHIOL POLYMERS

Thioesters.   Mercaptans, RSH, react with acids to give thioesters, $R'COSR$, that are probably at least as stable as ordinary esters. Little work has been done with dithiols, and none is commercially available.   If they should become of interest it seems likely that a few of them could be produced cheaply, particularly 1,4-butanedithiol, $HS(CH_2)_4SH$ (33).   Aliphatic polythioesters should have higher melting points than the oxygen compounds, and might possibly be film-forming.

In addition to forming polymers with dicarboxylic acids, dithiols should also polymerize with inorganic acids such as sulfuric acid and benzenephosphonic acid, $C_6H_5PO_3H_2$.   Another type of polythioester could be made from ω-thiolacids, especially β-thiolpropionic acid, $HSCH_2CH_2CO_2H$, which is easily obtained from acrylic acid or by the hydrolysis of the β-thiolnitrile.

Hydrogen disulfide, H_2S_2, is the simplest dithiol and should also produce polymers with dibasic acids, but none has been reported.

Polydisulfides. As mercaptans are readily oxidized to disulfides, dithiols should produce polymeric disulfides, $[(-CH_2)_x-S-S-]_n$.

Metal salts. Thiols are very weak acids and form hydrolytically stable salts only with heavy metals. A possibility exists that a dithiol might produce such a polymer as the zinc or tin salt. The former should be especially interesting since covalent bonds to zinc are at an angle of 180° and cyclization therefore would be unlikely.

Polythioacetals. Thioacetals are very stable thermally and hydrolytically. Their polymers, such as the formaldehyde derivative of butanedithiol, should also be very stable. As with many other sulfide types of compounds, the stability and stiffness could both be increased, if desired, by oxidizing to a polysulfone.

H. Polymers from Dihalides

Organic halides can be converted directly to a large number of other compounds by substitution reactions. Polymers could at least theoretically be made by the reaction of dihalides with difunctional reagents: for instance, polyethers from metal alkoxides (glycoloxides), polyesters from salts of dicarboxylic acids, polyamines from diamines or primary monoamines, polysulfides from sodium sulfide or dithiols, polydisulfides from sodium disulfide, polyamides from diamides, and polysulfonamides from the disodium salt of a disulfonamide.

Many of these products are made more easily by other methods. The amine-forming reaction is complicated by the possibility of forming tertiary amines and quaternary ammonium salts. Compounds of the type $Cl-(CH_2)_x-NR_2$ polymerize spontaneously to polysalts, $[-(CH_2)_x-NR_2^+]_nCl^-$.

Polyethers in general are low-melting materials, but the *polythioethers* (sulfides), $(-R-S-)_n$, seem to have more rigid structures and might have sufficiently high melting points for their use in preparing films. Oxidation to *polysulfones* $(-RSO_2-)_n$ produces high-melting polymers (see Section III-C).

Thiokol, often represented as $[-CH_2-CH_2-S(S)-S(S)-]_n$, may derive some of its rubbery character from the fact that the sodium polysulfide, Na_2S_x, from which it is made, is a mixture, and the polymer is actually a copolymer of $C_2H_4S_2$, $C_2H_4S_3$, $C_2H_4S_4$, etc.   The homopolymer of $C_2H_4S_2$ may be crystalline.

I. Carbon–Carbon Condensations

There are several condensation reactions that result in the formation of carbon chains.   Among these are the many base-promoted condensations of *active-hydrogen compounds*, such as the aldol condensation, the Perkin reaction, the Claisen condensation, and many others.   These are not particularly promising for polymers, however, because the necessary activating groups—ester, aldehyde, etc.—would be found as substituents in the final product and would interfere with crystallization.

Phenol–formaldehyde polymers may be considered as formed by the reaction of active-hydrogen compounds, the *ortho* and *para* hydrogen atoms of phenol being replaced   If the *para* position is blocked, as in *p*-cresol, *p*-$CH_3C_6H_4OH$, or in *p*-chlorophenol, there is the possibility of forming a linear polymer through *ortho*-methylene bridges.   It is possible to draw on paper a planar model of the polymer which seems as if it would pack well in a crystalline arrangement.   No film-forming polymer has yet been made from *p*-cresol however.

The *Wurtz reaction* (elimination of halogen between two molecules of alkyl halides) is subject to too many side reactions to be practical for polymers, although Carothers was able to produce carbon chains as long as 70 (possibly 90) by the reaction of decamethylene dibromide with sodium (17).   The *Wurtz-Fittig reaction* between aromatic and aliphatic halides by elimination of halogen proceeds somewhat more smoothly, but has not yet been used for polymers.   *p*-Dichlorobenzene possibly might give a useful product with sodium and ethylene chloride or tetramethylene chloride if the proper conditions for condensation were found.

The *Friedel-Crafts reaction* can lead to polymers, but generally, branched structures result.   For instance, the polymer from benzyl chloride, $(-C_6H_4CH_2-)_n$, is a black tar.   One possibility for obtaining linear polymers is durene, 1,2,4,5-tetramethylenebenzene. Only two positions are open and condensation reactions should produce linear polymers.   The monochloromethylation product should also give a linear polymer if rearrangement of the methyl groups can be avoided.

Diazoalkanes. Diazomethane, CH_2N_2, produces polymethylene by the loss of nitrogen. Diazoethane, CH_3CHN_2, produces a crystalline polyethylidene in the presence of colloidal gold (34). The polymers have some unique properties, but the monomers are too expensive to be of practical use.

II. Addition Polymers from Ethylenic Monomers

Polyethylene, the prototype of olefin polymers, is made by the addition polymerization of ethylene, $CH_2{=}CH_2$, the simplest olefin (35–38). Many other compounds containing a carbon–carbon double bond can be polymerized in a similar manner. Conjugated diolefins such as 1,3-butadiene, $CH_2{=}CH{-}CH{=}CH_2$, may polymerize by 1,2-addition to give a polymer with the structure $[-CH_2{-}CH(CH{=}CH_2){-}]_n$, or by 1,4-addition to give $[-CH_2{-}CH{=}CH{-}CH_2{-}]_n$. Some other diolefinic compounds polymerize by a cyclopolymerization process. All of these types of polymers are discussed in this section. Because of the enormous number of known and possible polymers in this classification only a few members of the class are mentioned.

A. POLYETHYLENE

Polyethylene is today probably the best-known film-forming polymer. When made by one of the low-pressure processes, it is almost completely linear, with a high degree of crystallinity and a melting point of up to 136°C. High-pressure processes give branched polymers, containing up to three branch points per 100 carbon atoms. The branched polymers are less crystalline and have lower melting points—in the range of 105–120°C; their softening points are still lower. As concerns mechanical properties, the linear polymers are denser, stiffer, harder, and have higher yield strength, while the branched polymers are more flexible and have higher impact strength (see Volume II).

B. FLUORINATED ETHYLENES

Polytetrafluoroethylene molecules, $(-CF_2{-}CF_2{-})_n$, have the same symmetry as polyethylene, although the crystal structure is not the same. Its very high melting point (over 400°C) is attributable in part to a more complete filling of space by the fluorine atoms, as compared with hydrogen atoms (radius 1.35 Å versus

1.2 Å), and there is consequently less freedom of motion in the polymer.

Vinyl fluoride, $CH_2\!\!=\!\!CHF$, *vinylidene fluoride,* $CH_2\!\!=\!\!CF_2$, and *trifluoroethylene,* $CHF\!\!=\!\!CF_2$, all give crystalline film-forming polymers. The melting points, approximately 175, 190, and 350°C, respectively, are higher than that of polyethylene even though the polymers have lower symmetry. The fluorines probably fit very tightly in the place of the hydrogens in polyethylene, and lack of symmetry is more than compensated for by an increase in rigidity.

Substituted fluoroethylenes may give crystalline polymers in instances in which the analogous hydrogen compounds do not. Thus, *trifluorochloroethylene,* $CF_2\!\!=\!\!CFCl$, gives a more crystalline polymer than does vinyl chloride. Hexafluoropropylene, $CF_2\!\!=\!\!CF\!\!-\!\!CF_3$, is known (39), but the polymer has not been described as crystalline.

The only potentially cheap fluorinated monomer is vinyl fluoride, made from acetylene and hydrogen fluoride. The polymer is more stable than polyvinyl chloride.

C. Vinyl Compounds

Vinyl compounds have the general formula $CH_2\!\!=\!\!CHX$. They give polyethylenes having a single substituent on every other carbon atom. If the substituent is large, it will not fit well in place of hydrogen in the lattice of crystalline polyethylene. Furthermore, the carbon atom bearing the X group becomes an asymmetric center in the polymer. If the D and L configurations are randomly distributed along the chain, the polymer is said to be *atactic,* and no regular crystalline arrangement is possible. *Isotactic* polymers, prepared with stereospecific catalysts, are described in a later section.

Only the smallest substituents are likely to permit crystallinity in an atactic polymer. The probable order of increasing size of various substituents is F, OH, NH_2, Cl, Br, CH_3, CN, OCH_3, C_2H_5, C_6H_5. Pauling (40) gives data on van der Waals (nonbonding) radii of atoms and groups as follows (in Angstrom units): H, 1.2; F, 1.35; Cl, 1.80; Br, 1.95; CH_3, 2.0; I, 2.15.

Polyvinyl fluoride, with the smallest substituent, is crystalline, as was mentioned above.

Polyvinyl alcohol, $(-CH_2\!\!-\!\!CHOH-)_n$, made by the hydrolysis of polyvinyl acetate, gives evidence of crystallinity when stretched.

Polyvinyl chloride, $(-CH_2CHCl-)_n$, is slightly crystalline. In this case, the repeat distance along the chain, according to x-ray information, corresponds to two monomer units, possibly indicating a tendency of chlorine atoms to alternate from one side of the chain to the other.

Polyvinylamine has been made by the hydrolysis of polyvinyl-succinimide and polyvinylphthalimide (41). The preparation of polyvinylamine perchlorate has also been reported (42). Succinimide and phthalimide polymers have also been subjected to partial hydrolysis (43).

Polyvinyl pivalate. Polymers and copolymers of vinyl pivalate have been described (44,45). A satisfactory method of preparation of the monomers is by ester exchange between vinyl acetate and pivalic acid (46), or from pivalyl chloride and acetaldehyde in pyridine (47). No simple way of making pivalic acid is available. The polymer is probably atactic and noncrystalline like other vinyl polymers, but the branched pivalate group imparts considerable rigidity to the chain, indicated by a softening point of 65–70°C as compared with about 35°C for polyvinyl acetate (48).

Polyvinyl bromide may be similar to the chloride in structure, but it is too unstable to be useful.

Monomers containing groups as large as methyl or larger give amorphous polymers. Thus, atactic polypropylene (prepared in solution, bulk, or emulsion), polyacrylonitrile, polyacrylic acid and its esters, polyvinyl ethers and esters, polystyrene, and many others, which together comprise the great majority of all synthetic polymers, are all amorphous. The same is true of polymers from disubstituted monomers of the type $CH_2=CXY$, such as methyl methacrylate, methacrylonitrile, itaconic acid, and α-methylstyrene.

Vinyl compounds, however, cannot be completely dismissed from consideration in making film. *Polystyrene,* $[-CH_2-CH(C_6H_5)-]_n$, for instance, is capable of a limited degree of orientation. A film of oriented polystyrene, Polyflex I, made by Plax, is clear and fairly stiff, with a metallic ring; it is as strong as cellophane in some respects, but is brittle, having low impact strength and stress–flex resistance. It also has a rather low softening point of about 90°C.

Polyvinyl chloride is extrudable into film when properly stabilized and plasticized. The films are rather limp, however, and have low softening points.

Polyacrylonitrile, $(-CH_2-CHCN-)_n$, is capable of a high degree of orientation. X-ray diffraction shows a one-dimensional *lateral* order, indicating definite spacing between chains, but not along the chains. This is attributed to the formation of hydrogen bonds between the nitrogens of one chain and the alpha hydrogens of another. No commercial films are made of polyacrylonitrile, but it forms a good fiber (Orlon, Acrilan) and has at least a theoretical potential for film.

Polyvinyl alcohols, made by the partial or complete hydrolysis of polyvinyl acetate, have been used for a variety of purposes, including film (49). They are all hygroscopic; those that are more than 75% hydrolyzed are soluble in water, although water uptake is slow. For most purposes they are plasticized with glycerol. The films are good gas and oil barriers, and they may be heat sealed at 160–200°C without a coating. They have desirable properties of toughness, stretch, and light resistance, all of which are needed for some applications. They have some specialty applications, since an infinite variety of derivatives can be obtained, most of which are soft and sticky, of which the best known is polyvinyl butyral, which is used in safety glass laminations.

Generally speaking, atactic vinyl polymers as a class probably lack the strength, flexibility, solvent resistance, and high softening points required for a packaging film. If any should be considered, they are probably those having polar groups to give added strength. *Polyacrylonitrile* and *polyacrylamide* are possibilites.

1. Vinyl Polymers: A Summary of Types

The preceding remarks upon what kinds of vinyl polymers may be suitable for film fails to give an inkling of the truly tremendous number of known vinyl polymers. For the sake of completeness a list has been compiled of monomers of the type $CH_2=CHX$ or $CH_2=CXY$ that have been polymerized. It should be pointed out that many of the entries are merely classes that include a great many specific examples. Furthermore, monomers which will homopolymerize will generally copolymerize, so that if there are x different monomers there will be almost x^2 binary copolymer systems (each having an infinite variety of compositions), x^3 ternary systems, and so forth, not to mention block and graft copolymers, as well as polymers modified chemically after formation, e.g., by oxidation, reduction, chlorination, sulfonation, or hydrolysis. To enumerate all of the possibilities is an endless task.

Hydrocarbons. *Alpha-olefins.* This category includes the straight-chain olefins, propylene, 1-butene, etc., as well as branched ones such as 5-methyl-1-butene. Olefins of the type $CH_2\!=\!CR_2$, where R is an alkyl group larger than methyl, have received little attention.

Vinyl aromatics. The prototype is styrene, upon which a monograph has been written (50). Innumerable substituted styrenes containing halogens, and alkyl, nitro, and other groups have been polymerized. 2,5-Dichlorostyrene, for instance, gives a polymer with a higher softening point than that of styrene. Sulfonated polystyrene, usually crosslinked, is used as an ion-exchange resin. Vinylnaphthalene and vinylbiphenyl are illustrations of other kinds of vinyl aromatics.

Vinyl Heterocyclics. The vinylpyridines are used in copolymers, especially for anion-exchange resins. Polymers from 2-vinylfuran are of little interest because of the reactivity of the furan ring. On the other hand, polymers from 2-vinyldibenzofuran have a high softening point, and have been used in capacitors, although they are somewhat brittle. Vinylferrocene is red, and has been converted to a red polymer. Other vinyl heterocyclics are also possible.

Halogen Compounds. In addition to the vinyl halides, mixed vinylidene halides, e.g., $CH_2\!=\!CFCl$, have been polymerized. Polymers formed from allyl chloride or longer-chain unsaturated alkyl halides are not known. Several polymers containing the trifluoromethyl group are known (48).

Vinyl Oxygen Compounds. *Vinyl ethers.* Many compounds of the type $CH_2\!=\!CHOR$ have been polymerized. The R group may be a hydrocarbon radical or a radical containing various functional groups, as in vinyl β-aminoethyl ether and the vinyl ether of methyl glycolate.

Vinyl esters. Polyvinyl acetate and its derivatives, e.g., polyvinyl alcohol, and the polyvinyl acetals, are well known (49). A wide variety of other vinyl esters has been polymerized.

Vinyl Nitrogen Compounds. *Vinyl amines.* Vinylamine is not known, but several of its derivatives have been made and polymerized, of which the best known is *N*-vinylcarbazole. The polymer is a useful dielectric material.

Vinyl amides. Polyvinylpyrrolidone is useful in synthetic blood plasma. *N*-Vinylcaprolactam has also been polymerized.

Vinyl imides. Vinyl phthalimide forms a polymer of interest as a possible source of polyvinylamine (41).

Nitroethylene. Nitroethylene gives a high-melting, but not very stable, polymer.

Vinyl Sulfur Compounds. *Vinyl alkyl sulfides* do not polymerize well except in copolymerization.

Vinyl alkyl sulfoxides and sulfones polymerize readily. Divinyl sulfone has been used as a crosslinking agent.

Vinylsulfonic acid. This compound and its salts are useful ingredients in cation-exchange resins. They are useful also as emulsion stabilizers in copolymerization.

Compounds of Vinyl Radicals with Miscellaneous Elements. Stable compounds are known with a vinyl group attached to boron, silicon, tin, lead, phosphorus, or arsenic, as well as polymers from vinylsilicon and vinylphosphorus compounds, but they are not of much current interest.

Allyl Compounds. A large number of allyl compounds is known, but few, if any, are of interest in film-forming polymers. Compounds such as allylbenzene, allyl chloride, allyl alcohol, allyl alkyl ethers, allyl esters of monobasic acids, and allyl amines will undergo addition polymerization, but since they all contain an allylic hydrogen, activated by another group, they are subject to chain transfer, with the formation of low polymers. This situation is more than counteracted in diallyl esters such as diallyl phthalate, which is useful in crosslinked polyesters resins. Allylstarch can be pressed to a film capable of crosslinking for increased stiffness. It does not appear possible to make linear high polymers from monoallyl compounds.

Acrylics. Acrylics comprise all those polymers that are related to acrylic acid, $CH_2{=}CH{-}CO_2H$, and methacrylic acid, $CH_2{=}C{-}(CH_3){-}CO_2H$. The most important examples are polymers of methyl methacrylate and acrylonitrile (vinyl cyanide). Copolymers of acrylonitrile with almost 100 other monomers are known (51). In addition to acrylic acid, its nitrile and amide, innumerable esters of acrylic acid and methacrylic acid have been used in polymers (52). Unusual ones include esters of hydroxymethylpenta-methyldisiloxane and long-chain polyethylene glycol monoesters. The softening points of the polymers generally decrease with increasing monomer size. Other monomers related to acrylic acid include acrolein, itaconic acid, maleic anhydride, fumaric acid, esters of maleic and fumaric acids (used in copolymers), vinylidene cyanide, and many vinyl ketones.

2. Vinylidene Compounds

The symmetry that is lost in vinyl polymers is partly restored in vinylidene polymers. Thus, *polyvinylidene chloride*, $(—CH_2—CCl_2—)_n$, is much more crystalline than polyvinyl chloride, and has a melting point of nearly 200°C. The structure is somewhat similar to that of polyethylene, but it is found that the chlorine atoms on adjacent monomer units in the same chain interfere with each other in the planar zigzag structure. Accordingly, the monomer units are twisted alternately from right to left so that the repeat distance in the crystalline polymer corresponds to the length of two monomer units.

Copolymers of vinylidene chloride are used for paper, cellophane, and polyolefin coatings; Dow sells Saran Wrap, a heavily plasticized film of similar composition. Although the latter is rather limp, its stiffness can be regulated by varying the composition of monomers and the amount of plasticizer. The pure homopolymer is very stiff and softens only at about 200°C; it is not film-forming. Its main disadvantage has been thermal instability; even copolymers with much lower softening points require stabilizers. It is possible that a saran film can be made with the required stiffness and stability of a good packaging film, but as yet none has been developed. Block copolymers of vinylidene chloride would be of interest.

Isobutylene, $CH_2\!\!=\!\!C(CH_3)_2$, is structurally similar to vinylidene chloride, but there is enough difference in size between a chlorine atom (radius 1.8 Å) and a methyl group (radius 2.0 Å) that the polymer molecules cannot exist in an extended form unless the polymer as a whole is under tension. Polyisobutylene is therefore rubbery.

Vinylidene cyanide, $CH_2\!\!=\!\!C(CN)_2$, makes a crystalline, fiber-forming polymer (53,54). The cyanide group has a radius of about 1.6 Å and cyanide groups in a 1,3-position along a chain do not interfere with one another, permitting the polymer to exist as extended chains.

Where the X groups in $CH_2\!\!=\!\!CX_2$ are as large as methyl there is little hope of getting a crystalline polymer, as the molecule will of necessity be coiled. This is true of all 1,1-dialkylethylenes, as well as of cyclic compounds such as methylenecyclohexane.

Ketene acetals, $CH_2\!\!=\!\!C(OR)_2$, generally give low polymers. The cyclic ethylene glycol acetal of ketene, however, gives a high-melting high polymer (55). A model shows that this polymer can exist in an extended form, there being less 1,3-interaction than in polyisobutylene.

3. Vinylene Compounds

1,2-Disubstituted ethylenes, $CHX\!\!=\!\!CHX$, generally do not undergo homopolymerization. Exceptions are the cyclic *vinylene carbonate* (obtainable from ethylene carbonate by chlorination and dehydrohalogenation) (56), *dialkyl fumarates* ($ROOC\!-\!CH\!\!=\!\!CH\!-\!COOR$), *indene* (1,2-benzocyclopentadiene), and coumarone (2,3-benzofuran). The only polymer of possible interest is polyvinylene carbonate, which is not as yet economically attractive.

Maleic anhydride does not homopolymerize, but forms copolymers readily—with styrene, for instance. With stilbene ($C_6H_5\!-\!CH\!\!=\!\!CH\!-\!C_6H_5$), which also does not homopolymerize, an interesting $1:1$ alternating copolymer is obtained. This kind of polymer is probably crystalline.

4. Isotactic and Syndiotactic Polymers

The polymerization of a vinyl monomer in bulk, solution, or emulsion normally results in a random stereochemical configuration of the asymmetric carbon atoms. As a result the polymer is atactic, and cannot crystallize. It may be reasoned, however, that in any growing polymer chain the configuration of the terminal asymmetric monomer unit will have some influence on the way in which a new monomer unit will add. Indeed, if the side chain is very bulky, it might be expected to direct the next side chain to the opposite side of the molecule, and a syndiotactic structure (alternating D and L units) would result. This is an intriguing possibility, and one that has been little investigated.

The difference in properties between polystyrenes prepared at different temperatures has been considered evidence for a certain degree of alternation from D to L configurations at the low temperatures, and more randomness at higher temperatures (57). There also is evidence for alternation in poly-α-methyl-*p*-methylstyrene (58) and in polyvinyl chloride (59).

The first truly crystalline vinyl polymers were prepared from vinyl ethers by Schildknecht in 1947 (60). Whereas vinyl ethers undergo a rapid "flash" polymerization in the presence of boron fluoride, boron fluoride etherate at −80°C leads to a slow "proliferous" polymerization. The flash polymer is amorphous, rubbery, tacky, and nonorientable, while the proliferous polymer is crystalline, fiber-forming, and orientable. It was thought at first that the polymer from *vinyl isobutyl ether*, $CH_2=CH-O-CH_2-CH(CH_3)_2$, was syndiotactic, but x-ray diffraction data seem to favor the isotactic form (61). Nothing has been reported on melting points or film-forming potential. If the melting points of the polyvinyl ethers are too low for film, those of the analogous thioethers might be high enough.

The first isotactic hydrocarbon polymers were reported by Natta (62) as a result of heterogeneous catalysis on a solid Ziegler catalyst. The effect is particularly striking in the case of propylene and styrene, which give crystalline polymers melting at 165°C and 230°C, respectively, whereas atactic polypropylene is a viscous liquid and polystyrene is an amorphous solid with a low softening point.

Straight-chain olefins higher than propylene also give isotactic polymers, but their melting points are lower than that of polypropylene. Isotactic poly-1-butene, for instance, melts at 126°C. Some of the branched olefins, however, give polymers with remarkably high melting points: polyisobutylethylene, mp 235°C; polyisopropylethylene, 300°C; polyneopentylethylene, >320°C; polycyclohexylethylene 322°C (63). Poly-4-methyl-1-hexene melts at 183–188°C (64). None of these olefins is readily available. Isobutylethylene can perhaps be isolated in the dimerization of propylene; it might also be possible to isomerize the 4-methyl-2-pentene, which is the major constituent of dipropylene.

Isotactic polypropylene is capable of extrusion and gives a very clear, tough, stable, orientable film. While the melting point of the pure isotactic polymer (165°C) is probably high enough for any packaging film, a rather difficult separation from atactic and partly isotactic polymer is required, in addition to thorough washing to remove catalyst residues, which tend to degrade the polymer. These problems have been solved in a sufficiently economical manner that an ultimate cost not much greater than that of linear polyethylene is reasonable.

It is difficult at present to envision any of the higher olefins competing in price with propylene. For specialty purposes, where a higher melting point is desired, isotactic polymers from isobutylethylene or styrene might be considered. *Styrene* is a practicable starting material, although it has given isotactic polymer only in low yields ($<30\%$).

The first definitely established syndiotactic polymer was that from butadiene (see Section II-D). In this case the syndiotactic polymer melts at a temperature 34°C higher than the isotactic isomer. If this difference is general, it might be expected that syndiotactic polypropylene, polybutylene, and polystyrene would have melting points of about 200, 160, and 260°C, respectively, which would be extremely interesting from a practical standpoint. Research directed toward obtaining such polymers, as by the use of catalysts based on chromium and vanadium complexes, might result in an entirely new class of polymers.

An interesting polymer which possibly belongs in this category is the crystalline fraction obtained from norbornene (ethylene–cyclopentadiene adduct) with a Ziegler catalyst. The polymer is believed to have the following structure (65):

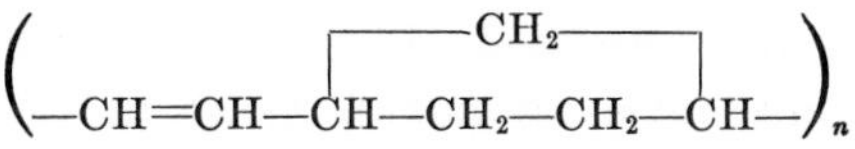

$$\left(\underset{-CH=CH-CH-CH_2-CH_2-CH-}{\overset{\lceil---CH_2---\rceil}{}} \right)_n$$

Apart from olefins, very interesting possibilities exist in the preparation of isotactic and syndiotactic polymers from substituted olefins. Those possibilities below serve merely to illustrate the size of the area that remains to be explored.

1. The preparation of isotactic polyvinyl acetate by a proliferous cationic polymerization like that of Schildknecht.

2. The preparation of isotactic polyvinyl alcohol by the hydrolysis of isotactic polyvinyl ethers or esters.

3. The preparation of syndiotactic or isotactic polyvinyl chloride, polyacrylonitrile, polystyrene, etc. by a free-radical process, using as low a temperature as possible.

4. The preparation of syndiotactic or isotactic polyacrylonitrile by an anionic process using a solid catalyst such as sodium amide or lithium dispersions, keeping the temperature as low as possible.

5. The preparation of syndiotactic polyolefins, from propylene and 1-butene in particular, with the aid of a special catalyst such as that for syndiotactic polybutadiene (see the following Section D).

D. Dienes

Isoprene. Most synthetic rubbers are diene polymers or copolymers. Natural rubber consists almost entirely of *cis*-1,4-polyisoprene.

$$\left[\begin{array}{c} \underset{\displaystyle -CH_2}{\overset{\displaystyle H}{\diagdown}} C = C \underset{\displaystyle CH_2-}{\overset{\displaystyle CH_3}{\diagup}} \end{array} \right]_n$$

The naturally occurring *trans* isomer, gutta percha, is crystalline rather than rubbery; it has not been investigated for film or fiber formation. Hydrogenation of either introduces an asymmetric carbon atom and makes an amorphous, atactic polymer, hydro-rubber. Most synthetic rubbers are mixtures of *cis* and *trans* forms.

Butadiene. With a Ziegler catalyst, butadiene, $CH_2=CH—CH=CH_2$, may be converted to either form at will, as well as to two crystalline 1,2-polymers (66). The *cis* form, amorphous and rubbery, results from a 1:1 ratio of trialkylaluminum to titanium tetrachloride. The *trans* form, mp 130°C, results from a 2:1 ratio; the *trans* form also results with titanium trichloride or vanadium trichloride and any ratio of aluminum alkyl. When the metal halide is replaced by oxygen or nitrogen complexes of chromium or vanadium, e.g., chromium acetylacetonate, the syndiotactic 1,2-polymer, mp 154°C, is formed, unless the ratio of aluminum to transition metal is greater than 6:1; in this case the isotactic 1,2-polymer, mp 120°C, is formed. (For the melting points see ref. 67.) It is particularly interesting that the melting point of the *trans* form is 6°C lower than that of the hydrogenated product, linear polyethylene, and that isotactic polybutadiene melts at a temperature 6°C lower than isotactic poly-1-butene, mp 126°C. It may be guessed, therefore, that hydrogenation of the syndiotactic polybutadiene will produce syndiotactic poly-1-butene (as yet unknown) with a melting point of about 160°C—a very satisfactory melting point for film.   For that matter, syndiotactic polybutadiene may be satisfactory for film.

p-Xylylene, which has the structure

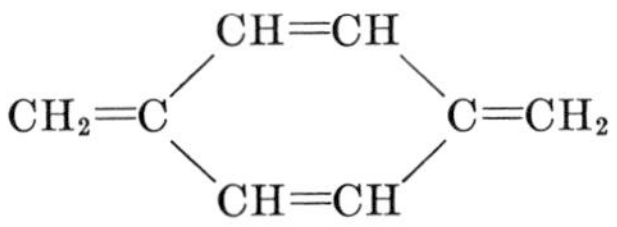

$$CH_2=C \overset{\displaystyle CH=CH}{\underset{\displaystyle CH=CH}{\Big\langle \qquad \Big\rangle}} C=CH_2$$

is obtained when *p*-xylene, $p\text{-}CH_3C_6H_4CH_3$, is pyrolyzed at 1000°C and the vapors are quenched at −78°C.   On warming to room temperature it polymerizes spontaneously to poly-*p*-xylylene by a 1,6-addition.   The polymer melts at 380°C; related polymers all have high melting points (68,69).

Other Dienes.   Bailey has prepared a series of dienes related to 1,2-dimethylenecyclohexane.   Polydimethylenecyclohexane is a crystalline material melting at 165°C; on hydrogenation it gives a syndiotactic (?) polymer melting at 110°C (70).

E.   CYCLOPOLYMERIZATION

Diallyl compounds normally give crosslinked polymers.   If the allyl groups are separated by a single atom, however, as in diallyl ether, $(CH_2=CH—CH_2—)_2O$, a new kind of linear polymer forms, with the following structure:

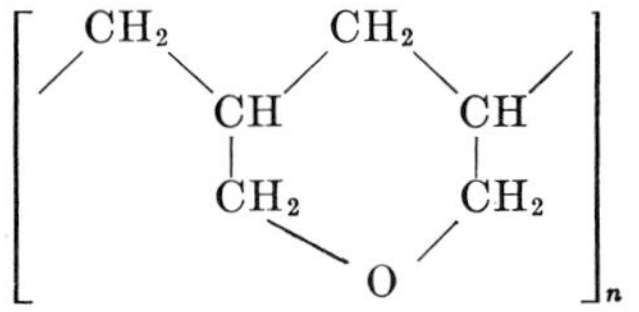

$$\left[\begin{array}{c} CH_2 \qquad CH_2 \\ / \quad \backslash \ / \quad \backslash \\ CH \qquad CH \\ | \qquad\quad | \\ CH_2 \qquad CH_2 \\ \backslash \qquad / \\ O \end{array} \right]_n$$

The reaction is best known in the case of diallylammonium salts (71). Similar reactions probably occur when the oxygen or nitrogen is replaced by sulfur, sulfoxide, sulfone, methylene, etc.

The term *cyclopolymerization* was coined for a more recent reaction, the polymerization of diacrylylmethane, $(CH_2=CH—CO—)_2CH_2$, under alkaline catalysis, to a brilliant yellow polymer of a structure similar to that of the diallyl polymers (72).

Divinyl acetals, such as divinyl butyral, $(CH_2=CH—O—)_2\text{-}CH—C_3H_7$, might be expected to behave similarly because the double bonds are in the same spatial relationship as in the previous examples. The polymers are all crosslinked (73), however, as would be expected of ordinary divinyl compounds.

Divinyl carbonate has not been prepared, but it perhaps could be made from ethylene oxide by way of bischloroethyl carbonate. This monomer also has structural similarities to diallyl ether, and might give a linear polymer.

III. Nonolefinic Addition Polymers

A variety of polymers containing elements other than carbon in the chain can be made by addition reactions. These may be classified in the following groups: (*1*) polymers from monomers containing C=O, C=S, or C=N double bonds; (*2*) polymers from cyclic monomers; (*3*) polymers requiring two dissimilar monomers in a 1:1 ratio; and (*4*) copolymers between olefins and nonolefins in nonstoichiometric ratios.

A. Polymers from Carbonyl, Thiocarbonyl, and Imino Compounds

Formaldehyde. Formaldehyde, CH_2O, as well as its trimer, trioxane, are converted by acid catalysts to polyoxymethylenes of high molecular weight. When suitably stabilized against depolymerization, the polymer has been found useful as a molding compound, sold by du Pont as Delrin (74). It is very tough and resilient.

Acetaldehyde, CH_3CHO, is converted by acids to a cyclic trimer (paraldehyde) and tetramer (metaldehyde) (75). Rubbery high polymer is formed with an alumina catalyst, and a crystalline polymer with organometallic catalysts (76,76a,76b).

Ketones do not form any polymers of the polyoxymethylene type; here the carbonyl group appears to be stabilized in monomeric form by the two alkyl groups.

Thioaldehydes and *thioketones* are of a different order of reactivity (77). The aliphatic compounds are known only in the form of polymers (cyclic and linear). Alkyl aryl thioketones are generally stable only as trimers. The monomer of thioacetophenone, $C_6H_5C(S)CH_3$, is obtained as a blue oil by heating the trimer at 140°C, but at room temperature it slowly reverts to the trimer. Diaryl thiones are known only as highly colored monomers.

Linear polymers of CH_2S, and copolymers with CH_2O, are formed by the reaction between formaldehyde, hydrogen sulfide, and hydrochloric acid.   They are analogous to polyoxymethylene and might be useful film formers.   In addition, they could be oxidized to polysulfones if it were desired to raise the melting point.

The isolation of a low polymer (decamer) of *thiobenzaldehyde*, C_6H_5CHS, makes it appear likely that higher polymers could also be formed.   The —S—$CH(C_6H_5)$— unit lacks lateral symmetry, however, and probably does not form a crystalline polymer.

Selenoketones and telluroketones appear to exist mainly in the trimeric form.

Imines, $R_2C{=}NH$, may be formed by the reaction between Grignard reagents and nitriles.   Corresponding trimers (hexahydrotriazines) are known, but no linear polymers have been reported.

B. Polyethers

Ethylene oxide, CH_2CH_2O, reacts with water and catalysts to give polymers of the type $HO{-}(CH_2{-}CH_2{-}O)_n{-}H$. The higher members of the series are crystalline, but the melting points are only about 66°C (78) and the polymers do not appear to be suitable for film.   They are soluble in water.

Propylene oxide, CH_3CHCH_2O, is a mixture of D and L isomers. With most catalysts it produces a random polymer which is amorphous and not film-forming.   The pure isomers, however, in the presence of potassium hydroxide or ferric chloride retain their optical configuration and give crystalline, isotactic, optically active polymers, which melt at 70–75°C (79).   Certain catalysts, e.g., one made from ferric chloride and propylene oxide, are stereospecific, each site reacting preferentially with one isomer.   As a result, the racemic mixture can be converted to a crystalline mixture of poly-D-propylene oxide and poly-L-propylene oxide, rather than to an amorphous copolymer.   The melting point is too low for film, but in principle, analogous compounds could be polymerized similarly. *Styrene oxide*, for instance, might give a crystalline polymer with a suitably high melting point.

Trimethylene oxide, oxetane, $CH_2CH_2CH_2O$, is the prototype of the four-membered cyclic ethers.   Like ethylene oxide (oxirane) it is strained, and polymerizes readily under the influence of acid catalysts (79a). It is not readily obtainable, however. Its 3,3-disubstitution products are better known. Penton, mp 180°C, the polymer obtained by the Hercules Research Center from 3,3-bischloromethyloxetane (ultimately from pentaerythritol), is tough, solvent-resistant, and noninflammable, and can be extruded as a strong transparent film (80). 3,3-Dimethyloxetane is perhaps more readily available by way of neopentylene glycol; its polymer has not been described.

Tetrahydrofuran, $CH_2CH_2CH_2CH_2O$, is only slightly strained, but it too will polymerize; the polymer has a low melting point (81). Substituted tetrahydrofurans, such as the 2-methyl derivative, are perfectly stable and do not polymerize (79a). A somewhat analogous compound, 1,4-cyclohexene oxide, is definitely strained, and polymerizes to a polymer with a very high melting point, 450°C (82).

Neither tetrahydropyran, nor dioxane, nor any other six-membered ether except trioxane has been polymerized, indicating a maximum stability in a six-membered ring. Polydecamethylene oxide was made by Carothers by the pyrolysis of polydecamethylene carbonate. Presumably, therefore, large-ringed ethers generally will polymerize.

C. Polythioethers and Polysulfones

The carbon–sulfur link is generally less reactive than the carbon–oxygen link, and should produce stable polymers in the form of thioethers, RSR, thioesters, RCOSR, sulfoxides, RSOR, and sulfones, RSO_2R. Futhermore, the melting points of sulfides, sulfoxides, and sulfones are higher than those of the corresponding ethers. Therefore, the possibility of using sulfur-containing polymers for film is good, although none has been reported for this purpose.

The preparation of polysulfides using dithiols and diolefins has been thoroughly explored (83–92); this is a free-radical reaction which may be performed in emulsion or solution to give polysulfides of high molecular weight. These in turn are oxidized to the polysulfone (84).

More recently polysulfones have been prepared by interaction of a dimercaptan with a dibromide in the presence of a base (cf. Section I-H) (93–95). The polysulfides obtained were oxidized with hydrogen peroxide in formic acid to the corresponding sulfones

$[—SO_2—(CH_2)_x—SO_2—(CH_2)_y—]$. For values of x and y, respectively 4-4, 4-5, 4-6, 5-5, 5-6, and 6-6, the melting points were 271 (dec), 246, 243, 243, 223, and 220°C respectively. [The melting point of hexamethylene sulfone had previously been reported by Carothers (96) as 196–8°C and by Marvel (84) as 205°C.] All of these polysulfones decomposed severely at, or slightly above, the melting temperatures. From the studies of Dainton and associates, it appears to be most improbable that stable melt-extrudable polysulfones will be made (97–103).

Ethylene sulfide, CH_2CH_2S, was many years ago reported to polymerize readily to a crystalline polymer melting at 170–175°C (104). It should be film-forming. The monomer is not commercially available. A crystalline polymer probably could not be made from propylene sufide unless it were isotactic.

Trimethylene sulfide has been reported to give low polymers (105). Possibly the same polymer is obtainable more readily from allyl mercaptan; this reaction is discussed later. The sulfur analog of Penton is not known, but it doubtless could be made, and it should have rather interesting properties.

Cyclic disulfides such as β,β'-oxydiethyl disulfide,

$$SCH_2CH_2OCH_2CH_2S$$

polymerize by an anionic cleavage of the S—S link (106).

D. Polyamines

Ethylenimine, CH_2CH_2NH, polymerizes readily, but is very expensive. Trimethylenimine, a four-membered ring, may polymerize, but little is known about it. Pyrrolidine and piperidine, five- and six-membered rings, are not known to polymerize. Linear polyamines can not be made by condensation reactions because of the formation of tertiary branched amines. Therefore, polyamines will probably remain unexplored.

E. Polymerization of Cyclic Lactones, Lactams, Formals, etc.

The formation of polyesters, polyamides, polyformals, etc. has already been discussed in Section I. This does not mean that an addition polymerization of a cyclic monomer may not sometimes be preferable.   6-Nylon, for instance, can be prepared far more rapidly by a catalytic polymerization of caprolactam than by the condensation of ϵ-aminocaproic acid.   Also, polyglycine may be prepared by the polymerization of the cyclic dimer, diketopiperazine, but not from glycine itself, $NH_2CH_2CO_2H$.

For most classes of ring compounds there is a characteristic ring size of maximum stability, corresponding to a minimum tendency to polymerize (17,79a).   For lactones the maximum occurs at 5 atoms, as in γ-butyrolactone, $\overline{OCH_2CH_2CH_2CO}$.   For lactams the number is probably 6, as in piperidone, $\overline{NHCH_2CH_2CH_2CH_2CO}$. For silicones the number seems to be 8 or 10, although all cyclic polydimethysiloxanes have a strong tendency to polymerize.

Little is known of ring–chain equilibrium in sulfur compounds. It may be possible to polymerize cyclic sulfides, sulfones, and sulfonamides.

F. Addition Polymers Requiring Two Monomers

In this category are those polymers which require two dissimilar monomers or groups because neither one polymerizes readily alone. One of the monomers has to be unsaturated in the sense that it contain a double bond or a reactive three-membered ring.

Isocyanate polymers.   The isocyanate group, $-N{=}C{=}O$, adds readily to compounds containing OH or NH groups to form urethanes, $-NH-CO-O-$, or ureas, $-NH-CO-NH-$.   With carboxylic acids an amide group, $-NH-CO-$, is formed, with loss of carbon dioxide; this is important in the formation of foam plastics. Diisocyanates react with diols and diamines to give linear polymers of considerable interest.

Polyurethanes have been used for several years in Germany for films and fibers.   They melt at lower temperatures than the corresponding polyamides.   Diisocyanates are based on diamines and are therefore subject to the same considerations of availability and cost.   The cheapest diisocyanate is tolylene diisocyanate, used mainly for polyurethane foams (107).

Epoxies.   1,2-Epoxides, such as ethylene oxide, add to alcohols and amines.   Diepoxides and diols or diamines might be expected to give linear polymers.   Actually, commercial "epoxy" resins

(108) are crosslinked resins. This crosslinkage results from the fact that the linear chains which might be formed contain OH groups which are capable of reacting with additional epoxide to form cross-links. Even without excess epoxide a branched, rather than linear, structure would result. It is likely that a diamine in place of a bisphenol or other dihydroxy compound would react so much faster than the hydroxyl groups produced that the crosslinking reaction would be virtually suppressed and a linear polymer would result—a polyaminoether. Such a polymer has not yet been reported.

Addition products of olefins. Olefins add to a great many substances and functional groups. Thus, under conditions where an alcohol will add to a monoolefin, diolefins and diols will give a polyether. In such a case the oxygen would not be joined to the terminal atom of the diolefin (Markownikoff's rule), and the polymer would not have the most desirable structure for a film. The only reactions seriously proposed for film-forming polymers involve the addition of S—H and Si—H bonds. Thus, a series of linear polymeric sulfides has been prepared by the addition of dithiols to diolefins (e.g., 1,5-hexadiene, $CH_2{=}CH{-}CH_2CH_2{-}CH{=}CH_2$) (83); benzoyl peroxide is a satisfactory catalyst. Butadiene cannot be used because it polymerizes by itself. Hydrogen sulfide probably can be used as the dithiol.

There is some evidence that *allyl mercaptan*, $CH_2{=}CH{-}CH_2SH$, will react similarly, producing polytrimethylene sulfide. It can be made cheaply from allyl chloride and sodium hydrosulfide. An analogous reaction of allyldimethylsilane, $CH_2{=}CH{-}CH_2SiMe_2H$, appears to be definitely possible.

G. Copolymers of Olefins

Copolymers with inorganic compounds. Olefins have been known to copolymerize with oxygen (109), sulfur dioxide (110), and carbon monoxide (111,112). The last two give polysulfones and polyketones. The ratio of monomers is not critical except that it is not possible to incorporate more inorganic monomer than olefin in the polymer, which indicates that adjacent carbonyl or sulfone groups are not produced. Polysulfones prepared in this manner have high melting points, but decompose rapidly above their melting points (see Section III-C).

Copolymers with cyclic monomers. The only example in this category is the anionic copolymerization of caprolactam and styrene mentioned by D'Alelio (113).   Since the temperatures required for the two monomers to polymerize are very different, it seems unlikely that the "copolymer" was other than a mixture of homopolymers. The possibility of further developments along this line has not been completely eliminated, however.

References

1. G. D. Buckley and N. H. Ray, U.S. Pat. 2,550,767 (May 1, 1951).
2. L. B. Sokolov, T. V. Kudim, and L. V. Turetskii, *Vysokomolekul. Soedin.*, **3**, 1369 (1961).
2a. H. K. Hall Jr. and J. W. Berge, *J. Polymer Sci. B*, **1**, 277 (1963).
3. T. Lieser, H. Gehlen, and M. Gehlen-Keller, *Ann.*, **556**, 114 (1944).
4. D. E. Hudgin and F. Brown, Abstracts of Papers, 132nd Meeting Am. Chem. Soc., New York, September 1957, 25T.
4a. M. Katz, *J. Polymer Sci.*, **40**, 337 (1959).
5. E. F. Carlston and F. G. Lum, *Ind. Eng. Chem.*, **49**, 1239 (1957).
6. A. E. Barkdoll, H. W. Gray, W. Kirk Jr., D. C. Pease, and R. S. Schreiber, *J. Am. Chem. Soc.*, **75**, 1238 (1953).
7. D. S. Breslow, G. E. Hulse, and A. S. Matlack, *J. Am. Chem. Soc.*, **79**, 3760 (1957).
8. J. Lincoln, U.S. Pat. 2,500,317 (March 14, 1950).
9. W. O. Ney, W. R. Nummy, and C. E. Barnes, U.S. Pat. 2,638,463 (May 12, 1953).
10. C. E. Barnes, W. O. Ney, and W. R. Nummy, U.S. Pat. 2,809,958 (Oct. 15, 1957).
11. W. E. Hanford and R. M. Joyce, *J. Polymer Sci.*, **3**, 171 (1948).
12. G. H. Berthold, U.S. Pat. 2,727,017 (Dec. 13, 1955).
13. M. Genas, U.S. Patent 2,462,855 (Mar. 1, 1949).
14. R. Aelion, U.S. Pat. 2,600,953 (June 17, 1952).
15. W. E. Hanford, U.S. Pat. 2,312,967 (Mar. 2, 1943).
16. K. Rössler and P. Schlack, Ger. Pat. 753,569 (Aug. 24, 1953); *Chem. Abstr.*, **52**, 4244 (1958).
17. H. Mark and G. S. Whitby, Eds., *Collected Papers of Wallace Hume Carothers on High Polymeric Substances*, Interscience, New York, 1940.
18. Goodyear, Brit. Pat. 766,290 (January 16, 1957); see also F. G. Lum and E. F. Carlston, Abstracts of Papers 121st Meeting Am. Chem. Soc., Milwaukee, March 1952, p. 4L (a preprint appeared in the papers of the Division of Paint, Varnish, and Plastics Chemistry).
19. R. Hill and E. E. Walker, *J. Polymer Sci.*, **3**, 609 (1948).
20. J. R. Caldwell, U.S. Pat. 2,799,666 (July 16, 1957).
21. J. Lincoln, U.S. Pat. 2,799,665 (July 16, 1957).
22. T. L. Gresham, J. E. Jansen, and F. W. Shaver, *J. Am. Chem. Soc.*, **70**, 998 (1948).

23. T. Alderson, U.S. Pat. 2,658,055 (Nov. 3, 1953).

24. C. A. Bischoff and P. Walden, *Ber.*, **26**, [2], 262 (1893).

25. D. D. Reynolds, K. R. Dunham, and J. van den Berghe, U.S. Pats. 2,789,509 and 2,789,964–972 (Apr. 23, 1957).

26. H. Schnell, *Angew. Chem.*, **68**, 633 (1956); see also Bayer, Brit. Pat. 772,627 (Apr. 17, 1957).

26a. C. E. Sroog, A. L. Endrey, S. V. Abramo, C. E. Berr, W. M. Edwards, and K. L. Oliver, *J. Polymer Sci.*, **A3**, 1373 (1965).

26b. H. Vogel and C. S. Marvel, *J. Polymer Sci.*, **50**, 511 (1961).

27. R. W. Moncrieff, U.S. Pat. 2,512,667 (June 27, 1950); J. W. Fisher, E. W. Wheatley, and H. Bates, U.S. Pat. 2,476,968 (July 26, 1949); U.S. Pats. 2,512,599–601, 2,512,624–634, and 2,512,891 (June 27, 1950).

28. *Mod. Packaging*, **38**, No. 7, 144 (March 1965); *Modern Plastics Encyclopedia* 1967, **44**, No. 1A, 17 (Sept. 1966).

29. K. Fukazawa, Japan. Pat. 2,137 (1952); *Chem. Abstr.*, **47**, 6697 (1953).

30. J. R. Wright, R. O. Bolt, A. Goldschmidt, and A. D. Abbott, *J. Am. Chem. Soc.*, **80**, 1733 (1958).

31. A. D. F. Toy, U.S. Pat. 2,435,252 (Feb. 3, 1948).

32. W. E. Hanford, U.S. Pat. 2,386,793 (Oct. 16, 1945); *Chem. Abstr.*, **40**, 604 (1946).

33. C. Colange and P. Stuchlik, *Bull. Soc. Chim. France*, **1950**, 832.

34. G. Saini, *Gazz. Chim. Ital.*, **87**, 342 (1957).

35. S. L. Aggarwal and O. J. Sweeting, *Chem. Rev.*, **57**, 665 (1957).

36. J. K. Stille, *Chem. Rev.*, **58**, 541 (1958).

37. R. A. V. Raff and J. B. Allison, *Polyethylene*, Interscience, New York, 1956.

38. A. Renfrew and P. Morgan, *Polythene*, 2nd ed., Interscience, New York, 1960.

39. D. A. Nelson, U.S. Pat. 2,758,138 (Aug. 7, 1956); J. Waddell, U.S. Pat. 2,759,983 (Aug. 21, 1956).

40. L. Pauling, *The Nature of the Chemical Bond*, 2nd ed., Cornell University Press, Ithaca, 1940, p. 189.

41. D. D. Reynolds and W. O. Kenyon, *J. Am. Chem. Soc.*, **69**, 911 (1947).

42. M. L. Wolfrom and A. Chaney, Tech. Rept. 675-5, Ohio State Research Foundation, May 1, 1957.

43. W. E. Hanford and H. B. Stevenson, U.S. pat. 2,365,340 (Dec. 19, 1944).

44. W. E. Hanford, J. R. Roland, and W. E. Mochel, U.S. Pat. 2,473,996 (June 21, 1949).

45. W. R. Cornthwaite and N. D. Scott, U.S. Pat. 2,381,338 (Aug. 7, 1945).

46. R. L. Adelman, *J. Org. Chem.*, **14**, 1057 (1949).

47. A. M. Sladkov and G. S. Petrov, *Zh. Obsch. Khim.*, **24**, 450 (1954); *Chem. Abstr.*, **49**, 6093 (1955).

48. C. A. Schildknecht, *Vinyl and Related Polymers*, Wiley, New York, 1952, p. 377.

49. F. Kainer, *Polyvinylalkohole*, Ferdinand Enke Verlag, Stüttgart, 1949.

50. R. H. Boundy, R. F. Boyer, and S. M. Stoesser, Eds., *Styrene*, Reinhold, New York, 1952, 1245 pp.

51. *The Chemistry of Acrylonitrile*, American Cyanamid Company, Beacon Press, New York, 1951.

52. E. H. Riddle, *Monomeric Acrylic Esters*, Reinhold, New York, 1954.

53. A. E. Ardis, *J. Am. Chem. Soc.*, **72**, 1305 (1950).

54. A. E. Ardis and H. Gilbert, U.S. Pat. 2,535,827 (Dec. 26, 1950).

55. S. M. McElvain and M. J. Curry, *J. Am. Chem. Soc.*, **70**, 378 (1948).

56. M. S. Newman and R. W. Addor, *J. Am. Chem. Soc.*, **77**, 3789 (1955).

57. M. L. Huggins, *J. Am. Chem. Soc.*, **66**, 1991 (1944); ref. 48, pp. 36–37.

58. Ref. 48, pp. 139, 145.

59. Ref. 48, p. 416.

60. Ref. 48, pp. 609–614.

61. C. W. Bunn and E. R. Howells, *J. Polymer Sci.*, **18**, 307 (1955).

62. G. Natta and P. Pino, *J. Am. Chem. Soc.*, **77**, 1708 (1955).

63. F. P. Reding, *J. Polymer Sci.*, **21**, 547 (1956).

64. G. Natta, P. Corradini, and I. W. Bassi, *Atti Accad. Nazl. Lincei, Rend. Classe Sci. Fis. Mat. Nat.*, **19**, 404 (1955).

65. W. L. Truett, D. L. Johnson, I. M. Robinson, and B. A. Montague, *J. Am. Chem. Soc.*, **82**, 2337 (1960).

66. G. Natta, *Rubber Plastics Age*, **38**, 495 (1957).

67. G. Natta and P. Corradini, *Atti Accad. Nazl. Lincei, Rend. Classe Sci. Fis. Mat. Nat.*, **19**, 229 (1955); **20**, 560 (1956).

68. M. Szwarc, *J. Polymer Sci.*, **6**, 319 (1951).

69. L. A. Errede and B. F. Landrum, *J. Am. Chem. Soc.*, **79**, 4952 (1957).

70. W. J. Bailey and M. J. Stanek, Abstracts of Papers 130th Meeting Am. Chem. Soc., Atlantic City, Sept. 1956, p. 20S; *Dissertation Abstr.*, **17**, 1459 (1957).

71. G. B. Butler and R. J. Angelo, *J. Am. Chem. Soc.*, **79**, 3128 (1957).

72. J. F. Jones, *J. Polymer Sci.*, **33**, 7 (1958).

73. Ref. 48, p. 618.

74. R. L. Craven, U.S. Pat. 2,481,981 (September 13, 1959).

75. E. C. Craven, *J. Soc. Chem. Ind.*, **63**, 251 (1944); *Chem. Abstr.*, **39**, 494 (1945).

76. J. Furukawa, T. Saegusa, H. Fujii, A. Kawasaki, H. Imai, and Y. Fujii, *Makromol. Chem.*, **37**, 149 (1960).

76a. J. Furukawa, T. Saegusa, and H. Fujii, *Makromol. Chem.*, **44–46**, 398 (1961).

76b. J. Furukawa and T. Saegusa, *Polymerization of Aldehydes and Oxides*, Interscience, New York, 1963.

77. E. Campaigne, *Chem. Rev.*, **39**, 1 (1946).

78. K. L. Smith and R. Van Cleve, *Ind. Eng. Chem.*, **59**, 12 (1958).

79. C. C. Price, M. Osgan, R. E. Hughes, and C. Shambelan, *J. Am. Chem. Soc.*, **78**, 690 (1956).

79a. P. A. Small, *Trans. Faraday Soc.*, **51**, 1716 (1955).

80. A. C. Farthing and R. J. Reynolds, *J. Polymer Sci.*, **12**, 503 (1954).

81. K. Hamann, *Angew. Chem.*, **63**, 231 (1951).

82. E. L. Wittbecker, H. K. Hall, Jr., and T. W. Campbell, *J. Am. Chem. Soc.*, **82**, 1218 (1960).

83. C. S. Marvel and R. R. Chambers, *J. Am. Chem. Soc.*, **70**, 993 (1948).
84. C. S. Marvel and P. H. Aldrich, *J. Am. Chem. Soc.*, **72**, 1978 (1950).
85. C. S. Marvel and G. J. Nowlin, *J. Am. Chem. Soc.*, **72**, 5026 (1950).
86. C. S. Marvel and A. H. Markhart, *J. Am. Chem. Soc.*, **73**, 1064 (1951).
87. C. S. Marvel and A. H. Markhart, *J. Am. Chem. Soc.*, **73**, 481 (1951).
88. C. S. Marvel and H. E. Baumgarten, *J. Polymer Sci.*, **6**, 127 (1951).
89. C. S. Marvel and A. H. Markhart, *J. Polymer Sci.*, **6**, 711 (1951).
90. C. S. Marvel and P. D. Ceaser, *J. Am. Chem. Soc.*, **73**, 1097 (1951).
91. C. S. Marvel and C. W. Roberts, *J. Polymer Sci.*, **6**, 717 (1951).
92. C. S. Marvel and H. N. Cripps, *J. Polymer Sci.*, **8**, 313 (1952).
93. Wingfoot Corp., Brit. Pat. 630,625 (Oct. 18, 1949); *Chem. Abstr.*, **44**, 3741 (1950).
94. H. D. Noether, *Textile Res. J.*, **28**, 533 (1958).
95. H. D. Noether, U.S. Pat. 2,534,366 (Dec. 19, 1950); British Celanese, Ltd., Brit. Pat. 661,811 (Nov. 28, 1951). *Chem. Abstr.*, **45**, 3652 (1951); **46**, 3800 (1952).
96. W. H. Carothers, U.S. Pat. 2,201,884 (May 21, 1940); *Chem. Abstr.*, **33**, 6389 (1939).
97. F. S. Dainton and K. J. Ivin, *Proc. Roy. Soc. (London)*, **A212**, 96 (1952).
98. F. S. Dainton and K. J. Ivin, *Proc. Roy Soc. (London)*, **A212**, 207 (1952).
99. G. M. Bristow and F. S. Dainton, *Proc. Roy. Soc. (London)*, **A229**, 509 (1955).
100. G. M. Bristow and F. S. Dainton, *Proc. Roy. Soc. (London)*, **A229**, 525 (1955).
101. F. S. Dainton, K. J. Ivin, and D. R. Sheard, *Trans. Faraday Soc.*, **52**, 414 (1956).
102. K. J. Ivin, *J. Polymer Sci.*, **25**, 229 (1957).
103. F. S. Dainton, J. Diaper, K. J. Ivin, and D. R. Sheard, *Trans. Faraday Soc.*, **53**, 1269 (1957).
104. M. Delepine and S. Eschenbrenner, *Bull. Soc. Chim. France*, **33**, 703 (1923); *Beilsteins Handbuch der Organischen Chemie*, Vol. E II 1, Springer, Berlin, 1941) p. 347; *Chem. Abstr.*, **17**, 3161 (1923).
105. R. W. Bost and M. W. Conn, *J. Elisha Mitchell Sci. Soc.*, **50**, 182 (1934); *Chem. Abstr.*, **29**, 1350 (1935).
106. F. S. Dainton, J. A. Davies, P. P. Manning, and S. A. Zahir, *Trans. Faraday Soc.*, **53**, 813 (1957).
107. J. H. Saunders and K. C. Frisch, *Polyurethanes*, Interscience, New York, 1962.
108. H. Lee and K. Neville, *Epoxy Resins*, McGraw-Hill, New York, 1947.
109. G. V. Schulz and G. Henrici, *Makromol. Chem.*, **18–19**, 437 (1956).
110. F. S. Dainton and K. J. Ivin, *Nature*, **162**, 705 (1948).
111. M. M. Brubaker and D. D. Coffman, *J. Am. Chem. Soc.*, **74**, 1509 (1952).
112. D. D. Coffman, P. S. Pinkney, F. T. Wall, W. H. Wood, and H. S. Young, *J. Am. Chem. Soc.*, **74**, 3391 (1952).
113. G. F. D'Alelio, *Fundamental Principles of Polymerization*, Wiley, New York, 1952, p. 345.

CHAPTER 4

NATURAL POLYMERS AS FILM FORMERS
A. Cellulose for Film Manufacture

DONALD R. WALTON

Rayonier, Inc., New York, New York

I. Cellulose

Cellulose is a polymer, constructed of anhydroglucose units, which occurs in nature in abundance. It is the building block which forms the supporting framework of trees and other plants. Its carbohydrate structure is related to other components of plants, namely, sugars and starches, but, unlike these materials, cellulose is not used for food by man. Although synthesis of cellulose is being investigated in the laboratory (1), it is a natural polymer, not a synthetic, and man depends on plant sources for cellulose supplies.

Nevertheless, cellulose is an important and versatile substance in great demand as a chemical raw material, as well as a useful fiber. It is inexpensive and can be grown on a sustained yield basis to be harvested as an agricultural crop. The crop intervals range from a year or less for cotton to many years for long-lived species of trees. Because of its wide distribution, cellulose is available locally in many areas of the world. Successful harvesting of cellulose, separation from other plant components, and economical purification involve large investments in land and equipment and a high degree of technical capability.

Purified cellulose, such as that used in the manufacture of films for flexible packaging materials, is termed *chemical cellulose* or *dissolving cellulose*. These names are also applied to similar products employed in the manufacture of regenerated cellulose fibers and chemical derivatives of cellulose.

Other types of cellulose used in fiber form as components of papers, textiles, and similar structures will not be discussed here.

Cotton, which has been known by man for thousands of years as a useful fiber, provides raw material for chemical cellulose in the form of "linters." These are the relatively short seed hairs which remain attached to the cotton seed after ginning has removed the longer fibers for textile operations. Cotton linters, like the staple cotton fiber, are a relatively pure form of cellulose, and it is not difficult to prepare linters in a form suitable for dissolving end uses. Historically, cotton linters were the first major source of chemical cellulose, and for many years they were generally accepted as the standard raw material of the dissolving cellulose industry.

Major efforts by wood fiber producers, backed by intensive research, have developed economical and effective methods for separation, purification, and bleaching of wood cellulose. Today wood is the dominant source of dissolving cellulose, having displaced cotton linters from most of their traditional markets. Linters are used for portions of the dissolving field, including certain cellulose acetate plastics, high-viscosity ether derivatives, high-viscosity cellulose nitrates, and some special viscose and cuprammonium applications. In terms of volume, linters have been almost completely displaced from the viscose industry. In some parts of the world, the availability of cotton linters as an agricultural by-product dictates their use in a broader range of applications. Sources other than linters or wood are under development. Reeds, bamboo (a grass), and agricultural wastes have been proposed for this purpose and serious efforts have been made and will continue to be made to adapt these materials for dissolving uses. A variety of technical problems exists with these sources, and, when their total volume is considered, they are of only minor importance.

A. CHEMICAL STRUCTURE

Investigation of cellulose is a complex and involved subject. Many early investigators thought that cellulose was chemically combined with lignin. Payen (1795–1871) is credited with the first separation of wood into cellulose and other components, and was the first to use the term *cellulose*. Over a long period of research, evidence has accumulated to show that cellulose consists of D-glucose units linked together into long chains, reaching high molecular weights. Bonding has been shown to be of the β-1,4-glycosidic type into cellobiose repeat units.

Hydrolysis produces primarily glucose, but it is possible that other sugars are present in the chain to a small degree. Studies of chemical derivatives confirm the existence of one primary hydroxyl group and two secondary hydroxyl groups for each anhydroglucose unit. It is these hydroxyl groups which provide sites for the important chemical reactions of cellulose. Structural representations of cellulose indicate a rather rigid configuration, which is associated with insolubility in conventional solvents and difficulty in plasticization. Because of its structure, cellulose is not thermoplastic. At elevated temperatures cellulose decomposes but does not melt. Conversion to useful fibers and films usually requires special techniques to attain the desired mobility and solubility. In most such systems cellulose is first treated to introduce substituent groups. Frequently, the process also involves breaking of the cellulose chains by oxidation or acid hydrolysis in order to reach convenient solution viscosity levels.

The generally accepted structure of cellulose is shown in Figure
1. More recent representation of the cellulose molecule is the
"chair form" (Fig. 2).

Not only are the cellulose chains themselves rigid and relatively
insoluble, but they also exhibit a chemical affinity between neighboring chains through hydrogen bonding (2). The resulting ordered
arrangement gives rise to crystalline areas which can be readily
demonstrated by x-ray techniques. The presence of these tightly
bonded regions within the natural cellulose fiber introduces additional problems in swelling the fiber and inserting substituent groups
in a uniform manner. Such ordered regions, however, provide an
effective means of transmitting stress from one chain to neighboring chains. Stress can be transmitted readily along the axis of the
chain, and the mechanical strength of a cellulosic fiber or film is
largely determined by the intermolecular structure, the length of
the chains, and the location and type of substituent groups, if
present. Interruption of the structure, rupture of the chains, or the
presence of short chains substantially reduces mechanical strength.
Substituent groups can also result in a disordered arrangement and
poorer physical properties. Introduction of crosslinks, as would
be expected, produces increased stability, reduced swelling, greater
rigidity, higher strength, and lower elongation. In commercial
practice crosslinking is usually limited to an aftertreatment of the
cellulosic product which has previously been prepared in the desired
physical form.

As cellulose exists in nature (3), it comprises a relatively broad

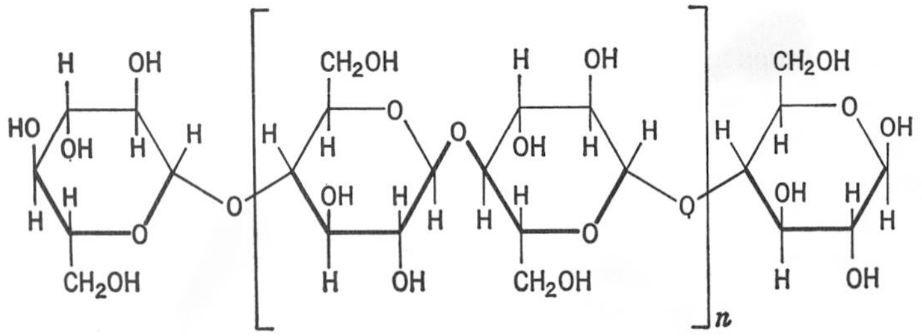

Fig. 1.   Conventional representation of cellulose, showing the cellobiose repeat
unit.

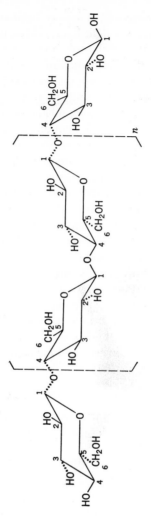

Fig. 2.   Cellulose representation in the chair form.

spectrum of chain length (4), not only from plant species to plant species but also within a single fiber. Measurement of molecular weight and separation of cellulose into various molecular weight fractions have received a great deal of attention (5). Several solvent systems have been employed, including cuprammonium hydroxide, cupriethylenediamine (6), and, more recently, complexes with other metals, such as cobalt, zinc, nickel, cadmium (7,8), and iron. Although the structure of such complexes has been investigated repeatedly, their precise nature is still not clear. They are, nevertheless, useful in describing cellulose, and the viscosity of solutions of cellulose assists in classifying cellulose types.

Cellulose nitrate is frequently used for separating cellulose into fractions. This derivative offers the advantage of fractionation based on both solubility and precipitation. Graphs of chain-length distribution constructed on the basis of cellulose nitrate are highly useful in estimating the degree of uniformity of cellulose chain length, a most important characteristic of cellulose for dissolving and regenerating operations.

Less precise estimates of fractions of low, medium, and high molecular weight are based on the partial solubility of cellulose in alkalis (9). Alkali metal hydroxides of sodium, potassium, and lithium at various temperatures and concentrations have been proposed for this purpose.

B. Physical Structure

Since cellulose occurs primarily as the wall of plant cells, its shape and physical structure are determined by the species of plant, the type and function of the cell, and the growth situation during which the cell was formed. Major differences exist between deciduous (commonly termed "hardwood") trees and coniferous ("softwood") trees. Historically, the softwoods, with their longer fibers, were the basis for bleached paper pulps. From these fibers were developed the first wood pulps used for dissolving celluloses. More recently, the shorter-fibered hardwoods have also been used successfully, but in smaller volume and in a narrower range of end uses. Typical physical dimensions of some cellulose fibers are listed in Table 1.

The physical arrangement of cellulose in natural fibers consists of three general types: (1) the grouping of molecular chains into ran-

TABLE 1
Length of Wood Fibers

	Length, mm (approx.)
Southern pines	
Loblolly	3
Longleaf	3.7
Western softwoods	
Western hemlock	2.8
White fir	3.5
Eastern softwoods	
Hemlock	2.9
White spruce	3.1
Deciduous woods (hardwoods)	
Black gum	1.7
Beech	1.2

dom or ordered areas, (2) the distribution and interconnection of these areas in a fibrillar structure, and (3) the association of fibrils to form the cellulose fiber.

X-ray diagrams and kinetic studies have been used to differentiate between ordered areas and amorphous regions. The size of the crystallites and the proportion of these regions are difficult to establish, since their borders are indistinct, and since the amorphous areas appear to have some degree of order. The chemical behavior of cellulose is influenced by the physical structure of the fiber.

The crystalline forms found in natural cellulose from many different sources appear to be identical in x-ray pattern. This form of cellulose is known as cellulose I, or "native" cellulose. Treatment with strong swelling agents, such as cold, strong alkalis, causes shifting to a different arrangement known as cellulose II. This is the form present in regenerated cellulose, such as rayon or cellophane. Other crystalline forms are known and have been described, but they will not be discussed here.

The distribution and interconnection of ordered and random regions within a fibril have been only partially clarified. Little is known regarding the basic mechanism of production of orientation in a growing fiber. It is certain, however, that cellulose crystallites in natural fibers are oriented along the axis of the fibril.   In light microscopy and electron microscopy studies, the patterns of fibrils have been well established.   In the case of ramie, for example, with the fibrils parallel to the fiber axis, the cellulose crystallites are oriented parallel to the fiber, producing high tensile strength, but low transverse properties.   In both wood and cotton celluloses, the fibrils assume a spiral pattern (10).   In cotton there are layers with S-twist or Z-twist directions.   In the same layer, in the same cotton fiber, the twist has been observed to reverse direction repeatedly.

Detailed studies of wood cells describe three principal layers: S_1, the outer or "primary" wall; S_2, the middle layer or "secondary" wall; and S_3, the inner or "tertiary" wall.   In the outermost layer, S_1, fibrils are arranged in a spiral pattern almost perpendicular to the axis of the fiber.   This layer is thin and is located adjacent to the middle lamella in which large amounts of lignin are deposited.   The S_2 layer is composed of fibrils arranged in spirals nearly parallel to the axis of the fiber.   The S_2 layer may be extremely thin or very thick, depending on the growth situation when the cell was formed. Physical tests of individual fibers show greatest tensile strength for cells with thick S_2 layers.   These results are attributed not only to the favorable alignment of the fibrils in the S_2 layer but also to the greater content of alpha cellulose and to the presence of highly ordered areas in this region.

Recent work on the S_3 layer indicates it is quite thin, but somewhat resistant to chemical attack.   Its fibrils are arranged in an interwoven structure.   Investigators report that the S_3 layer is removed by the sulfite pulping process, but remains almost intact after conventional sulfate pulping.

The electron microscope has revealed the presence of still another layer, the "warty" layer (11).   It consists of distinct lumps or warts and is present as the extreme inner surface of wood cells, lining the cavity or lumen.   It has been suggested that this layer is the residue of protoplast.   Chemical tests show the warty layer to be resistant to sulfuric acid, sodium hydroxide, and mild oxidizing agents, and this layer has been separated for study by dissolving other layers away.

It has been suggested that the resistance of the warty layer to chemical attack may account for the presence of undissolved particles in solutions of cellulose and cellulose derivatives, such as viscose and acetate. It appears to be too early to draw a conclusion linking such particles with the warty layer. Inhomogeneities in such solutions are of many types, and the size and frequency of the warts do not correspond to known particles occurring in solutions. Early studies of the swelling of cellulose in various media produced a great deal of evidence that fibers tend to swell nonuniformly, leaving narrow, annular, less swollen rings or "gels." These gels were not readily accessible to reagents and had not reacted sufficiently to go into solution. These gels are inherently much larger than any particles which could be derived from warts, and the influence of gels and fiber fragments on filtration, spinning, and casting operations would logically be expected to be of greater importance. Smaller particles derived from the warty layer may be of concern, particularly with respect to optical properties of solutions, but more work needs to be done to establish the relationship between the warty layer and inhomogeneities in solution.

II. Development of Cellulose for Films

A. CELLULOSE SOLUTIONS

With the exception of cuprammonium solutions, current commercial practice for converting cellulose to useful fibers or films involves addition of substituent groups to cellulose (12), followed by solution or dispersion in appropriate media, and subsequent recovery of the cellulose or cellulose compound in the desired physical form. Commercial utilization of this principle has produced an endless variety of useful products. This discussion will be limited to those subjects which are of interest in the manufacture of films.

Since cellulose, when heated, decomposes before it can soften, it is not thermoplastic. Its derivatives, aided by the presence of solvents, swelling agents, or plasticizers, can be used commercially as thermoplastics. In a broad sense, a plastic condition may be achieved by dissolving or dispersing the cellulose or derivative in a suitable medium, and concentrating the dispersion by partial removal of the solvent or swelling medium or substituent groups. Although these systems are quite specific, rather delicately balanced, and extremely complex in their interrelation of physical and chemical phenomena, they form the basis for commercial manufacture of cellulosic products of great utility. In the case of fibers, draw ratios of more than 100% are common. This ease of stretch is largely responsible for anisotropy of films made from similar systems. Tension forces required to pull the film along the desired path tend to lengthen the film along the machine direction and shrink it in the transverse direction. These effects are reflected in differences in product physical properties between machine direction and transverse direction and are serious enough to have stimulated efforts to avoid these differences. Current manufacturing conditions involve compromises designed to reduce these differences. More radical solutions to the problem have been proposed, including efforts to develop machinery to equalize the lateral and longitudinal forces and effects.

The three hydroxyl groups available for substitution form the basis for key reactions in preparation of solutions of cellulose. These hydroxyls give cellulose a distinct affinity for water and other polar liquids. Cellulose does not dissolve in water, although its monomer, D-glucose, is readily soluble. As moisture is adsorbed on cellulose, swelling occurs, and more sites become available for adsorption. Conversely, as water is removed by drying, the reverse occurs. Excessive rates of drying at high temperatures cause an irreversible shrinkage, with resultant reduced accessibility and lowered reactivity.

Aqueous strong alkalis cause a high degree of swelling of cellulose, and are capable of dissolving a small amount of cellulose, particularly the shorter chain-length material, depending on temperature and concentration of the alkali. This partial solubility forms the basis of several empirical systems for describing cellulose (13). Selection of two or more concentrations provides meaningful quantitative data helpful in differentiating between cellulose types and grades. Techniques for estimating the accessibility, crystallinity, and size of crystalline areas have been proposed, based on alkaline swelling and partial solubility. The swelling of cellulose in alkali increases the accessibility to other reagents, and this behavior permits the use of carbon disulfide, ethylene oxide, or other reagents to introduce substituents to the cellulose.

Strong mineral acids, such as sulfuric, phosphoric, and hydrochloric acids are capable of dissolving cellulose but severe degradation of cellulose occurs, even at room temperatures. Low temperatures are required to slow the rate of hydrolysis. Acids cause swelling of cellulose, although to a lower degree than do alkalis.

Acid systems are successfully used in the formation of commercially important cellulose esters, including the acetates, butyrates, nitrates, and mixed esters.

Reduction of viscosity and control of viscosity level are important factors in making solutions of cellulose. As found in nature, most celluloses are too high in degree of polymerization and too highly ordered to permit uniform accessibility for reaction or convenient handling of derivative solutions. Controlled reduction of chain length (14) is of major importance to the dissolving industries. Viscosity reduction is accomplished both in the pulp mill and the fiber or film plant. In the pulp mills oxidizing agents are commonly used. In the viscose process and in ether manufacture, alkaline oxidation with air is commonly used, while in acid systems used in the preparation of organic and inorganic esters, acid hydrolysis is employed. These mechanisms have been studied in detail. Oxidation in the presence of alkali appears to be complex and may be dependent on a free-radical mechanism. In any case, all glycosidic bonds seem equally accessible when the cellulose is highly swollen by the alkaline medium, and cleavage occurs at random. Alkaline oxidation results in more uniform reduction of chain length than does acid hydrolysis. Chain cleavage in acid media involves hydrolysis of the glycosidic bond, forming a new reducing end group.

Commercial control of these reactions depends upon accurate measurement and close regulation of temperature and time. If present, catalysts are of extreme importance. In alkaline oxidation, catalysts consisting of salts of manganese or cobalt have important commercial application (15,16). The term *catalyst* is also applied to mineral acids in esterification reactions. These acids swell and hydrolyze the cellulose and reduce the activation energy of esterification.

Control of chain length is essential, not only because of the dependence of solution viscosity on degree of polymerization but also because physical properties of the product are influenced by the level and uniformity of chain length. Concentration of cellulose in solution is sometimes limited by solution viscosity.

B. Cellophane

The thin, clear, flexible packaging material known as cellophane was first patented in England in 1898 by Stearn (17). Cast from viscose, cellophane was first made from cotton cellulose, but wood

cellulose was also used at an early date.   Manufacture of cellophane
began in France in 1908 and in the United States in 1924.   Commercial viscose films, therefore, were developed more recently than
fibers, and they profited from prior developments in selecting suitable cellulose sources.   In 1927, development of successful coatings
to impart moistureproof characteristics to cellophane overcame serious limitations of the early film, and the manufacture of cellophane
soon began a rapid growth, requiring large amounts of cellulose.
Wood cellulose was quickly adopted as the standard source for cellophane manufacture (18), and currently accounts for almost the
entire needs of this major industry (19,20).   Wood cellulose has
proved to be particularly well suited to the requirements of cellophane (21).   Major factors in the selection of cellulose for cellophane are uniformity, reasonable price, price stability, low viscosity,
low aging requirement, good yield of regenerated cellulose, purity,
chain length sufficient to provide adequate strength, and adapt-
ability to various types of process and equipment.   The versatility
of wood is shown by the fact that such widely different species as
western softwoods, southern pine, and some hardwoods are successfully used in the industry.

Uniformity of characteristics is the most important consideration
in selecting cellulose for cellophane manufacture.   Almost equally
important are cost and price stability.   Cost should be measured in
terms of total final cost of first-quality film.   This figure comprises
performance factors which are markedly influenced by the cellulose.
Cost penalties in terms of poor processing and operating problems
can more than counterbalance the lower price of an inferior, nonuniform source of cellulose.

At the other end of the spectrum of cellulose quality, efforts to
utilize highest-quality celluloses have produced satisfactory film.
These attempts have not yet been successful, however, in attaining
meaningful improvements in properties which can be translated into
increased earnings for the film maker.   Current indications are that
new technology and modified equipment may be needed to take full
advantage of the potential of the highest-quality celluloses in cellophane manufacture.   It is believed that this approach will eventually lead to improved film properties.

As a result of extensive work on the part of film manufacturers
and chemical cellulose producers, specially designed celluloses have

been developed for cellophane. These celluloses are characterized by high reactivity, low viscosity, a high degree of uniformity, and the capability of high productivity in economical viscose compositions. Special efforts have been directed toward adjusting sheet characteristics to provide rapid and easy disintegration in slurry steeping. Also, wood species are selected to attain rapid, uniform pressing on continuous presses with maximum throughput (22–24). The same fiber species are easily shredded after pressing, yielding an alkali cellulose which is open and accessible to the xanthation reaction.

In the interest of economy, aging of alkali cellulose should take place rapidly with a minimum of space and equipment. Introduction of low-viscosity celluloses and catalytic aging has resulted in important increases in capacity and reductions in working capital requirements. Viscose quality (25,26) is also improved by the mild conditions made possible by these tools. As a group, celluloses designed for cellophane manufacture have the lowest viscosity in the dissolving industry.

Good reactivity is extremely important to the economics of cellophane production. As reactivity is increased and made more uniform, the quality of the viscose solution is improved (27), and the cost of filtering the solution declines. Of even more importance is the fact that as reactivity increases, the composition of the viscose can be adjusted to reduce the consumption of chemicals. Lower usage of sodium hydroxide per pound of cellophane can be attained by such changes in viscose formulation. Higher cellulose concentration in viscose can be employed, and some improvement in film properties can thereby be realized. The high-cellulose, low-caustic, low-carbon disulfide viscose compositions commonly used for cellophane are dictated by economic factors, but are made feasible by the development of wood celluloses of the desired characteristics.

Yield of regenerated cellulose is important, not only from the standpoint of direct economics but also from the view of the amount of material which must be discarded from the system. Control of the quantity of material soluble in steeping caustic is an important concern of cellophane plant operation and pulp production. Cellulose producers have met the problem by offering grades with a higher yield and lower solubility in caustic.

Although it is not always recognized as important, the adapt-

TABLE
Typical Analyses of Celluloses

	Cellophane		Acetate			
Plant source	Western soft- woods	Southern pine	Western soft- woods	Western soft- woods	Southern pine	Southern pine
Process	Sulfite	Sulfite	Sulfite	Sulfite	Sulfite	Sulfite
Analyses, %						
S_{10}	11.3	11.5	5.1	4.4	4.3	3.4
R_{10}	88.7	88.5	94.9	95.6	95.7	96.6
S_{10}–S_{18}	6.6	6.5	2.4	1.4	1.3	1.8
S_{18}	4.7	5.0	2.7	3.0	3.0	1.6
10 % KOH solubility	7.0	9.1	6.3	5.0	5.8	10.4
Ash	0.04	0.19	0.06	0.05	0.12	0.05
Calcium	0.012	0.007	0.032	0.003	0.003	0.001
Iron	0.0003	0.0004	0.0002	0.0003	0.0004	0.0003
Copper	0.0003	0.0002	0.0001	0.0003	0.0002	0.0002
Manganese	0.00002	0.00001	0.00001	0.00001	0.00001	0.00001
Ether extract	0.19	0.10	0.08	0.11	0.08	0.04
Brightness, Elrepho	93.0	94.0	94.4	93.5	92.6	95.5
Intrinsic viscosity, dl/g	4.35	3.8	8.1	9.0	9.3	7.3
Nitrate DP	755	645	1460	1620	1675	1315

ability of cellulose to equipment and process is a key to the successful operation of a cellophane plant. In the early history of cellophane, conventional steeping presses required sheet characteristics which would permit good contact of steeping caustic with all portions of the sheet. The sheets had to be stiff, self-supporting, and capable of sufficient swelling to allow caustic to flow freely between the sheets in the passages opened by the swelling. With the introduction of slurry-steeping systems, sheet characteristics were changed markedly in order to favor rapid and complete disintegration of the sheets in the slurry. Selection of fiber species to favor rapid draining and ease of pressing was also a key factor in reaching maximum productivity of the equipment. More recent methods of steeping, such as the force-through principle, require more uniform sheet characteristics, high porosity, and low pressure drop for liquid flowing at right angles to the plane of the sheet. Such systems are not yet in general use, but their advantages in uniformity of processing, ease of removing caustic solubles, low operating cost, and ease of continuous operation are expected to be recognized in

2
for Film Manufacture

Ethers			Nitration		Meat casings, caps, bands		
Southern pine	Southern pine	Western soft-woods	Western soft-woods	Southern pine	Southern pine	Southern pine	Cotton linters
Sulfite	Sulfite	Sulfite	Sulfite	Sulfite	Sulfite	Sulfate	Sulfate
11.5	11.7	13.2	8.5	9.1	1.7	4.4	0.8
88.5	88.3	86.8	91.5	90.9	98.3	95.6	99.2
6.5	1.8	1.4	2.7	2.1	0.6	1.5	0.5
5.0	9.9	11.8	5.8	7.0	1.1	2.8	0.3
9.1	16.1	18.7	8.0	11.8	3.2	4.7	2.0
0.19	0.16	0.08	0.05	0.17	0.05	0.06	0.04
0.007	0.003	0.044	0.012	0.005	0.002	0.002	0.006
0.0004	0.0004	0.0002	0.0004	0.0004	0.0003	0.0002	0.0005
0.0002	0.0002	0.0001	0.0003	0.0002	0.0001	0.0001	0.0001
0.00001	0.00001	0.00001	0.00001	0.00001	0.00001	0.00001	0.00001
0.10	0.14	0.17	0.15	0.09	0.01	0.01	0.01
94.0	93.3	93.2	92.5	94.3	88.	89.5	89.5
3.8	8.6	10.8	7.5	7.0	6.1	6.0	11.
645	1550	1945	1350	1260	1100	1075	1980

the future.   Typical celluloses used for cellophane manufacture are listed in Table 2.

Cellulose is also used in certain coatings for cellophane.   Cellulose for this purpose is discussed in Section II-D-1.

The viscose process is also used to produce continuous tubes of regenerated cellulose.   These tubes are used for meat casings, caps, bands, and bottle closures.   Since these products are often shipped and used in a wet condition, they must have high wet strength and resistance to bursting or tearing.

Cellulose for this purpose is selected for maximum physical-property potential, and is characterized by high purity and uniformity of chain length.

C. CELLULOSE ACETATE

Cellulose acetate is an organic ester of cellulose which is broadly used in fibers, plastics, and films.   Since cellulose is a polyhydric alcohol, esterification can be carried out with organic acids, assisted by the presence of anhydrides and mineral acids.

The earliest source of cellulose for acetate manufacture was cotton linters, and for many years purified linters were the preferred source of cellulose for acetylation. Acetate production increased rapidly in the decade from 1930 to 1940 and further heavy demands on the available linters supply during World War II led to serious efforts to use wood cellulose in acetate manufacture.

In the first attempts to develop wood cellulose for acetate, efforts were made to approach the characteristics of cotton linters. Although cotton linters were not precisely duplicated, success in adapting wood cellulose was attained. Wood cellulose continued to penetrate the market at the expense of linters, and today wood is the major source of cellulose for acetate and other esters. Linters are still used to a limited extent for specialized uses, such as plastics requiring exceptional whiteness. Improvements in characteristics of wood cellulose have continued, important economies have been realized, and special celluloses have been developed for specific applications within the ester field. At present, a range of grades is available for this industry.

For acetate film manufacture, wood cellulose is the dominant source. Up to this time, sulfite pulping processes are the favored method of preparation. Other pulping systems have encountered technical difficulties, and they have only limited application.

As in the case of cellophane, uniformity is the most important consideration in selection of cellulose for acetate production. While price is also important, chemical purity is generally rated as of more concern than the cost of pulp. A comparison of acetate pulps with cellophane grades shows an important difference in purity (28). As a consequence, the price of acetate pulps as a group is slightly higher than that of cellophane pulps. The first successful wood celluloses were of the highest quality then obtainable. As knowledge of the interrelationships of cellulose and manufacture of acetate accumulated, it became possible to effect economies in the pulping process, with little or no loss in quality of the acetate product performance. At the present time, manufacturers of acetate products select from a spectrum of wood cellulose types, choosing the grades which best suit their requirements for quality, performance, and cost, consistent with the end product to be made (29).

For film manufacture, the nuclear age has introduced the problem of man-made radioactive material. In the form of fallout from the testing of nuclear devices, radioactive particles represent a severe and intolerable contamination for products used in photosensitive applications.   The amount of fallout and the location of these particles vary widely, depending on the testing programs and the movement of air currents.   When there is a possibility of contamination, elaborate precautions are taken.   Producers of wood cellulose have adopted similar measures to protect the cellulose from such particles.   Both water and air supplies are carefully monitored and specially cleaned to insure quality of cellulose for these critical applications.

D. Other Cellulose Derivatives

1. Cellulose Nitrate

Cellulose nitrate was the first commercially successful cellulosic film and was an important factor in the early years of the film industry.   Its strength, durability, clarity, dimensional stability, and resistance to moisture were outstanding.   The serious problem of inflammability, however, particularly in thin sections, has favored newer films, including the acetates.   Nitrocellulose is no longer of importance as a base film for packaging or photographic purposes and is limited to certain specialty applications in the field, usually as sheeting rather than as thin film.

Cellulose nitrate can be readily dissolved and formed into a desired shape; then the nitrate groups can be split off by saponification, leaving regenerated cellulose.   As would be expected, such products are expensive, and are of limited use.

Since cellulose nitrate was the first cellulosic film, numerous sources of cellulose have been investigated, including cotton, cotton waste, cotton linters, linen rags, cotton rags, creped paper, wood fibers of many types, and other materials.   In the early days, cotton linters were the preferred source for celluloid, film, and smokeless powder.   The subsequent discovery of nitrocellulose lacquers and and requirements of war for cotton linters resulted in such a demand for cellulose that wood cellulose suitable for nitrocellulose was developed.   Wood sources have continued to displace other cellulose types, and today cellulose nitrates for only a few specialized end uses are made from cotton linters.

While nitrocellulose films are no longer of importance as base films, a key discovery of Charch and Prindle (30) resulted in the use of nitrocellulose as a component of coatings for cellophane. Its desirable properties include moisture resistance, transparency, strength, toughness, resilience, high density, high luster, and resistance to oils. Its solubility in various mixed solvents has been an asset to coating operations.

The same properties have also contributed to the success of nitrocellulose in protective coatings of infinite variety. Lacquers of high durability and desirable esthetics (the properties of transparency, high luster, and clarity, primarily) are a major end use for nitrocellulose. A floor, furniture, wall, or other item coated by such a formulation is, in effect, wrapped in a protective film of nitrocellulose.

In selection of wood cellulose for nitration, chemical characteristics and physical characteristics (31) are equally important. In fact, a breakthrough in the physical preparation of shredded wood cellulose made possible the economical use of wood cellulose (32) in large volume for nitration. Still further developments in sheeting and shredding cellulose have resulted in specialized celluloses adapted to the commercial equipment now in service. Important sheet properties include density, porosity, drainage rate, and fiber type. Chemical characteristics of concern are uniformity, viscosity, purity, and color. Uniformity of reactivity is highly important, since clarity of solutions and smoothness of coatings are necessary in today's markets.

2. Cellulose Ethers

There are several cellulose derivatives, classed as ethers, which are capable of forming films (33) or sheets. As a group their volume is small, compared with cellophane or the esters. Nevertheless, they are in commercial use and several have considerable future potential. The ethers are noted for their variety and versatility, and the volume is growing. Films made from them range from water-soluble through alkali-soluble types, and include edible compounds with interesting applications. Even the water-soluble types can be rendered insoluble after casting by suitable aftertreatment with crosslinking agents. Many of the markets for cellulose ethers involve formulations for coatings, thickeners, finishes, sizes, and similar applications. In many cases, the effectiveness of these formulations depends on the type of film which remains after the solution or dispersion is dried. In comparison with natural products, such as resins, starches, and gums, the cellulosics have advantages in uniformity, solubility, stability, viscosity control, and compatibility.

Cellulose sources for cellulose ethers (34) range from bleached paper pulps to the highest-quality wood and cotton linters. Special precautions are taken in the case of products to be used in applications that involve foods.

E. Cuprammonium Films

Cuprammonium solutions of cellulose have been known for many years and have been used for both fiber and film production. The film obtained is regenerated cellulose, not a derivative, and it is produced commercially in Germany. An effort in the United States after World War II to manufacture cuprammonium film was not commercially successful.

Cellulose sources for cuprammonium processes have traditionally been of high quality. Cotton linters have been used and are still considered a preferred raw material, although economics favor wood cellulose. Wood cellulose has been prepared for this industry and is well liked for this process, particularly for fiber manufacture. In a solution process, such as the cuprammonium system, the chemical reactivity so important in other systems involving formation of derivatives is of less importance. The solution process has more flexibility in selection of cellulose, and can choose the quality and characteristics most suited to the requirements of the end product and process economics. While low-quality celluloses can be dissolved in a cuprammonium solution, there is no purification step or extraction stage to remove short-chain material and noncellulosics. The end product, therefore, reflects the quality of the raw material. Also, the costly ingredients (copper and ammonia) of the solvent system dictate efficient practices, maximum yield, and minimum losses. Recovery of both copper and ammonia are essential features of the cuprammonium process.

In order to obtain the desired combination of maximum physical properties of film and low cost, good to high-quality celluloses are always used. High-quality celluloses are capable of producing high wet-strength films, particularly in the early stages of casting, thus permitting rapid casting speeds. The uniformity of chain length of such cellulose assists physical properties, particularly wet strength, tear, and toughness. The structure of cellulose film prepared from cuprammonium solution results in a somewhat higher moisture content than cellophane. This higher moisture content permits lower softener content with adequate flexibility and fewer problems with brittleness and shelf life.

F. Cellulose Additives

In the long history of cellulose for film and fiber manufacture, an extensive list of materials has been suggested for addition to the system at various points in the process. Of the many possibilities, only those items which may be added to the cellulose and which are of commercial importance are discussed here. Broadly these additives may be classed as catalysts or as wetting agents. Their usefulness is described in the patent literature (35,36). For catalysis of alkaline oxidation of cellulose, manganese(II) salts are preferred. The metal is applied in controlled amounts to cellulose sheets, and the aging time of alkali cellulose can readily be reduced approximately 50% by the addition of 10 parts of manganese per million parts of cellulose. Large tonnages of cellulose for film have been treated with manganese.

Cobalt(II) is even more effective, since 1 ppm of cobalt is equivalent in effect to 10 ppm of manganese. Cobalt is only infrequently used at present, since its effectiveness is greater than usual commercial needs.

Wetting agents and lubricants can be grouped together since their functions overlap to a considerable degree. In extrusion of viscose into coagulating baths, suitable agents assist in preventing deposits in the orifices. Fiber makers have used such materials for some years. The clearance of certain agents by the U.S. Food and Drug Administration late in 1964 has removed a serious obstacle to commercial use in film manufacture. Like catalysts, these materials can be added to the cellulose in controlled amounts at the pulp mill. These same agents also have other beneficial effects in processing cellulose into viscose (37).

III. Characteristics of Dissolving Cellulose

In other sections of this chapter, mention has been made of the use of cellulose in a variety of films and sheet products. The characteristics of cellulose which are of importance to selection of cellulose types and adapting them to the equipment and process for making film are discussed in this section.

There is no single test or set of tests which is capable of characterizing cellulose and predicting its performance in a film-forming process. Consequently, in addition to the analyses mentioned below, which are made on the cellulose itself, manufacturers of film customarily use laboratory versions of the commercial solution-preparation steps to verify the solution characteristics of the type of cellulose. These steps are invariably checked before adopting or selecting a new grade for commercial use, and are sometimes made in a routine manner for quality control purposes.

A. PURITY

Cellulose purity is difficult to define. It has different meanings for various industries. In this section it will be used in a broad sense to indicate the proportion of material insoluble in strong alkali and the absence of noncellulosics (38). The term *alpha cellulose,* used in the past, has obvious limitations. Strong alkalis are capable of dissolving carbohydrates other than cellulose, as well as short-chain fractions of cellulose. Since the fraction insoluble in alkali also includes such noncellulosics as inorganic ash constituents, other analyses must be considered in assessing the purity of the material. For example, viscosity or chain-length measurements should be examined in conjunction with alkali insolubles, since the average chain length of a cellulose obviously has an influence on the proportion of chains which will be soluble in alkali. A useful measure can be obtained by using two concentrations of alkali, such as 10 and 18% sodium hydroxide (39).

Celluloses for acetate are generally of higher purity than those for cellophane.

B. VISCOSITY

Viscosity of cellulose refers to the viscosity of a solution of cellulose prepared under specific conditions. There is a series of such methods and several are in use. Of great utility are those methods which can be related to the degree of polymerization (DP) of cellulose (40). For this purpose the intrinsic viscosity (41) is highly useful. It should be recognized that an average DP can be estimated by such methods, but the distribution of chain length requires more elaborate techniques.

The pulping processes used to prepare celluloses of high purity also tend to produce more uniform chain-length distributions. For film manufacture, the uniformity of cellulose viscosity is important, although DP distribution can be rather broad. The celluloses of low viscosity commonly used for cellophane viscose film manufacture at present do not require as uniform a DP distribution as do corresponding viscose fiber processes.

C. Ash

Although at one time considered extremely important in describing cellulose, ash has declined in concern in the light of more sophisticated information on the influence of individual constituents of ash. Sodium salts, which appear as ash, are considered of negligible effect on xanthation and other processes which involve addition of caustic. Alkaline earth metals are considered objectionable in both cellophane and acetate processes, but for different reasons. Silica has received much attention, and large amounts are considered harmful to viscose filtration, although the size and physical condition of the silica particles is a factor. A small amount of sulfates is not objectionable in viscose, but chlorides and carbonates are avoided. Relatively minute amounts of metals are important and are discussed in the next section.

D. Metals

Iron, copper, and manganese occur in minute quantities in cellulose. Usually they are held to very low levels, primarily because each has an influence on the rate of depolymerization during the aging of alkali cellulose. This rate is increased by cobalt, manganese, and iron and is decreased by copper. As long as the content of these metals can be held uniform, there is usually no problem, but variations of a few parts per million of manganese or cobalt can upset viscose viscosity, unless appropriate corrections are made.

Celluloses intended for films to be used in contact with foods are periodically analyzed for heavy metals, such as lead, arsenic, and mercury. These metals do not occur naturally in cellulose and maximum precautions are taken in cellulose manufacturing plants to eliminate any possibility of contamination by heavy metals. It is common practice to exclude completely such useful instruments as mercury thermometers in order to avoid contamination from accidental breakage.

E. Resins

Extraction with organic solvents shows that commercial cellulose commonly contains from 0 to 0.2% extractables. While empirical evidence indicates they can play a useful role in certain viscose processes, they are considered objectionable in ester manufacture. Recent trends include efforts to reduce the amount of such extractables, followed by addition of selected agents directly to the cellulose for those industries which require the lubricating effect and the benefits in viscose performance.

F. Brightness

Whiteness of cellulose has an influence on the color of films and fibers made from the cellulose. Other factors in manufacture of the final product may mask this effect. Nevertheless, pulp whiteness can be an important consideration in selection of cellulose. Fiber end uses are often more sensitive to pulp whiteness than film products.

IV. Manufacture of Chemical Cellulose

A. Cotton Linters

Cotton linters are the short fibers remaining on the cotton seed after the staple cotton has been removed by ginning. The seeds, with the short fibers adhering to them, are screened and cleaned to remove trash and sent to the delinting step. The short seed hairs are removed in successive cuts. The seeds are cracked, the hulls removed, and the seed is pressed to yield oil and meal products. These are of more economic importance than the linters.

First-cut linters find uses in paddings, upholstery, surgical dressings, and other outlets. Second-cut linters are the usual source of fiber for preparation of chemical cellulose for dissolving uses. When removed from the seed, these linters comprise fuzzy fibers or true linters, some immature fibers, small amounts of cotton staple or fragments of cotton staple, and trash. Linters are contaminated with dirt and various plant fragments, including hulls, leaves, and stems. In the case of linters, contamination accounts for most of the noncellulosics entering the purification process.

Mechanical separation of trash by air elution is the first step. Then the linters are digested under pressure in dilute sodium hydroxide with a detergent. After washing, the fibers are bleached to improve whiteness, increase reactivity, and reduce chain length. Bleaching is less complex than in the case of wood cellulose, but may be a multistage process. An acid treatment to remove inorganic material may be inserted before sheeting and drying. Linters may be dried in bulk form and baled. Shipments are made in bales, bales of cut sheets, or rolls.

B. Wood Cellulose (42–44)

1. Sulfite Process (45)

Wood cellulose prepared by the sulfite process was used in England to manufacture viscose products over 65 years ago, and this process is still generally considered to be the most versatile method of preparing dissolving cellulose. New technology and new techniques have changed all phases of the sulfite system, but the basic steps remain the same. They are harvesting, debarking, and cleaning the wood, cutting the wood into chips, cooking (digestion under pressure in a hot solution of sulfurous acid and bisulfite salts), alkaline extraction, bleaching, washing, drying, and packaging. Suitable screening and cleaning operations are also essential parts of the overall operation. Recent advances in technology have introduced important modifications to sulfite systems, with emphasis on the quality of the product and the recovery of chemical and heat values.

Manufacture from trees of materials other than cellulose has received a great deal of attention, and commercial products are appearing in such widely diversified markets as drilling mud additives, dispersants, agricultural micronutrients, boiler water chemicals, flavorings, and solvents. These products will not be discussed here.

The preparation of dissolving cellulose (46,47) requires a precise control of all manufacturing conditions, beginning with the selection of wood species, and ending with the physical form of the final cellulose package. All of the processing steps are interrelated, and the entire system must be in proper balance to insure chemical cellulose of uniform quality.   To add complexity to a difficult operation, the requirements of various industries differ in important respects, and a change from one type of cellulose to another requires, in turn, changes in operating conditions throughout the process.

As manufacture and use of dissolving cellulose has become more sophisticated, there has been a distinct trend toward development of a range of types designed for each industry.   Thus, consuming industries are able to select the type which most nearly suits their needs for quality, purity, reactivity, property potential, and price.   Modern technology has developed and is continuing to contribute important advances in pulping processes.   Although the proliferation of process modifications makes impossible a detailed description here, typical steps for manufacture of dissolving cellulose are discussed in principle in this section.

Wood Preparation. Selected species are harvested and transported to the mill where the bark is removed by hydraulic or mechanical barkers, the logs are washed, and the wood is reduced to chips of controlled size.   The chips may be stored before charging to the digesters.   Species, age of wood, size of chip, and moisture content are important.

Cooking. At present, batch digesters are commonly used and offer important advantages in flexibility and control.   Continuous equipment offers other advantages and will grow in importance in the future.

In the batch system, a solution of sulfurous acid and bisulfite salts is introduced into the digester.   Temperature, pressure, and time are precisely controlled.   When cooking is completed, the batch is discharged while under pressure.   The rapid release of pressure aids in separating the fibers from the original chip structure.   Corrosion can be a serious problem, and the digesters are lined with resistant brick or stainless steel.

After digestion, the spent liquor is removed from the fibers, which are washed, screened, and sent to the bleach section.   Spent liquor returns to the recovery area.   Evolved sulfur dioxide gases are also recovered.

Bleaching. In recent years the technology of bleaching (48,49) and purification of sulfite wood pulp has become so complex that as many as six stages may be used.   Among these are chlorination, alkaline extraction, hypochlorite treatment, and chlorine dioxide treatment (50). Some stages may be repeated. The number of stages, the sequence of treatment, the chemical usage, time, temperature, pH, and the efficiency of intermediate washing determine the characteristics of the bleached fiber. The bleached cellulose is usually screened, cleaned, sheeted, and dried. The cellulose is collected in large rolls, which are cut to the desired final roll or sheet size. Additives may be applied, and the packages are prepared to fit customer handling requirements, weighed, and shipped. Rail cars or ocean-going vessels are generally used, although barges and trucks carry some shipments. Inflatable dunnage in rail cars has materially improved the physical condition of the packages at destination.

2. Prehydrolyzed Kraft Process

One of the reasons for the dominant position of the sulfite process for manufacture of dissolving cellulose in the early years of the industry was the difficulty of purifying and bleaching pulps made by the conventional Kraft process. This process uses a strongly alkaline cooking medium. A key discovery of the research on this problem was that an initial acid prehydrolysis stage before the hot alkaline digestion could overcome this problem. Prehydrolyzed Kraft mills can produce celluloses of high quality for many dissolving uses, although they are not favored for ester manufacture. Cooking temperatures are somewhat higher than in the sulfite process, and the subsequent bleaching and purification systems are more complex.

Highly efficient recovery systems for heat and inorganic chemicals are an integral part of the Kraft process (51). The heart of this operation is the recovery boiler which burns the organic matter in concentrated spent liquor. This boiler produces valuable heat while reducing sulfate salts to sufides. Elaborate precautions are taken to insure safe operation of the recovery unit, since occasionally explosions have occurred. Resinous woods in the Kraft process produce turpentine and tall oil, which are collected and sold for a variety of uses.

The recovery systems of the Kraft process have assisted in the development of modifications in the sulfite process. Variations of these methods have been in plant service for a number of years. They involve soluble "bases," such as ammonium, sodium, and magnesium ions in place of the less soluble calcium. These bases are more soluble and permit a broader range of cooking and recovery conditions. Without recovery, they are more expensive than calcium, and recycling of chemicals is an essential consideration. Economic studies (52) are optimistic that these new systems will not only improve the economics of the traditional sulfite process but also permit greater flexibility in operation, while making important contributions to reduction of wastes.

V. Economic Aspects

In the economic use of man's material resources, wood cellulose is outstanding because of its wide availability, renewable character, economic value, versatility, and reasonable cost (53,54). Its chemical stability and absence of hazard allow long storage periods and transportation over long distances in commercial carriers.

The abundance of the resource and its variety of forms provide the essential elements for price stability. Accurate forecasting of manufacturing, transport, and raw material needs is possible, both for the cellulose producer and the cellulose consumer.

The cost of dissolving cellulose as a polymer for the manufacture of film is sufficiently low to permit successful competition in the marketplace with synthetic polymers. These materials are derived primarily from fossil sources of chemicals which are not renewable. Consequently, in the long run, synthetic polymers are expected to increase in cost as exploration and production expenditures increase. While cellulose is grown on land, it is not limited to arable land. On the contrary, it is currently produced largely either in areas unsuited to agriculture or on lands considered marginal for agricultural use. Tree farms are displacing other crops on the sound basis of better economic use and higher return to the grower. Tree farms and related industries have contributed markedly to the economy of large areas, notably Canada, Scandinavia, and northwestern and southeastern United States. Beneficial effects extend into other areas, including recreation, conservation, wildlife resources, and other concerns of vital human interest. Scientific research on genetics, nutrition, insect infestation, and disease control have already increased the growth of healthy timber and reduced losses. Wise forest management and fire protection have aided in securing these benefits.

Currently, cellulose production capacity is increasing (55,56).   In the five years between 1965 and 1970 a total of 45 ventures in the United States and about 50 in Canada are forecast.   While a large portion of this capacity is intended for fibrous uses, such as paper, some capacity will be available for dissolving cellulose.

In the future, as world population increases, the growing of cellulose will increase in importance in the economic-use pattern of man's resources.   Additional land may be converted from fiber crops to wood, or ways may be found to produce cellulose by synthesis. Laboratory studies on synthesis of cellulose show limited promise, and commercial application appears to be reserved for the distant future.   Nevertheless, it is of interest to note that the raw materials used by plants are carbon dioxide and water.   Both of these materials are cheap and abundant.

Research on grafting various groups on a cellulose molecule is arousing interest, and futher work along this line is expected.   Modification of the cellulose polymer would permit even broader ranges of base film properties, coating anchorage, and end-use performance. As the trend continues toward films tailored to specific end-use requirements, films made from cellulose derivatives may become increasingly important.

General References

E. Heuser, *The Chemistry of Cellulose*, Wiley, New York, 1941.
L. E. Wise and E. C. Jahn, *Wood Chemistry*, Vol. 1 and 2, 2nd ed., Reinhold, New York, 1952.
E. Hagglund, *Chemistry of Wood*, Academic Press, New York, 1951.
J. N. Stephenson, *Pulp and Paper Manufacture*, Vol. 1–4, McGraw-Hill, New York, 1950–1955.
E. Ott, H. M. Spurlin, and M. W. Grafflin, *Cellulose and Cellulose Derivatives*, (Parts I, II, and III, Vol. 5, *High Polymers*, 2nd ed.), Interscience, New York, 1960.
J. P. Casey, *Pulp and Paper*, Vol. 1 and 2, Interscience, New York, 1960.
W. D. Paist, *Cellulosics*, Reinhold, New York, 1958.
F. D. Miles, *Cellulose Nitrate*, Interscience, New York, 1955.

Analytical Methods

R. L. Whistler, *Methods in Carbohydrate Chemistry*, Vol. 3, Cellulose, Academic Press, New York, 1963.
ASTM Book of Standards, American Society for Testing and Materials, Philadelphia, Part 15, issued annually in April.

TAPPI Routine Control Methods and Technical Information Sheets, Technical Association of the Pulp and Paper Industry, New York, issued periodically.

References

1. J. R. Colvin, *Can. J. Biochem. Physiol.*, **39**, 1921 (1961).
2. L. Hunter, W. C. Price, and A. R. Martin, *Report of Symposium on Hydrogen Bond*, the Royal Institute of Chemistry, London, 1950.
3. J. K. Hamilton and N. S. Thompson, *Pulp Paper Mag. Can.*, **59**, (10), 233 (1958).
4. T. E. Timell, *Tappi*, **40**, 25, 30, 568, 749 (1957).
5. "Molecular Weights of Cellulose Symposium," *Ind. Eng. Chem.*, **45**, 2482 (1953).
6. R. Levy, P. Moffat, and W. D. Harrison, *Paper Trade J.*, **118**, (6), 29 (1944).
7. G. Jayme, *Papier*, **12**, 624 (1958).
8. D. K. Smith, R. F. Bampton, and R. L. Mitchell, *Ind. Eng. Chem., Process Design and Develop.*, **2**, 57 (1963).
9. R. Bartunek, *Papier*, **6**, 120 (1952).
10. M. L. Rollins, A. T. Moore, W. R. Goynes, J. H. Carra, and I. V. de Gruy, *Am. Dyestuff Reptr.*, **54** (14), 36 (1965).
11. W. Liese, *J. Polymer Sci.*, **2C**, 213 (1963).
12. D. B. Mutton, *Pulp Paper Mag. Can.*, **65**, 2 T-41 (1964).
13. F. R. Charles, *Tappi*, **37**, 148 (1954).
14. B. B. Thomas and W. J. Alexander, *J. Polymer Sci.*, **15**, 361 (1955).
15. R. L. Mitchell, U.S. Pat. 2,542,285 (Feb. 20, 1951); U.S. Pat. 2,602,536 (July 20, 1954).
16. D. K. Smith, W. J. Alexander, and R. L. Mitchell, *Ind. Eng. Chem.*, **52**, 905 (1960).
17. C. H. Stearn, Brit. Pat. 1020 (Jan. 13, 1898).
18. R. L. Mitchell, *Ind. Eng. Chem.*, **47**, 2370 (1955).
19. H. H. Sineath, *Tappi*, **46**, 142A (1963).
20. *Chem. Eng. News*, **42**, No. 42, 29 (1964).
21. G. C. Inskeep and P. Van Horn, *Ind. Eng. Chem.*, **44**, 2511 (1952).
22. J. W. More, *Svensk Papperstid.*, **65**, 1 (1962).
23. E. Treiber and R. Pennell, *Svensk Papperstid.*, **65**, 149 (1962).
24. R. Bartunek, *6th Symposium Viscose, Technical Questions*, Section I, Stockholm (Oct. 1961).
25. H. L. Vosters, *Svensk Papperstid.*, **53**, 613 (1950).
26. D. K. Smith, R. F. Bampton, and R. L. Mitchell, *Ind. Eng. Chem. Process Design Develop.*, **2**, 223 (1963).
27. O. Samuelson, *Svensk Papperstid.*, **29**, 866 (1953).
28. A. J. Rosenthal and B. B. White, *Ind. Eng. Chem.*, **44**, 2693 (1952).
29. R. J. Conca, J. K. Hamilton, and H. M. Kircher, *Tappi*, **46**, 644 (1963).
30. W. H. Charch and K. E. Prindle, U.S. Pat. 1,737,187 (Nov. 6, 1929); U.S. Pat. 1,826,696 (Oct. 6, 1931).
31. B. Miller and T. E. Timell, *Textile Res. J.*, **26**, 255 (1956).
32. R. L. Stern, U.S. Pat. 2,028,080 (Jan. 14, 1936).

33. D. R. Erickson, U.S. Pat. 2,469,764 (May 10, 1949).
34. A. W. Anderson and R. W. Swinehart, *Tappi*, **39,** 548 (1956).
35. J. J. Polak and J. G. Weeldenburg, Netherlands Pat. 39,956 (1935).
36. R. L. Mitchell, U.S. Pat. 2,805,169 (Sept. 3, 1957).
37. W. J. Alexander and R. D. Kross, *Ind. Eng. Chem.,* **51,** 535 (1959).
38. I. Croon and A. Donetzhuber, *Tappi*, **46,** 648 (1963).
39. J. Gordy, *Can. Pulp Paper Ind.,* **14,** No. 5, 28–30, 32, 36 (1961).
40. W. J. Alexander and R. L. Mitchell, *Anal. Chem.,* **21,** 1497 (1949).
41. W. J. Alexander, O. Goldschmid, and R. L. Mitchell, *Ind. Eng. Chem.,* **49,** 1303 (1957).
42. B. B. Thomas, *J. Chem. Ed.,* **35,** 493 (1958).
43. J. K. Hamilton and N. S. Thompson, *Pulp Paper Mag. Can.,* **61,** T-263 (1960).
44. J. K. Hamilton, *Pure Appl. Chem.,* **5,** 197 (1962).
45. *Ray-O-Sun,* **22,** No. 11-12, Rayonier, Inc. (Nov.–Dec., 1964).
46. G. A. Richter, *Tappi*, **38,** 129 (1955).
47. G. A. Richter, *Tappi*, **40,** 429 (1957).
48. *The Bleaching of Pulp,* Tappi Monograph No. 10, Badger Printing Co., Appleton, Wis. (1953).
49. J. D. Rue, *Pulp Bleaching,* Bulletin No. 200, Hooker Electrochemical Co., Niagara Falls, New York (1957).
50. W. H. Rapson, *Chlorine Dioxide Bleaching,* Bulletin No. 203, Hooker Electrochemical Co., Niagara Falls, New York (1957).
51. "Cut-Price Pulping," *Chem. Week,* **96,** No. 6, 85 (Feb. 6, 1965).
52. J. J. Voci and F. D. Iannazzi, *Chem. Eng. Progr.,* **61**(5), 110 (1965).
53. *Pulp and Paper International,* Annual World Review (September issue, annually).
54. "Cellulose," *Chem. Week,* **85,** No. 9, 53 (August 29, 1959).
55. C. W. Heckroth, *Pulp Paper,* **39,** 32, (1965).
56. "Pulp, Perilous Market for Chemicals," *Chem. Week,* **99,** No. 22, 57 (November 26, 1966).

CHAPTER 4

NATURAL POLYMERS AS FILM FORMERS
B. Cellophane Viscose and Its Conversion into Film

VERNON C. HASKELL

Film Department, E. I. du Pont de Nemours & Co., Richmond, Virginia

I. Introduction

The conversion of wood pulp into a transparent film of regenerated cellulose has become a sizable industry giving direct employment to thousands of people in North and South America, Europe, and Japan. Commercial production of cellophane in the United States was begun in 1925 and its initial use was mainly as a decorative wrap on candy boxes and the like. The development of a moisture-proof, heat-sealing coating for cellophane extended its usefulness greatly and applications mainly on food and tobacco products

115

grew to a point where over 400 million pounds were sold annually in the U.S. in the early 1960's.   World production is at an annual rate of over one billion pounds.

Many methods have been described for making a film of regenerated cellulose.   E. Weston (1) described as early as 1882 a process for making a film of regenerated cellulose by the denitration of nitrocellulose. A substantial quantity of regenerated cellulose film is made in Europe by the cuprammonium process.   Deacetylation of cellulose acetate film (2) yields a regenerated cellulose film having exceptional toughness. A film of similar structure and properties to that made from cellulose acetate is produced by drying a film of viscose on a heated support and regenerating cellulose under conditions which do not permit reswelling before regeneration is complete (3).   Wood pulp has been dissolved in concentrated calcium thiocyanate solution (4) and wet cast into a regenerated cellulose film.   In principle, all of these processes are similar—most of them involve first putting the cellulose into solution.   This can be accomplished only by reagents which interact strongly enough with the cellulose hydroxyls to overcome the powerful intermolecular attraction of the native cellulose structure. Examples include the formation of a stable derivative at the hydroxyl site as in cellulose acetate or nitrocellulose, an unstable derivative which persists long enough to dissolve the cellulose and which is decomposed at a later stage after a film is formed as in the viscose process or a relatively stable coordination compound to be reconverted to cellulose later, as in the cuprammonium process. Formation of a film from the cellulosic solution is achieved by extruding the solution through a slot or spreading a layer on a flat support. In the case of xanthate or of other unstable derivatives, conversion back to cellulose before the film loses its shape is the next step. The freeing of the cellulose hydroxyls results in hydrogen-bond interaction between structural elements of the film which preserves the form of the film and renders it insoluble in water.

II. Viscose for Cellophane

Since the viscose process is so well known and so completely reviewed in the literature concerning viscose rayon, only those phases of viscose preparation and properties which are specific to film processing will be discussed here.

Viscose for cellophane differs substantially from that used to manufacture either textile or tire cord rayon because the film and the fiber have different property requirements and different process economics.   In the fiber such properties as tensile strength, abrasion resistance, and the ability to withstand repeated stresses without failure are important.   In the film, the amount of work absorbed before failure under conditions like those of film usage is a better index of its performance.   Other properties which are necessary or desirable in the film but less so in the fiber are transparency and surface smoothness.   In this discussion, it is proposed to show the relationship which exists between the composition and properties of cellophane viscose and the properties of the resulting film.

A. $\overline{DP}$ and DP Distribution

As in other polymeric systems, the strength and toughness of the regenerated cellulose film increase as the degree of polymerization (DP) of the cellulose increases.   In Figure 1 (5) are shown data of

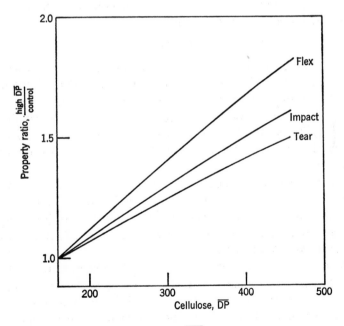

Fig. 1.   Effect of cellulose $\overline{DP}$ on film properties.

physical properties versus degree of polymerization of the cellulose for a series of films made in the laboratory by spreading a thin layer of the viscose on a glass plate and coagulating the film by immersing the plate in a coagulating bath similar in composition to that used for the commercial manufacture of cellophane. At still higher levels of $\overline{DP}$, the rate of improvement in properties is lower, but practical difficulties interfere with production of uniform film when using extremely high-viscosity viscoses. The slope and shape of the curves of physical properties versus $\overline{DP}$ depend upon coagulating and regenerating conditions and upon the composition of the viscose.

The above data would suggest making commercial cellulose film with the highest $\overline{DP}$ cellulose available from the pulp suppliers. Economic considerations prevent this. The extremely high viscosity solutions could not be handled without the expensive replacement of pumps, pipelines, filters, and other equipment.

Hopkins and Whatley (6) have studied the steady-state flow of viscose in a capillary tube with varying concentration and molecular weight of the cellulose. Viscose was assumed to obey the empirical power law for shear stress τ as follows:

$$\tau = K \, (du/dr)^n \qquad (1)$$

K being the fluid consistency index, du/dr the rate of shear in a cylindrical tube, and n the flow behavior index. For a Newtonian fluid, $n = 1$ and K becomes the coefficient of viscosity.

The effect of cellulose content at constant $\overline{DP}$ of the cellulose is shown in Table 1. In the measurements, n remained constant while K increased fourfold as the cellulose content was increased from 9.58 to 15.94%.

Similar data showing the effect of $\overline{DP}$ at constant cellulose content on the flow behavior of viscose are given in Table 2. Here a variation in both n and K is seen, n decreasing to two-fifths of the original value and K increasing by a factor of 27 as the $\overline{DP}$ is increased from 250 to 725.

These data confirm that viscose is a non-Newtonian fluid and that the non-Newtonian behavoir increases with increasing $\overline{DP}$ and with increasing rates of shear. In the cellophane process this is of significance in two steps where high pressure is required to force the viscose through small openings: during filtration in which

TABLE 1

Rheological Properties of Constant-$\overline{DP}$ Cellulose Viscose
at Varying Cellulose Contents ($\overline{DP}$ = 725)

Conditions	Viscose sample No.[a]			
	1-1	1-2	1-3	1-4
Cellulose, %	9.58	11.97	13.23	15.94
NaOH, %	5.54	6.99	8.30	10.61
n	0.267	0.276	0.273	0.272
K, (lb force)(sec)^n/in.^{2}	0.172	0.292	0.431	0.689

[a] All samples contained 45% CS_2 based on cellulose.

undissolved fibers and foreign particles are removed and at the extrusion die where the sheet is formed. At both of these points in the process, pressures of the order of 100 psi are required to maintain satisfactory flow rates. Without the reduction of the apparent viscosity at high shear rates, filtration and extrusion of viscose would either be very slow or would require expensive high-pressure equipment.

The effect of DP distribution on the physical properties of cellophane has received little, if any, attention in the published literature.

TABLE 2

Rheological Properties of Constant-Cellulose-Content Viscose at Varying $\overline{DP}$

Conditions	Viscose sample No.[a]		
	1-1	1-5	1-6
Cellulose, %	9.58	9.03	8.85
NaOH, %	5.54	5.38	5.33
Alkali cellulose aging time, hr	0	16	32
$\overline{DP}$ of alkali cellulose	725	480	250
n	0.267	0.343	0.666
K, (lb force)(sec)^n/in.^{2}	0.172	0.068	0.00634

[a] All samples contained 45% CS_2 based on cellulose.

If cellophane is similar to rayon and other polymeric systems in this respect, a narrow molecular weight distribution should be beneficial to properties. Solution viscosities depend closely upon weight-average DP, but the physical properties of the condensed polymeric products depend more upon the number-average DP. Granting that a narrower molecular weight distribution should be beneficial to the properties of cellophane, the question of economics must again be introduced. The first step might be to utilize a wood pulp having a higher α-cellulose content. Such a pulp is produced by caustic extraction of low molecular weight fractions in the pulping process and the attendant loss in yield necessitates a higher price for the pulp. In the cellophane process some aging of the alkali cellulose is required so that the solution viscosity may be controlled by means of variations in the aging time. This aging is carefully controlled with respect to average time and temperature, but temperature gradients in large masses of alkali cellulose are inevitable and some variations in the holdup time must occur, particularly in continuous aging processes. The effects of these variations would be to broaden the molecular weight distribution for a given average DP. If the steeping caustic is recycled without dialysis to remove hemicellulose and if waste steeping caustic is used as the caustic source for dissolving the cellulose xanthate in the mixing operation, a more economical process will result, but the molecular weight distribution of the product will be broader.

B. Cellulose Concentration

When the cellulose content of viscose is increased by reducing only the water content of the recipe, the resulting cellophane has improved mechanical properties. This has been observed in casting viscoses of normal degree of substitution of carbon disulfide into the regular acid sulfate coagulating bath as well as when casting a more highly substituted viscose into a two-bath system of concentrated ammonium sulfate followed by the acid sulfate bath. Data are shown in Table 3 (7).

An extreme example of this improvement in toughness is observed when a film of viscose is evaporated to a self-supporting state in which the cellulose concentration is 35–40% (3). In this case, some physical properties are increased several fold although, as

TABLE 3
Viscose Composition vs. Film Properties

Conditions	Control	High cellulose	High $\overline{DP}$	High cellulose and CS_2
Cellulose, %	8.5	14.0	8.5	12.0
NaOH, %	5.25	8.9	5.25	10.3
CS_2, %	26.5	26.5	26.5	50.0
Cellulose, $\overline{DP}$	595	535	865	684
Viscosity, poise	49.6	620	350	496
Film properties	195 Gauge at 75°F—35% R.H.			
Tear, g M.D.[a]	2.3	5.4	3.1	9.0
Tear, g T.D.[b]	5.7	8.9	8.6	15.9
Pendulum impact, kg-cm	2.9	5.6	3.4	10.9
S-1,[c] %	19.8	15.3	19.0	16.0

[a] Lengthwise direction of the casting machine.
[b] Transverse direction.
[c] S-1 refers to the softener, glycerol.

will be seen later in discussing dry cast cellophane, not all of this improvement can be attributed to high solids content alone. It is of interest to inquire into the structural differences between films made from high cellulose and normal cellulose viscose.

One consequence of increasing the cellulose content of the viscose is that a more compact gel film is produced, as shown in Table 4 (8).

This decrease in gel swelling corresponds with a drop in moisture content from 370 to 180% on a dry basis.

TABLE 4
The Effect of Cellulose Concentration in the Viscose on the Degree of Swelling of Cellulose Films

Cellulose in viscose, %	Gel swelling, $\dfrac{\text{wet wt.}}{\text{dry wt.}}$
9.0	4.72
12.0	3.02
16.0	2.77

Two possible effects on the fine structure of the film could arise from these differences in swelling, differences in molecular and crystallite orientation and differences in the degree of crystallinity. The effect of the amount of shrinkage during drying, and hence of the degree of swelling, on the orientation of regenerated cellulose fibers and films is well known (9). Crystallinity as measured by x-ray diffraction seems to be determined by the conditions under which coagulation and regeneration occur and is much less sensitive to deformations of the gel or dry sheet.

C. Alkalinity

The choice of viscose alkalinity for cellophane is based on the principle of minimizing costs without sacrificing film quality. The presence of sodium hydroxide in viscose is necessary for its solubilizing action on the cellulose xanthate. Since it is completely removed from the cellulose in the cellophane casting process, its presence adds to the material cost per pound of cellophane produced. It is an additional expense because sulfuric acid is consumed in neutralizing it. If the caustic concentration is too low, the cellulose is incompletely dissolved and the filtering capacity of plate and frame presses or other filtering devices may become too low for economical operation. Treiber (10) has pointed out that it is only the free sodium hydroxide in the viscose which actively solubilizes the cellulose.

D. Degree of Substitution

Since reaction with carbon disulfide is the basis for solubilization of cellulose in the viscose process, it is to be expected that the degree of substitution of carbon disulfide on the cellulose chains has a profound effect on the process and on the product. The degree of substitution depends upon the amount of carbon disulfide per unit weight of cellulose which is charged into the reactor, on the efficiency by which the CS_2 is converted to xanthate, in competition with the reaction of carbon disulfide with sodium hydroxide to form sodium trithiocarbonate, and on the extent of ripening or dexanthation which has occurred between xanthation and coagulation of the viscose. In viscose for cellophane, economic pressure keeps the level of carbon disulfide used to make the viscose at the minimum which will yield a satisfactory product and an efficiently operable process.   Because of the lower degree of polymerization of the cellulose pulps compared with rayon pulp and because of the high reactivity of the cellophane grade pulps, a lower level of carbon disulfide is used in cellophane viscose than in textile rayon or tire cord viscoses.   This is usually between 25 and 30% carbon disulfide on a pulp basis.

Besides being more costly, highly xanthated viscoses are more difficult to convert into film of good transparency.   When the usual acid sulfate bath is used for coagulation of a highly xanthated viscose, excess gas is liberated within the film, forming microscopic bubbles which scatter light and reduce transparency.   When this can be avoided, however, by the use of special bath or other coagulating and regenerating techniques, it is found that a tougher cellulose sheet can be produced from the more highly xanthated viscoses.   One such system (11) consists of an initial bath buffered to a pH of 5–7 for coagulation and a second bath containing sulfuric acid and sodium sulfate for regeneration.   With such a bath, viscoses containing 40% or more of carbon disulfide have been cast dynamically into tough, strong, clear cellulose film.   It appears that the buffered bath not only coagulates the viscose but also leaches out enough of the gas-forming by-products so that excessive gas formation is avoided within the film.

Chemically, the ripening of viscose can be thought of as a reaction between the cellulose xanthate and sodium hydroxide to form sodium trithiocarbonate and alkali cellulose.   Dunbrant and Samuelson (12) have recently shown that the dexanthation reaction is the net result of two decomposition reactions having markedly different rates, and of the reverse rexanthation reactions.   These were studied separately by eliminating the rexanthation by stirring viscose with an excess of an ion-exchange resin charged with hydroxyl ion.   The resin particles absorbed trithiocarbonate and carbonate ions as quickly as they were formed, thus preventing rexanthation on other hydroxyl sites.   The fastest dexanthation reaction was ascribed to the decomposition of 2,3-xanthate and the other slower ones to the decomposition of 6-xanthate.   Thus, the xanthate group on the 6-position of the anhydroglucose group appeared to be more stable.   It was found that the concentration of 2,3-xanthate decreases continuously during the ripening period, but the 6-xanthate concentration passes through a maximum (see p. 89).

E. Viscose Additives

The purpose of adding other ingredients to viscose used for making cellophane would be to change the properties of the cellophane in some advantageous way, to improve the processing characteristics of the viscose, or to reduce the cost of the product. For example, attempts have been made to reinforce cellophane with fibrous materials of various kinds. Attempts to do this by adding fibrous materials or mineral particles to viscose resulted in decreased clarity of the film. While the tear resistance was improved, the resulting loss of transparency was a serious disadvantage. This resulted not only from light scattering from the interface between cellulose and the fibers, but also because bubbles and voids were formed around the particles during coagulation and regeneration.

Incompatibility with cellulose is exhibited so generally that it is extremely difficult to find additives which do not produce hazy cellophane. In one trial, starch was xanthated and dissolved in caustic solution. When this was blended with viscose in various proportions, the resulting films were hazy if the starch constituted more than 2–3% of the film.

Certain viscose additives have the effect of clarifying both the viscose and the resulting cellophane. Examples of these are formaldehyde and other aldehydes (13) and certain specific surface-active agents. The clarifying action on viscose has not been explained, but it may result from improved dispersion of the xanthated cellulose molecules and the breaking up of clusters which scatter incident light. Formaldehyde has been found to interact with the xanthate during the regeneration reaction in such a way that the liberation of gas is slowed. Among the surfactants which have been found to have a clarifying action on viscose and cellophane are sulfated castor oil, certain amine-substituted fatty acids, and sulfonamides. These probably function by keeping gas bubbles dispersed or solubilized.

Another large class of viscose additives termed *modifiers* are used along with modifications to the coagulation and regeneration baths to produce high tenacity rayon. The applicability of these agents to the cellophane process will be discussed in Section III-D.

III. Film Formation from Viscose

For purposes of discussion, conversion of viscose to cellophane can be broken down into several steps, the first of which is to produce a film of viscose. This is accomplished by extrusion through a long, narrow slot of carefully controlled thickness. The second step is coagulation or changing the viscose from a liquid to the solid state so that it will hold its shape. The third step, regeneration, is a chemical process in which the sodium cellulose xanthate in the viscose is decomposed by acid to convert the film to highly swollen cellulose hydrate. The subsequent steps of purification, softening, and drying are outside the scope of this chapter. In the following discussion an attempt will be made to show the influence of the variables of extrusion, coagulation, and regeneration on the structure and morphology of the cellophane sheet.

A. Viscose Extrusion

Viscose extrusion into film is accomplished by a mechanical assembly which includes a viscose metering pump supplying viscose at a required rate for the casting operation, a pressure dome partly filled with air which helps to smooth out surges of pressure, and the extrusion die which shapes the film. The extrusion die or "hopper," as it is called in the cellophane plant, contains a screen to remove gel particles and other solid material which might plug the hopper lips. The lip assembly at the bottom of the hopper forms the long, narrow slot of carefully controlled width which shapes the viscose stream into the form of a thin sheet. In the designing of the hopper and the extrusion channel, it is desirable that viscose be fed uniformly to the full length of the film-forming slot. This problem has been complicated by the fact mentioned earlier that viscose exhibits non-Newtonian behavior at high shear rates. At a point between the lips of the extrusion die, the shear rate is so high that the apparent viscosity is less than 1% of the viscosity of viscose as measured at low shear rates. In order to achieve uniform flow in the die, it is desirable that most of the pressure drop in the flowing viscose be across the die lips. Another practical problem in hopper design is to eliminate, as much as possible, recesses where stagnant viscose can collect during operation. These would eventually form gel and continuously feed particles into the viscose stream, which would rapidly plug the screen or become caught between the lips, causing the film to split.

Control of thickness uniformity of the sheet is one of the most important problems of casting. If gauge variations were distributed randomly across the sheet, they would be less troublesome.

Instead, a thick lane piles on top of itself, giving a visible gauge band in the roll and even stretching the film locally, if severe enough. Optimizing the gauge is not a simple matter of providing a sheet of viscose at the die having uniform thickness. Because of nonuniform shrinkage in width from point to point across the sheet, it is necessary to compensate at the extrusion die for the resulting variations in thickness in order to produce a dry sheet of uniform thickness at the other end of the machine. This is achieved by providing closely spaced adjusting bolts across the die so that localized adjustments can be made to the slot thickness.

B. EXTRUSION THEORY

A phenomenological theory of viscose extrusion through cellophane hopper lips was developed independently in this country by Hinkle (14) and in Japan by Hatakeyama (15) from the same empirical flow equation. Using Hinkle's nomenclature, and considering the forces acting on a plug of fluid passing between parallel plates a distance a apart, shearing stress S is given by

$$S = (\Delta P/L)h \tag{1}$$

where ΔP is the pressure drop across the length L in the direction of flow and h is the half-height of the fluid volume element. For the non-Newtonian, pseudoplastic flow of viscose, the shearing stress and the rate of shear can be accurately represented over a large range by an equation of the form

$$du/dh = AS + BS^b \tag{2}$$

where A, B, and b are empirical constants. Substitution of eq. (1) into eq. (2) and rearrangement leads to

$$du = A(\Delta P/L)h\,dh + B[(\Delta P/L)^b]h^b\,dh \tag{3}$$

Integration and evaluation of the integration constant by placing $U = 0$ at $h = a/2$ gives

$$U = A\,\frac{\Delta P}{L}\left(\frac{h^2}{2} - \frac{a^2}{8}\right) + B\left(\frac{\Delta P}{L}\right)^b \frac{h^{b+1} - (a/2)^{b+1}}{b+1} \tag{4}$$

Now, the average velocity can be calculated:

$$U_{\mathrm{av}} = \frac{\displaystyle\int_0^{a/2} U\,dh}{a/2} \tag{5}$$

Combination of eq. (4) with eq. (5), integration, and rearrangement gives

$$\frac{U_{av}}{a} = \frac{A}{6}\frac{a\Delta P}{2L} + \frac{B}{2(b+2)}\frac{(a\Delta P)^{b}}{2L} \tag{6}$$

Comparison of eq. (6) with flow data of viscose through cellophane hopper lips shows that A is the reciprocal of the viscosity coefficient measured at low shear rates. If the appropriate value of A is substituted, the factors B and b can be obtained from a logarithmic plot of the data as shown in Figure 2. The linearity of this plot

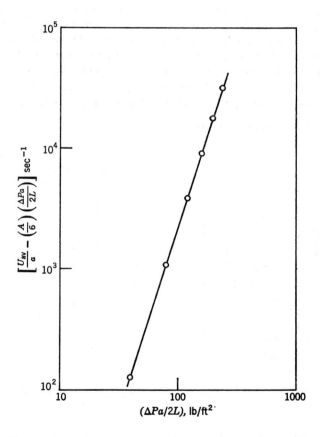

Fig. 2.   Application of the extrusion equation to viscose data.

confirms the validity of eq. (6) for interpolation and limited extrapolation of viscose flow data.

C. Shear-Induced Orientation in Viscose

Viscose, like solutions of other high polymers, exhibits optical anisotropy when examined in a shear gradient. Since a very high shear gradient exists in the viscose as it passes through the extrusion die, it is important to determine if the anisotropy introduced at this point persists long enough to be permanently set into the film by coagulation and regeneration. To investigate this phenomenon, an extrusion die (Fig. 3) was constructed from Lucite* acrylic resin, having an extrusion cross section similar to that of the cellophane casting hopper (16). Commercial cellophane grade viscose was pumped through the die at various pressures. Plane polarized light was passed perpendicularly through the extrusion channel at three positions A, B, and C and by means of a graduated quartz wedge compensator, readings of optical retardation were obtained. In Figure 4 are shown data of optical retardation versus block pressure

* Du Pont registered trademark.

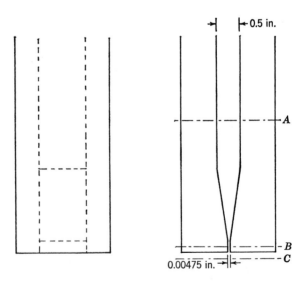

Fig. 3.   Transparent extrusion die.

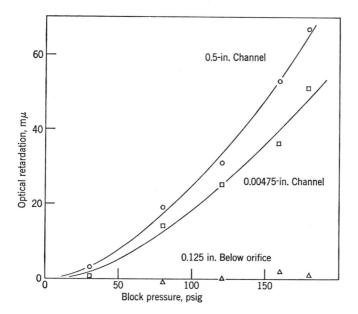

Fig. 4.   Shear-induced orientation at different points in extrusion die.

for the three positions *A*, *B*, and *C*.   Within the extrusion channel the optical retardation did not vary greatly with channel thickness, but upon leaving the channel the flow birefringence was dissipated very rapidly.   At an extrusion velocity of 162 m/min, the optical retardation was zero $\frac{1}{8}$ in. from the extrusion orifice.   This shows complete relaxation of shear-induced orientation in less than 1 msec.   Since coagulation times for cellophane are of the order of several seconds, as will be shown later, this result suggests that shear-induced orientation at the hopper lips is dissipated too rapidly to be frozen into the cellophane structure by coagulation.

D. Coagulation and Regeneration

The two main processes involved in formation of a self-supporting sheet of cellulose from the viscose solution are coagulation and regeneration.   Coagulation is the process by which the fluid viscose is converted to a semisolid structure or gel.   Coagulation occurs, for example, when a film of viscose is immersed in a solution of sodium chloride. A sheet of gel is formed which can be peeled cleanly from its support. The coagulated film will redissolve, if immersed in dilute caustic. Regeneration, on the other hand, is an irreversible chemical change in which the sodium cellulose xanthate reacts with acid (typically, sulfuric acid) with the formation of the sodium salt, carbon disulfide, hydrogen sulfide, and regenerated cellulose. In the commercial preparation of cellophane, viscose is extruded into a bath containing sodium sulfate and sulfuric acid so that these two mechanisms operate simultaneously. Since the structure and the properties of the cellulose film are profoundly affected by the conditions under which coagulation and regeneration are carried out, these processes will be discussed in detail.

Hermans (17) has suggested that coagulation is initiated by the formation of coagulation nuclei which subsequently grow until they intersect one another and form a three-dimensional random network.

In a study of the mechanism of film formation from viscose (5), the coagulation and regenerating requirements for obtaining film of improved strength and toughness were determined. Since tough cellulose films were found to have a low degree of swelling in the gel state, the rates of dehydration and the caustic removal were studied using various viscose and coagulating bath compositions. A known quantity of viscose was spread uniformly on a glass plate and immersed in the coagulating bath for various times and at various bath temperatures. Films were then analyzed for moisture content and alkalinity. Viscoses having different degrees of xanthate substitution lost caustic as well as moisture at different rates. The highly substituted viscose lost sodium hydroxide less rapidly, but water more rapidly than viscose with a lower xanthate substitution when both were exposed to 40% ammonium sulfate at 45°C. This is explained by earlier gel formation in the "leaner" viscose leading to slower dehydration and higher ultimate swelling. The difference between high and low xanthate viscoses was less pronounced when the coagulating bath was 30% sodium sulfate. In a series of viscoses having varying cellulose contents, the rate of water loss was greatest with viscoses of low cellulose content, but the high cellulose viscoses reached a lower final moisture content. It was also found that with the ammonium sulfate bath, temperature had only a minor effect on the coagulation process. In the neutral bath, the rates of removal of both sodium hydroxide and water were increased with increasing bath temperature.   Thus, it appeared that the rate of coagulation, as measured by removal of sodium hydroxide and water, was much more affected by the buffering and neutralizing capacity of the bath than by thermal or osmotic effects.   Acidifying the bath with sulfuric acid increased the rate of removal of sodium hydroxide, as expected.   The effect was much more pronounced, however, with the sodium sulfate bath than with the ammonium sulfate bath.   The extra acidity introduced by the sulfuric acid caused regeneration before dehydration had been completed and consequently the film had a higher degree of gel swelling after coagulation.

Rates of acid penetration into a viscose film were studied by measuring the rate of movement of the neutralization boundary in a tapered layer of viscose containing an acid–base indicator (0.2% bromcresol purple) (5).   The movement of the color boundary was converted to the depth of acid penetration into the viscose film and, as shown in Figure 5, an almost linear relationship was obtained when a plot was made of thickness squared versus time. A comparison of the rates of water loss and the rate of acid penetration with the bath consisting of sulfuric acid and sodium sulfate showed that there was very little, if any, dehydration of the viscose

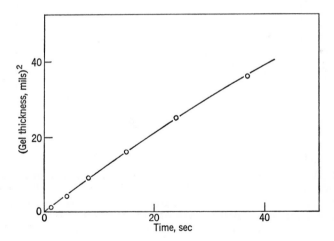

Fig. 5.   The rate of diffusion of acid into viscose film: Bath composition, 12:18 H_2SO_4/Na_2SO_4; temperature, 45°C.

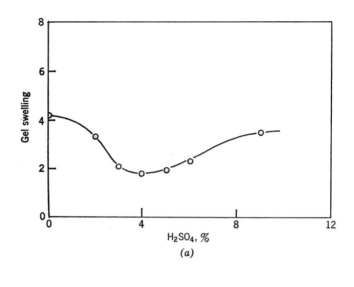

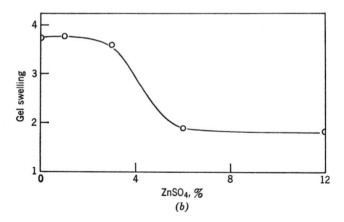

Fig. 6(a). The effect of coagulating bath composition on the degree of swelling of cellulose gel film: sulfuric acid concentration. Bath 1 was $XH_2SO_4/18Na_2SO_4/6ZnSO_4$ and Bath 2 was $12H_2SO_4/18Na_2SO_4$. (b) The effect of coagulating bath composition on the degree of swelling of cellulose gel film: zinc sulfate concentration. Bath 1 was $4H_2SO_4/18Na_2SO_4/XZnSO_4$ and Bath 2 was $12H_2SO_4/18Na_2SO_4$ as before. The viscose used was 9.0 cellulose/5.5NaOH.

at a site inside the film until the acid has penetrated to that point. Continued exposure produced continued dehydration, but part of this deswelling was reversible and reswelling occurred during washing with water.

For many years, high-tenacity yarns from viscose rayon have been made for tire cord applications using a well-dispersed viscose in combination with a coagulating bath containing zinc salt. Further improvements have been made in the properties of tire cord yarns using certain organic compounds as modifiers in the bath and in the viscose. The combination of zinc and viscose modifiers acts to control the rates of diffusion of bath components into the fiber and of viscose components out of the fiber. The applicability of these processes to the preparation of stronger and tougher cellophanes has been explored (8). The effect of bath composition on the degree of swelling of films made with a cellophane type of viscose is shown in Figure 6a and b. In Figure 6a the critical effect of concentration of sulfuric acid is shown. The gel swelling drops sharply as the acidity increases over 2%, reaching a minimum value at 4% and then increasing more gradually to the upper limiting value at around 8%. With 4% sulfuric acid in the bath, which is the optimum acid concentration, an increase in the zinc sulfate concentration, as in Figure 6b, does little to lower gel swelling until the level is more than 3%. There is a sharp drop in gel swelling as the zinc sulfate concentration is increased from 3 to 6%, and higher concentrations yield no further benefits. A completely satisfactory explanation of the role of zinc sulfate in its action on viscose has not yet been advanced.

While a zinc-modified coagulating bath produced a film with improved toughness and reduced sensitivity to moisture, it caused a four- or fivefold reduction in coagulation rate and it imparted an undesirable cloudiness to the film.

E. The Dynamics of Casting Cellophane

Morphological studies (16) of the cellophane base sheet have shown that the molecular orientation as shown by birefringence measurements varies considerably from point to point within the sheet. The variation in birefringence in different machine-direction lanes as one moves across the sheet is shown in Figure 7. The fact that the middle portion of the sheet shows less machine-direction

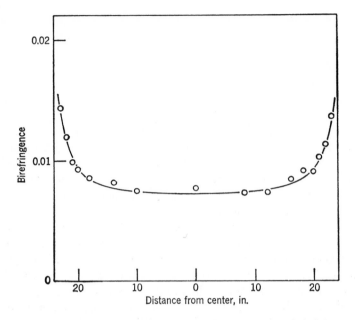

Fig. 7.   Variation of birefringence across a sheet of cellophane.

orientation than the outer lanes can be explained as resulting from excessive transverse shrinkage of the sheet near the edges. This in turn seems to result from the less effective tentering of the lanes near edges on the rolls of the casting machine.

Cross sections cut from the edge of machine-cast cellophane (16) show optical retardation or birefringence profiles which vary from the surface layers to the center layer in an almost parabolic curve. Plate-cast film shows no such variation. Since it was shown earlier that molecular orientation induced by shear in the viscose as it passes through the extrusion slot fades too rapidly to be incorporated into the film structure, it appears that the parabolic orientation profile indicates a corresponding profile of stretching applied in differing amounts to different layers. This suggests that the zone where this profile is established is between the hopper lips and the point where coagulation is complete. Stretching after this point should merely shift the curved profile to higher levels without significantly altering its shape.

In Figure 8 is shown a schematic diagram of viscose being extruded through an orifice into an acid–sulfate coagulating bath. A two-phase web is shown, consisting of unchanged viscose in the interior surrounded by an envelope of regenerated cellulose gel. At a distance D from the hopper lips, all of the viscose has been converted to regenerated cellulose. The following is a list of variables which seem to play a role in the critical zone between the hopper and the point of complete coagulation and descriptions of their probable effects.

1. Bath Drag

The tension required to draw the film away from the extrusion die is supplied from the first driven roll over which the film passes. One component of this tension is the viscous drag of the coagulating bath through which the film is drawn. Factors contributing to bath drag are the surface area of the film from hopper to the point of complete coagulation, the viscosity of the coagulating bath, and a shear rate in the coagulating bath adjacent to the film. Not much can be done to reduce bath viscosity, but the other factors can be minimized by proper choice of coagulating variables. Faster coagulation will shorten the D distance and hence reduce bath drag in the critical zone. The shear rate in the bath layer depends on the difference in the speed of the film and the average speed of the surrounding coagulating bath. Methods have been described for casting into a cocurrent stream of coagulating bath or supporting the film of viscose until coagulation is complete (18).

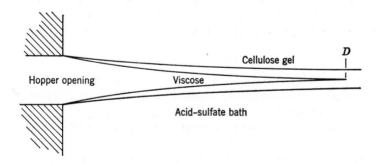

Fig. 8.   Extrusion and coagulation of viscose film.

2. Viscose Deformation

Control of the unit weight of the film is obtained by matching the viscose delivery to the hopper with the machine speed. Film of a given unit weight can be obtained within a wide range of hopper slot openings, the slot opening being usually adjusted so that the extrusion velocity is considerably below the draw-off velocity. The contribution of viscose deformation to tension in the partly coagulated film depends on the viscosity of the viscose at low rates of shear and upon the hopper opening. For viscoses of normal viscosity (40–100 poises), viscose deformation does not appear to contribute appreciably to the orientation profile.

3. Gel Stretching

The contribution of gel stretching to the tension in the partly coagulated film depends upon the amount of stretching required and upon the tensile modulus of the gel film. The amount of stretching is a function of the hopper slot opening and of the solids content of the viscose. The tensile modulus increases with increasing compactness of the gel. Any process variable which changes the degree of swelling of the gel would be expected to affect film tension resulting from gel stretching.

4. Coagulation Rate

It was shown in Figure 5 that acid penetration, and hence coagulation, is most rapid in the early stages of the process, i.e., immediately after extrusion. In addition, the greatest amount of stretching is applied to the film in the early stages of coagulation where the gel envelope is thinnest and weakest. These facts lead to the conclusion that the zone immediately following extrusion is the most critical one in obtaining the optimum orientation profile in the film.

F. Dry-Cast Cellophane

An alternate method of film formation from viscose is the dry casting process which gives a product whose structure and properties are markedly different from those of conventionally cast cellophane (3). The process involves extrusion or spreading of a thin layer of viscose on a heated support, such as a casting wheel or belt, evaporation of the water in the viscose film in a stream of hot air, regeneration by heat or by acidic decomposition of the xanthate

TABLE 5
Comparison of Physical Properties of Conventional and Dry-Cast Cellophane

Film, type[a]	Softener content,[b] %	Tear, g	Pendulum impact, kg-cm	Stress-flex, cycles
Conventional[a]	0	2.5	2	2
Dry-cast	0	25	18+	25
Conventional[c]	16	6	3	5
Dry-cast	14	30	18+	30

[a] All samples were 1 mil thick.
[b] Glycerol was used as softener.
[c] Laboratory cast.

in a nonswelling medium, washing, and final drying. As shown in Table 5, the physical properties of dry-cast cellophane prove it to be exceptionally tough, even in the absence of a softener. The explanation for these exceptional properties is to be found in the fine structure of the dry-cast cellophane as compared with regular cellophane and as revealed by measurements of lateral order, crystallinity, or accessibility of the cellulose to chemical reagents. The structure and properties of the dry-cast cellophane are similar to those of a cellulosic film made by deacetylation of cellulose acetate film in a nonswelling medium (2). The reason for the unusual structure and properties of these experimental films as compared with the conventional wet-cast cellophane is to be found in the mechanism by which the film is formed from the solution of cellulose. In the conventional process, the acid–sulfate bath simultaneously coagulates and regenerates the cellulose. Consequently, hydroxyl groups on the cellulose molecules are freed while the structure is in a swollen state. This apparently causes localized deswelling, forming ordered structural elements surrounded by water-filled voids so that moisture content on a dry basis is about 300%. In contrast with this, the two dry-casting processes involve, first, a deswelling step and, second, a regeneration of the cellulose under conditions which do not permit reswelling. Under these conditions, the cellulose molecules do not form highly ordered structures characteristic of the wet-casting method and also the degree of swelling is much lower (moisture content 80–120% on a dry basis).

While dry-cast cellophane has excellent toughness and strength, its reduced transparency and greater sensitivity to moisture, as well as less favorable process economics, have kept it from being commercialized.

References

1. E. Weston, U.S. Pat. 264,987 (Sept. 26, 1882).
2. V. C. Haskell and D. K. Owens, *Textile Res. J.*, **30**, 993 (1960).
3. C. R. Price and V. C. Haskell, *J. Appl. Polymer Sci.*, **5**, 635 (1961).
4. M. F. Bechtold (to E. I. du Pont de Nemours & Co.), U.S. Pat. 2,737,437 (March 6, 1956).
5. M. Miller and V. C. Haskell, *J. Appl. Polymer Sci.*, **5**, 627 (1961).
6. T. E. Hopkins and J. W. Whatley, *J. Appl. Polymer Sci.*, **6**, 600 (1962).
7. Data of Table 3 are taken from an unpublished report by F. N. Bowman, R. L. Burton, K. C. Hodges, H. B. Sanford, and W. W. Story, Film Dept., E. I. du Pont de Nemours & Co., Wilmington, Del.
8 Data of Table 4 are taken from an unpublished report by V. C. Haskell, Film Dept., E. I. du Pont de Nemours & Co., Wilmington, Del.
9. E. Ott, H. M. Spurlin, and M. W. Grafflin, *Cellulose and Cellulose Derivatives*, Interscience, New York, 1954, Part I, pp. 302–310.
10. E. Treiber, *Tappi*, **46**, 595 (1963).
11. H. G. Ingersoll, U.S. Pat. 2,991,510 (July 11, 1961).
12. S. Dunbrant and O. Samuelson, *Tappi*, **46**, 520 (1963).
13. F. M. C. Corp., Brit. Pat. 963,072 (July 8, 1964).
14. Unpublished report by B. L. Hinkle, E. I. du Pont de Nemours & Co. (1955).
15. S. Hatakeyama, *Chem. High Polymers (Japan)*, **13**, 231 (1956).
16. V. C. Haskell and D. K. Owens, *J. Appl. Polymer Sci.*, **4**, 225 (1960).
17. P. H. Hermans, *Physics and Chemistry of Cellulose Fibers*, Elsevier, New York, 1949, p. 74 ff.
18. W. Bender (to E. I. du Pont de Nemours & Co.), U.S. Pat. 2,254,203 (Sept. 2, 1941); J. A. Mitchell (to E. I. du Pont de Nemours & Co.), U.S. Pat. 3,073,733 (Jan. 15, 1963); N. V. Onderzoeking, Brit. Pat. 922,655 (Apr. 3, 1963); B. L. Hinkle and F. C. Stults (to E. I. du Pont de Nemours & Co.), U.S. Pat. 2,962,766 (Dec. 6, 1960).

CHAPTER 4

NATURAL POLYMERS AS FILM FORMERS
C. Properties of Regenerated Cellulose

ERIC WELLISCH

*Film Division, Olin Mathieson Chemical Corporation,
Pisgah Forest, North Carolina*

I. Introduction

Cellophane is primarily cellulose regenerated from viscose and cast into sheet form. It usually contains a plasticizer or softening agent such as glycerol or other polyhydroxy alcohols which contribute to the physical properties exhibited by the cellophane base sheet. In addition, the coating process which moistureproofs the

139

film also has a considerable effect on the physical properties of the final product.

The structure and therefore the properties of cellophane are influenced by the process of manufacture. The viscose process, which solubilizes and then regenerates cellulose, largely disperses cellulose chains until, in a viscose solution, the cellulose chains are no longer held together by strong hydrogen bonding, although some association of chains and some cellulose crystallites doubtless exist in viscose solution (1). After regeneration, none of the original pulp fibers can be found in the film and both its degree of polymerization and its degree of crystallinity have been decreased by the process of dissolution and regeneration.

The dissolving of the original wood pulp, that is, the methods used in the preparation of viscose beginning with alkali cellulose, its xanthation and the dissolution, blending, and ripening conditions, influence film structure and film properties (2). In particular, the degree of polymerization of the alkali cellulose, the extent and uniformity of xanthation, and the cellulose and caustic content of the viscose will determine its state of solution and its viscosity and thus contribute to the properties of the regenerated product (3–5).

The film casting process and particularly the initial coagulation and regeneration steps, as well as the final drying step, which are all carried out on a casting machine running at a considerable speed and exerting a continuous stress on the cellulosic web during its formative stage, will influence molecular structure and orientation and therefore affect the durability of the regenerated cellulose.

This chapter is concerned with the physical properties of regenerated cellulose film as determined and controlled by the interaction of viscose preparation, film casting, plasticizing, and drying. It attempts to show that the influence of each of these steps on the cellulosic fine structure contributes to those properties that ultimately determine its performance in use.

II. The Fine Structure of Cellophane

A. Casting Process and Structure

In a viscose solution, cellulose chains are not oriented except for some association of incompletely dispersed chains or crystallites which are held together by hydrogen bonds and are, therefore, not subjected to the solubilizing effect of the xanthation reaction. During coagulation in an acid–salt bath, the viscose solution—crystallites plus dispersed xanthated regions of the cellulose chains—forms a gel. The rate of gel formation or loss of water and the extent of gelation or final equilibrium water content depend on the ionic strength of the regenerating (acid–salt) bath and on the state of solution of cellulose (viscose) as determined by the degree and uniformity of xanthation; they also depend on the onset and rate of regeneration, as will be discussed further. This gel is subjected to the pull of the casting machine in the machine direction and becomes deformed and oriented by mechanical stresses until regeneration is complete and reformation of hydrogen bonds restricts the movement of neighboring chains. The degree of orientation and of lateral order (crystallinity) of the regenerated film thus depends on the rate of both the coagulation and the regeneration process (5–7).

The degree of lateral order of cellulose film as reported by Price and Haskell (8) is known to have a pronounced effect on its physical properties. A low lateral order appears to be the most significant feature controlling physical properties in a film. Highly durable films with a very low degree of lateral order were prepared by dry casting (on a polished, heated surface), resulting in a random cellulose structure of unusual toughness and extensibility. For the same reason, i.e., a very low degree of order, films produced by deacetylation of cellulose acetate also show radically improved physical properties (9).

With the aid of birefringence measurements, Haskell and Owens also investigated the orientation of cellophane (6). Films hand cast on glass plates exhibit primarily uniplanar orientations while machine-cast films exhibit a superimposed uniaxial orientation in the machine direction in addition to uniplanar orientation. This orientation differs both as a function of film width and cross section. The degree of uniaxial orientation is greater at the edges (edge effect) than at the center of the sheet while the cross section of the film shows an orientation profile which indicates higher orientation at the film surface than within the sheet. This difference in orientation in thickness can be correlated with differences in crystallinity as determined by both dye uptake and electron microscopy measurements. This uniaxial orientation can be related to the pull of the casting machine on the gel film in the machine direction, and is a function of casting speed, machine-direction tensions, and transverse-direction shrinkage; these forces influence the molecular orientation of the cellulose chains primarily during the coagulation and regeneration step while the gel film is most subject to facile deformation by the pull of the casting machine (10). Additional deformation and orientation occur when film is dried, as the removal of large amounts of water causes the gel to shrink mostly in the thickness and transverse direction while under machine-direction tension (11).

The degree of lateral order is largely responsible for the physical properties of the regenerated cellulose film and can be directly related to its anisotropic characteristics (imbalance of properties in the machine and transverse directions) and the dimensional instability which results when the film absorbs moisture and relaxes at high ambient humidities (11,12).

The degree of uniaxial orientation and the degree of crystallinity of the regenerated cellulose film are a function of the competing rates of the coagulation and regeneration process. Thus, while viscose prior to coagulation is subject to little effective distortion, the coagulated gel becomes highly oriented during regeneration (H-bond formation) (6,7); this is desired in the case of viscose fibers where retarding additives are used to produce and maximize orientation (13). On the other hand, a hydroxyethylcellulose film which regenerates approximately 300 times faster than viscose exhibits considerably less orientation and has much greater isotropic character than the corresponding viscose film; however, because of this rapid regeneration, the coagulation and gel formation are incomplete and both the wet gel film and the regenerated dry film are inherently weaker in most of their physical properties than is film regenerated from viscose solutions; the best answer to proper balance of strength and isotropy lies in permitting maximum coagulation under tensionless conditions, such as by casting on an endless belt or drum (8,14–20) followed by rapid regeneration, and delaying the application of casting machine stress until regeneration has been completed. A similar procedure should be followed during drying where shrinkage resulting from water removal should be controlled in proportion to the loss of water, rather than uniformly restricted on the drier rolls and in effect enhanced by the pull of the casting machine (72).

B. X-Ray Studies and Birefringence

The question of how molecules are oriented in a polymer film such as cellophane is important in relation to the resulting directional variation of strength characteristics of the film. In order to determine the nature and extent of molecular orientation in a film, it is necessary to measure the change with direction of some property of the material that depends on the molecular orientation within the sheet. In general, mechanical properties are measured, and therefore defined, in terms of molecular responses to external stimuli. Elasticity, for example, is measured in terms of an elastic modulus, which is the ratio of the force acting on the material to the elongation produced. If the degree of elasticity is the same regardless of the direction of the applied force and resulting stretch, the material is said to be isotropic with respect to this property. If, however, there is variation of elasticity with direction, the material is anisotropic. Because the elasticity in any one direction depends on the nature and intensity of forces in that direction, the change in values of elastic modulus can be taken as a measure of molecular orientation.

Birefringence measurements can be used to determine the degree of molecular orientation present in a piece of film without making a distinction between the crystalline and amorphous areas. Such measurements alone, however, do not completely define molecular orientation, for there are several distinct types of orientation that may occur. The most general case is random orientation, in which there is no relative orientation among molecules. If the molecules have their chain axes parallel to a common direction, this type of orientation is referred to as uniaxial (although "uniaxial" is actually a term borrowed from crystallography, and has a precise definition within that field). If the chain axes lie parallel to a given plane, this orientation is called uniplanar. If, in addition to having uniplanar orientation, the molecules also have a fixed rotation about their chain axes, this case constitutes selective uniplanar orientation. Finally, a combination of selective uniplanar and uniaxial orientation is called selective uniaxial orientation.

In order to distinguish among these types of orientation, some method, in addition to birefringence measurement, must be employed. The question to be answered is the following: What are the exact positions of the molecules and bonds in a cellulose film with respect to the length, width, and thickness coordinates of the film?   The answer depends first of all on a knowledge of the unit cell structure for the material.   The most important features of the unit cell for cellulose II, which represents the structure of regenerated cellulose, are given by the Meyer and Misch model (21), shown in Figure 1.   This model has recently been the subject of some dispute regarding its more detailed features, but the main

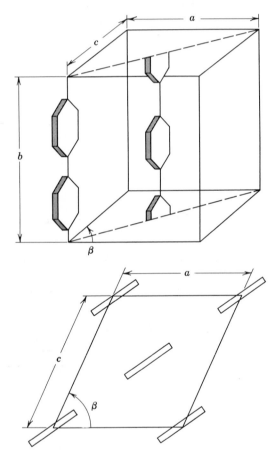

Fig. 1.   Unit cell for cellulose II structure: $a = 8.14$ Å; $b = 10.3$ Å, $c = 9.14$ Å; $\beta = 62°$.

points of the model are well established (22). Once the molecular structure within the unit cell is known, a method of establishing the orientation of the unit cell with respect to the film coordinates will give, as well, the orientation of molecules with respect to the film coordinates. Such a method is x-ray diffraction. By taking x-ray pictures of the films and studying the resulting patterns, the several types of molecular orientation described above can be distinguished. Thus, x-ray diffraction gives information about the type of orientation present in the crystalline areas of the film and birefringence measurements give a quantitative estimate of the orientation in the film as a whole. By use of these two measurements, the selective uniaxial orientation of cellophane, which is a combination of selective uniplanar and uniaxial orientation, has been confirmed (6,9,23–25).

On the basis of birefringence measurements, it appears that viscose films have their chain axis in the plane of the film. The existence of such orientation of chain axis parallel to the plane of the film is important in that it may be related to the degree of orientation, or orientability, of chain axes into the machine direction in commercial films. It is probable that the degree of orientation of molecules parallel to the plane of the film influences its dimensional stability. The lower this kind of orientation is in a given film, the better are its dimensional (or deformation) characteristics since, with little orientation present initially, molecules are already in a relaxed state and the film will return to its original dimensions after swelling under conditions of high humidity.

X-ray diffraction has established that the orientation of the molecules about their chain axes is such that the anhydroglucose rings and the intermolecular hydrogen bonds are in the thickness direction, giving predominantly selective uniplanar orientation which is responsible for the imbalance of physical properties in the machine and transverse directions and produces considerable difference in the birefringence associated with these two directions. It is suggested that the optimum orientation from the point of view of strength properties would be such that all of the hydrogen bonds be parallel to the plane of the film.   The complete description of the optimum orientation in cellulose film—to what extent it should be uniplanar, selective uniplanar, or uniaxial—must depend on the relative importance and desirability of each of the resulting properties of the film. Birefringence measurements and x-ray photographs were made of deacetylated cellulose acetate films; the very low degree of molecular order of these films appears to be associated with increased extensibility and greater toughness in comparison with viscose films (9).

C. Fine Structure and Dye Uptake

An investigation of the fine structure of regenerated cellulose film by absorption of Solophenyl blue green dye distinguished two distinctly different crystalline structures in the cross section of the cellophane sheet (26–29). A thin dense outer skin comprising only 5–15% of the total cross section whose thickness depends on the regenerating conditions and a more porous inner structure of different crystalline order can be distinguished by differential dyeing and by absorption measurements at 630 mμ (5,10,30). The two regions of quite different responses to dye absorption observed may be analogous to the skin and core regions that have been observed in rayon fibers (13,29,31). The thin outer "skin" layer is of low crystalline order and of a more dense structure while the inner "core" has a loose structure and is of a relatively high degree of lateral order.

In general, it appears that the fine structure of today's cellophane films is similar to a "core" structure which is covered by a thin cuticle of "skin" structure whose thickness is a function of the type of regeneration used and regenerating solution employed. A rapid regeneration using high acid concentrations, as found in commercial use (32), produces a thin outer skin type of structure of low crystalline order. This outer dense structure facilitates the formation of an inner core type of structure of high lateral order since further acid penetration is slowed and becomes diffusion-controlled (33,34). On the other hand, the regeneration of viscose from an acid–salt bath containing zinc sulfate results in complex formation of zinc xanthate which will make the film more skinlike in structure resulting in a decrease in dye penetration (29,30,35–37). The addition of zincate to viscose (38–40), or the addition of small amounts of formaldehyde to viscose, or to the regenerating bath, will also assist in obtaining an all-skin type of film (41).

At present, the principal difference in fine structure of different films is the existence of these outer skin barriers of different thickness affecting the accessibility to moisture and softener which might control the rate of softener uptake and restrict the use of softener types to those of small molecular volume and high polarity. On the other hand, the existence of a loose core type of structure of high lateral order (which at present represents 85–90% of the cellulose sheet) confirms the existence of large accessible amorphous regions which make softener absorption possible and thus facilitate the plasticization of the cellulose base sheet, a necessary condition for durability requirements and commercial acceptability as a useful packaging material. All-skin, high-tenacity rayon fibers do not require similar plasticization.

Recently, a radically new model for the fine structure of natural cellulose has been proposed (42,43) according to which cellulose microfibrils are composed of flat ribbons wound in the form of a tight helix. The cellulose molecule is arranged parallel to the fibril axis and forms flat ribbons folding back and forth with its axis tilted at some preferred angle to the ribbon axis. With the ribbon wound as a helix, the molecular chain becomes parallel to the fibril axis. If this idea should be applicable to regenerated cellulose chains, we must reinterpret crystalline order and skin and core structure in terms of the number of folds and the tightness of the helical structure of the microfibrils and think of accessibility in terms of imperfections and distortions at the helix fold rather than in terms of molecular order or disorder (44–46).

III. Cellophane Additives

A. Softeners

Regenerated cellulose is brittle at low relative humidities without softener or plasticizer. The dry film is rigidly crosslinked and immobilized by hydrogen bonds along the chain length. Water molecules break these interchain bonds by interaction with the cellulosic hydroxyl groups and cause swelling and plasticizing; flexibility and deformability result from this interaction, as is desirable in a packaging material. Since water is too volatile a plasticizer, however, commercial practice requires less fugitive, less volatile softeners. The commonly employed commercial softeners are the polyhydroxy alcohols of high hydroxyl-to-hydrocarbon ratio which are not only capable of interaction with cellulose chains but also are highly hygroscopic and thus can absorb and hold a large proportion of water at a given relative humidity, which further aids plasticization (47).

The control of the proper softener level and the use of the proper softener type are critical factors in commercial cellophane production. Too little softener results in brittle film and too high a softener concentration makes a film too tacky and too soft. On the other hand, final conditions in use must also be considered. At high relative humidity, less softener will be required since a high water content will compensate for the softener. High ambient temperatures usually require less softener than do low-temperature conditions. Special softeners may be needed for special low-temperature performance; thus, winter and summer films require different softener types and films required to perform in the arctic regions may be softened with softeners of much higher vapor pressure than films shipped to regions of high relative humidities and high ambient temperatures. In addition to softener volatility, hygroscopicity, and cellulose-softener interaction, other factors such as taste, odor, and toxicity will determine the acceptability of a specific softener.

Softeners such as glycerol or other polyhydroxy alcohols are absorbed into gel film by diffusion of softener solution into the accessible free volume of the gel cellulose during casting. The simplest process of softener uptake by gel film involves the replacement of water by softener solution from the impregnating bath of the casting machine by establishment of a dynamic diffusion equilibrium. If no enrichment in softener by interaction with cellulose is assumed, the amount of softener taken up by the film is the result of two variables: the concentration of the softener bath and the volume of the free space in the cellulosic structure to be filled. Volume changes of the gel film during softener uptake with 10% glycerol solution were found to be negligible. Preferential softener uptake or enrichment within the gel film was found to be very slight at the softener concentration used. It appears that at 10% concentration glycerol and other organic compounds are taken up by gel film by a diffusion process in which the free water in films is rapidly replaced by softener solution without appreciable enrichment or change in volume (48,49a,50).

The free volume of the film available for water–softener exchange is determined by the cellulose structure, its crystalline-to-amorphous ratio and by the water content of the gel film as measured by the gel swelling index. The gel swelling index is the water-to-cellulose ratio of the gel film (51) which depends on the extent of gel densification occurring during the coagulation–regeneration process which is a function of several processing variables, including the degree of polymerization, the cellulose content of the viscose, and the regenerating conditions (5,7,33,52).

On the basis of this mechanism of softener uptake, the final softener concentration in the film after drying is related to and can be calculated from the original water-to-cellulose ratio of the gel film and the softener bath concentration in the casting machine (49b,50).

B. Cellulose–Softener Interaction

It is of considerable theoretical as well as practical interest to understand the interactions of cellulose chains with each other and with small molecules such as water, glycerol, or glycols which are used as commercial softeners for cellophane and are a major factor in determining the final mechanical properties.

We have already mentioned that softener is taken up by the gel cellophane principally by an exchange of water for softener solution without volume change or enrichment in softener concentration. Only when water is expelled and the film begins to collapse during drying does the softener begin to interact with cellulose. This interaction begins at almost 95% relative humidity corresponding to 100–80% moisture content and increases as the softener displaces water down to a moisture content of 6–9%. The dried base sheet then contains tightly held water and softener, solvating and screening the cellulosic hydroxyl groups. The interaction between softener and cellulose prevents interchain hydrogen bonding and is therefore believed to be a necessary condition for proper plasticization of cellulose.

Depending upon the affinity of the softener for cellulosic hydroxyls and upon the conditions of drying, some change in the cellulose structure may occur (10); essentially, however, the softening process can be considered as one of competitive exchange of water for softener at the cellulosic hydroxyl sites without any appreciable change in the amorphous-to-crystalline ratio of the base sheet. The assumption that softeners do not penetrate the crystalline regions is supported by volume measurements of gel cellophane softened with softener solutions ranging from 100% water to 80% glycerol; a constant volume was found equal to more than 99% of the added volumes of glycerol solutions plus dry cellulose measured separately (49a). This constant contraction suggests that interactions between cellulose and water and between cellulose and glycerol are similar, without appreciable change in the accessibility of cellulose (50). Further confirmation of this assumption has been found in the work of McKeown and Lyness (53) who have shown that no change occurs in the ratio of crystalline-to-amorphous cellulose even under more drastic conditions of hydrolysis.

If, therefore, the accessibility of cellulose is considered constant while in a state of collapse (between 70 and 5% water content), the cellulose–softener–water system may be regarded as a binary system of water and softener which should obey Raoult's law. Deviations from Raoult's law should give an indication of the inter-action of softener and water with cellulosic hydroxyls. This interaction can be measured quantitatively by measuring the equilibrium moisture content of the systems cellulose–water, softener–water, and cellulose–water–softener such that the deviation in additivity in equilibrium moisture content represents a measure of cellulose–softener interaction. By use of this method of inter-action measurement, it was found that glycerol and ethylene glycol, which are considered effective softening agents, exhibit a positive deviation from additivity over a range of 0–30% softener in the film and at relative humidities from 15 to 93%; similarly, a much smaller positive deviation was measured for tetramethylene sulfone while a negative deviation was observed for ethylene carbonate. The magnitude of these deviations was found to be in direct proportion to softener effectiveness under film performance conditions. The deviations are therefore a useful measure of cellulose–softener interaction and of softener efficiency. The absolute values of interaction depend on the softener concentration and relative humidity conditions. At 35% relative humidity, in a range where the reduction in water content caused by softener-cellulose interaction was at a maximum, the measured deviations from additivity (in moles of water per mole of softener and cellulose) were: ethylene glycol, 0.26; glycerol, 0.21; tetramethylene sulfone, 0.08; and ethylene carbonate, −0.08. Thus, ethylene carbonate, the poorest softener, shows no interaction with cellulose and in effect slightly increases the equilibrium moisture content of the system cellulose–softener–water in comparison to unsoftened film; this moisture increase is probably a result of a slight increase in the overall volume of the system cellulose–water–inert softener, as discussed in the beginning of this section. On the other hand, a very large positive deviation of 0.60 was found for ethylenediamine which is completely and preferentially held by the cellulose hydroxyls; this extreme interaction results in a crosslinking effect which embrittles the cellulose base sheet rather than plasticizing it at low relative humidities (54,55).

C. Anchoring Agents

Anchoring agents, such as urea– or melamine–formaldehyde resins, are added in small concentrations to cellophane to increase the adhesion, especially under conditions of high humidity, between the cellulose base sheet and the hydrophobic nitrocellulose- or polymer-based lacquer coating (56–63). These resins are usually added in small concentrations to the softener bath and are picked up by the wet gel film with the softener just before drying. The resin must be in a primary condensate form and be soluble or dispersible in the aqueous softener bath; a small amount of curing agent, usually an organic acid, is also added at the same time. Since the molecular volume of these prepolymers is much larger than the water or softener molecules, they cannot diffuse into the voids of the gel film; they are screened out during the water–softener solution exchange and are adsorbed on the film surface followed by possible chemical bonding during curing. Partial curing of the resin occurs on the casting machine during drying and is completed in the coating towers during coating. Care must be taken to avoid adhesion of base sheet to base sheet (called blocking) if too much time is allowed to elapse between casting and coating of anchored stock. For this reason, anchoring resins have been applied alternatively to the cast film as a subcoat or by spray or roll-on techniques just prior to coating; addition of solvent-based resin to the coating bath has also been used (64).

In addition to melamine– and urea–formaldehyde anchoring agents, other resins have been used, including polyethyleneimine or polyacrylamide and modifications, and other materials such as quaternary ammonium compounds whose cationic character improves adhesion to the polymeric electronegative base sheet (65). Recently, polymeric coatings have also been developed which are self-anchoring by addition of small amounts of a copolymer of styrene to the vinylidene chloride coating lacquer (66).

In addition to their primary use to improve coating adhesion under conditions of high relative humidity, anchoring agents were found to retard and reduce the moisture regain of coated films and thus assist in improving the dimensional stability of the base sheet (66a).

IV. Drying of Cellophane and Physical Properties

A. The Drying Process

When the regenerated cellulose base sheet enters the drying section of the casting machine, the gel film contains approximately 300% water based on its cellulose content and about 15–25% softener present as an aqueous solution equal in concentration to that of the softener bath the film is leaving just prior to drying. The gel film is nearly of the same volume before and after softening since the softening process involves neither appreciable swelling nor enrichment (49a,b,50). On water removal during drying, the gel is reduced in volume and shrinks mostly in the thickness direction to one-fourth of its original thickness. During this shrinkage and reduction in thickness, while water is removed by heat down to about 6–8% moisture content, tensions are being built up in the machine direction against the pull of the casting machine and also in the transverse direction where it is prevented from major shrinkage by a tight wrap on the drier rolls and by special surface coatings of these rolls. Thus, a certain amount of transverse film tentering is achieved which decreases the anisotropy of the film. Nevertheless, drying imparts further distortion to the film in addition to the strain imparted during the coagulation–regeneration process. This affects the imbalance of the physical properties of the base sheet in the machine and transverse directions including its "pullout" (uniformity of tension across the sheet) and its dimensional stability (10,12,33).

Experiments have shown that when drying is carried out under controlled conditions of shrinkage and tension (12,67–70), or when removal of water is carried out in discrete stages in accordance with the difference in position and in interaction of the water present and its effect on the cellulosic sheet during such removal (71), the isotropic characteristics of the base sheet are increased and all durability properties including dimensional properties are improved.

B. DRYING, ACCESSIBILITY, AND SHRINKAGE

As indicated in the above discussion, the drying of cellophane can impart tensions and further enhance the anisotropy of the cellophane base sheet and increase the dimensional instability of the finished product under conditions of actual use.   Thus, when a film on a packaged product is exposed to high relative humidity, it absorbs moisture and swells; interchain bonds are broken by interaction with the entering water molecules (72).   Under these conditions, any stresses set up during the casting and drying process, as discussed earlier, are relaxed and the film shrinks irreversibly and anisotropically, resulting in a wrinkled or distorted package which becomes less protective, less attractive, and less salable.

To obtain a cellophane base sheet of maximum dimensional stability, it is necessary to produce film of minimum and, if possible, isotropic shrinkage characteristics.   This requires that we not only recognize and control the factors contributing to film anisotropy during the early stages of coagulation and regeneration but also that we understand the process of water removal and its effect on film shrinkage during drying.   It might then be possible not only to avoid additional anisotropy during drying but to use the drying process to compensate for and thus reduce the orientations and tensions produced on the wet end of the casting machine before the film enters the driers.

Shrinkage measurements as a function of water content carried out with films softened with glycerol showed (1) that shrinkage in the thickness direction is directly proportional to the water content, (2) that in both the machine and transverse directions, little shrinkage occurs initially down from 300 to 80% water, (3) that shrinkage in the transverse direction is greater than in the machine direction, and (4) that this anisotropic shrinkage is related to the molecular orientation of the film imparted during wet-end processing (6,12,73). Therefore, shrinkage and tensions produced during drying appear to depend on the orientation of the film as it enters the drier, i.e.,

on the processing conditions at the wet end of the casting machine
and on the volume changes of the film during drying, as determined
by the rate and extent of water removal. Volume measurements
of gel film as a function of water removal during drying have shown
that little volume changes occur as water is removed from 370%
down to 100%, indicating that this water is located interstitially
and is removed by creating empty capillaries without affecting the
gel film volume. As more water is removed, volume changes begin
to occur between 100 and 80% water content which are related to
adsorbed water in the accessible region of cellulose. Further drying
and water removal causes increased film shrinkage in the transverse
direction as water held by adsorption to cellulose hydroxyls is
removed; this shrinkage corresponds to a water content down to
15% moisture. At this moisture level and below, cellulosic hydroxyl
groups begin to hydrogen bond, irreversibly replacing water of
hydration. Water removal at this moisture level is difficult and the
rate of drying becomes very slow even at high drying temperatures.

Thus, we essentially have three phases of drying: removal of
absorbed, adsorbed, and hydrated water. Only during removal
of the latter two types of water does appreciable shrinkage occur
and only during the last phase of moisture removal does the shrinkage become irreversible. On the basis of this shrinkage pattern,
it is possible to control the dimensional stability of cellophane by
drying the film from 400 to 100% water content under sufficient
tension to prevent shrinkage, and controlling the shrinkage that
occurs below 100% water content in such a manner as to impart a
minimum amount of tension to avoid the buildup of stresses which
will cause shrinkage on rewetting and thus result in dimensional
instability.

In practice, improvement in dimensional stability has been
achieved by combinations of stretching and relaxation during drying
as indicated above (71) and by proper adjustment of drying rate
and temperature (12,67,69). Other methods of improving the
dimensional stability are based on stress relaxation by rewetting of
dry film in hot water (74) or by crosslinking of the cellulose base
sheet to prevent relaxation on exposure to high relative humidities
(75–77).

This last method usually causes deterioration of the durability
of the base sheet without resulting in effective dimensional stabilization of the base sheet (78).

V. Cellophane–Moisture Equilibrium

A. Effect of Softener Content

When regenerated cellulose film softened with a standard softener such as glycerol or a glycol is exposed to the atmosphere, the equilibrium moisture content will vary with relative humidity. If this moisture content is plotted as a function of vapor pressure of water, a sorption isotherm is obtained whose general sigmoidal shape is similar for both softened and unsoftened films. It is generally believed that the first portion of the curve between 0 and 15% relative humidity corresponds to very strongly bonded water "chemisorbed" to the primary hydroxyl groups. The fairly flat portion between 15 and 65% relative humidity represents water *adsorbed* by free cellulosic hydroxyl groups in the amorphous region of cellulose and the steep portion of the curve between 65 and 100% relative humidity represents water *absorbed* by capillary action and condensed water which enters the film as it swells at high relative humidity. In the presence of a softener the water content is usually reduced at the low humidity as softener is adsorbed on cellulosic hydroxyls in preference to water. At high vapor pressure of water, the equilibrium moisture content of the film increases beyond that found for unsoftened film as some of the softener molecules are dislodged from the cellulosic hydroxyl sites by water under high vapor pressure and the softener exerts its own hygroscopicity. This is essentially the qualitative picture of softener action as we know it today and is represented in Figure 2 for a glycerol-softened film and for unsoftened film.

The softening action of various softeners on cellophane can best be compared, at least qualitatively, at high relative humidity where softeners must compete for cellulosic hydroxyl sites with large amounts of water, whereas at low humidities little water is present for competitive interaction. This is the basis of quantitative measurements of cellulose–softener interaction discussed in an early section of this chapter. A softener which shows some interaction with cellulose in the presence of large amounts of moisture will certainly interact with cellulose in the absence of water. The most effective softener will have a positive interaction with cellulose in the sense that the moisture regain of the cellulose–softener system is less than the combined regain of the separate components (54,55).

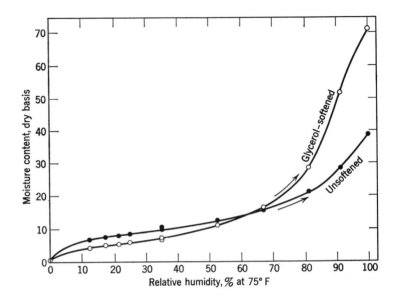

Fig. 2. Comparison of cellophane sorption isotherms: glycerol-softened films
(○); unsoftened films (●).

As mentioned above, this positive interaction is limited by cross-linking effects, as found in the case of ethylenediamine.

B. Drying Condition and Moisture Equilibrium

A difference exists between the equilibrium moisture content of dried unsoftened film and undried gel film which at high relative humidity is greater than the normal difference between absorption and desorption of moisture in dried films caused by relief of tension and by swelling (79).

When unsoftened film is dried, secondary hydroxyl sites are gradually vacated by water (80) and interact with each other, which results in crosslinking by hydrogen bonding.   The dry, brittle film that results obviously will be less hygroscopic than an undried gel film which has been allowed to reach equilibrium moisture content by slow desorption at any given relative humidity and whose secondary hydroxyl groups are therefore not so likely to form hydrogen bonds.   When softened film is dried, the softener replaces the water at the secondary hydroxyl sites and hydrogen bonding between cellulose chains is prevented.

Depending on the extent of drying, three different conditions may result: (1) The softened film is dried to a moisture content of 10–12% (based on cellulose) which results in an open cellulose structure with water hydrogen-bonded to primary and many secondary hydroxyl groups and with some softener still unbonded. (2) The softened film is dried to a moisture content of 4–6% without loss of softener; this will probably free most secondary hydroxyl groups of cellulose for interaction with the softener. Not much hydrogen bonding between cellulose hydroxyls will occur under this condition. (3) During prolonged drying at high temperatures (100–150° C), as during coating, most of the moisture (even that held by primary hydroxyl groups) may be expelled and some of the softener may be lost. Under these conditions, we would expect the film to collapse further and additional hydrogen bonds to be formed between adjacent cellulosic hydroxyl groups. The resulting film should be of much lower hygroscopicity and of poor durability.

Experimental evidence supports the above hypothesis. When unsoftened films were dried at 115° C to different moisture levels by variation of the drying time and allowed to reach equilibrium under different relative humidity conditions, the moisture sorption and desorption data shown in Table 1 were obtained. These data show that the hygroscopicity of a film decreases considerably as the unsoftened film is dried to lower and lower initial moisture content. If we assume a 40–50% average accessibility for regenerated cellulose (81,82), a cursory calculation shows that approximately 14% moisture content (per 100 g of dry cellulose or 0.62 mole of anhydroglucose units) is equivalent to the three hydroxyl groups in cellulose being fully hydrated in the accessible region. On the other hand, if the film is dried below 5% moisture, none of the remaining water will be held by *secondary* hydroxyls but only by primary hydroxyls where it is believed to be most firmly associated (80). It is interesting to note from Figure 3 that the desorption curve for gel film is not very different from those of films A and B while films C and D (Table 1) dried to very low moisture content show sorption data which appear to be linear between 0 and 65% relative humidity. The elimination of the usual S shape of the sorption isotherms for overdried films can be explained by assuming effective inactivation

TABLE 1
Equilibrium Moisture Content of Unsoftened Films
Moisture Hysteresis

Relative humidity, %	Equilibrium moisture content, %				
	Film A	Film B	Film C	Film D	Gel film
Initial[a]	12.0	5.9	3.1	1.3	261
15	7.0	5.3	1.6	1.5	
35	9.3	7.3	4.8	3.9	
65	13.9	12.6	9.6	10.0	
81	20.9	19.2	13.5	15.5	
93	33.3	30.5	27.8	24.8	36.3
81	22.5	19.8	15.1	16.3	27.4
65	17.1	14.6	10.3	11.6	17.9
35	10.9	8.6	4.8	5.4	9.2
15	7.0	5.3	0.8	2.3	9.4

[a] Initial moisture content on a dry cellulose basis.

of secondary hydroxyls by interchain bonding, thus preventing their participation in the adsorption of water. The desorption curves also retain this linearity, indicating the irreversibility of the hydrogen bonding between secondary hydroxyls despite the capillary water and swelling of the film which occurred at the high humidities. This indicates that the accessibility of a collapsed film remains unchanged by the capillary and swelling action of water. Similar evidence supporting this theory of adsorption in terms of primary and secondary hydroxyl groups is found in the work by Urquhart and Williams (83) who showed that the S-shaped sorption isotherm of cellulosic materials becomes a straight line up to 50% relative humidity when the temperature is increased to 80°C, and Ayer in studying the primary adsorbed water in cotton fibers by drying techniques was able to show that never-dried cotton fiber has about three times as many available hydroxyl groups for the formation of hydrate as a cotton which has been completely dried (84). The linear relationship obtained in our experiments corresponds to the Langmuir adsorption isotherm on a surface containing only one elementary space (85).

Similar results are obtained when films of a given softener content are initially dried to different moisture levels and permitted to

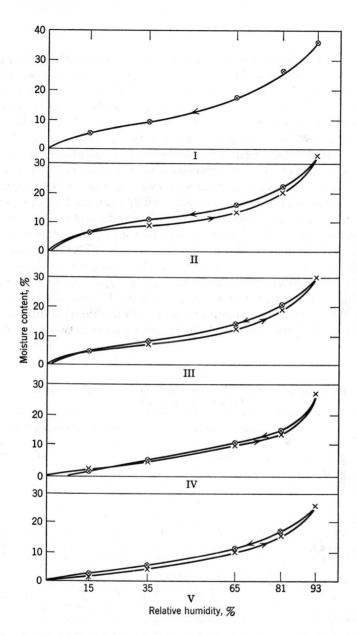

Fig. 3. Moisture hysteresis of unsoftened films: I, gel film desorption; II, film A, 12% moisture; III, film B, 5.9% moisture; IV, film C, 3.1% moisture; V, film D, 1.3% moisture.

reach equilibrium moisture content as a function of relative humidity. The data of Table 2 show that drying conditions strongly affect the equilibrium moisture content of softened film except that here the equilibrium moisture content depends not only on the interaction of cellulosic hydroxyls with each other but also on the cellulose–softener interaction and on the hygroscopicity of the particular softener used. The differences in equilibrium moisture content indicate that less interaction between glycerol or ethylene glycol and the cellulosic hydroxyls exists at high initial moisture content than at low initial moisture content when the film is properly dried. When film is dried below specification moisture content, as may sometimes occur during casting and coating, the following may happen: (1) some softener will be lost, especially a volatile softener such as propylene glycol; (2) softener will replace more water at the cellulosic hydroxyl sites; (3) additional hydrogen bonding between cellulose chains may occur. Each of these factors will lower the equilibrium moisture content and decrease the base sheet durability.

This additional irreversible hydrogen bonding between cellulose

TABLE 2
Equilibrium Moisture Content of Softened Films as a Function of
Initial Drying Conditions

Softener in film	Initial analysis[a]		Equilibrium moisture content, %[a]				
	H_2O, %	Softener, %	15% RH	35% RH	65% RH	81% RH	93% RH
A. *Underdried films*							
Glycerol	9.3	35.8	7.9	11.2	19.6	26.8	77.5
Glycerol	18.8	33.2	12.6	15.6	23.6	35.6	83.6
Ethylene glycol	9.2	17.5	6.5	10.0	22.0	27.1	54.4
Ethylene glycol	14.3	18.5	8.8	12.6	24.4	30.9	62.5
B. *Overdried films*							
Glycerol	4.5	24.6	5.4	9.1	17.6	26.1	60.2
Glycerol	7.5	21.9	8.2	10.3	18.3	35.3	67.5
Ethylene glycol	5.1	21.8	6.9	11.7	21.9	25.7	44.3
Ethylene glycol	7.8	24.7	11.4	13.2	24.9	30.4	51.1

[a] Dry cellulose basis.

chains on overdrying of softened films is indicated by the differences in the equilibrium moisture data of standard and overdried films softened with glycerol or with ethylene glycol, as shown in Table 2. These data indicate that overdrying results in irreversible changes in the cellulose structure believed to be caused by additional inter-chain hydrogen bonding. Drying conditions during casting and coating may therefore have an important effect on the mechanical properties of coated cellophane.

VI. Mechanical Properties of Cellophane

The durability of cellophane is measured at the present time by several physical tests such as by impact strength, flexing under stress, tenacity, elongation, tear, and others, and no correlation between these test data and the structure and interaction of the polymer chains exists. Several studies have indicated that there is little relation, if any, among the physical tests mentioned, and that therefore none of these tests when measured separately is indicative of the performance of cellophane in use. To relate the performance of cellophane with its molecular structure, the deforma-tion of regenerated cellulose under stress in terms of a mechanical model has been evaluated and the elastic and viscous flow portions of this model have been correlated with the interactions between cellulose chains in the absence and presence of softener molecules of different types and concentrations.

The interaction between cellulose and various softeners in terms of the relative affinity of cellulose and of the softener molecule for water (54,55) has been discussed in Section III.

A. ELASTIC MODULUS

The effect of the softener molecule on the elastic modulus of cellophane was studied using a dynamic modulus apparatus which measured the resonant frequency of vibration of a given film specimen from which the elastic dynamic modulus could then be calculated.

The data for glycerol-softened films indicate that each mole of glycerol is twice as effective as a mole of water in softening the cellulose structure. This would seem to indicate that a molecule of glycerol is able to cause twice as much disruption in the cellulose structure as a water molecule because of its larger size and the presence of hydrogen bonding groups at each end of the molecule. At each relative humidity, the modulus of film in equilibrium with that humidity decreases with increasing glycerol content. Since at low relative humidities (up to 35%) the equilibrium moisture content remains nearly constant with varying glycerol content, the point of view that softening action results from partial disruption or prevention of hydrogen bonding of cellulosic hydroxyls has been strengthened (86).

In another study, the viscoelastic behavior of cellophane was evaluated using a simple mechanical model consisting of Maxwell and Voigt elements of springs and dashpots in series (87,88). The effect of added softener molecules on the elastic and viscous parameters of the model was determined and compared with the elastic modulus of the same films when measured separately. If the logarithms of these moduli for glycerol-softened films and for the unsoftened film are plotted as a function of the total effective molar concentration (the concentration multiplied by an appropriate weighting factor), it is found that the data fall on a straight line. The modulus data for other softeners also fit this line if their molar concentrations are weighted by suitable factors. Since the abscissa of this curve is a function of the number of hydrogen bonds broken, the modulus may be considered to be a direct measure of softener effectiveness relative to water (86). These results are shown in Figure 4, and the effective molar concentrations of several softeners are given in Table 3. The factor weighting each softener with respect to water indicates that a mole of glycerol is twice as effective in interrupting hydrogen bonds as is a mole of water. Although all of these other factors were arbitrarily assigned to each softener to secure fit to the glycerol–water curve they agree with our experience regarding softener effectiveness on a weight basis, i.e., three times as much softener is required by weight as water to produce cellophane of the same flexibility at 35% relative humidity (88).

In another investigation (89), the effect of film thickness on the dynamic elastic modulus of cellophane has been studied to inquire into structural differences that may exist in cellophane as a result of different conditions of regeneration. The moduli values were found to be independent of film thickness for all types of cellophane, including plant and hand-cast laboratory films. Therefore, cellophane by this method of study showed no indication of a variation in structure in the thickness direction *when prepared by the same*

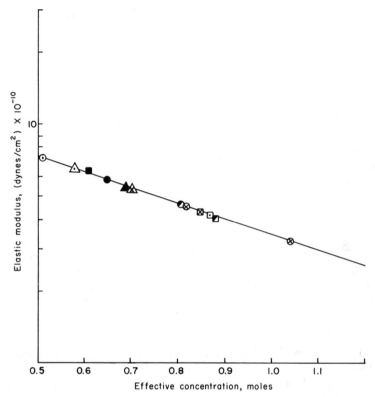

Fig. 4.   Elastic modulus of cellophanes as a function of the effective molar concentration of softener.   Beginning at the left, the points represent, respectively: water, ethylenediamine, ethylene glycol, propylene glycol, ethylene carbonate, 1,3-butanediol, diethylene glycol, glycerol (two points), triethylene glycol, diethanolamine, glycerol.

method.   Different types of cellophane showed small differences in modulus undoubtedly as a result of slight but not basic structural differences arising from different methods of preparation.

B. Delayed Elasticity

Strain–time curves evaluated in terms of the mechanical model were used to obtain the elastic, inelastic, and flow components of viscose in the presence of added softener molecules (88).

TABLE 3
Effective Molar Concentrations of Softeners and Water

Softener	Softener concentration, moles	Factor	Water, moles	Effective concentration, moles
From Young's modulus				
Water			0.51	0.51
Glycerol	0.22	2	0.38	0.82
Glycerol	0.32	2	0.40	1.04
Triethylene glycol	0.16	3	0.39	0.87
Ethylenediamine	0.30	0.75	0.36	0.58
From mechanical model				
Glycerol	0.20	2	0.44	0.84
Ethylene glycol	0.21	1	0.40	0.61
Propylene glycol	0.16	1.5	0.41	0.65
1,3-Butanediol	0.17	1.5	0.45	0.70
Diethylene glycol	0.15	2.5	0.43	0.81
Diethanolamine	0.14	2.5	0.53	0.88
Ethylene carbonate	0.17	1.75	0.39	0.69
Water			0.51	0.51

The delayed elasticity is considered a function of the amorphous structure of the system and represents short-range motion of chain segments. The activation energy of this movement is expected to be related to the difference between the internal energies of polymer and softener. If the free energy of the process is proportional to the difference in internal energy of the participants, we would expect that the logarithm of either the modulus or the recoverable creep viscosity (pseudoelastic components of strain) should vary as a linear function of the heat of vaporization. On this basis, the total heat of vaporization of the amount of softener (including water) contained in the cellophane was calculated. When the logarithm of either the viscosity η_2 or the modulus E_2 was plotted against this value, a linear relationship was obtained, as shown in Figure 5. The calculated energy values and the viscosities and moduli are given in Table 4.

For these particular softeners, the heat of vaporization depends almost entirely on the energy necessary to break hydrogen bonds.

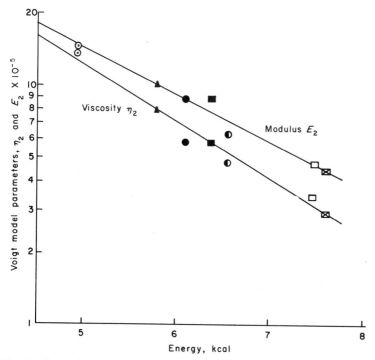

Fig. 5. Dependence of the Voigt model parameters E_2 and η_2 on the heat of vaporization of softener plus water present in the films; ($\boxed{X}$) glycerol; ($\blacksquare$) ethylene glycol; ($\bullet$) propylene glycol; ($\bullet\!\!\!\bigcirc$) diethylene glycol; ($\square$) diethanolamine; ($\blacktriangle$) ethylene carbonate; ($\odot$) water.

We might suspect, therefore, that the deformation process related to delayed elasticity requires breaking of interchain hydrogen bonds in contrast to the purely elastic deformation which involves only the stretching of hydrogen bonds. Both the modulus E_2 and recoverable creep viscosity η_2 decrease with increasing energy, indicating that the stronger the interaction between softener molecules, the stronger is the interaction with cellulose. The cellulose-cellulose interaction must be correspondingly weakened. This implies that softener efficiency can be increased by any regeneration and casting process which results in a less ordered cellulose structure of decreased cellulose–cellulose interaction.

TABLE 4
Heat of Vaporization of Softeners (Including Water)
in Regenerated Cellulose Sheets[a]

Softener	Heat of vaporization, Kcal			$E_2 \times 10^{-5}$, lb/in.^{2}	$\eta_2 \times 10^{-6}$, lb-min/in.^{2}
	Softener	Water	Total		
Glycerol	3.34	4.25	7.59	4.42	0.294
Ethylene glycol	2.48	3.88	6.36	8.82	0.585
Propylene glycol	2.12	3.98	6.10	8.82	0.585
Diethylene glycol	2.38	4.17	6.55	6.25	0.475
Diethanolamine	2.32	5.15	7.47	4.69	0.346
Ethylene carbonate	2.02	3.78	5.80	10.00	0.795
Unsoftened film		4.95	4.95	15.00	1.47

[a] These films are completely described in ref. 88.

C. VISCOUS FLOW

The unrecoverable flow viscosities η_3 were obtained from the strain–time curves as related to the mechanical model (88). When log η_3 was plotted as a function of the log E_1, the pure elastic modulus, a linear plot was obtained, as shown in Figure 6, indicating that η_3 is functionally dependent on the modulus E_1. Therefore, elastic modulus and viscous flow are not independent but depend on the molar concentration of the softener in the same manner (88).

These results indicate a new approach to the measurement of the mechanical properties of cellophane based on hydrogen bond energies as influenced by cellulose–softener interaction. Pure elasticity and plastic flow depend on stretching and finally on the breaking of hydrogen bonds, whereas delayed elasticity depends on the movement of polymer chain segments and requires breaking of interchain bonds. Meaningful durability data can be obtained for cellophane from these measurements by correlating these mechanical property parameters to end-use performance. A single durability coefficient relating process conditions to cellulose–softener interaction and cellulose–cellulose bonding may thus replace present durability measurements. This should lead to tailor-made film types of superior mechanical capabilities (89a).

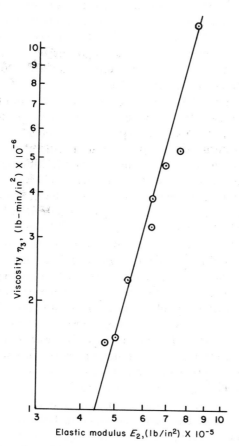

Fig. 6.   Interdependence of elastic modulus E_1 and flow viscosity η_3.

VII. Conclusion

This chapter has sought to show that the physical properties of cellophane depend on its structure which is largely determined by the process of manufacture, and specifically by the viscose process used in cellophane technology.   Cellophane technology differs from rayon technology to a considerable extent.   Beginning with differences in pulp grades with respect to molecular weight and α-cellulose content and with considerable differences in viscose preparation in terms of CS_2 concentrations, viscose alkalinity, and cellulose content, the major differences are in the regeneration processes of spinning and casting. It is here when the viscose solution is again transformed to a cellulose network where major processing differences produce major property differentials as required by the end use in the textile or packaging industry.

While optimum properties of rayon fibers are obtained by high orientation and low lateral order at very high degrees of polymerization achieved by viscose regeneration in the presence of additives and modifiers permitting slow viscose regeneration under high stretch at moderately low cellulose content, different conditions are required to achieve the desirable structure and properties of cellophane. Here the desired low degree of lateral order and high degree of isotropy are achieved by rapid regeneration with minimum orientation, which avoids the use of additives and modifiers, while attempting to increase gel densification at the wet end by balancing degree of polymerization and cellulose content as a means to resist orientation during casting. In particular, the addition of softeners and their interaction with the regenerated cellulose base sheet, known to be of much lower skin-to-core ratio than may be desirable, represents a vast difference in present cellophane technology, and leads to an end product of much lower order of average durability characteristics as compared with present-day high-strength rayon fibers where an all-skin cellulose produces high strength and flexibility without plasticization (90).

During the last decade under constant pressure from new synthetic polymer fibers and films available in large quantity and at low cost, we have already seen how the rayon industry has made a comeback and regained a large portion of the tire cord market in competition with nylon and polyester fibers. Here a new technology of converting wood pulp to a filament under conditions of minimum degradation has produced super tire cord and high-strength "polynosic" fibers competitive with, if not superior to, the synthetic counterparts (91,92). The cellophane industry at present is engaged in a similar competitive challenge. High-density polyolefins obtainable at lower cost and at lower unit weights are a serious threat to regenerated cellulose film in many packaging applications. In response to this challenge, a new cellophane technology based on old facts and new theories is emerging, as indicated in this chapter.

Belt casting methods are now in use which permit increased gel densification and greater film isotropy, resulting in stronger and more dimensionally stable films of thinner gauges and, therefore, of lower cost per square meter of coverage (5,93). Dry casting on heated drums has been found to produce isotropic cellophane of low lateral order and very high strength even though economical casting speeds have yet to be extablished (8). Tubular casting has been claimed to produce film of high strength by isotropic stretching during regeneration (94,95); this may be an approach which should permit the application of fiber technology to film casting. Most recently, studies in viscose particle content and distribution and in viscose filtration (96,97) and new theories relating viscose rheology to its behavior during regeneration and to plasticizer efficiency (88) indicate that cellophane may yet be capable of competing successfully with its synthetic counterparts.

References

1. E. Treiber, *J. Polymer Sci.*, **51**, 297 (1961).
2. E. Treiber, *Lenzinger Ber.*, **16**, 5, (1964); *Tappi*, **46**, 594 (1963).
3. K. Hess, *Holzforschung*, **9**, No. 3, 65 (1955).
4. W. Koblitz, *Papier*, **9**, 262 (1955).
5. J. Voss, *Svensk Papperstid.*, **64**, 863 (1961).
6. V. C. Haskell and D. K. Owens, *J. Appl. Polymer Sci.*, **4**, 225 (1960).
7. M. Miller and V. C. Haskell, *J. Appl. Polymer Sci.*, **5**, 627 (1961).
8. C. R. Price and V. C. Haskell, *J. Appl. Polymer Sci.*, **5**, 635 (1961).
9. V. C. Haskell and D. K. Owens, *Textile Res. J.*, **30**, 993 (1960).
10. H. A. Pearse, D. J. Priest, R. J. Shimell, and A. G. White, *Tappi*, **46**, 622 (1963).
11. A. J. Pennings, W. Prins, R. D. Hale, and B. G. Rånby, *J. Appl. Polymer Sci.*, **5**, 676 (1961).
12. P. Görling, *Papier*, **16**, 9 (1962).
13. W. A. Sisson, *Textile Res. J.*, **30**, 153 (1960).
14. E. Czapek and R. Weingand, U.S. Pat. 1,745,247 (Jan. 28, 1930).
15. M. Koenig, U.S. Pat. 1,861,701 (June 7, 1932).
16. J. B. Nichols, U.S. Pat. 2,445,333 (July 20, 1948).
17. J. B. Nichols, U.S. Pat. 2,451,768 (Oct. 19, 1948).
18. Brit. Pat. 852,609 (Sept. 29, 1958).
19. Brit. Pat. 852,614 (Oct. 1, 1958).
20. B. L. Hinkle and F. C. Stults, U.S. Pat. 2,962,766 (Dec. 6, 1960).
21. E. Ott, H. M. Spurlin, and M. W. Grafflin, *Cellulose and Cellulose Derivatives*, Interscience, New York, 1954, p. 339.
22. D. W. Jones, *J. Polymer Sci.*, **32**, 371 (1958).
23. J. Voss and O. Meyer-Berge, *Papier*, **16**, 724 (1962).

24. C. J. Heffelfinger and R. I. Burton, *J. Polymer Sci.*, **47**, 289 (1960).
25. J. Honeyman, *Recent Advances in the Chemistry of Cellulose and Starch*, Interscience, New York, 1959, p. 142 ff.
26. W. R. Berry, *Textile Res. J.*, **24**, 397 (1954).
27. K. J. Heritage, *Trans. Faraday Soc.*, **54**, 1902 (1957).
28. B. Carroll and H. C. Cheung, *J. Phys. Chem.*, **66**, 2585 (1962).
29. E. Zehmisch and V. Jancarik, *Faserforsch. Textiltech.*, **13**, 62 (1962).
30. E. Wellisch, M. K. Gupta, L. Marker, and O. J. Sweeting, *J. Appl. Polymer Sci.*, **9**, 2591 (1965).
31. R. J. E. Cumberbirch, J. E. Ford, and R. E. Gee, *J. Textile Inst. Trans.*, **52**, T330 (1961).
32. P. Van Horn, *Ind. Eng. Chem.*, **44**, 2512 (1952).
33. G. Jayne and K. Balser, *Papier*, **18**, 746 (1964).
34. G. Jayne and K. Balser, *Papier*, **19**, 572 (1965).
35. H. Klare and A. Groebe, *Z. Chem.*, **1962**, 15.
36. A. Groebe, H. Klare and H. Jost, *Faserforsch. Textiltech.*, **11**, 210 (1960).
37. M. Levine and R. H. Burroughs, Jr., *J. Appl. Polymer Sci.*, **2**, 192 (1959).
38. F. Kebl and L. Kudlacek, *Chem. Prumysl*, **11**, 324 (1961); *Chem. Abstr.*, **56**, 3696f (1962).
39. D. Vermaas, *Textile Res. J.*, **32**, 353 (1962).
40. A. M. van de Ven, *Papier*, **19**, 757 (1965).
41. R. L. Mitchell, J. W. Berry, and W. H. Woodman, U.S. Pat. 3,018,158 (Jan. 23, 1962).
42. R. St. J. Manley, *J. Polymer Sci.*, **2A**, 4503 (1964).
43. R. St. J. Manley, *Nature*, **204**, 1155 (1964).
44. R. St. J. Manley, and S. Inoue, *J. Polymer Sci.*, **3B**, 691 (1965).
45. I. L. Wadhera and R. St. J. Manley, *Makromol. Chem.*, **94**, 42 (1966).
46. O. Ellefson and K. Kringstad, *Faserforsch. Textiltech.*, **15**, 582 (1964).
47. J. H. Young, *Paper Maker* and *Brit. Paper Trade J.*, **139**, 48 (1960).
48. S. N. Danilov, M. A. Sokolovskiĭ, and A. E. Evdokimova, *J. Gen. Chem. USSR English Transl.*, **17**, 507 (1947).
49a. R. Mykolajewycz, O. J. Sweeting, and E. Czapek, *J. Appl. Polymer Sci.*, **7**, 251 (1963).
49b. O. J. Sweeting, R. Mykolajewycz, E. Wellisch, and R. N. Lewis, *J. Appl. Polymer Sci.*, **1**, 356 (1959).
50. R. Mykolajewycz, E. Wellisch, R. N. Lewis, and O. J. Sweeting, *J. Appl. Polymer Sci.*, **2**, 236 (1959).
51. Method developed by Rayonier, Inc., Eastern Research Division, Whippany New Jersey, n. d.
52. H. Schmiedeknecht and H. Klare, *Faserforsch. Textiltech.*, **14**, 219 (1963).
53. J. J. McKeown and W. I. Lyness, *J. Polymer Sci.*, **47**, 9 (1960).
54. E. Wellisch, L. Hagan, L. Marker, and O. J. Sweeting, *J. Appl. Polymer Sci.*, **3**, 331 (1960).
55. E. Wellisch, L. Hagan, L. Marker, and O. J. Sweeting, *J. Polymer Sci.*, **51**, 263 (1961).
56. W. J. Jebens, U.S. Pat. 2,280,829 (Apr. 28, 1942).
57. W. M. Wooding, U.S. Pat. 2,546,575 (March 27, 1951); U.S. Pats. 2,688,570

and 2,688,571 (Sept. 7, 1954); U.S. Pat. 2,987,418 (June 6, 1961); W. M. Wooding, Y. Jen, and E. H. Sheers, Can. Pat. 583,443 (Sept. 15, 1959).

58. W. Berry, C. R. Oswin, and J. Boyd, U.S. Pat. 2,684,919 (July 27, 1954).
59. G. I. Keim, U.S. Pat. 2,864,724 (Dec. 16, 1958).
60. E. Wellisch, U.S. Pat. 2,728,688 (Dec. 27, 1955).
61. G. Pitzl, U.S. Pat. 2,432,542 (Dec. 16, 1947).
62. R. T. K. Cornwell, U.S. Pat. 2,575,443 (Nov. 20, 1951).
63. L. M. Ellis, U.S. Pat. 2,523,868 (Sept. 26, 1950).
64. V. C. Haskell, "Cellophane," Encyclopedia of Polymer Science and Technology, Vol. 3, Interscience, New York, 1965, p. 66.
65. L. Hagan and G. R. Mitchell, U.S. Pat. 2,823,141 (Feb. 11, 1958).
66. M. C. Raes, U.S. Pat. 3,192,066 (June 29, 1965).
66a. W. P. Ericks, U.S. Pat. 2,891,019 (June 16, 1950); U.S. Pat. 2,977,331 (March 28, 1961).
67. E. M. Hicks, Jr., Furman Univ. Bull. Furman Studies Number, 3, No. 4, 67 (1956).
68. E. M. Hicks, Jr., Furman Univ. Bull. Furman Studies Number, 8, No. 1, 13 (1960).
69. E. M. Hicks, Jr., Furman Univ. Bull. Furman Studies Number, 9, No. 1, 16 (1961).
70. P. E. Wrist and C. A. Schiel, U.S. Pat. 3,257,268 (June 21, 1966).
71. W. P. Kane, U.S. Pat. 3,068,529 (Dec. 18, 1962).
72. Paper Trade J., 134, 30 (May 30, 1952).
73. A. Groebe, H. J. Purz, R. Maron, and H. Klare, Faserforsch. Textiltech., 14, 347 (1963).
74. Brit. Pat. 905,735 (Sept. 12, 1962).
75. W. P. Ericks, U.S. Pat. 3,074,833 (Jan. 22, 1963).
76. W. P. Ericks, U.S. Pat. 3,098,760 (July 23, 1963).
77. N. R. Eldred and J. C. Spicer, Tappi, 48, 608 (1963).
78. Tech. Rept. No. 2299, "Cellulose Viscose Process Studies," International Cellulose Research, April, 1966, pp. 139–146.
79. P. H. Hermans, Physics and Chemistry of Cellulose Fibers, Elsevier, New York, 1949, Chapter II, p. 180 ff.
80. A. G. Assaf, R. H. Haas, and C. B. Purves, J. Am. Chem. Soc., 66, 66 (1944).
81. Ref. 21, p. 267 ff.
82. H. Tarkow and A. J. Stamm, J. Phys. Chem., 56, 266 (1952).
83. A. R. Urquhart and A. M. Williams, J. Textile Inst. Trans., 15, 559 (1924).
84. E. Ayer, J. Polymer Sci., 21, 445 (1956).
85. I. Langmuir, J. Am. Chem. Soc., 40, 1361 (1918).
86. O. C. Hansen, Jr., L. Marker, and O. J. Sweeting, J. Appl. Polymer Sci., 5, 655 (1961).
87. T. Alfrey, Jr., Mechanical Behavior of High Polymers, Interscience, New York, 1948, p. 103 ff.
88. E. Wellisch, L. Marker, and O. J. Sweeting, J. Appl. Polymer Sci., 5, 647 (1961).
89. E. Gipstein, L. Marker, E. Wellisch, and O. J. Sweeting, J. Polymer Sci., 3A, 2313 (1965).

89a. W. J. Hamburger, *Textile Res. J.*, **18**, 102 (1948).
90. D. K. Smith, *Textile Res. J.*, **29**, 32 (1959).
91. R. H. Braunlish, *Am. Dyestuff Reptr.*, **54**, 160 (1965).
92. K. Goetze, *Chemiefasern*, **12**, 936 (1965).
93. French Pat. 1,435,701 (March 7, 1966).
94. R. O. Osborn, U.S. Pat. 3,121,761 (Feb. 18, 1964).
95. J. Hafstad and K. R. Wilson, U.S. Pat. 3,121,762 (Feb. 18, 1964).
96. D. F. Durso and L. R. Parks, *Svensk Papperstid.*, **64**, 853 (1961).
97. D. F. Durso and L. R. Parks, *Tappi*, **48**, 594 (1965).

CHAPTER 4

NATURAL POLYMERS AS FILM FORMERS
D. Other Natural Organic Polymers

ORVILLE J. SWEETING*

Yale University, New Haven, Connecticut

I. Introduction

Of the multitude of natural organic polymers, only cellulose (Chapter 4A, B, and C) has been extensively exploited as a self-supporting transparent film. In this concluding section, we shall survey very briefly the domain of natural organic polymers (other than cellulose) and discuss briefly one natural polymer that may have some promise of commercial application, amylose, and one example of another group simply derived from cellulose, the hydroxy-alkyl celluloses. Regenerated hydroxyethyl cellulose has properties that are sufficiently different from those of cellophane to make

* Formerly Associate Director of Research, Film Division, Olin Mathieson Chemical Corp., New Haven, Connecticut.

it of interest. Biological polymers include proteins, shellac, poly-saccharides, hydrocarbons, and nucleic acids. The polysaccharides include (besides cellulose) starch, chitin, isinglas, and amylose. The biological hydrocarbons include rubber, balata, and guttapercha as well as the various naturally altered substances such as peat, coals, asphalts, graphite, and diamond. Though all of these sub-stances may have had their origins in living organisms, they are as diverse as the organisms and tissues from which they are derived, even without the modifying effects of heat and pressure in the earth's crust.

Proteins. Interest in regenerated or synthetic protein has resulted from the fine qualities of wool as a textile material, and wool has been the primary standard against which these materials have been judged. The chief sources from which fibers have been prepared are milk (casein), peanuts (peanut protein), soybeans (soy-bean protein), and corn (zein). Little or no attention has been given any of these proteins for use as transparent self-supporting films.

The industrial development of regenerated protein fiber covers a period of about 70 years, but little effort has been expended to make a useful transparent film. The first work was done with Vandura silk in 1894, a material based on gelatine.

Casein fiber was first prepared commercially in 1935; Fibrolane in England and Merinova in Italy were produced commercially, but no films based on casein were manufactured.

Soybean protein for fibers has been studied in Japan, Italy, USSR, England, and the United States and was produced commercially for a time in the United States in 1943 by the Drackett Co. Peanut protein has been used by Imperial Chemical Industries, Ltd., in the production of Ardil in commercial quantities.

The four regional laboratories of the Department of Agriculture have investigated fibers from various proteins. The Eastern Regional Research Laboratory has worked on improving casein fiber, the Southern has worked on peanut and cottonseed protein fibers, the Western laboratory has produced fibers from egg albu-men and chicken feather keratin and the Northern laboratory has developed zein fibers. The Virginia-Carolina Chemical Corp. pro-duced zein fiber under the trade name Vicara in the early 1950s, but the fiber was not sufficiently strong, and manufacture has been discontinued.

In preparing fibers from proteins, the process embodies the following steps: (*1*) solubilization, or spinning solution preparation; (*2*) regeneration, or coagulation; (*3*) orientation and stretching; and (*4*) insolubilization or hardening. These steps have their counterparts in the manufacture of films from natural sources, e.g., cellophane from cellulose, and it is not improbable that commercial modifications to produce transparent protein films could be developed. Plasticized zein has been cast into film, but the product has found no commercial use.

Polysaccharides are most abundant in the higher orders of land plants and in seaweed where they constitute approximately three quarters of the dry weight.

Polysaccharides, as the term implies, are polymerized saccharides and may be considered to be condensation polymers in which monosaccharides or their derivatives (such as the uronic acids, $HO_2C(CHOH)_4CHO$, or the amino sugars) are glycosidically joined by the elimination of water. The molecular size may vary markedly; however, most polysaccharides found in nature are of high molecular weight and usually contain from a hundred to several thousand monomer units. Those polymers below ten units are termed oligosaccharides.

Polysaccharides are deposited in tissues along with numerous other substances, and consequently occur as mixtures. In some cases the purity may be high, as in cotton fiber (approximately 98% cellulose). Starch in seeds and tubers is likewise free of contaminants. Some plant exudates, or gums, and many slimes of bacteria are nearly pure carbohydrate. Where intimate mixtures of polysaccharides and other biological materials occur, separation and purification of a particular polysaccharide may be difficult.

Polysaccharides containing only one kind of polymerized sugar unit occur in greater tonnages than polysaccharides which contain two or more kinds of sugar units, although there is a greater number of the latter. No polysaccharide is known which contains more than five types of monomer unit. Cellulose is the most abundant polysaccharide. Also abundant are laminarin from seaweed, and amylose and amylopectin from starch. Xylan is abundant and widely distributed in plants. Chitin of insects and crustacea is also abundant.

Of approximately 130 known polysaccharides, one-third to one-half contain up to 50% uronic acid. A few polysaccharides are composed solely of uronic acid units: alginic acid (a polymer of mannuronic acid) from seaweed and galacturonan (a polymer of galacturonic acid) of the pectic group of substances. Two polysaccharides, mesquite gum and slippery elm mucilage, contain some units with a methyl ether group.

The properties of polysaccharides vary with their structure, size of molecule, and constituent monosaccharides. Structural polysaccharides such as cellulose and chitin are almost always linear. Usually, branched polysaccharides are easily soluble in water and have immense thickening power. Linear polysaccharides have low solubility and show a marked tendency to associate. For example, amylose, a linear glucan found in most starches, slowly precipitates from starch solution (Section II).

II. Amylose Films

In 1951, the Northern Regional Research Laboratory at Peoria, Illinois reported the preparation of thin transparent films from corn amylose (1). These films were prepared by a butanol precipitation (2), though it is possible and more desirable, in fact, to cast films from hot aqueous solution (3).

TABLE 1

Physical Properties of Unplasticized Corn Amylose Film

Refractive index	1.53
Density, g/cm³ at 25°C	1.46
Tensile strength, psi	10,000–12,000
Young's modulus, psi	350,000–500,000
Elongation at break, %	10–15

The dry films are strong and transparent; they are not thermoplastic or heat sealing. They burn slowly with a luminous flame. Heating for 4 hr at 170°C does not appreciably discolor amylose films, but some degradation takes place under these conditions (1). Table 1 lists some of the film properties. Compatible plasticizers include water, glycerol, ethylene glycol, propylene glycol, urea, sorbitol—in short, all of those that are effective in softening cellophane.

TABLE 2

Properties of Corn Amylose for Film Casting

Physical form	Crystalline powder
Color	Colorless
Particle size, μ	10–100
Moisture, %	10
Starch carbohydrates,[a] %	97
Amylose,[a] %	90
Amylopectin,[a] %	0
Noncarbohydrates,[a] %	1.5
Cellulose,[a] %	1.0
Ash,[a] %	0.5
Cold water solubility, %	5
Intrinsic viscosity,[b] $(g/dl)^{-1}$	1.4
Degree of polymerization	700
Density, g/cm^3 at 25°C	1.46
Refractive index at 25°C	1.53

[a] Dry basis
[b] Determined in $1N$ KOH at 35°C, concentration 0.4 g/dl.

Starch is an abundant raw material, but inexpensive methods to separate its constituent polymers have not been found. Geneticists have raised the amylose content of corn starch from 25 to 75%—even to 82% in one instance—and since a high-amylose starch could probably be produced for approximately the same cost as normal starch, new genetic varieties are promising as a low-cost source of amylose (4,5).

The presence of amylopectin (a concomitant polysaccharide in corn starch, similar to amylose but with frequent crosslinks between anhydroglucose chains) is seriously detrimental to the properties of amylose films, reducing flex life, tensile strength, and elongation. The manufacture of an acceptable amylose film will therefore require effective separation of the amylopectin fraction. In Table 2 are presented the properties of a typical commercially available amylose (6).

III. Hydroxyethyl Cellulose Films

The preparation of hydroxyethyl cellulose is not new (7,8), but in recent years, renewed interest has been shown in this product that can be readily derived from cellulose. The principal reasons for the renewed activity are the possibilities for simpler processing than the current viscose process for cellophane (Chapter 4B), and the indication that certain film properties may be for some applications more desirable than those of cellophane, particularly dimensional stability. A commercial operation based on the patents of Erickson (8) is producing in California a film known as Cal-Wrap.

In recent years, Rayonier, Inc. has manufactured a grade of hydroxyethyl cellulose called Ethynier for a variety of applications, among them film (9,10), but this product is no longer commercially available, although production can be resumed at any time if the market permits. Ethynier as produced consists of $\frac{1}{4} \times \frac{1}{2}$-in. cylindrical pellets, with a moisture content of about 7%, which are readily converted to a clear casting solution containing preferentially 8% Ethylose and 7% sodium hydroxide. The material has a bulk density of 30–35 lb/ft^3. Ethynier differs from previous varieties of hydroxyethyl cellulose in being soluble in as low a concentration of sodium hydroxide as 6%, a characteristic said to be the result of uniform hydroxyethylation. The product contains only 4% ethylene oxide, with a degree of substitution of 0.15; i.e., one hydroxyl on every seventh anhydroglucose unit, approximately, has been substituted.

Since the solubility in alkaline solution depends on the conversion of the hydroxyethyl groups to the sodium salt, one should expect that both the coagulation–dehydration and the regeneration of sodium hydroxyethyl cellulose in an acid–salt bath will occur at different rates than with cellulose xanthate. Based on the more polar nature and basicity of the sodium salt of hydroxyethyl cellulose as compared with sodium cellulose xanthate the combined rates of coagulation and regeneration are faster in the case of the former. Therefore, one might expect some difference in final film properties. A faster regeneration (relative to its coagulation) might result in less densification of the gel, making it inherently weaker, but on the other hand, the rapid regeneration will increase the isotropic character of the regenerated film and result in a better balance of its physical properties. The higher water content of this film in the gel form results in increased softener pickup, and its less dense structure and hydroxyethyl side chains increase the cellulose-softener interaction (11,12).

A. Laboratory Preparation of Hydroxyethyl Cellulose Solution and Film

To obtain film in the laboratory by plate casting, a solution of hydroxyethyl cellulose* is prepared in a two-step process.   Ethynier pellets are first slurried in water with mild stirring.   Caustic solution to give a final 7% sodium hydroxide concentration is then added with stirring at high speed to effect colloidal dispersion. Complete solution is obtained by working below room temperature (5–10°C) and filtering out any undissolved particles under vacuum. The solution is then deaerated and is ready for casting.   The desired solution viscosity is obtained by varying the hydroxyethyl cellulose content of the solution between 6 and 12%.   A 6–8% content has been found to be most suitable for plate casting.   The solution is cast on glass plates using a casting bar with a gap setting of about 18–20 mils to obtain a final film thickness of 1 mil from a solution containing 8% hydroxyethyl cellulose and 7% sodium hydroxide.

Coagulation and regeneration of the viscous solution is achieved by neutralization in a salt–acid bath containing 15% sodium sulfate and 10% sulfuric acid.   A standard bath used for viscose (20% Na_2SO_4 and 10% H_2SO_4) may also be used.   After thorough washing (to free the film from sulfate) and adding softener, if desired, the film is dried on a glass plate in a forced-air oven at 180°F.

1. Solution Viscosity and Aging

Considerable latitude in solution viscosity is possible by selection of proper solution compositions.   As with most polymer solutions, viscosity increases rapidly with degree of polymerization and with increased solids content.   The sodium hydroxide concentration also affects the viscosity; the higher sodium hydroxide content increases the solubility of hydroxyethyl cellulose and lowers the viscosity. The degree of polymerization is controlled during the hydroxyethylation step, and can therefore be considered constant (9,10). The viscosity curves (Fig. 1) illustrate this relationship for Ethynier solutions at two different sodium hydroxide contents.   Solution vis-

---

* In these experiments, Rayonier, Inc. Ethynier brand of hydroxyethyl cellulose was used.   Grateful acknowledgment is made to Donald R. Walton, and to R. Logan Mitchell, General Manager of Research, for supplying the material.

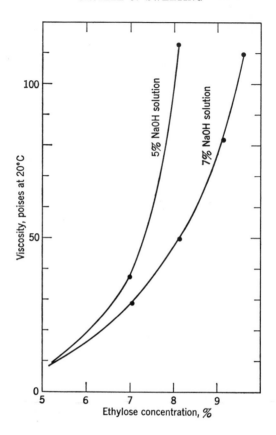

Fig. 1. Variation of viscosity of Ethynier solutions with concentration of
Ethynier in 5 and 7% NaOH solutions.   (Courtesy Rayonier, Inc.)

cosity also varies with temperature as shown in Figure 2, decreasing
with increasing temperature.

The viscosity of Ethynier solutions in 8% sodium hydroxide
measured immediately after preparation and filtration, and after
aging for a period up to one week at 5°C showed that the viscosity
is considerably lower than that of standard viscose, and it does not
change appreciably during aging at low temperature (Fig. 3).
Table 3 summarizes the results of several such tests (measurements
were made at 18°C with the Brookfield viscometer).   The viscosi-

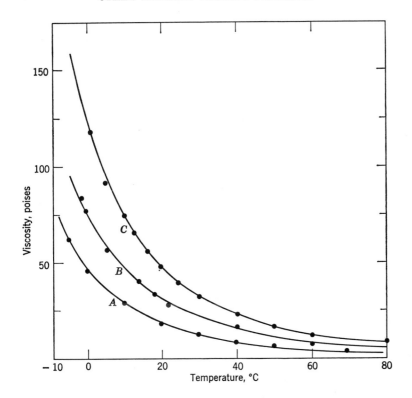

Fig. 2. Viscosity of hydroxyethyl cellulose solutions as a function of temperature: *A*, 6%; *B*, 7%; *C*, 8% Ethynier in 7% NaOH. (Courtesy Rayonier, Inc.)

ties of a 1:1 mixture of Ethynier and viscose have been included for comparison.

In Figure 4 are shown the regions in which objectionable gels are present that interfere with filtration of Ethynier solutions in aqueous sodium hydroxide of various concentrations.

2. Coagulation and Regeneration

Coagulation rates determined in sodium sulfate solutions of concentration ranging from 12 to 18% sodium sulfate in the absence of acid showed that hydroxyethyl cellulose coagulates at approximately half the coagulation rate of viscose. For the former, rates

TABLE 3
Solution Viscosity of Ethynier and of Viscose Measured at 18°C

Cellulose solution, type and concentration	Aging time, days at 5°C	Viscosity, Brookfield seconds	Specific gravity
Ethynier, 8%	0	40.5	1.100
	3	40.5	
	7	40.0	1.100
Ethynier, 8% and viscose, 9%, 1:1	0	51.8	1.108
	3	50.5	1.108
	7	50.0	1.110
Viscose, 9%	0	68.5	1.113
	3	68.0	
	7	68.0	1.114

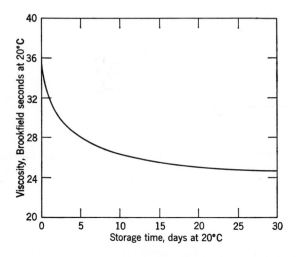

Fig. 3.   Effect of time of storage upon viscosity; 8% Ethynier, 7% NaOH in water.

were from 14.4 to 25.2 $\times$ 10^{-6} g H_2O/cm-sec, and for the latter, 22.9 to 36.3 $\times$ 10^{-6} g H_2O/cm-sec.   In ammonium sulfate solutions at concentrations ranging from 17 to 40%, the rates of coagulation were found to be about equal, ranging from 8.4 to 40.0 $\times$ 10^{-6} g H_2O/cm-sec.

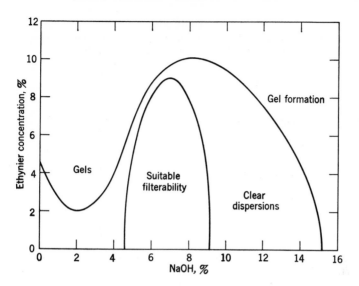

Fig. 4. Filterability diagram of solutions of Ethynier in aqueous sodium hydroxide.

On the other hand, the rate of regeneration of hydroxyethyl cellulose was found to be 15 times as fast as that of viscose in $1N$ sulfuric acid solution, and 300 times as fast in a standard acid–salt regeneration bath. Since the regeneration of sodium hydroxyethyl cellulose can be considered a neutralization of a strong base by a strong acid, this rapid regeneration is to be expected. In a single salt–acid bath, coagulation and regeneration can no longer be considered separately. One would expect that the presence of acid would speed up the rate of coagulation, while regeneration, which is diffusion controlled, would be somewhat slower than when measured in acid alone. This rapid completion of the coagulation–regeneration step affects the physical properties of both the gel film and the dry product.

3. Gel Swelling

The gel-swelling index, defined as the weight ratio of gel film to oven-dry film times 100, is a measure of the densification of the gel film during coagulation and regeneration. The extent of this

TABLE 4

Gel Swelling of Films Prepared from Ethynier, from Viscose,
and from Ethynier–Viscose Mixtures at 25°C

Film composition, %	Gel swelling index
Ethynier, 100	824
	797
Ethynier, 50	647
	621
Ethynier, 35	576
	547
Ethynier, 25	519
	538
Viscose, 100	444
	452

densification influences both the wet strength properties of the gel
and the toughness of the dried film.   The degree of densification
probably depends on the structure of the cellulose derivative having
a hydroxyethyl side chain, on the relative rates of coagulation and
regeneration as discussed above, and on the hydroxyethyl cellulose
concentration of the alkaline solution used for regeneration.

The gel-swelling index was determined for a series of films cast on
glass plates from hydroxyethyl cellulose and from viscose solutions
and regenerated from a standard acid–salt bath (containing 12.5%
sulfuric acid and 20% sodium sulfate) at room temperature.   Mixtures of hydroxyethyl cellulose and viscose containing 25, 35, and
50% Ethynier (based on cellulose content) were also prepared and
regenerated.   The gel-swelling data are given in Table 4 and are
plotted as a function of composition in Figure 5.   From these data
one may conclude that there is no interaction between hydroxyethyl
cellulose and viscose when regenerated in various compositions.   The
gel-swelling index of mixtures increases linearly with increasing
hydroxyethyl cellulose content and is nearly twice that of viscose
at 100% Ethynier concentration.

Experiments summarized in Table 5 have determined the effect
of different regenerating bath types and concentrations on gel swelling.   An acid bath alone containing 12.5% sulfuric acid caused a

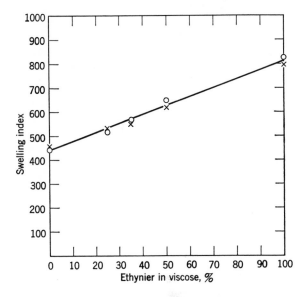

Fig. 5.   Gel-swelling index of viscose–Ethynier films.

TABLE 5

Gel-Swelling Data of Hydroxyethyl Cellulose Films Regenerated
from Different Regenerating Baths at Room Temperature[a]

	Regenerating bath composition, %	Gel swelling index
$\begin{cases} Na_2SO_4 \\ H_2SO_4 \end{cases}$	20 12.5	805 806
H_2SO_4	12.5	861
$\begin{cases} (NH_4)_2SO_4 \\ H_2SO_4 \end{cases}$	40 12.5	789 804
$\begin{cases} ZnSO_4 \\ Na_2SO_4 \\ H_2SO_4 \end{cases}$	10 16 8	706
$\begin{cases} Na_2SO_4 \\ H_2SO_4 \end{cases}$	15 10	1055[b]
$\begin{cases} H_3PO_4 \\ NaOH \end{cases}$	39.5 10	891[b]

[a] 8% Ethynier, 7% NaOH.
[b] 6% Ethynier, 7% NaOH.

small increase in gel swelling, while the addition of zinc sulfate (which slows the regeneration process) decreased the gel swelling by at least 10%. On the other hand, a salt–acid bath consisting of 40% ammonium sulfate and 12.5% sulfuric acid did not increase the gel-swelling index compared to the control. Regeneration with this bath at 60°C also was without effect. Since coagulation studies have shown that hydroxyethyl cellulose coagulates in ammonium sulfate at the same rate as viscose (see above), these results show that the rapid rate of regeneration in a single coagulation–regeneration bath determines the water content of the regenerated gel film rather than a comparatively small increase in the rate of coagulation resulting from differences in the dehydrating potential of various salts and their concentrations. In agreement with this argument, the use of a buffer solution consisting of phosphoric acid and its sodium salts as the regenerating bath may be of value, since such a buffered system should slow the regeneration process and thus permit the gel film to dehydrate to a greater extent (13). Thus, when an Ethynier solution was regenerated in a solution consisting of 39.5% phosphoric acid and 10% sodium hydroxide, the gel swelling index dropped by about 15% to 891, compared with an index of 1055 obtained when a regenerating bath consisting of 15% sulfuric acid and 10% sodium sulfate was used.

4. Softener Uptake

The softener uptake by gel film is related to its gel-swelling index, since softener uptake has been found to occur by direct exchange of water in the gel film for softener solution; thus, softener uptake depends on the available volume (14,15). The latter in turn is determined by the state of collapse of the gel structure after coagulation and regeneration. The softener pickup ratio, defined as the ratio of percentage of softener in the dry film (cellulose basis) to the percentage of softener in the softener bath, therefore depends on the gel-swelling index and is related to the hydroxyethyl cellulose concentration of the casting charge.

The pickup ratio of hydroxyethyl cellulose and of viscose films and of mixtures cast on glass plates and softened with glycerol show that the pickup ratio increases in proportion to the gel swelling index, which, in turn, is a linear function of composition as shown in Figure 5. The regenerated films were washed and softened in a

TABLE 6

Pickup Ratios for Hydroxyethyl Cellulose and Viscose Films
and Mixed Films Softened with Glycerol[a]

Composition, %	Softener content, %	Pickup ratio[b]
Viscose, 100	18.3	3.1
{ Viscose, 75 Ethynier, 25	25.5	4.3
{ Viscose, 50 Ethynier, 50	29.6	5.9[c]
{ Ethynier, 75 Viscose, 25	40.7	6.6
Ethynier, 100	51.5	8.6

[a] Glycerol bath 6%.
[b] Based on dry cellulose.
[c] Glycerol bath 5%.

6% glycerol bath. The pickup ratio calculated for these films is given in Table 6.

5. Softener Loss

The high gel-swelling index found for hydroxyethyl cellulose films which results in a high softer pickup ratio as discussed above would indicate a higher moisture regain than has been observed for viscose films. This higher moisture regain indicates the possibility of a higher degree of reswelling of hydroxyethyl cellulose at high-humidity conditions, which would result in excessive softener losses of softened films under these conditions. To evaluate the softener loss of hydroxyethyl cellulose films, a series was prepared containing between 20 and 60% glycerol (moisture-free basis). The softener content in these films was analyzed after 4, 7, 21, and 42 days of conditioning at 75°F and 81% relative humidity. The results in Table 7 show that the loss of glycerol at the 20% level in the film reaches 10% of the glycerol present. This loss increased to 15% for the 60% glycerol initially in the film, i.e., after 42 days of conditioning between 85 and 90% of the softener present initially was still in the film. This retention of softener is at least as good as that shown by cellulose base sheet regenerated from viscose.

TABLE 7
Softener Loss Study of Hydroxyethyl Cellulose Base Sheet Conditioned
at 75°F, 81% Relative Humidity

| | Softener content, glycerol, %, dry basis, averages | |
Sample films	Initial	Final
1	19.9	19.7 after 4 days
2	19.4	17.2 after 42 days
3	60.5	53.6 after 7 days
4	59.4	56.2 after 21 days
5	61.7	52.9 after 42 days

6. Properties of Hydroxyethyl Cellulose Film

Table 8 summarizes the composite results of tests made on a
large number of samples of hydroxyethyl cellulose film prepared by
plate casting in the laboratory and of samples made on continuous

TABLE 8
Properties of Hydroxyethyl Cellulose Films Compared with Films from Viscose

| | Hydroxyethyl cellulose[a] | | Cellophane, commercial (195 MST-54) |
	Laboratory	Commercial	
Unit weight, g/m²	34.5	36.8	36.0
Glycerol, % (dry basis)	15.0	18.5	19.0
Moisture		6.5	6
Tear, g			
MD	1.9	2.1	3.4
TD	2.1	2.4	5.3
Tenacity, lb			
MD	33.4	50.0	18.5
TD	30.8	46.0	25.3
Elongation			
MD	17.1	14.0	12
TD	22.6	18.0	51
Pendulum impact, kg-cm	6.4	7.1	9
Stress–flex, number		7	45

[a] Regenerated from Ethynier, a product of Rayonier, Inc.

casting machines of commercial size. A comparison is made with one of the commonly used cellophanes, 195 MST-54.

In general, plate-cast hydroxyethyl cellulose films are inferior in physical properties to viscose film at equivalent softener concentrations. Compared with hand-cast viscose controls, the hydroxyethyl cellulose films approach the properties of the former but are still somewhat lower. It is possible, however, to make hydroxyethyl cellulose film readily of much higher glycerol content than that of a film from viscose gel film under similar conditions. Thus a softener bath of 10–12% glycerol content gives a hydroxyethyl cellulose film containing 40–50% glycerol, while hand-cast viscose film contains only 30% glycerol. These differences result from different pickup ratios of the films, which in turn are related to differences in gel swelling, as discussed earlier. Ethynier films of such high softener content exhibit high stress-flex performance, and may be superior to viscose films at low-temperature conditions.

Values for tear, tenacity, and elongation show a better balance of physical properties in the properties in the machine and transverse directions (MD and TD, respectively) than is exhibited by regenerated viscose films. This result follows from the greater rate of coagulation and regeneration for hydroxyethyl cellulose, which results in less anisotropy of these films, as compared with viscose cellophane. The poor tear strength of Ethynier exhibited here may pose a serious problem during film slitting and lower the performance on high-speed packaging machines.

The balance of properties, which is related to internal structure, is reflected in the dimensional stability of hydroxyethyl cellulose films under changing conditions of relative humidity and temperature.

Deformation is defined as the percentage change in dimension resulting from conditioning for 24 hr at 100°F and 90% relative humidity (RH) and then drying the film for 24 hours at 103°F and 4% RH. Permanent set is defined as the percentage change between the initial and final dimension resulting from the conditioning cycle, 75°F—35% RH, to 100°F—90% RH, to 103°F—4% RH, and back to 75°F—35% RH. The dimensional stability values and permanent set are given in Table 9. Since deformation and permanent set are a measure of the built-in stress and orientation of a film imparted by the casting machine, this test is significant only when applied to machine-cast film. These data show

TABLE 9
Dimensional Stability of Hydroxyethyl Cellulose Films

	Hydroxyethyl Cellulose	195 MST-54 Cellophane
Deformation, %		
MD	1.8	1.6
TD	2.2	5.2
Permanent set, %		
MD	2.4	2.6
TD	0.2	3.0

that hydroxyethyl cellulose film shrinks less in the transverse direction during casting and stretches to about the same degree as viscose film in the machine direction.

Figures 6 and 7 are plots of more extensive data on dimensional stability as a function of ambient relative humidity for both regenerated viscose films and hydroxyethyl cellulose films (6a and 7a in the machine direction, 6b and 7b in the transverse direction of the samples). The atmospheres used were 8, 28, 68, and 94% relative humidity, at 70°F, and each exposure was for 22 hr. Figure 8 exhibits the variation of moisture content of the films as a function of relative humidity.

B. Semicommercial Production

In recent years many tons of hydroxyethyl cellulose have been produced on full-size casting machines such as are normally used for viscose film casting. It is certain that several commercial cellulose films at the present time include appreciable amounts of hydroxyethyl cellulose in their structure. Several inherent processing conditions contribute to simplicity of operation, but factors such as the greater gel swelling present problems in the driers.

No difficulty is encountered in dispersing hydroxyethyl cellulose in caustic solutions, but since the viscosity drops sharply with temperature (Fig. 2) a temperature of about 5°C is highly advantageous for filtration and pumping.

One important advantage is that hydroxyethyl cellulose solution may be allowed to stand in equipment for unlimited lengths of time without danger of gelation. Unlike viscose, the solution is not susceptible to hydrolysis and therefore coagulation susceptibility

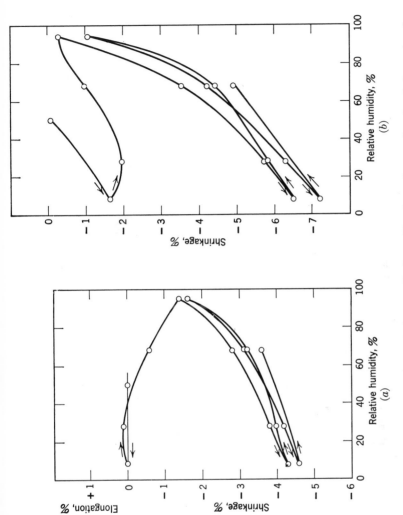

Fig. 6. Dimensional changes of viscose cellophane film through several cycles of change in relative humidity: (a) machine direction, (b) transverse direction of the films.

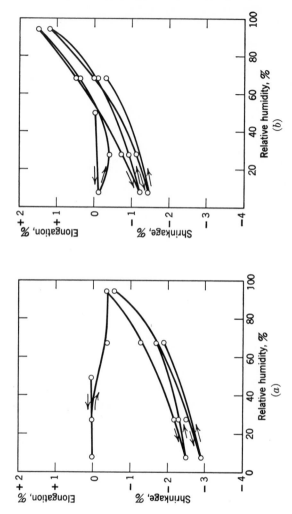

Fig. 7. Dimensional changes of Ethynier hydroxyethyl cellulose film through several cycles of change in relative humidity: (a) machine direction, (b) transverse direction of the films.

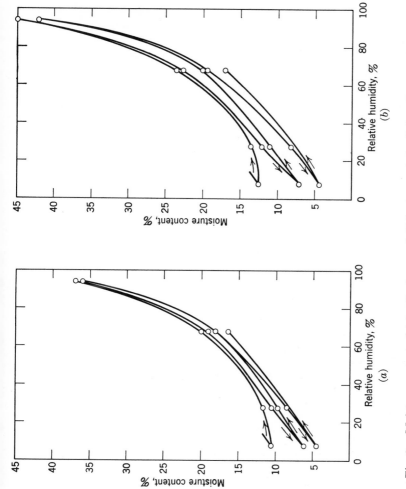

Fig. 8.   Moisture content of (a) cellophane and (b) hydroxyethyl cellulose films as a function of ambient relative humidity: temperature, 70°F; glycerol content (dry basis), 20% in all samples.

does not change with time at constant temperature. No trouble is encountered in deaerating the solutions, since no gas is present, as in the case of viscose.

The casting of hydroxyethyl cellulose solutions is very similar to that of viscose. Most of the film shrinkage occurs in the coagulating acid bath, and the shrinkage is less than that which occurs with viscose film. Since the gel swelling is greater than that of viscose film, it would be expected that the more open structure would wash more readily. No desulfuring or bleaching is required with hydroxyethyl cellulose film and it can therefore be passed directly from the wash water tanks to the softener tanks. The total absence of sulfur compounds, many of which are toxic, and all of which are odoriferous, is of great advantage in processing, as well as in product applications.

It may be expected that in the future, the use of hydroxyethyl cellulose, either alone as a casting material, or in combination with viscose, will continue in order to secure certain special properties of some commercial films.

References

1. I. A. Wolff, H. A. Davis, J. E. Cluskey, L. J. Gundrum, and C. E. Rist, *Ind. Eng. Chem.*, **43**, 915 (1951).
2. T. J. Schoch, in *Advances in Carbohydrate Chemistry*, W. W. Pigman and M. L. Wolfrom, Eds., Academic Press, New York, 1945, p. 259.
3. E. Muetgeert and P. Hiemstra, U.S. Pat. 2,822,581 (Feb. 11, 1958).
4. R. L. Whistler and G. N. Richards, *Ind. Eng. Chem.*, **50**, 1551 (1958).
5. Anon., *Agr. Food Chem.*, **6**, 641 (1958).
6. A. E. Staley Mfg. Co., Decatur, Illinois.
7. A. W. Scharger and M. J. Shoemaker, *Ind. Eng. Chem.*, **29**, 114 (1937).
8. D. R. Erickson, U.S. Pat. 2,469,764 (May 10, 1949).
9. R. L. Mitchell, A. R. Albohn, and S. G. Palmgren, to Rayonier, Inc., U.S. Pat. 3,045,007 (July 17, 1962).
10. R. L. Mitchell and D. K. Smith, to Rayonier, Inc., U.S. Pat. 3,054,705 (Sept. 18, 1962).
11. E. Wellisch, L. Hagan, L. Marker, and O. J. Sweeting, *J. Appl. Polymer Sci.*, **3**, 331 (1960).
12. E. Wellisch, L. Hagan, L. Marker, and O. J. Sweeting, *J. Polymer Sci.*, **51**, 263 (1961).
13. D. T. Milne, to American Viscose, U.S. Pat. 2,994,924 (Aug. 8, 1961).
14. O. J. Sweeting, R. Mykolajewycz, E. Wellisch, and R. N. Lewis, *J. Appl. Polymer Sci.*, **1**, 356 (1959).
15. R. Mykolajewycz, E. Wellisch, R. N. Lewis, and O. J. Sweeting, *J. Appl. Polymer Sci.*, **2**, 236 (1959).

CHAPTER 5

INORGANIC POLYMERS AS FILM FORMERS

ORVILLE J. SWEETING*

Yale University, New Haven, Connecticut

and

RICHARD N. LEWIS

Stauffer Chemical Co., Adrian, Michigan

* Formerly Associate Director of Research, Film Division, Olin Mathieson Chemical Corp., New Haven, Connecticut

I. Introduction

Inorganic polymeric compounds have been known for some time, but they have not been studied as polymers until recently. Many, of course, are three-dimensional and hence unsuitable for films. Only in the last few years have there been any serious attempts to control functionality and prepare tractable linear polymers analogous to the better known organic polymers.

Inorganic polymers might well be considered in any long-range research planning because of their variety and novelty, and because unusual properties are to be expected. Thus, if we consider mica as the prototype of an inorganic film, we might expect other inorganic films to have outstanding resistance to light and other radiations, high temperature, and oxidizing agents; high strength, stiffness, and dimensional stability; good adhesion to coatings; and very low permeability to moisture and gases.

The range of inorganic compounds includes the saltlike compounds between metals and nonmetals and the covalent compounds formed between nonmetals. Most linear inorganic polymers are to be found among the nonmetals and the metalloids, including boron and most of the elements of Groups IVA, VA, and VIA (Table 1). Most ionic compounds are three-dimensional because ionic forces extend in all directions. By the proper choice of complexing or chelating agents, however, it is sometimes possible to reduce the crosslinking tendency of metals so that linear polymers result.

In the discussion that follows, various sheetlike structures will occasionally be mentioned. One might think that if linear (one-dimensional) polymers are ideal for fibers, two-dimensional polymers should be best for film. Perhaps this will some day be the case, but we do not now know how to process such materials, as they are insoluble and cannot be melted without decomposition. They are included here only as a matter of interest.

A recent book on inorganic polymers (1) discusses the following classes of polymers in detail: phosphorus-based macromolecules; sulfur polymers; boron polymers; silicone polymers; organopolymers of silicon, germanium, tin, and lead; polymeric metal alkoxides, organometalloxanes, and organometalloxanosiloxanes; coordination polymers; and electron-deficient polymers. It is not the purpose of this chapter to condense this excellent work, but rather to elucidate the principles on which one might base a search for new materials.

Periodic table of the elements.

Main Periodic Table

	Representative Elements		Transition Elements—d										Representative Elements					Noble Gas Elements
Group:	IA	IIA	IIIB	IVB	VB	VIB	VIIB	VIII			IB	IIB	IIIA	IVA	VA	VIA	VIIA	0
Valence Shell	s^1	s^2	d^1s^2	d^2s^2	$(d^3s^2)^a$	$(d^4s^2)^a$	d^5s^2	$(d^6s^2)^a$	$(d^7s^2)^a$	$(d^8s^2)^a$	s^1d^{10}	s^2	s^2p^1	s^2p^2	s^2p^3	s^2p^4	s^2p^5	s^2p^6
$n=1$ ($1s$)	1 H 1.00797																	2 He 4.0026
$n=2$ ($2s2p$)	3 Li 6.939	4 Be 9.0122											5 B 10.811	6 C 12.01115	7 N 14.0067	8 O 15.9994	9 F 18.9984	10 Ne 20.183
$n=3$ ($3s3p$)	11 Na 22.9898	12 Mg 24.312											13 Al 26.9815	14 Si 28.086	15 P 30.9738	16 S 32.064	17 Cl 35.453	18 Ar 39.948
$n=4$ ($4s3d4p$)	19 K 39.102	20 Ca 40.08	21 Sc 44.956	22 Ti 47.90	23 V 50.942	24 Cr 51.996	25 Mn 54.9380	26 Fe 55.847	27 Co 58.9332	28 Ni 58.71	29 Cu 63.54	30 Zn 65.37	31 Ga 69.72	32 Ge 72.59	33 As 74.9216	34 Se 78.96	35 Br 79.909	36 Kr 83.80
$n=5$ ($5s4d5p$)	37 Rb 85.47	38 Sr 87.62	39 Y 88.905	40 Zr 91.22	41 Nb 92.906	42 Mo 95.94	43 Tc 99	44 Ru 101.07	45 Rh 102.905	46 Pd 106.4	47 Ag 107.870	48 Cd 112.40	49 In 114.82	50 Sn 118.69	51 Sb 121.75	52 Te 127.60	53 I 126.9044	54 Xe 131.30
$n=6$ ($6s4f5d6p$)	55 Cs 132.905	56 Ba 137.34	57–71 ☆	72 Hf 178.49	73 Ta 180.948	74 W 183.85	75 Re 186.2	76 Os 190.2	77 Ir 192.2	78 Pt 195.09	79 Au 196.967	80 Hg 200.59	81 Tl 204.37	82 Pb 207.19	83 Bi 208.980	84 Po 210	85 At 210	86 Rn 222
$n=7$ ($7s5f6d7p$)	87 Fr 223	88 Ra 226	89–103 ★															

Inner Transition Elements—f

☆ Lanthanide Series	57 La 138.91	58 Ce 140.12	59 Pr 140.907	60 Nd 144.24	61 Pm 145	62 Sm 150.35	63 Eu 151.96	64 Gd 157.25	65 Tb 158.924	66 Dy 162.50	67 Ho 164.930	68 Er 167.26	69 Tm 168.934	70 Yb 173.04	71 Lu 174.97
★ Actinide Series	89 Ac 227	90 Th 232.038	91 Pa 231	92 U 238.03	93 Np 237	94 Pu 242	95 Am 243	96 Cm 247	97 Bk 247	98 Cf 249	99 Es 254	100 Fm 253	101 Md 256	102 No 254	103 Lw 257

II. Polymeric Structure Related to Bonding

Among ionic and covalent compounds there are all degrees between the extreme types of bonds. The variation is based on differences in electronegativity, or electron-holding power, of atoms. Pauling's original table of electronegativities (2) has been extended to cover the entire periodic table, with characteristic values for each element in each valence state (3) (Table 2).

An electon-pair bond between *like* atoms in a symmetrical molecule, e.g., Cl_2, is purely covalent; that is, the probability is equal for finding the electrons associated with either atom. For two atoms of different electronegativity, however, e.g., HCl, the probability of finding the electrons associated with one atom of the two is greater for the more electronegative atom, and the bond has a certain degree of polarity or ionic character. Bond polarity is readily measured in diatomic molecules by determining the dipole moment of the molecule and the distance between the atoms. Based on the dipole moments of the hydrogen halide molecules, Hannay and Smyth (4) have developed the equation

$$\text{amount of ionic character} = 0.16(x_A - x_B) + 0.035(x_A - x_B)^2$$

x_A and x_B being the electronegativities of the two atoms. Some representative values are calculated in Table 3.

According to this table, bonds in which the electronegativity difference is greater than 2.1 are predominantly ionic. Thus, almost all of the metallic fluorides and most of the metallic oxides are ionic.

There is a spatial restriction in covalent bonds not found in ionic bonds. The force of attraction between two ions is governed only by the ionic charges and the distance between them, in accordance with Coulomb's law. Covalent bonds, however, are formed only within a limited range of distances, and are formed only at nearly fixed angles from other covalent bonds. Compounds of the sodium chloride type, where the angles are unfavorable for covalent bonding, have little or no covalent character. The commonest covalent bond orbitals, sp^3, formed by the hybridization of one s and three p orbitals, cause the bonds to be directed toward the corners of a regular tetrahedron, as in the bonds of a tetracoordinated carbon atom. This makes angles between the bonds of 109° 28', the so-called tetrahedral angle. Other types of bonds will be discussed later under coordination compounds.

TABLE 2

Electronegativities of the Principal Elements

H 2.1_5							
Li 0.9_5	Be 1.5	B 2.0	C 2.5	N 3.0	O 3.5	F 3.9_5	
Na 0.9	Mg 1.2	Al 1.5	Si 1.8	P 2.5	S 2.8	Cl 3.0	
K 0.8	Ca 1.0	Sc 1.3	Ti(IV) 1.6	V(III) 1.4 V(IV) 1.7 V(V) 1.9	Cr(II) 1.4 Cr(III) 1.6 Cr(VI) 2.2	Mn(II) 1.4 Mn(III) 1.5 Mn(VII) 2.5	Fe(II) 1.7 Fe(III) 1.8 Co 1.7 Ni 1.8
Cu(I) 1.8 Cu(II) 2.0	Zn 1.5	Ga 1.5	Ge 1.8	As(III) 2.0 As(V) 2.2	Se 2.0	Br 2.8	
Ag(I) 1.8	Cd 1.5	In 1.5	Sn(II) 1.7 Sn(IV) 1.8	Sb(III) 1.8 Sb(V) 2.1	Te 2.1	I 2.5_5	

TABLE 3
Ionic Character of Covalent Bonds

Electronegativity difference	Per cent of ionic character
0.5	9
1.0	20
1.5	32
2.0	46
2.1	49
2.5	62

A. The Graphite Structure

Of purely covalent substances, we may consider the different forms of carbon. The stable form is graphite, which is composed of sheets of carbon atoms arranged in a chicken-wire pattern of fused hexagons. Each atom is bonded to only three other atoms in the sheet, but because of the possibility of formation of double bonds, as in condensed aromatic rings, there is almost no bonding between sheets. Consequently, graphite has pronounced cleavage planes and is very soft.

The only other substance with the graphite structure is the ordinary form of boron nitride, BN. This so-called white graphite is isoelectronic with graphite and has similar properties, except that it is noncombustible. Although the electronegativity difference between boron and nitrogen is 1.0, the bonds are not essentially different from those of graphite.

B. Diamond and The Structures of 1:1 Binary Compounds

In diamond, all of the bonds are tetrahedral. While the structure is isotropic, we may consider it to be built up of puckered layers, with bonds between the high points of one layer and the low points of the layer above it. In making a model of such a structure it is possible to arrange the layers so that every second layer is superimposable. This would result in a crystal with hexagonal symmetry, which is never found in diamond. Instead, every third layer is superimposable, with the result that diamond belongs to the cubic system. The extreme hardness of diamond, 10 on the Moh scale, results in part from the fact that it is held together by rigid covalent bonds. Bonds with high ionic character may be stronger, in the sense that they have a high heat of formation, but they do not impart such a high degree of hardness.

Other elements which crystallize with the diamond structure are silicon, germanium, and the gray form of tin.

There are several binary compounds that have essentially the same structure as diamond, both cubic and hexagonal symmetries being found. Zinc sulfide exhibits both types in different natural minerals. Sphalerite is cubic, every third layer repeating, while wurtzite is hexagonal, every second layer repeating. The density and hardness of these are almost identical, indicating that the bond lengths are the same. In all such compounds each atom is surrounded tetrahedrally by four others, as in diamond.

Just as ordinary boron nitride has the same structure as graphite, the high-pressure form of boron nitride, "borazon," has the same structure as diamond (5). It is the only compound that equals diamond in hardness. It has the cubic, i.e., sphalerite, structure.

Proceeding in the same direction we find that beryllium oxide is isoelectronic with diamond, except that it has the wurtzite structure. With a hardness of 9 it is almost as hard as diamond. Even with an electronegativity difference of 2.0 the covalent structure is maintained.

The last member of the series is lithium fluoride, in which the electronegativity difference is 3.0. This has the sodium chloride structure and there is no indication of covalence. All of the lithium halides, in fact, are ionic and crystallize with the same structure, even lithium iodide, where the electronegativity difference is 1.55.

C. OTHER BINARY COMPOUNDS

So far we have considered only compounds of the type AB. The same kind of discussion may be given for compounds of the types AB_2, A_2B, AB_3, A_3B, A_2B_3, and A_3B_2 (6). The structures are not simple, but a brief statement of the structures may be useful.

AB_2. Most of the halides of bivalent metals are ionic, with three-dimensional structures, as in fluorite, CaF_2. Some, e.g., magnesium chloride, have layer structures (see AB_3 below). Interesting flat chain structures are found in palladium(II) chloride, copper(II)

chloride, and copper(II) bromide; if the chains could be separated, flexible linear polymers might result.

Oxides of the type MO_2 are ionic, with the exception of silicon dioxide, which will be discussed later. While sulfides are generally covalent, many of them involve six coordination (octahedral) about the metal atoms, as in ionic compounds. Germanium disulfide has a three-dimensional structure like silicon dioxide. Silicon disulfide has a linear structure like palladium(II) chloride, except that there is tetrahedral coordination about the silicon, rather than planar.

A_2B. Oxides and sulfides are mostly ionic, with the antifluorite structure.

AB_3. Most trihalides, like many dihalides, are ionic and exhibit a layer structure. In these layer structures there are sandwiches, two layers of halide ions with the appropriate number of cations in the middle, two thirds as many in the trihalides as in the dihalides. Aluminum bromide is better regarded as a molecular crystal. Fluorides such as FeF_3, and oxides such as CrO_3, have three-dimensional structures. The layer structures in both dihalides and trihalides are rather weak, as shown by the very low hardness for these materials.

A_3B. Compounds of this type are all three-dimensional. The type compound, sodium arsenide, Na_3As, has an unusually complex structure (6).

A_2B_3, A_3B_2. These compounds are essentially ionic, three-dimensional compounds, except where covalent bonds are to be expected, as in B_2O_3 (crosslinked), P_4O_6 (molecules), Sb_2O_3 and Sb_2S_3 (chains). In corundum, Al_2O_3, each aluminum atom is surrounded by six oxygens and each oxygen by four aluminums. γ-Alumina has a defect structure related to the spinels, which have the general formula MAl_2O_4.

It might appear from the foregoing discussion that ionic compounds are not promising as polymeric materials, and that polymers should rather be sought among the nonmetallic compounds, which may be expected to be covalent.   In general, this is true.   Nevertheless, a number of interesting possibilities involving metals, where the bonds have some degree of ionic character, will be presented in the following sections.

D. CHELATE POLYMERS

In metallic compounds the metal atoms or ions tend to be surrounded by negatively charged ions.   If a metal atom were to be found in a polymer, it would be a potential branch point, and one might not anticipate any linear polymers from metallic compounds. A possible way of circumventing this difficulty is to reduce the functionality of the metal to two by combining it with a monofunctional complexing agent.   As an example consider nickel(II) ion.   It has a coordination number of four, and with ammonia it gives the complex ion $Ni(NH_3)_4^{+2}$, which is not polymer-forming. On the other hand, the complex with a diamine such as hexamethylenediamine would be expected to be a crosslinked polymer. If, however, each nickel ion could first be combined with two molecules of ammonia and then with a molecule of hexamethylenediamine we would have a polymer-forming reaction to give the product:

$$-NH_2(CH_2)_6NH_2-\overset{+2}{Ni}(NH_3)_2-NH_2(CH_2)_6NH_2-\overset{+2}{Ni}(NH_3)_2-$$

The difficulty is that the complexing bonds tend to be in mobile equilibrium.   Some of the nickel ions would therefore become attached to, say, three ammonia molecules and only one diamine, while others would be attached to three diamine molecules, and the polymer would have a branched structure.

It is therefore desirable to have all of the complexing molecules of the same species.   But since almost all metal ions have coordination numbers greater than two, how is it possible to maintain bifunctionality?   One answer is to use chelating agents.

A chelating agent is a polyfunctional complexing agent which is capable of forming two or more bonds to the same metal atom. Although some complexes are rather loosely held, many are extremely stable (7,8).

What kinds of organic molecules form chelates? We have seen that a bifunctional molecule such as hexamethylenediamine is likely to give polymeric complexes. Ethylenediamine, on the other hand, readily forms monomeric chelates, thus:

$$\left[\begin{array}{ccc} CH_2-NH_2 & & NH_2-CH_2 \\ | & \diagdown \diagup & | \\ & Ni & \\ | & \diagup \diagdown & | \\ CH_2-NH_2 & & NH_2-CH_2 \end{array}\right]^{2+}$$

In general it is found that a chelate will result if a ring of five or six atoms can be formed. This is an extension of a principle of organic chemistry that five- and six-membered rings are the most stable. Four-membered chelate rings are known, being found in sulfate and other complexes, and in salts of dithio acids. Seven-membered rings of low stability are found in succinate complexes and a few others. A few larger rings have been reported, including an unstable copper complex with hexamethylenediamine, which contains a nine-membered ring.

Acetylacetone, $CH_3-CO-CH_2-CO-CH_3$, is one of the best known chelating agents. It is a weak acid, and the active hydrogen atom can be replaced by metal ions to give chelates having structures as shown:

$$\left[\begin{array}{c} CH_3 \\ \diagdown \\ C=O \\ HC \diagdown \diagdown \\ C-O \diagup \\ \diagup \\ CH_3 \end{array}\right]_2 M(II) \qquad \left[\begin{array}{c} CH_3 \\ \diagdown \\ C=O \\ HC \diagdown \diagdown \\ C-O \diagup \\ \diagup \\ CH_3 \end{array}\right]_3 M(III)$$

Acetylacetone gives neutral chelates with most bivalent and trivalent ions, and some quadrivalent ions. For most polymer purposes electrically neutral polymers are to be desired. There are of course many synthetic and natural polyelectrolytes, and those with only a small proportion of charged atoms, e.g., silk and cellulose, are sometimes fiber-forming. A highly charged polymer, however, must have a large number of counterions, which have a crosslinking effect.

1. Tetraketone Polymers

Chelate polymers from tetraketones have been described (9). These tetraketones are really pairs of β-diketones linked together. The examples given are the beryllium complexes of terephthalyldiacetone, $CH_3COCH_2COC_6H_4COCH_2COCH_3$ (melting above 295°C); 4,4'-bis(acetoacetyl)diphenyl ether, $CH_3COCH_2COC_6H_4OC_6H_4-COCH_3$ (stick temperature 197°C); 1,8-bis(benzoylacetyl)octane (melt-spun into fibers which were somewhat brittle); and β, β, β',β'-tetraacetyldiethylbenzene (also forms somewhat brittle fibers). The polymerization was carried out by heating the tetraketone with beryllium acetylacetonate, acetylacetone being removed by distillation; this is a "trans-chelation" reaction.

2. Other Known Chelate Polymers

Dithiooxamide forms polymers with nickel and copper (10) with the following structure:

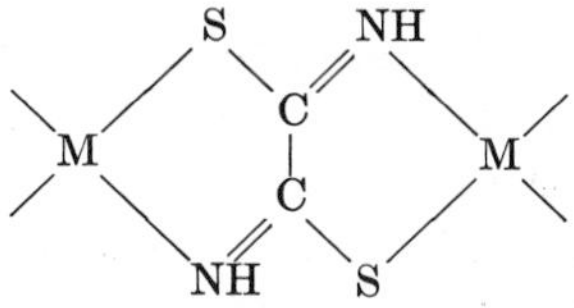

In a search for synthetic oxygen carriers Bailes and Calvin (11) reported a number of cobalt(II) chelates, including neutral polymeric chelates of 4-hydroxy-5-formylsalicylaldehyde and naphthazarin, with the following structures:

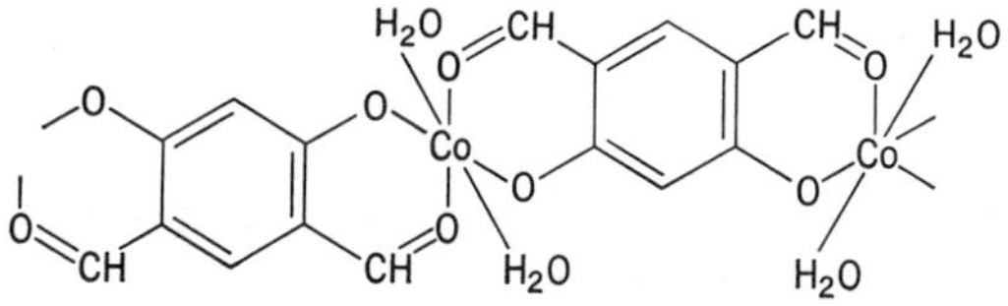

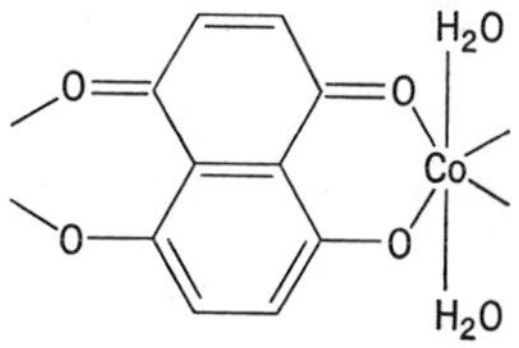

Since cobalt(II) has a coordination number of four as well as six, the water molecules are readily removed by heating.

Whereas ethylenediaminetetraacetic acid, $[—CH_2N(CH_2CO-OH)_2]_2$, is a sexadentate chelating agent, e.g., for calcium ion, homologs with more methylene groups between the nitrogens tend to coordinate more than one metal ion and therefore produce poly-chelates. These are anionic polymers associated with uncomplexed alkali metal cations.

Bis-α-amino acids of the type $HOOC—CR_2—NH—(CH_2)_n—NH—CR_2—COOH$ are capable of forming either monomeric or polymeric chelates. Schlesinger (12) studied the copper(II) complexes where $n = 2, 3, 5, 7$, and 10. For $n = 2$ or 3, the complexes are soluble, blue, and monomolecular. For $n = 5$ or 7, there are both soluble, blue, monomolecular forms and insoluble violet forms of unknown molecular weight. For $n = 10$, only the insoluble violet form is known. The author attributed the two forms to geometrical isomerism [copper(II) complexes are planar]. It would seem, however, that the insoluble forms are more likely polymeric.

Other chelates in which rings of eight atoms or more have been formulated are very probably polymeric.

Recent emphasis has been on the development of polymeric fluids, elastomers, and plastics of exceptional thermal stability for jet aircraft applications. To this end a series of copper(II), nickel(II), and zinc chelates of bis-α-thiopicolinamides has been studied (13). The most stable is a zinc chelate which can be heated to 390°C without appreciable decomposition or melting. It has the structure shown below.

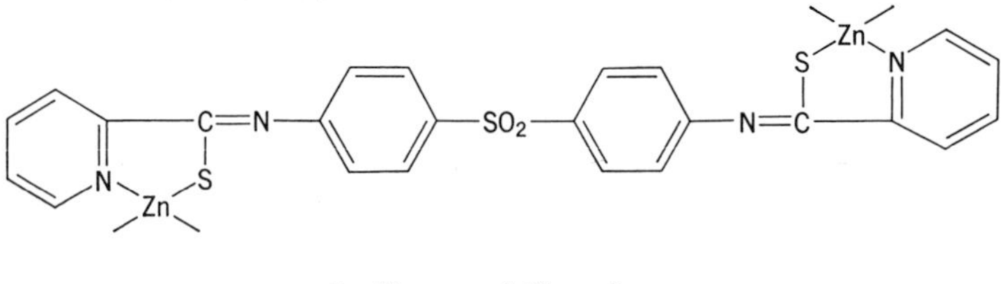

3. Types of Complexes

Another factor enters when some degree of covalent bonding is possible, as with the more electronegative metals. The strongest bonds are formed when d orbitals can be used. Whereas the elements in the two short rows of the Periodic Table use mainly the

sp^3 tetrahedral orbitals, transition metals hybridize the d orbitals, forming d^2sp^3 bonds (octahedral), or dsp^2 bonds (square planar). Square complexes occur with the ions of metals near the end of the transition periods, where only one d orbital is available, as in Ni(II), Pd(II), Pt(II), Ag(III), and Au(III). Square complexes are also formed with Cu(II) and Ag(II), in which the odd d electron is promoted into a higher orbital. With the other transition metals, tetrahedral or octahedral coordination is favored by both covalent and ionic bonding.

We may consider the following types of complexes: 2-4, 2-6, 3-4, 3-6, 4-6, 4-8, in which the first digit represents the oxidation state of the metal and the second its coordination number. Most of the chelate polymers known involve complexes of the 2-4 type. It would be expected that chelates with a larger number of bonds, as in the 2-6, 3-6, or 4-8 types, would have greater strength, both mechanically and in terms of thermal resistance.

4. Types of Chelating Agents

To make a linear polymer, the nature of the chelating agent depends on the metal ion that is used. With a 2-4 metal the chelating agent must have two bidentate ends. Furthermore, it must have a net charge of -2, if it is desired to make an electrically neutral polymer, i.e., it must be derived from a dibasic acid. Symmetrical chelating agents, such as the bis-α-amino acids will give an unambiguous polymer.

An unsymmetrical agent, such as a bis-amino acid with two amino groups at one end and two carboxyl groups at the other, would give not only the head-to-tail type of structure, in which each metal ion is attached to two amino and two carboxyl groups and is thus electrically neutral, but also the head-to-head and tail-to-tail structures, in which charged metal atoms would be surrounded by four amino groups or four carboxyl groups. This would be a polymer with both positive and negative charges along the chain. It might have rather unusual properties, perhaps similar to those of proteins at their isoelectric points. Symmetry considerations make it difficult to construct an uncharged polymer based on tervalent metals, unless the effective valence can be reduced to two by a strong complexing agent such as acetylacetone.

In general, a polymer-forming chelating agent must have on either end one or two anionic groups and one or two neutral coordinating groups. The anionic groups may be derived from carboxylic acids, alcohols, amides, their sulfur analogs, or diketones. The neutral donor atoms may be oxygen in ethers, alcohols, carbonyl groups, etc.; nitrogen in amines, amides, or nitriles; sulfur in mercaptans, thioethers, thioacids, etc.; or, theoretically, phosphorus or arsenic in phosphino or arsino compounds, as in dimethylphosphinoacetic acid. The various combinations represent a considerable challenge to the synthetic organic chemist. It can hardly be doubted that the number of possible chelate polymers is greater than that of all known synthetic polymers.

One other point should be kept in mind. For a film-forming polymer a reasonable degree of flexibility is required; this means that the molecular chains themselves must be flexible. In the polychelates there are of necessity at least two rings joined together at the metal atom, and this gives a rather stiff place in the chain. To counteract this it is probably necessary to have a fairly long aliphatic chain between the chelating ends of the organic reagent, or else to incorporate in the chain flexibilizing groups such as ether or ester groups. An existing example of a polymer in which rigidifying groups are separated by a long hydrocarbon chain is 11-nylon, "Rilsan," which makes a very fine transparent film.

5. Possible Chelating Agents

One polymer-forming chelating agent that would not be expensive is the xanthate of ethylene glycol, $NaSSCOCH_2CH_2OCSSNa$, made from sodium, ethylene glycol, and carbon disulfide. The dithio acid end groups act as chelating agents, forming four-membered rings, as in the nickel complex made from ethyl xanthates (14).

$$C_2H_5OC \underset{\diagdown S \diagup}{\overset{S \diagdown \diagup}{}} Ni \underset{\diagdown S \diagup}{\overset{\diagdown S}{}} COC_2H_5$$

Polyethylene sulfide of low molecular weight and having mercaptan end groups would very likely form polymeric chelates with either

2-4 metals or 2-6 metals which make covalent bonds to sulfur. Possible structures are as follows:

$$CH_2\!\!-\!\!CH_2$$
$$|\qquad\quad|$$
$$S\qquad S\!-\!(CH_2\!\!-\!CH_2\!\!-\!S\!-\!)_x CH_2\!\!-\!CH_2\!\!-\!S \qquad \underset{Zn}{\diagup} \quad S$$
$$\diagdown Zn \diagup \qquad\qquad\qquad\qquad\qquad\qquad\qquad | \qquad\quad |$$
$$\qquad\qquad\qquad\qquad\qquad\qquad\qquad\qquad\qquad\qquad CH_2\!\!-\!CH_2$$

$$CH_2\!\!-\!\!CH_2 \qquad CH_2\!\!-\!\!CH_2$$
$$|\qquad\qquad\qquad|$$
$$S \diagdown \qquad S \diagup \qquad S\!-\!(CH_2\!\!-\!CH_2\!\!-\!S\!-\!)_x CH_2\!\!-\!CH_2\!\!-\!S \diagup \underset{Fe}{} \quad S \quad S$$
$$\diagdown Fe \diagup \qquad\qquad\qquad\qquad\qquad\qquad\qquad CH_2\!\!-\!CH_2 \quad CH_2\!\!-\!CH_2$$

Triglycol dichloride, Cl—CH_2—CH_2—O—CH_2—CH_2—O—CH_2—CH_2—Cl, as well as its homologs, are potential intermediates. With the sodium derivative of acetylacetone, for instance, it will give a tetraketone, or bis-β-diketone. With glycine, NH_2CH_2COOH, it will give a bis-α-amino acid. With iminodiacetic acid, $HOOCCH_2NHCH_2COOH$, it will give a diaminotetracarboxylic acid, which will be particularly useful for polymerizing 4-6 metals such as Ti(IV) and Sn(IV). With ammonia it is possible to get a diamine, which can then be mercaptoethylated on both ends, giving the chelate-forming groups HS—CH_2—CH_2—NH—. Mercapto-ethylation of amines may be carried out with ethylene sulfide (15) or with Eastman's ethylene monothiocarbonate (16).

6. Choice of Metal Ion

For most film purposes it is necessary to eliminate elements that are colored in all of their forms. The choice is then narrowed to the following ions: Be, Mg, Ca, Zn, Cd, Sn(II), Pb(II), Al, Ce(III), Ce(IV), Sn(IV), Ti(IV), Zr, and Th. Mn(II) and Fe(II) might be included if a slight color were not objectionable, although both give strongly colored ions on oxidation. Calcium can probably be eliminated as not giving very stable complexes. Beryllium, cadmium, and thorium can probably be eliminated on grounds of cost, and lead on grounds of health hazard. This leaves three bivalent ions: magnesium, zinc, and tin(II); two tervalent ions: aluminum and cerium(III); and four tetravalent ions: cerium(IV), tin(IV), titanium, and zirconium. Of these, only zinc has previously been reported in polychelates. All of them warrant further investigation.

E. OTHER METAL-CONTAINING POLYMERS

1. Oxygen-Bridged Complexes

Among the earliest known of the bridge complexes are the chromium "olation" complexes in which two or more chromium atoms are joined by pairs of hydroxyl groups (17). The bonding is somewhat similar to that in the aluminum bromide dimer, where two of the bromine atoms are joined to both metal atoms. In the olates, however, chromium is octahedral.

Similar coordination complexes are formed with Be, Al, Sc, Mn(II), Fe(III), Co(III), Zr, Th, and Pu(IV) (18). By loss of protons, the hydroxide bridges are converted to oxide bridges, and in general such complexes have limited stability.

More stable bridge structures are formed in the metal alkoxides, and many well characterized low polymers have been described (19). Aluminum *ter*-butoxide is a dimer, apparently with the aluminum bromide structure, i.e., the aluminum is 4-coordinated. Other aluminum alkoxides are typically tetrameric, at least one of the aluminums being 6-coordinated.

Similarly, zirconium tetra-*ter*-butoxide is monomeric (4-coordinated), whereas the ethoxide is a trimer (6-coordinated). Titanium tetra-*n*-butoxide is normally a trimer (all Ti 6-coordinated), but dissociates in benzene to a monomer (all 4-coordinated) without forming any dimer. In the trimer, the three metal atoms are in a straight line, with six bridging alkoxy groups and six terminal ones.

Ceric tetra-*n*-butoxide is a polymer with both 6-coordinate and 8-coordinate cerium. Thorium alkoxides approach the theoretical cyclic octamer, with every metal atom having a possibly cubic 8-coordinate structure (20).

Beryllium ethoxide is insoluble and apparently polymeric, and is very probably a linear polymer (21). Lithium methoxide is a polymer with a two-dimensional layer structure (22). Bradley (19) has predicted infinite linear polymers for 2-5 and 2-6 metal alkoxides (bivalent metals with coordination numbers 5 and 6), but these have not yet been studied. Possibilities include the alkoxides of Fe(II) and Co(II).

The metal alkoxides are susceptible to hydrolysis, with partial or complete replacement of alkoxide with oxide. An insoluble polymer of the composition $Ti_3O_4(OEt)_4$ has been prepared (23); an infinite ribbonlike structure has been proposed for it, but not definitely proved. A zirconium oxide ethoxide of similar composition has also been prepared.

The structure of a given composition is often difficult to predict, because of uncertainties regarding the coordination number of the metal and the steric effect of the alkyl group. Further developments can be expected in this area, however.

2. Other Bridged Complexes

Basic beryllium acetate, $Be_4O(CH_3COO)_6$, consists of tetrahedral molecules, with oxygen at the center, beryllium atoms at the four corners, and acetate bridges joining each pair of beryllium atoms and forming a series of 6-membered rings.

Similar acetate bridges appear to be formed in chromium complexes (24), but no clearly polymeric acetates have been described.

Definitely polymeric structures have been obtained with diphenylphosphinate as the bridging group. Thus, chromium(III) acetylacetonate bis(diphenylphosphinate) is a one-dimensional high polymer with film-forming properties (25). Flexibility in this polymer is attributed to a coiled structure resulting from the acetylacetonate ligand, which prevents the adoption of a *trans* configuration in the other groups.

Similar polymers have been made with aluminum; analogous polymers, without the acetylacetone, have been made with zinc and cobalt. Other phosphinic acids, including dimethyl- and methylphenylphosphinic acid may also be used. Zinc methylphenylphosphinate has been obtained as a stable polymer melting below 100°C with a molecular weight of over 35,000 (26).

F. POLYMERS FROM FERROCENE

Ferrocene is a very stable substance and might be expected to form thermally stable polymers. Various disubstituted ferrocenes have been prepared and converted to polymers (27). Thus, bis-hydroxyalkylferrocenes have been converted to polyurethanes, and ferrocene-1,1'-dicarboxylic acid chloride has been converted to a polyamide. These were no more thermally stable, however, than organic polyurethanes and polyamides.

All ferrocene derivatives have an orange-yellow color, and by replacing the iron with other transition metals it is possible to obtain a wide variety of colors (28). Examples include titanium(IV), red; titanium(III), green; vanadium(III), green; chromium(II), scarlet; nickel(II), dark green; ruthenium(II), pale yellow; uranium(IV), dark red.

G. CARBORANE POLYMERS

Attempts have been made to make polymers based on the very stable o-carborane, an icosahedral molecule, $C_2B_{10}H_{10}$. It was found, however, that with substituents in the *ortho* position, only cyclic products could be obtained.

m-Carborane has been converted to the bis-dimethylchlorosilyl derivative and thence to a polymeric siloxane,

$$\left[\begin{array}{c} CH_3 \\ | \\ -Si-CB_{10}H_{10}C-Si-O- \\ | \\ CH_3 \end{array} \quad \begin{array}{c} CH_3 \\ | \\ | \\ CH_3 \end{array} \right]_n$$

and to polymers of the series

$$\left[\begin{array}{c} CH_3 \\ | \\ -Si-CB_{10}H_{10}C-Si-O- \\ | \\ CH_3 \end{array} \quad \begin{array}{c} CH_3 \\ | \\ | \\ CH_3 \end{array} \left(\begin{array}{c} CH_3 \\ | \\ Si-O- \\ | \\ CH_3 \end{array} \right)_x \right]_n$$

Very stable high polymers were obtained, with properties ranging from high-melting solids through rubbers to fluids as x varied from 0 to 3 (29).

H. COVALENT POLYMERS FROM THE NONMETALS

One problem with a great many elements is to reduce the functionality to two, in order to get a linear polymer. This is done by combining a multivalent element with univalent radicals, such as alkyl, aryl, or alkoxy groups. Halogen atoms may sometimes be used, although they are often too easily hydrolyzable to be useful. For instance, the hydrolysis of silicon tetrachloride leads to silica, SiO_2, which is highly crosslinked. If each silicon atom bears two methyl groups, however, a linear silicone polymer can be obtained.

Occasionally linear polymers are formed in potentially crosslinking reactions through the joining together of rings. An example is silicon disulfide, SiS_2. It might have been expected to be similar to the dioxide, but it is a linear polymer having the structure

Chains of this type are hard to predict and thus far a search for polymer-forming reactions among the nonmetals other than carbon has not been fruitful.

This discussion will be closed with a few remarks about elements that have been considered and a brief mention of some areas of phosphorus and nitrogen chemistry that may be worthy of another look.

1. Boron-Containing Polymers

Boron–Oxygen Systems.   Boron oxide, B_2O_3, is usually obtained as a glass, presumably a three-dimensional network.   Organoboron oxides, such as methylboron oxide, $(CH_3BO)_3$, occur as cyclic trimers (30–32).   No larger rings are formed, and no linear polymers, indicating unusual stability in the six-membered boron–oxygen ring.

One of the few commercial boron-containing polymers is the peculiar substance known as "bouncing putty," which is a highly viscoelastic silicone made by the addition of five percent of boron oxide or a borate.   It appears to be a linear copolymer in which every boron atom bears a hydroxyl group in an otherwise nonpolar material (33).

Polymeric chain ions are found in such compounds as calcium metaborate, $Ca(BO_2)_2$.   The structure of the ion is represented as

$$
\begin{array}{ccccccc}
O^- & & O^- & & O^- & & O^- \\
| & & | & & | & & | \\
-B-O-&B-&O-&B-&O-&B-&O-
\end{array}
$$

The calcium ions in this case bind the whole structure together so that it does not have the properties of a linear polymer.

Boron–Nitrogen Compounds.   Boron nitride, BN, was mentioned before as existing in both diamond and graphite forms.   There are several substances related to borazine (34) ("borazole," "inorganic benzene"), $B_3N_3H_6$, in which part or all of the hydrogens are replaced by methyl or chlorine.

There seems to be no tendency for the ring to open to give linear polymers.   Borazine itself is rather reactive chemically because of

its B—H bonds. There are some theoretical possibilities of including it or its derivatives in a polymer, replacing the benzene ring in,
say, polystyrene, but there would appear to be no particular advantage as a film-forming material (35).

The obvious question arises as to whether a complete boron–
nitrogen chemistry exists analogous to carbon chemistry, particularly with reference to polymers. The answer would seem to be
definitely no. Whereas the hydrogen attached to nitrogen is pro-
tonic, the hydrogen attached to boron is hydridic, and when they
occur together in the same molecule, they tend to react to split out
hydrogen. Thermally stable compounds are not achieved until all
of the hydrogen has been removed from either boron or nitrogen,
or until the borazine structure has been reached.

It is theoretically possible that polymers such as $-BH_2-NR_2-$
BH_2-NR_2- or $-BR_2-NH_2-BR_2-NH_2-$, structurally analogous to polyisobutylene, could exist, although ring formation is
more probable. A linear polymer of the composition $C_3H_6NBH_2$,
in which the amine is part of a trimethylenimine ring, has been
made and is apparently stable (36).

By analogy with borazine, inorganic naphthalene and anthracene
analogs should be possible, but none has yet been reported. It is
doubtful in any case that long strings of six-membered rings could
be produced.

Boron–Phosphorus Polymers. Boron phosphide, BP, is a solid of
unknown structure, but probably crosslinked. The borine–phosphine complex, PH_3BH_3, is somewhat more stable than the corresponding ammonia complex, although it loses hydrogen at 65°C.
The compound $(CH_3)_3PBH_3$ is stable to almost 200°C. A trimeric
compound, $[(CH_3)_2PBH_2]_3$, is thermally stable to 350°C, and is
extremely resistant to acids and bases; the B—H bond is thus
unusually inert (37). A cyclic tetramer, also quite stable, is known.
A linear polymer has also been made. It must be stabilized by
strong electron-donor end groups, but even when stabilized it reverts
to the trimer at 330°C (36).

Other polymeric phosphinoborines include the following.
Dimethylphosphinodimethylborine, $(CH_3)_2PB(CH_3)_2$, occurs only
as the cyclic trimer. Methylphosphinodimethylborine, CH_3-
$PHB(CH_3)_2$, and phosphinodimethylborine, $PH_2B(CH_3)_2$, are
monomeric at first (prepared from the addition products of the

appropriate borine and phosphine) and change into fairly high polymers that are viscous oils. Phosphinoborine, PH_2BH_2, made by the removal of hydrogen from the addition product PH_3BH_3, is a very high polymer, insoluble in all solvents and inert to hydrochloric acid. Analysis indicates the formula $PBH_{3.75}$, which would imply some degree of unsaturation or crosslinking.

It can be seen that the various polymers are structural analogs of polyethylene, polypropylene, polyisobutylene, and poly-2-methyl-2-butene, in which the chain carbons have been replaced alternately by boron and phosphorus. The inertness of these materials, in contrast with the boron–nitrogen compounds, is quite remarkable. An unusual type of boron–phosphorus bond has been suggested to account for it (37), but whatever the explanation it would appear that boron–phosphorus polymer chemistry has not been fully exploited.

Boron–Carbon Polymers. These are not known, although polymers containing alternate boron and carbon atoms ought to be possible. Also polymethylene or phenylene bridges might be built between boron atoms. Although one organic radical attached to boron, as in phenylboronic acid, $C_6H_5B(OH)_2$, is very inert, it is found that organoboron compounds containing two or three alkyl groups are somewhat susceptible to hydrolysis and oxidation. This would limit the usefulness of any boron–carbon polymer.

Semiorganic Boron Polymers. There are a few instances where boron has been incorporated in an essentially organic polymer. One of these is a polymeric borate of diethylene glycol, suitable for adhesives. Another, made from phenylboronic acid and hexamethylene diisocyanate was supposed to make a polyamide, but was later found not to contain boron (35). Phenylboronic acid and hexamethylenediurea give a film-forming polymer containing 5.8% boron. The compound butylboronimine, BuBNH (38) is probably the cyclic trimer, a derivative of borazine.

2. Silicates

The silicates have been classified in various ways, but from a polymer standpoint only the chain and layer structures are of interest here.

Chain Silicates.   The simplest chain ion is found in the pyroxene minerals, e.g. $MgSiO_3$, which have the structure:

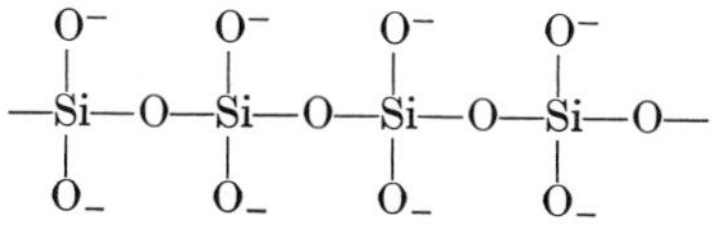

$$
\begin{array}{ccccccc}
O^- & & O^- & & O^- & & O^- \\
| & & | & & | & & | \\
-Si-O- & Si-O- & Si-O- & Si-O- \\
| & & | & & | & & | \\
O_- & & O_- & & O_- & & O_-
\end{array}
$$

The bivalent magnesium cations may be replaced by other bivalent cations, or by alternate univalent and tervalent cations, as in spodumene, $LiAl(SiO_3)_2$. In any case the cations hold the chains firmly together, and the polymers have no film-forming properties. It is conceivable to replace the cations with covalent radicals, forming siliconelike polymers, but this has never been done. Sodium silicate, Na_2SiO_3, in the form of water glass, may also be a linear polymer, and would certainly be more amenable to modification.

The amphibole minerals contain double-chain ions of the structure:

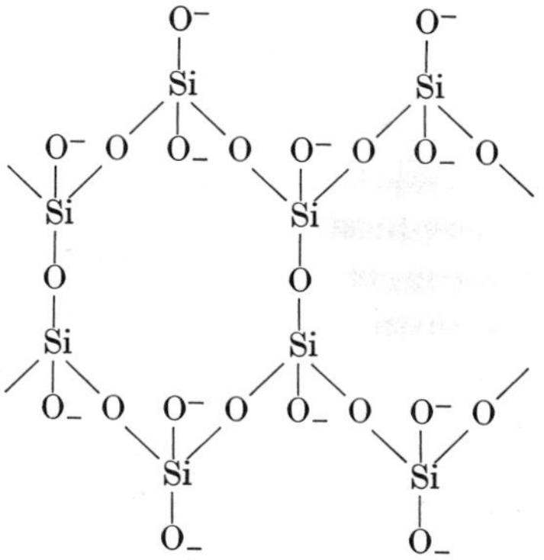

A typical example is tremolite, $Ca_2Mg_5(Si_4O_{11})_2(OH)_2$. Like the pyroxenes, the amphiboles are strongly held together by the cationic metal ions.

The asbestos minerals are exemplified by chrysotile, $Mg_3Si_2O_5$-$(OH)_4$. They have the same type of double chains as the amphiboles, but these are packed to give layer structures rather similar to those of kaolin and mica (39).

Layer Silicates.   Silicates in which the polymeric unit is represented by the formula $Si_2O_5^{-2}$ form characteristic layers (39). In apophyllite, $KF \cdot Ca_4Si_8O_{20} \cdot 8H_2O$, there are fused rings of four and eight silicon atoms, with anionic oxygen atoms protruding on both sides, so that the calcium ions can bind the sheets together.

The other layer silicates consist of fused rings of six silicon atoms, in which all the anionic oxygen atoms lie on one side, the other side being electrically neutral. The minerals talc, $Mg_3Si_4O_{10}(OH)_2$, and pyrophyllite, $Al_2Si_4O_{10}(OH)_2$, have sandwich structures with the neutral faces on the outside and the cations and attendant hydroxide ions inside. The outside of the sandwich is inert, so that the forces between the double layers are very small, and the minerals are soft, 1–2 on the Moh scale.

Kaolin, $Al_2Si_2O_5(OH)_4$, is an open sandwich with a similar structure, but lacking the top layer of Si_2O_5. The other clay minerals are similar, and because of the exposed layer of hydroxyl groups, they all have in common the properties of water absorption and cation exchange.

The micas, e.g., phlogopite, $KMg_3(Si_3AlO_{10})(OH)_2$, and muscovite, $KAl_2(Si_3AlO_{10})(OH)_2$, are somewhat similar to talc and pyrophyllite, except that one fourth of the silicon atoms in the sheets have been replaced by aluminum. This results in a negative charge on the sandwich as a whole, balanced by the positive charge of the alkali metal ions, which lie outside the sandwiches and bind them together. As a result the micas are harder minerals than talc.

In the brittle micas, e.g., margarite, $CaAl_2(Si_2Al_2O_{10})(OH)_2$, there is further replacement of silicon by aluminum. The charge on the sheet is therefore doubled and the alkali metal ions are replaced by bivalent calcium ions.

None of these minerals has a melting point low enough to permit extrusion. On heating they lose the elements of water and decompose. Nevertheless, the film-forming properties of mica are well known. The General Electric Company makes a mica tape for insulating purposes by bonding thin sheets of mica to a paper backing by means of a low-melting resin.

Synthetic Mica. During World War II synthetic mica was developed in Germany, and several improvements have been made since. The synthetic material is characterized by the substitution of fluoride for hydroxide, so that there is no loss of water on heating. One form, with the formula $KMg_3AlSi_3O_{10}F_2$ (fluorophlogopite), is said to be stable to above its melting point of about 1250°C (40). Other formulas include $K_2Mg_5LiAlSi_7O_{20}F_4$ (13) and $KMg_3AlSi_3O_{10}F_2$ (41). Slow cooling is apparently necessary to get large crystals.

Other Synthetic Silicates. An extension of the idea above for synthetic mica is to make other silicates in which hydroxide ions are replaced by fluoride ions. Thus one might make synthetic talc, amphibole, or asbestos. As in the case of mica, the fluoride substitutes should be very stable, and if they could be handled, some of them might prove to have suitable properties for film or fiber. They would have a unique advantage in thermal stability over everything but glass fiber, and might have better mechanical properties than glass.

3. Silicones (Polysiloxanes)

The name silicone was derived for compounds of the type $(R_2SiO)_n$ by analogy with ketones, R_2CO, which have a similar empirical formula. The silicon–oxygen double bond does not exist, however, and all silicones are polymeric, containing the alternating silicon–oxygen chain. The name has now been extended to include all organosilicon oxides, including hexamethyldisiloxane, $Me_3SiOSiMe_3$, the hydrolysis product of trimethylchlorosilane, Me_3SiCl, as well as the gel polymer $MeSiO_{1.5}$ obtained from methyltrichlorosilane, $MeSiCl_3$.

Linear polymers are formed from the cohydrolysis of trimethylchlorosilane and dimethyldichlorosilane. They have the general formula $Me_3SiO(SiMe_2O)_xSiMe_3$, MD_xM in silicone shorthand (42), where M is the monofunctional end unit $Me_3SiO_{0.5}$ and D is the difunctional chain unit Me_2SiO. Branches arise from the incorporation of T units (trifunctional—from methyltrichlorosilane) or Q units (tetrafunctional—from silicon tetrachloride). As long as there is a chain end for every chain branch the molecular weight is finite and the compounds are liquids, of varying degrees of viscosity. Oils have been made with molecular weights of over 100,000; silicone gums have molecular weights of the order of 1,000,000. Although these are regular polymers, they are not crystalline at room temperature, because of the very high flexibility of the chain.

Some or all of the methyl groups may be replaced by higher alkyl groups without changing the basic character of the polymer. Diphenylsiloxane low polymers are crystalline, but no corresponding high polymers are known.

The only crystalline high polymer in the silicone series is the double-chain, or ladder, polymer of phenylsilsesquioxane, which has a double syndiotactic structure (43).   Molecular weights as high as 4,000,000 have been obtained.

$$
\begin{array}{cccc}
\text{Ph} & \text{Ph} & \text{Ph} & \text{Ph} \\
| & | & | & | \\
-\text{Si}-\text{O}-\text{Si}-\text{O}-\text{Si}-\text{O}-\text{Si}-\text{O}- \\
| & | & | & | \\
\text{O} & \text{O} & \text{O} & \text{O} \\
| & | & | & | \\
-\text{Si}-\text{O}-\text{Si}-\text{O}-\text{Si}-\text{O}-\text{Si}-\text{O}- \\
| & | & | & | \\
\text{Ph} & \text{Ph} & \text{Ph} & \text{Ph}
\end{array}
$$

Orientable films have been cast from benzene solution.

4. Phosphorus

The oxides and sulfides of phosphorus are all tetrahedral molecules, but the existence of phosphate glasses shows that polymeric phosphorus–oxygen compounds can be made.   Linear polymers are also theoretically possible having empirical formulas such as RPO, RPO_2, RPS, RPOS, RPS_2, RPNH, RPONH, RPNR, and RPONR.

Polymers of Unsaturated Phosphorus Compounds.   In the case of phosphorus the simplest unsaturated compound is phosphoryl monochloride, POCl, which has been described only once many years ago (44).   If it were monomeric, it would be expected to be a gas, like phosgene, $COCl_2$, or a volatile soluble liquid like PCl_3. It is, however, a paraffinlike solid, insoluble in ordinary sovents, and obviously polymeric.   On the other hand, phosphorus oxychloride, $POCl_3$, is a volatile liquid, which may be written as $Cl_3P{=}O$.*

* It may be questioned whether the compounds of pentavalent phosphorus should be written with a covalent double bond $-\overset{|}{P}{=}O$, or with a semipolar bond $-\overset{|}{\underset{|}{P}}{}^+\!-O^-$ or $-\overset{|}{\underset{|}{P}}-O$.   Phosphorus, unlike nitrogen, is capable of expanding its valence shell to form five covalent bonds as in PCl_5.   The dipole moment of the phosphine oxides, R_3PO, indicates about 50% ionic structure.   The properties of the phosphine oxides are very different from those of the amine oxides, which definitely have the ionic structure; e.g. the phosphine oxides are only feebly basic whereas the amine oxides are very strong bases, and in water they give the ions R_3NOH^+ and OH^-.   For these and other reasons the double-bond structure of the phosphine oxides and related compounds is preferred (45).   It may be noted that the carbon–oxygen double bond has a considerable degree of polarity, yet is usually written $\diagdown\!\!\!\diagup C{=}O$.

It is concluded that double bonds may exist when the phosphorus is attached to four atoms, but not when it is attached to fewer atoms. Metaphosphoryl chloride, PO_2Cl, is an interesting case. It is a syrupy material, probably having a polymeric structure.

$$\left[\begin{array}{c} O \\ \| \\ -P-O- \\ | \\ Cl \end{array} \right]_n$$

In this formulation one double bond per phosphorus atom remains.

When the hypothetical monomer has a triple bond, as in phosphonitrilic chloride, $Cl_2P{\equiv}N$, a third type of polymer having the units

$$\begin{array}{c} Cl \\ | \\ -P{=}N- \\ | \\ Cl \end{array}$$

is possible. These will be discussed below.

Polymers of Trivalent Phosphorus. As mentioned above, phosphoryl monochloride, POCl, appears to be polymeric. Theoretically, similar compounds could be prepared in which chlorine is replaced by alkyl or alkoxy radicals, NH_2, NHR, or NR_2. With the possible exception of some esters of metaphosphorous acid, ROPO (46), no such compounds have been described. The only related materials are a few in which the monomer has a phosphorus-nitrogen double bond, as in "imides of phosphonous acids," e.g., $C_6H_5P{=}N-NHC_6H_5$, and "imidophosphites," as in $C_6H_5N{=}PCl$, $C_6H_5N{=}P(OEt)$, $C_6H_5N{=}P(OC_6H_5)$, $C_6H_5N{=}P(OCH_2C_6H_5)$, and $C_6H_5N{=}P(NHC_6H_5)$. The last compound was shown to exist as a dimer (47). This would imply an unusual four-membered nitrogen-phosphorus ring.

Polymers of Pentavalent Phosphorus. From the six-membered rings which are found in P_4O_{10} one would infer that other anhydrides of pentavalent phosphorus acids, $ROPO_2$ and RPO_2, might also be composed of six-membered rings. These have been postulated only in the case of amides of metaphosphoric acid, i.e., Et_2NPO_2, Pr_2NPO_2, and i-Bu_2NPO_2 (48). The esters of metaphosphoric acid, $ROPO_2$, are syrupy liquids, apparently low polymers. Analogous sulfur compounds, e.g., $C_6H_5SP_2$, are crystalline solids. The phosphonoanhydrides, RPO_2, are crystalline, but have not been investigated in over 60 years.

It has been found that phosphonamides of the general formula $RP(O)(NHR')_2$ lose a molecule of amine on heating and are converted to phosphonimides, $(—RPO—NR—)_n$, that are linear high polymers with film-forming properties (49). In view of this discovery it would be well to reinvestigate other anhydrides and imides. It is possible that many will prove capable of existence as high polymers. In addition to the polyphosphonimides, those of particular interest for film might be one of the imidophosphates, $(—RNHPO—NR'—)_n$, or a polythioanhydride, e.g. $(—C_6H_5PO—S—)_n$. The former are easily made from a primary amine and phosphorus oxychloride, and the latter could be made from benzenephosphonyl chloride, $C_6H_5POCl_2$.

Salts of metaphosphoric acid, HPO_3, are all polymeric. A number of low polymers of sodium metaphosphate have been characterized, although the hexamer, $Na_6P_6O_{18}$, corresponding to the well known name, sodium hexametaphosphate, has not. Potassium metaphosphate is quite different; it forms stable high polymers on heating. Ultracentrifuge measurements in aqueous salt solutions indicate molecular weights of 150,000–2,500,000 (39). It is conceivable that potassium metaphosphate could be used as such for a film. Alternately, part or all of the potassium could be replaced by organic radicals, for example by treatment with ethyl chloride, and a linear metaphosphate ester would result:

$$\left[\begin{array}{c} O \\ \parallel \\ —P—O— \\ | \\ O \\ | \\ R \end{array} \right]_n$$

Phosphorus–Oxygen–Carbon Polymers. Polymeric phosphonous and phosphonic esters, $[—RP—O—(CH_2)_x—O—]_n$ and $[—RPO—O—(CH_2)_x—O—]_n$, have been described (50,51); they are all liquid or low-melting materials of no interest for film.

Phosphonitrilic Chloride (39,52,53). By heating a mixture of phosphorus pentachloride and ammonium chloride one obtains phosphonitrilic chloride, $PNCl_2$. As formed, it consists of a cyclic trimer and tetramer, both of which are solid, and various homologous low-melting polymers. If the mixture is heated slowly it is converted to a linear, rubbery high polymer, called "inorganic rubber," best represented as follows:

$$\left[\begin{array}{c} Cl \\ | \\ -P=N- \\ | \\ Cl \end{array}\right]_n$$

The very existence of such a compound is remarkable; its inertness, thermal stability, and physical properties are even more so. It is best likened to a polydimethylsiloxane rubber. Both compounds are stable as polymers up to about 350°C, and are reversibly depolymerized at higher temperatures. The chains have the same resonance structures:

$$-\overset{|}{P}=N- \;\leftrightarrow\; -\overset{|}{P}\overset{+}{-}N\overset{-}{-}$$

$$-\overset{|}{Si}-O- \;\leftrightarrow\; -\overset{|}{Si}\overset{-}{=}O\overset{+}{-}$$

The difference is only that the phosphonitrilic compound is uncharged in the double-bonded state, while the silicone is uncharged in the single-bonded state. In both compounds there is probably resonance between the two forms. One might have predicted that the double bonds would stiffen the chain and make the polymer crystalline rather than rubbery, but this is not the case. It is interesting, however, that the phosphonitrilic chloride becomes crystalline on stretching, whereas the silicone does not. Despite the apparently conjugated double bonds, the compound is colorless.

Other Phosphonitrilic Compounds. A rubbery chlorofluoride, $P_4N_4Cl_2F_6$, has been made by the treatment of the chloride with lead fluoride; it is somewhat less stable than the chloride. While the chlorides are hardly attacked by water alone, when dissolved in a solvent they can be hydrolyzed stepwise to a phosphinic acid, $[-P(OH)_2-N-]_n$, or more likely, $[-P(O)OH-NH-]_n$. Likewise, the chlorines may be replaced by NH_2 or NHR on treatment with ammonia or amines, or by OR or SR on treatment with alcohols or mercaptans; it is not known whether the products have any useful properties. Treatment with phenylmagnesium bromide gives low yields of trimeric and tetrameric diphenylphosphinic nitride. A better synthesis is from diphenylphosphorus trichloride and ammonium chloride (54).   The low polymers have been converted to a high polymer, but its properties have not been described.   Other alkyl and aryl radicals have been introduced similarly.   It seems that further investigation of derivatives of the phosphonitrilic chlorides would be worthwhile.

The substitution of phosphorus by silicon in the P—N backbone of a polymer chain has been accomplished, producing silazanes (1) that are analogous to the phosphonitrilic compounds.   The polymers so far made are mostly cyclic trimers or tetramers, and are quite susceptible to hydrolysis.   By catalytic rearrangement, however, some waxy or rubbery products also have been isolated, with molecular weights greater than 10,000.   It is not clear whether the latter materials consist of linear chains, or are cyclolinear polymers consisting of hexagons of alternating Si and N atoms, with the rings linked by Si atoms between N atoms of adjacent rings.   Though somewhat less water-sensitive than the trimers and tetramers, they are rapidly hydrolyzed, melt at low temperatures, and are thermally decomposed at temperatures slightly above 250°C.

Phosphorus–Phosphorus Bonds.   One might expect linear polymers from a monosubstituted phosphorus such as phenylphosphorus, $(C_6H_5P)_n$, or phosphorus monochloride, $(PCl)_n$, but no such compounds have been found.   The only compounds with a phosphorus-phosphorus bond are P_2I_4, tetramethyldiphosphine, $(CH_3)_2PP(CH_3)_2$, tetraphenyldiphosphine, $(C_6H_5)_2PP(C_6H_5)_2$,   and   rather   poorly characterized compounds formulated as $C_6H_5P=PC_6H_5$, $C_6H_5P=POH$, and $C_6H_5P_4H$.   They are all very susceptible to oxidation and of little interest.

References

1. F. G. A. Stone and W. A. G. Graham, *Inorganic Polymers*, Academic Press, New York, 1962.
2. L. Pauling, *The Nature of the Chemical Bond*, 2d ed., Cornell University Press, Ithaca, 1940, p. 64.
3. W. Gordy and W. J. O. Thomas, *J. Chem. Phys.*, **24**, 439 (1956).
4. N. B. Hannay and C. P. Smyth, *J. Am. Chem. Soc.*, **68**, 171 (1946).
5. R. H. Wentorf, Jr., *J. Chem. Phys.*, **26**, 956 (1957).
6. A. F. Wells, *Structural Inorganic Chemistry*, 3d ed., Oxford University Press, 1962.

7. A. E. Martell and M. Calvin, *The Chemistry of the Metal Chelate Compounds*, Prentice-Hall, New York, 1952.
8. J. C. Bailar, Jr., *The Chemistry of the Coordination Compounds*, Reinhold, New York, 1956.
9. J. P. Wilkins and E. L. Wittbecker, U.S. Pat. 2,659,711 (Nov. 17, 1953).
10. (a) K. A. Jensen, *Z. Anorg. Allgem. Chem.*, **252**, 227 (1944); (b) P. Rây, *Z. Anal. Chem.*, **79**, 94 (1929).
11. R. H. Bailes and M. Calvin, *J. Am. Chem. Soc.*, **69**, 1886 (1947).
12. N. Schlesinger, *Chem. Ber.*, **58**, 1877 (1925).
13. K. V. Martin, *J. Am. Chem. Soc.*, **80**, 233 (1958).
14. R. Rigamonti and M. T. Cereti Mazza, *Ann. Chim. (Rome)*, **53**, (10), 1453 (1963); *Chem. Abstr.*, **60**, 7664 (1964).
15. G. I. Braz, *Zh. Obshch. Khim.*, **21**, 688 (1951).
16. D. D. Reynolds, *J. Am. Chem. Soc.*, **79**, 4951 (1957).
17. C. L. Rollinson, *The Chemistry of the Coordination Compounds*, J. C. Bailar, Jr., Ed., Reinhold, New York, 1956, Chap. 13.
18. A. Brunstad, *Hanford Works* 54203; *Chem. Abstr.*, **52**, 13506a (1958).
19. D. C. Bradley, "Polymeric Metal Alkoxides, Organometalloxanes, and Organometalloxosiloxanes," *Inorganic Polymers*, ref. 1, Chap. 7.
20. D. C. Bradley, A. K. Chatterjee, and W. Wardlaw, *J. Chem. Soc.*, **1956**, 2260.
21. N. Ya. Turova, A. V. Novoselova, and K. N. Semenenko, *Zh. Neorgan. Khim.*, **4**, 997 (1959); *Chem. Abstr.* **54**, 8397 (1960).
22. P. J. Wheatley, *J. Chem. Soc.*, **1960**, 4270.
23. D. C. Bradley, R. Gaze, and W. Wardlaw, *J. Chem. Soc.*, **1955**, 721, 3977.
24. F. B. Hauserman, *Metal-Organic Compounds*, Advances in Chemistry Series No. 23, American Chemical Society, Washington, D.C., 1959, p. 338.
25. B. P. Block, J. Simkin, and L. R. Ocone, *J. Am. Chem. Soc.*, **84**, 1749 (1962).
26. S. H. Rose and B. P. Block, *J. Polymer Sci.*, **(A-1)4**, 573, 583 (1966).
27. M. Okawara, Y. Takemoto, H. Kitaoka, E. Haruki, and E. Imoto, *Kogyo Kagaku Zasshi* **65**, 685 (1962); *Chem. Abstr.* **58**, 577 (1963).
28. E. G. Rochow, D. T. Hurd, and R. N. Lewis, *The Chemistry of Organometallic Compounds*, Wiley, New York, 1957, Chap. 10.
29. S. Papetti, B. B. Schaefer, A. P. Gray, and T. L. Heying, *J. Polymer Sci. A-1*, **4**, 1623 (1966).
30. A. B. Burg, *J. Am. Chem. Soc.*, **62**, 2228 (1940).
31. H. C. Mattraw, C. E. Erickson, and A. W. Laubengayer, *J. Am. Chem. Soc.*, **78**, 4901 (1956).
32. J. C. Perrine and R. N. Keller, *J. Am. Chem. Soc.*, **80**, 1823 (1958).
33. R. R. McGregor, *Silicones and Their Uses*, McGraw-Hill, New York, 1954.
34. For the name "borazine" see D. T. Haworth and L. F. Hohnstedt, *J. Am. Chem. Soc.*, **81**, 842 (1959).
35. W. L. Ruigh, *Congr. Intern. Chim. Pure Appl. 16ᵉ Paris*, **1957**, *Mem. Sect. Chim. Minerale* 1958; *Chem. Abstr.*, **54**, 10914 (1960).
36. A. B. Burg, Abstracts of Papers, 133rd Meeting, American Chemical Society, San Francisco, April 1958, p. 33L.
37. A. B. Burg and R. I. Wagner, *J. Am. Chem. Soc.*, **75**, 3872 (1953).

38. R. B. Booth and C. A. Kraus, *J. Am. Chem. Soc.*, **74**, 1415 (1952).
39. K. H. Meyer, *Natural and Synthetic High Polymers*, 2d ed., Interscience, New York, 1950.
40. J. M. Stevels, *Chem. Weekblad*, **44**, 608 (1948).
41. A. Van Valkenburg and R. G. Pike, *J. Res. Natl. Bur. Std.*, **48**, 360 (1952).
42. R. N. Meals and F. M. Lewis, *Silicones*, Reinhold, New York, 1959.
43. J. F. Brown, Jr., L. H. Vogt, Jr., A. Katchman, J. W. Eustance, K. M. Kiser, and K. W. Krantz, *J. Am. Chem. Soc.*, **82**, 6194 (1960).
44. A. Besson, *Compt. Rend.*, **125**, 771 (1897).
45. G. M. Phillips, J. S. Hunter, and L. E. Sutton, *J. Chem. Soc.*, **1945**, 146.
46. G. M. Kosolapoff, *Organophosphorus Compounds*, Wiley, New York, 1950.
47. H. W. Grummel, A. Guenther, and J. F. Morgan, *J. Am. Chem. Soc.*, **68**, 539 (1946).
48. A. Michaelis, *Ann. Chem.*, **326**, 129 (1903).
49. J. B. Dickey and H. W. Coover, U.S. Pat. 2,666,750 (Jan. 19, 1954).
50. V. V. Korshak, I. A. Gribova, and M. A. Andreeva, *Izv. Akad. Nauk SSSR Odt. Khim. Nauk*, **1957**, 631.
51. V. V. Korshak, *Angew. Chem.*, **70**, 85 (1958).
52. D. M. Yost and H. Russell, Jr., *Systematic Inorganic Chemistry of the Fifth-and-Sixth-Group Nonmetallic Elements*, Prentice-Hall, New York, 1946.
53. L. F. Audrieth, R. Steinman, and A. D. F. Toy, *Chem. Rev.*, **32**, 109 (1943).
54. C. P. Haber, D. L. Herring, and E. A. Lawton, *J. Am. Chem. Soc.*, **80**, 2116 (1958).

CHAPTER 6

MELT RHEOLOGY OF FILM-FORMING POLYMERS

HENRY J. KARAM

The Dow Chemical Company
Midland, Michigan

227

I. Introduction

A basic understanding of film technology requires some knowledge of rheological concepts. Rheology is defined as the science concerned with deformation of materials subject to an external stress. It is quite obvious from the definition that a knowledge of rheology is important, if we wish to understand the fabrication of film. We shall attempt in this chapter to introduce the different concepts of rheological behavior, and its measurement and significance in film technology.

II. Terminology

An understanding of the behavior of idealized materials is important if we wish to understand the rheological properties of polymeric materials. Properties of polymeric materials are defined in terms of deviation from ideal material behavior.

A. Hookean Solid

The properties of an ideal solid will be considered primarily because the molten polymer exhibits solid or elastic behavior as well as fluid or viscous behavior. An ideal solid is one which obeys Hooke's law (1,2), which states that

$$\sigma = E\delta \tag{1}$$

where σ is stress, or the force per unit area; δ is strain or the fractional change in length (it is a dimensionless quantity); and E is the modulus of elasticity of the material, a physical property of the solid. The modulus can be defined in terms of a tensile stress or shearing stress. The two moduli are related to each other, but not necessarily equal (1), and the units in the cgs system are dynes/cm^2 (see Chapter 14 for additional details).

Rheologists will sometimes use a spring to describe the behavior of an elastic body. A spring, like an elastic body, will respond linearly

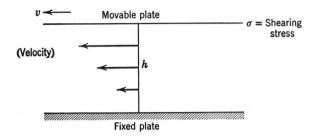

Fig. 1.   Graphic representation of laminar flow in a fluid.

with stress and will instantaneously recover when the stress is removed. The spring, because of its instantaneous recovery, is said to have perfect memory.

B. Newtonian Liquid (1,2)

Assume, as illustrated in Figure 1, that a liquid is confined between two parallel plates. When a shearing stress, σ, is applied to the upper plate, the latter does not move instantaneously to a new position. Instead, the top plate will move at a constant velocity, v, relative to the bottom fixed plate. Furthermore, when the stress is removed, the fluid does not recover its original shape. Since the shearing stress is transmitted uniformly through the liquid, each layer of fluid within the space, h, will slip relative to the next. There will be a uniform change of velocity with distance. Mathematically, the relationship between the shearing stress and the change of velocity with distance is the following:

$$\sigma = (v/h)\eta \qquad (2)$$

where σ is shearing stress in force/unit area; v is velocity of the top plate; h is the height or distance between plates; and η is the coefficient of viscosity of the fluid. Equation 2 is known as Newton's law for laminar flow of an ideal fluid.

The ratio of velocity to the distance between the plates is referred to as the shear rate, rate of strain, or shear gradient. A more rigorous definition of the shear rate is the change in velocity with change in distance, i.e., dv/dh.

Viscosity spectrum, Poises

10^{-4}	10^{-2}	10^{0}-10^{2}	10-10^{6}	10^{3}-10^{9}	10^{4}-10^{12} or greater
Gases	Water and alcohol	Oils, inks, paints, and varnishes	Fats and greases	Resins and gums	Pitches, asphalts, thermoplastics, and glass

Fig. 2. Spectrum of viscosities in fluids and plastic materials. Values are expressed as poises.

Newton's law for the deformation of a fluid is analogous to Hooke's law for deformation of a solid. The coefficient of viscosity is an elastic constant which describes a true physical property of a material. Furthermore, Newton's law for a fluid and Hooke's law for a solid are linear laws. This means that the stress varies in a linear manner with the strain in the case of a solid, or with shear gradient in the case of a fluid. Materials which obey Hooke's law or Newton's law are referred to as Hookean or Newtonian. Materials which respond to a stress in a nonlinear manner are non-Hookean or non-Newtonian.

The unit of viscosity in the cgs system is dyne-second/cm² or poise. Figure 2 shows the viscosity spectrum of materials which are encountered by film technologists.

A dashpot (1,2) is often used to depict mechanically the viscosity property. The dashpot consists of a piston moving in a container filled with a viscous Newtonian fluid. The rate of motion of the piston varies in a linear manner with the application of the force on the piston. Furthermore, if the force is suddenly removed, the piston has no tendency to return to its original position. The mechanistic model has all the features of a Newtonian liquid.

C. NON-NEWTONIAN LIQUID

Rheograms are used to describe deviations from ideal behavior. A rheogram is a plot of the shear rate as a function of the shear stress response of a material. How these curves are obtained will be discussed in the next section. Figure 3 shows rheograms of different materials. Newtonian materials are characterized by rheogram A of Figure 3. The inverse slope of the line is defined as the viscosity. Curve B is described as plastic or Bingham flow. It is characteristic of two phase systems such as a latex or plastisol.

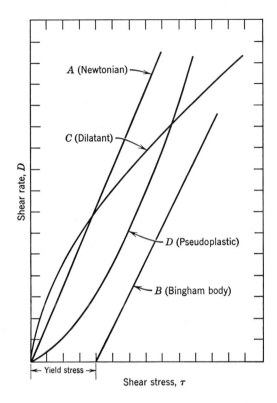

Fig. 3.   Typical flow curves of various types of deformable materials.

The classical equation of flow (Eq. 2), in which the shear stress is reduced by the yield stress, is used to describe plastic flow (1,3). The intercept on the x axis is defined as the yield stress. The inverse of the slope of the line is defined as the rigidity. Curve C is an example of a material which is dilatant. The viscosity of this class of materials increases with increasing shear. As far as we now know, polymer melts do not exhibit such rheological behavior. Curve D is defined as pseudoplastic flow, that is, the viscosity decreases with increasing shear. This type of flow is characteristic of high polymers. Much has been published in this area. Practically every issue of journals dealing with polymer technology contains data on this subject.

Some materials change in resistance to shear with time. A material which shows a decrease in viscosity with increasing shear and increasing time of application of that shear is defined as thixotropic (1,3,4). There are other materials which exhibit an increase of viscosity with increasing time of application of shear and they are defined as rheopetic (5). Polymer melts or solutions, as far as we know, fail to exhibit either time-dependent property, but latex printing inks and plastisols do exhibit such complex rheological behavior. When studying such problems as coating a film with latex, or printing films, one must consider such properties in the analysis of these problems.

The dependence of viscosity on shearing stress is not without a theoretical basis. The reaction rate theory of Eyring and Powell (6,7), the polymer entanglement theory of Bueche (8), Goodeve (9), Gillespie (10,11), Cross (12), and Spencer (13) all predict a decrease in viscosity at high shearing stress.

D. Orientation

1. Uniaxial

The term *orientation* is often used to describe a unit operation in film fabrication. It implies stretching a molten polymer as it is extruded from a film die, or cast from a polymer solution, and stretching a molten polymer over a mandrel or blowing a bubble (14). Orientation can be uniaxial as in the case of sheet extrusion of film or drawing of a filament. Here the polymer chains are aligned in one direction. This produces maximum strength in the direction of orientation. For example, unoriented nylon has a tensile strength of 8000 psi, but by proper orientation, this can be increased to over 80,000 psi. Although this type of orientation is desirable in a fiber, a uniaxially oriented film tends to crack and split along lines parallel to the direction of stretching (see Chapter 10).

2. Biaxial

Biaxial orientation (14), as the name implies, is orientation along two axes at right angles to one another. It should be emphasized that a balance of orientation is desirable for most purposes, but there are some applications in which a film of unbalanced orientation is preferable. Biaxial orientation can be achieved by blowing a bubble, or by use of a tentering frame (14).

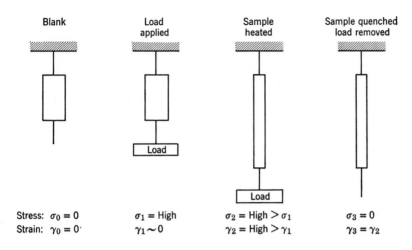

Blank	Load applied	Sample heated	Sample quenched load removed

Stress: $\sigma_0 = 0$ $\sigma_1 = $ High $\sigma_2 = $ High $> \sigma_1$ $\sigma_3 = 0$

Strain: $\gamma_0 = 0^{\cdot}$ $\gamma_1 \sim 0$ $\gamma_2 = $ High $> \gamma_1$ $\gamma_3 = \gamma_2$

Fig. 4.   Diagram illustrating stress–strain relationships in uniaxial orientation.

The polymer chains in the biaxially oriented case are in a web parallel to the plane of the film and not parallel to one another as in the case of oriented fibers. Mechanical properties such as tensile strength are not as high for biaxially oriented film as those for a uniaxially oriented fiber. As the degree of orientation in a biaxially oriented film becomes greater in one direction than in another, values of properties in the one direction increase at the expense of those in the other.

Orientation improves tensile properties, elongation, flexibility, toughness, chemical properties, weatherability, permeability, optical clarity, and increased shrinkability of polymer films.

E. Viscoelasticity

Before we discuss orientation on a molecular level, let us consider in a simple manner the deformation process which occurs when a polymeric material is stretched. We will restrict our discussion to a simple uniaxial case.

Suppose that a strip of rigid film is supported at one end (Fig. 4). At this stage, the stress and strain on the sample are zero. A large load is then applied to the free end of the film. In this case, the strain is nearly zero (the film has a high modulus) and the stress is high.   The film sample is heated above the softening point.   The film will immediately elongate, i.e., will undergo high strain.   The stress level will be much greater than during the previous stage since the cross-sectional area is greatly reduced.   Assume that we can quench the sample rapidly in this stage, and then remove the load.   The strain is now high when compared to its original length, and the external stress is zero.   If we reheat the sample, it will shrink.   A crosslinked polymer will recover, with time, its original dimension.   An amorphous or crystalline polymer will recover only partially.   The recovered portion of the deformation is referred to as the time-dependent elastic, or the high elastic deformation (1), and the irrecoverable deformation as the viscous deformation.

The theory of molecular orientation for polymers has been discussed in detail by many authors (1,8,15–17).   It will be briefly reviewed in a very elementary fashion here (cf. also Chapter 10).

The configurations of the polymer chains in the unoriented state are random.   When a constant stress is applied to a polymer, as in the case of hot stretching, three types of deformation take place. The total deformation, δ_T, is expressed by the following equation:

$$\delta_T = \delta_1 + \delta_2 + \delta_3 \tag{3}$$

Here δ_1 is the instantaneous elastic deformation.   It is caused by bond stretching or valence angle deformation.   It is independent of time and temperature of stretching.   This type of deformation is recoverable as soon as the stress is removed.

Also, δ_2 is the high elastic or time-dependent deformation associated with the uncoiling and orienting of molecular chains.   This is accomplished by rotation of molecular segments about carbon–carbon bonds.   The molecules have a tendency to align themselves in a direction parallel to the axis of stretch.   This orientation is frozen in when the film sample is quenched.   This deformation is sometimes referred to as *memory*.

Finally, δ_3 is a viscous deformation caused by molecules sliding past each other.   This deformation is not recovered when the stress is removed.

The mechanical behavior of a high polymer when sheared can be represented by the model shown in Figure 5. E_1 is the instantaneous elastic response modulus represented by a spring. E_2 and η_2 depict the retarded elastic response caused by uncoiling and ori-

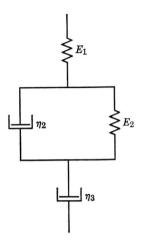

Fig. 5. Mechanical model representing viscoelasticity.

enting of the polymer chains.   In the mechanical model, it is represented by means of a spring whose elastic response is retarded by a
Newtonian dashpot.   The ratio

$$\eta_2/E_2 = \tau \tag{4}$$

is defined as the retardation or orientation time.   The flow viscosity of the polymer is designated η_3.   For simplicity, it is por-
trayed as a Newtonian dashpot.   Figure 6 indicates how the
mechanical model reproduces the deformation and recovery effects
just mentioned.

   In the preceding paragraph, a single orientation time was used to
represent the retarded elastic response of a polymer.   Actually, the
behavior can be best represented with a continuous distribution of
retardation time.   The model is replaced by the representation
shown in Figure 7.   The distribution of elastic properties is referred
to as the retardation time spectrum.   The distribution of elastic
properties is represented mathematically by a continuous function
$J(\tau)$.   This function $J(\tau)$ represents the amount of compliance that
has a retardation time, τ.   The short-time relaxation is probably
associated with unkinking of the molecular chains and the longer
time with uncoiling macromolecules.

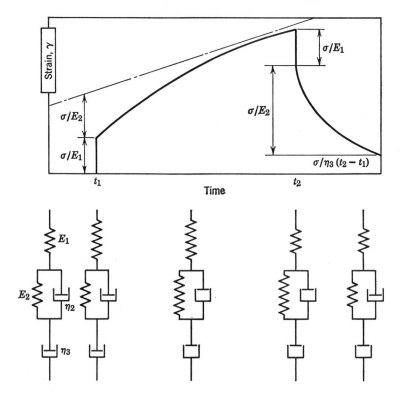

Fig. 6.   Mechanical model depicting deformation and recovery effects in terms
of springs and dashpots.

To represent the total deformation in terms of an orientation
time spectrum, Eq. (3) is rewritten in the following manner:

$$\delta_T = \frac{\sigma}{E_1} + \sigma \int_0^\infty J(\tau)[1 - e^{-t/\tau}] \, d\tau + \frac{\sigma}{\eta_3} t \qquad (5)$$

where σ is stress; t is time; η_3 is the apparent flow viscosity; E_1 is
the instantaneous elastic modulus; and τ is the retardation time.
    Also

$$\frac{1}{E_2} = \int_0^\infty J(\tau) d\tau \qquad (6)$$

where E_2 is the configurational or high-elastic modulus.

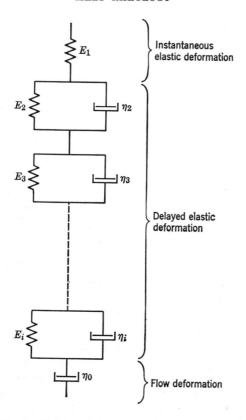

Fig. 7.   Voigt model representing viscoelasticity.

The portion and extent of the retardation time spectrum oriented is a function of fabrication conditions.   The resulting properties of the film are not only a function of the amount of orientation, but also the portion of the retardation time spectrum oriented.

In the previous discussion, we restricted the discussion to tension experiments.   Each spring in the model is characterized by a Young's modulus, instead of being characterized by shear modulus, and each dashpot in the model is represented by a tensile viscosity. If the material is nearly incompressible, which is the case with most molten high polymers, the relationship between tension modulus, E, and shear modulus, G, is a factor 3, i.e., $E = 3G$.   Similarly, the tension viscosity, η_3, is three times the shear viscosity.

The model discussed in the film stretching experiment is referred to as the linear Voigt model (1,8,15–17). For stress relaxation studies, a Maxwell model could be more convenient (1,8,15–17). The two are related. Details of the different models and their relationship are discussed elsewhere (1,8,15–18).

The foregoing discussion is based on the linear theory of viscoelasticity. In many cases of film fabrication, it is necessary to use a nonlinear theory (17,19,20), but these theories are beyond the scope of this chapter. We wish to emphasize that an understanding of linear theory is very useful for many problems encountered in film fabrication, and knowledge of the linear theory will provide one a better insight into the more involved nonlinear theory.

F. Birefringence

It is quite evident from the preceding discussion that orientation implies both magnitude and direction. The magnitude is dependent on the stress–time history, and the direction is dependent on the mode in which the stress is applied. The magnitude of the orientation can be obtained by measuring a parameter which is defined as birefringence.

The partial alignment of polymer chains during stretching in a uniaxial direction produces a difference between electronic polarizability in the direction of stress and that in the direction perpendicular to the stretch. This polarization anisotropy gives rise to a difference between the refractive indexes in these directions, which is by definition the birefringence. The axis in the direction of stretch is referred to as the principal optical axis.

Films are usually biaxially oriented. It is not exactly clear what is the best way to use the word birefringence in this case as compared with the uniaxial case. In biaxially oriented film, there are three principal values of the indexes of refraction, each associated with a principal optical axis. Two axes are in the plane of the film and the third is perpendicular to the plane. The birefringence in the plane of the film is usually small, and therefore the two birefringences are approximately equal. By definition, the birefringence in the case of biaxial orientation is the difference in the index of refraction along one axis in the plane of the film and that perpendicular to the plane.

The theoretical aspects of birefringence have been discussed by a number of authors (17,21,22). Muller (23) derived the following equation relating birefringence to structure from theoretical considerations.

$$\Delta n = -\pi \left(\frac{\bar{n}^2 + 2}{3\bar{n}} \right) a \frac{16}{15} N \left(\frac{\alpha_{11} + \alpha_{22}}{2} - \alpha_{33} \right) \frac{\Delta L}{L} \qquad (7)$$

In this equation, Δn is the birefringence value; $\bar{n}$ is the mean refractive index; α_{11}, α_{22}, and α_{33} represent polarizability of the monomer along the three principal directions, respectively; $\Delta L/L$ is the stretch ratio; N is the number of chains per unit volume; and a is a constant that characterizes the orientation of polymer chains.

The molecular mechanism that causes birefringence in an amorphous polymer such as polystyrene is different from the asymmetric molecular arrangement which causes birefringence in a crystal lattice. To distinguish between the two, the literature refers to them as orientation birefringence and intrinsic birefringence, respectively (24).

G. Crystallinity Parameters

Polymers are divided into two categories, those which are amorphous, and those which are partially crystalline. Polymers such as polystyrene are amorphous under all conditions. Partially crystalline polymers can be made amorphous under certain conditions.

Crystalline polymers are characterized by many parameters which are important from a film technology viewpoint. They include monomer crystal cell; size, shape, orientation, and aggregation of crystallites (15,25–27); crystalline induction time (28); crystallization rate (25,27,29); and crystalline melting point (30). The final three named are most important from a film fabrication point of view. The other parameters are important to the properties and end use applications of the film. It should be emphasized, however, that the orientation of the amorphous phase is, to a first approximation, more critical with respect to properties than is the orientation of the crystalline phase.

The *crystalline melting point* is defined as the temperature above which all crystalline order disappears; it is a function of pressure (25), and normally the crystalline melting point of polymers increases with pressure. For example, the increase in melting point for saran resin is 3.5°C per 1000 psi (31).

The increase in melting point with pressure is extremely important from a practical point of view. A heat-sensitive polymer such as saran, for example, should not be extruded at excessive back pressure. An increase in melting point of 7°C will double the degradation rate of saran (31).

A polymer which is heated above its crystalline melting point and quickly quenched remains amorphous. The time required to become again crystalline from the amorphous phase is defined as the *crystallization induction time* (28), and the rate at which it crystallizes is defined as the *crystallization rate* (29).

Factors which control crystallization induction time are many. The most important are molecular weight, molecular-weight distribution, copolymer composition, the addition of agents such as plasticizers, and rate of stretching of the polymer. When crystallization has started, several marked changes occur in the physical properties of the polymer. The most important are:

1. A gradual development of a typical x-ray diffraction pattern.

2. Increase in hardness.

3. Increase in force required to stretch a polymer.

4. Increase in density.

5. Evolution of heat, which is referred to as the heat of crystallization.

6. Variation in electrical properties.

7. Increase of resistance to solvents.

8. Change in gas permeability.

9. Change in optical properties.

H. Transition Temperature

The temperature at which a polymeric material changes from a glassy to a rubbery behavior is defined as the glass transition temperature, T_g. A polymer can exhibit various numbers of subgroup transitions below T_g. Boyer (30) extensively discusses the subject of transition temperatures and its implication regarding structure (see Section III-D).

III. Methods of Measuring Rheological Properties

A. Flow Properties

Rheometers to be used to measure rheological properties of high polymers must fulfill certain requirements, as follows.

1. Flow properties must be evaluated over a wide range of temperature, shear, and shear rate. This is important, since most polymers are non-Newtonian in nature; hence, measurements at one set of conditions give limited information.

2. The range of shear, shear rate, and temperature must be in the range in which the material is processed. This is important when one wishes to correlate processing information with rheological data.

3. Operation of the rheometer should be simple.

4. The data should be reproducible.

5. Samples must not degrade during the time interval over which the measurements are made.

Table 1 lists the various types of plastics rheometers that are used, their range, and a literature reference. The list is far from complete, but the most popular have been listed. For a more complete discussion of the subject, the reader should consult References 32 and 33.

TABLE 1
Summary of Types of Plastics Rheometers

Instrument	Range, poises	Reference
Capillary type		
Bingham high pressure	10^2–10^{12}	34
Caplastometer	10^2–10^6	35
Consistometer	10^2–10^6	36
Instron rheometer	10^2–10^{12}	37
Melt indexer	10^2–10^6	38, 39
Falling sphere type		
Falling coaxial cylinder	10^{-2}–10^5	40
Falling sphere	10^4–10^8	41
Rotational type		
Open-end rotating	10^3–10^9	42
Conicylinder	10^3–10^9	43
High-shear coaxial cylinder	10^3–10^9	44, 45
Cone and plate	10^{-2}–6×10^5	46
Coaxial cylinder	10^3–10^9	45
Compression or extension type		
Parallel plate	10^4–10^{12}	47
Penetration	10^4–10^{12}	48
Tensile	10^4–10^{12}	49

Capillary rheometers are ideal for control testing because of their simplicity, wide range, and capability of a high degree of precision. Falling sphere rheometers can be used for in-stream control, and are especially applicable where low shear viscosity is needed. Rotational viscometers can be used to study viscoelastic properties, yield stress of materials, and time–viscosity effects; they are useful in a wide range of viscosities. The compression or extension type of viscometer is useful for small samples, applicable over a wide temperature range, and can be used to study the viscoelastic properties of polymers; low shear viscosity data can be readily obtained.

B. Viscoelastic Properties

Much has been published in recent years on viscoelastic properties of polymers (1,8,15–17). Instruments and techniques that can be used to study viscoelastic properties include the following: rheogoniometer (50), viscoelastometer (51), coaxial cylinder viscometer (52), double transducer (16), vibrating plate (53), relaxation balance (16), creep experiment (8), and torsional experiments (16,54). The list, as in the case of viscometers, is far from complete. The reader is referred to the references cited for additional techniques.

C. Measurement of Orientation

There are three techniques to measure the level of orientation in oriented film: birefringence (17,55–57), orientation release stress (58,59), and shrinkage (55,60). In biaxial orientation, values are usually obtained in what is referred to as machine and transverse direction. Each of these tests measures a different aspect of orientation; hence, it is not necessary that all values from each of these tests correlate (cf. Chapters 10 and 14).

1. Birefringence

Birefringence as discussed previously can be measured by optical techniques. The most convenient way is to prepare a specimen in such a manner that it exhibits an interference fringe pattern when viewed between crossed Nicol prisms. By noting the order of fringes, the thickness of the specimen, and the wavelength of light being used, one can compute the birefringence directly. The technique for uniaxial orientation, such as in fibers, is discussed in detail elsewhere (55,57).

It is not necessary to have a fringe pattern to determine bire-fringence. In fact, for biaxially oriented film, the technique of tilting the film between crossed Nicol prisms is sometimes a more convenient way of determining birefringence. The tilting technique will also enable the investigator to determine the principal optical axis. Details of the technique are discussed by Spence (56).

Closely allied to the problem of birefringence is the study of orientation of crystallites, their size, shape, and orientation. X-ray and infrared dichroism are used for such details (26,61).

2. Orientation or Stress Release Test

This test (58,59) measures the shrinkage force of oriented film upon reheating. The test consists of restraining a sample between two clamps, and heating it to the desired temperature. The orien-tation release stress, calculated in pounds per square inch, is the maximum stress measured at the clamp. Many variations of the test are possible. It is often used as a quality control test.

3. Shrinkage

The test to determine shrinkage consists of measuring the change of dimensions of squares, circles, or strips of film which have been heated without restraint at various temperatures and times. The determination of shrinkage is often carried out between closely spaced plates to avoid curling.

Limited information can be obtained if one obtains shrinkage data at one set of conditions. The test, to be of use from a fundamental standpoint, should be conducted over a wide range of conditions. For example, by conducting a shrinkage test properly, one can deter-mine the portion of the retardation spectrum oriented and the amount of orientation for any one set of fabrication conditions. Reference 55 discusses details of such tests.

D. Crystalline Melting Point

The simplest technique for determining the crystalline melting point is by use of a petrographic microscope in conjunction with a hot stage. The sample is placed on the hot stage and slowly heated.

The temperature at which the observed spherulite melts out is defined as the crystalline melting point.

E. CRYSTALLINE INDUCTION TIME

Crystallization induction time is measured by placing a sample which is supercooled (below its melt temperature) between two crossed Nicol prisms. The time in which the light becomes transmitted is defined as the crystallization induction time. The experiment is repeated for various temperatures.

The crystallization rate is measured by a dilatometric technique (62), or by an infrared spectrophotometric technique (63).

F. TRANSITION TEMPERATURE

There are a number of techniques for studying transition temperatures, such as differential thermal analysis (64), dynamic mechanical testing (65) (sometimes referred to as mechanical spectroscopy), and nuclear magnetic resonance (66). There are numerous empirical techniques for studying this property, but the techniques noted are the most useful and fundamental.

G. DEGRADATION

Much has been published on the degradation of properties of polymeric materials. In film studies, tests do not have to be elaborate. Film technologists are usually concerned with the change of some physical property of the film as a function of time, for some environmental condition such as temperature or humidity. For example, the tests will consist merely of exposing the film or polymer to a time–temperature schedule and monitoring the property. More elaborate tests are needed if one is concerned with mechanisms of degradation. Degradation data coupled with rheological information are extremely valuable in the designing of extrusion dies, extruders, and other associated film fabrication equipment.

IV. Characterization of Rheological Properties

A. FLOW PROPERTIES

Flow properties of film-forming materials are pseudoplastic. At low shear stress they exhibit Newtonian flow. Similar behavior is observed at high shear stress and this is referred to as the secondary Newtonian region.

The theoretical equations describing the pseudoplastic behavior of polymers are not very useful for many problems encountered by film technologists. The so-called power law (2,67–69) has proved useful for many problems. The power law is the following:

$$\sigma = KD^n$$

where σ is shear stress; D is shear rate; and n is the flow behavior index. For a Newtonian material, $n = 1$. Generally n is a constant which has a value less than 1. K is defined as a consistency index. For a Newtonian material, K is equal to the viscosity. The constants, K and n, can be determined by plotting viscometric data (shear rate vs. shear stress) on log-log paper. The slope of the line is defined as n and the σ intercept at $D = 1$ is defined as K.

The power law is an empirical equation. The limitation of the equation is that it is dimensionally inconsistent and does not describe the flow behavior over a wide range. In spite of its limitations, the equation is extremely useful. References 67–69 discuss the application of the relationship to many engineering problems.

Flow viscosity of polymers is extremely sensitive to temperature. Crystalline polymers will exhibit a marked decrease in viscosity at the crystalline melting point. The change of viscosity with temperature is reported in terms of activation energy. The activation energy is proportional to the slope of the curve of the logarithm of viscosity as a function of the reciprocal of absolute temperature. The sensitivity of polymer viscosity to temperature stresses the importance of good temperature control in all film fabrication equipment.

Much has been published in recent years on how molecular parameters control flow viscosity of polymers. A good review of the subject is given in Reference 70. No further discussion will be given here since Chapter 8 treats these matters in detail.

The flow of polymer melts is considered to be laminar rather than turbulent. Physically, this means that the fluid velocity, at any point in the flowing mass, is in the flow direction (Fig. 1). At high shear rates there is evidence to show that the flow of the polymer melt becomes unstable. The polymer extruded from a film die may become wavy, bumpy, or even break up into fragments. Figure 8 shows a polystyrene parison extruded from a

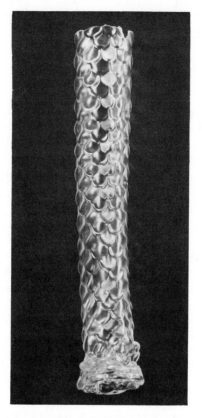

Fig. 8.   Melt fracture in polystyrene.

tubular film die. Note the unusual appearance of the surface. This phenomenon has been referred to as "melt fracture" and the stress at which it occurs is referred to as the critical stress. Melt fracture has been investigated thoroughly, but the causes are not yet well understood (71–77).

It is quite obvious that melt fracture is important from a practical standpoint, for the rate of extrusion at which it occurs represents the maximum rate of extrusion for any particular polymer. One of the most important considerations in film die design is to attain maximum rates without the occurrence of melt fracture.

The designer does not have any quantitative theory to guide him for proper design. This is largely the result of disagreement among rheologists as to the fundamental cause of this instability (78).

B. Viscoelastic Properties

Viscoelastic properties are usually characterized by evaluating the moduli of the elements of a viscoelastic model (1,8,15–17), in terms of normal stresses (79,80), in terms of the constants of a nonlinear theory (17,19,20), by the amount of swelling of an extrudate (13,68,81–83), or by end correction (84).

C. Characterization of Orientation

1. Birefringence

Birefringence is expressed as the difference in the indexes of refraction along the principal axes. Birefringence measurements provide a convenient index for measuring the extent of orientation. For example, Gurnee, by assuming a model for oriented polystyrene, calculated the maximum birefringence for uniaxial oriented polystyrene to be -0.3 and for biaxially oriented film to be -0.15 (57). Similar calculations have been done for rubber (17). The extent of orientation can be determined by comparing maximum birefringence and the measured value.

A knowledge of the three principal optical axes of an oriented film provides the investigator an insight into the way in which the film was fabricated. Related to this problem, it enables the investigator to determine if the stress applied was uniform when the film was being fabricated.

Birefringence enables one to determine quantitatively the homogeneity of orientation. Figure 9 illustrates this application, and represents the birefringent interference pattern of oriented polystyrene. It is quite evident that the skin is highly oriented compared to the core in Figure 9a, and the sample is uniformly oriented in Figure 9b and c.

The basic disadvantage of birefringence data is that they fail to indicate what portion of the retardation spectrum is being oriented. The second disadvantage is the inability to distinguish between the relative contribution of the birefringence from the crystalline and amorphous regions of the polymer. Both of these considerations

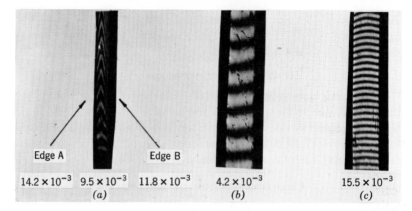

Edge A   Edge B

14.2 × 10⁻³ 9.5 × 10⁻³ 11.8 × 10⁻³   4.2 × 10⁻³   15.5 × 10⁻³
       (a)   (b)   (c)

Fig. 9. Birefringence patterns of biaxially oriented polytsyrene film: (a) Film that has been nonuniformly oriented; (b and c) film uniformly oriented, low and high birefringence, respectively.

are important if we wish to correlate physical properties of oriented films with birefringence measurements.

2. Stress Release

Stress release is reported in pounds per square inch. Sometimes data are reported in terms of a ratio of stress release in one direction to that perpendicular. An index of unity implies uniformly biaxially oriented film.

3. Shrinkage

Shrinkage data are given in terms of a curve or in terms of the ratio of the change in length to original length. Another convenient index is the ratio of shrinkage in the machine direction to that in the transverse direction. A shrinkage ratio of unity denotes uniform biaxial orientation.

D. CRYSTALLINITY PROPERTIES

1. Crystalline Melting Point and Glass Transition Temperature

Table 2 lists values of T_m, the crystalline melting point, and of T_g, the glass transition temperature (a second-order transition) for commercial film-forming polymers. The relation of these two values for a given film is directly related to the possibility of orienting a semicrystalline film, since in the process the film is heated to as low a temperature above the glass transition temperature as feasible and stretched. In the process, it is important that the film not approach closely the crystalline melting point, or orientation effects will be wiped out.

TABLE 2

Crystalline Melting Points and Glass Transition Temperatures

Polymer	T_g,°C	T_m,°C
Polyethylene		
Low-density	−25	98
High-density	−120	135
Polyoxymethylene	−50	185
Polyvinyl fluoride	−20	195
Polypropylene	−20–0	175
Saran	−17	175
Polyvinyl acetate	29	175
Polyhexamethylene adipamide	50	250
Polyethylene terephthalate	69	255
Polystyrene	100	100
Isotactic polystyrene	100	240
Polyvinyl chloride, unplasticized	87	212
Polyvinyl chloride, 15% plasticizer	60	
Cellulose acetate	70–120	160
Polymethyl methacrylate	105	160
Polytetrafluoroethylene	126	327
Polycarbonate	150	230

An alternate method of orienting a crystalline polymer is to heat it above the temperature of maximum crystallization after it has been oriented, with the film under constant restraint to prevent shrinkage during this operation. The film is then quenched to room temperature, and the restraint on the film removed. In both of these methods, the difference in temperature between the crystalline melting point and the glass transition point is crucial for successful commercial operations.

Theoretical discussion of the orientation of plastic films is beyond the scope of this chapter. A short discussion is included in Chapter 10. For details the reader should consult the published works on the subject (85–89).

2. Crystallization Induction Time

Data for this property will not be presented here, primarily because the crystalline induction time is a complex function of many variables. Anyone concerned with the theory of film manufacture and handling has to obtain data on his own particular formulation.

3. Other Crystallinity Properties

Other crystalline properties which are important for characterizing a polymer, such as size, shape, and number of monomer units in a unit crystal cell, and crystallization rates, should be determined for the particular polymer under consideration by standard methods described elsewhere (15,25–27,61).

Acknowledgment

The assistance of V. D. Floria in the preparation of this manuscript, and the permission of the Dow Chemical Company to publish are gratefully acknowledged.

References

1. T. Alfrey, Jr., *Mechanical Behavior of High Polymers*, Interscience, New York, 1948.
2. M. Reiner, *Deformation, Strain and Flow*, H. K. Lewis, London, 1960.
3. H. Green and R. N. Weltman, *Industrial Rheology and Rheological Structure*, Wiley, New York, 1949.
4. H. Green and R. N. Weltman, "Thixotropy," in *Colloid Chemistry*, Vol. 6, Reinhold, New York, 1946, p. 328.
5. H. Freundlich and F. Julesberger, *Trans. Faraday Soc.*, **31**, 920 (1935).
6. R. E. Powell and H. Eyring, *Nature*, **154**, 427 (1944).
7. T. Ree and H. Eyring, *J. Appl. Phys.*, **26**, 793 (1955).
8. F. Bueche, *Physical Properties of Polymers*, Interscience, New York, 1962.
9. C. F. Goodeve, *Trans. Faraday Soc.*, **35**, 342 (1939).
10. T. Gillespie, *J. Colloid Sci.*, **15**, 219 (1960).
11. T. Gillespie, *Trans. Soc. Rheol.*, **9**, 235 (1965).
12. M. M. Cross, *J. Colloid Sci.*, **20**, 417 (1966).
13. R. S. Spencer, *J. Polymer Sci.*, **5**, 591 (1950).
14. W. R. R. Park and J. Conrad, in *Encyclopedia of Polymer Science and Technology*, Vol. 2, Interscience, New York, 1965, p. 339.
15. A. V. Tobolsky, *Properties and Structure of Polymers*, Wiley, New York, 1960.
16. J. D. Ferry, *Viscoelastic Properties of Polymers*, Wiley, New York, 1961.
17. L. R. G. Treloar, *The Physics of Rubber Elasticity*, Oxford, London, 1958.
18. R. A. Wessling, *J. Polymer Sci.*, **A2**, 2001 (1964).

19. R. S. Rivlin, "Large Elastic Deformations," in *Rheology Theory and Applications*, Vol. 1, F. R. Eirich, Ed., Academic Press, New York, 1956, p. 351.
20. A. E. Green and J. E. Adkins, *Large Elastic Deformations*, Oxford, London, 1960.
21. W. Kuhn and F. Grun, *Kolloid-Z.*, **101**, 248 (1942).
22. E. F. Gurnee, *J. Appl. Phys.*, **25**, 1232 (1954).
23. F. H. Muller, *Kolloid-Z.*, **98**, 138, 306 (1941).
24. R. Houwink, *Elastomers and Plastomers, Part 1, General Theory*, Elsevier, New York, 1950.
25. A. Sharples, *Introduction to Polymer Crystallization*, Edward Arnold, London, 1966.
26. R. S. Stein, "Optical Methods of Characterizing High Polymers," in *Newer Methods of Characterizing High Polymers*, B. Ke, Ed., Interscience, New York, 1964, p. 156.
27. L. Mandelkern, *Crystallization of Polymers*, McGraw-Hill, New York, 1964.
28. J. Jack, *Brit. Plastics*, **34**, 312, 391 (1961).
29. M. Avrami, *J. Chem. Phys.*, **7**, 1103 (1939); **8**, 212 (1940).
30. R. F. Boyer, *Rubber Reviews for 1963*, **36**, No. 5 (Dec. 1963).
31. R. M. Wiley, unpublished data, Dow Chemical Company.
32. J. R. Van Wazer, J. W. Lyons, K. Y. Kim, and R. E. Colwell, *Viscosity and Flow Measurements. A Laboratory Handbook of Rheology*, Interscience, New York, 1963.
33. H. J. Karam, *Ind. Eng. Chem.*, **55**, 38 (1963).
34. H. K. Nason, *J. Appl. Phys.*, **16**, 338 (1945).
35. H. J. Karam, K. J. Cleereman, and J. L. Williams, *Mod. Plastics*, **32**, No. 7, 129 (March, 1955).
36. S. A. McKee and H. S. White, *J. Res. Natl. Bur. Std.*, **A46**, 18 (1951).
37. E. H. Merz and R. E. Colwell, *ASTM Bull.*, **232**, 63 (Sept., 1958).
38. J. P. Tordella and R. E. Jolly, *Mod. Plastics*, **31**, No. 2, 146 (Oct. 1953).
39. *1966 Standards*, Method D 1238-65T, "Measuring Flow Rates of Thermoplastics by Extrusion Plastometer," Am. Soc. Testing Materials, Philadelphia, Part 27, p. 455.
40. R. N. Traxler, J. W. Romberg, and H. E. Schweyer, *Ind. Eng. Chem. Anal. Ed.*, **14**, 340 (1942).
41. L. R. Bacon, *J. Franklin Inst.*, **221**, 251 (1936).
42. T. F. Ford and K. G. Arbran, *ASTM Proc.*, **40**, 1174 (1940).
43. M. Mooney and R. H. Ewart, *Physics*, **5**, 350 (1934).
44. E. W. Merrill, Symposium on Non-Newtonian Viscosity, ASTM Special Tech. Publ. No. 299.
45. H. Green and R. Weltman, *Ind. Eng. Chem. Anal. Ed.*, **15**, 201 (1943).
46. R. McKennell, *Proc. 2nd Intern. Congr. Rheol.*, p. 350 (July, 1953).
47. G. Dienes and H. F. Klemm, *J. Appl. Phys.*, **17**, 458 (1946).
48. P. A. Small, *J. Polymer Sci.*, **28**, 223 (1958).
49. H. J. Karam and J. C. Bellinger, *Trans. Soc. Rheol.*, **8**, 61 (1964).
50. W. F. O. Pollet and A. H. Cross, *J. Sci. Instr.*, **27**, 209 (1950).

51. R. J. Overberg and H. Leaderman, *J. Res. Natl. Bur. Std.*, **C65**, 9 (1961).
52. F. D. Dexter, *J. Appl. Phys.*, **25**, 1124 (1954).
53. J. G. Woodward, *J. Colloid Sci.*, **6**, 481 (1951).
54. D. J. Plazek, *J. Phys. Chem.*, **69**, 3840 (1965).
55. K. J. Cleereman, H. J. Karam, and J. L. Williams, *Mod. Plastics*, **30**, No. 9, 119 (May 1953).
56. J. Spence, *J. Phys. Chem.*, **43**, 865 (1939); **45**, 401 (1941).
57. E. F. Gurnee, *J. Appl. Phys.*, **25**, 1232 (1954).
58. C. P. Fortner, *Soc. Plastics Engrs. J.*, **9**, No. 5, 21 (1953).
59. *1966 Standards*, Method D 1504-61, Am. Society for Testing and Materials, Philadelphia, Part 26, p. 115.
60. R. D. Andrews, *J. Appl. Phys.*, **26**, 1061 (1955).
61. S. L. Aggarwal, G. P. Tilley, and O. J. Sweeting, *J. Appl. Polymer Sci.*, **1**, 91 (1959).
62. S. Newman, *J. Polymer Sci.*, **47**, 111 (1960).
63. W. H. Cobbs and R. L. Burton, *J. Polymer Sci.*, **10**, 275 (1953).
64. B. Ke, "Differential Thermal Analysis," in *Newer Methods of Polymeric Characterization*, B. Ke, Ed., Interscience, New York, 1964, p. 347.
65. L. E. Neilsen, *Mechanical Properties of Polymers*, Reinhold, New York, 1962, Chapter 7.
66. J. G. Powles, *Polymer*, **1**, 219 (1960).
67. E. C. Bernhardt, *Processing of Thermoplastic Materials*, Reinhold, New York, 1959.
68. J. M. McKelvey, *Polymer Processing*, Wiley, New York, 1962.
69. W. L. Wilkinson, *Non-Newtonian Fluids*, Pergamon Press, London, 1960.
70. T. G Fox, S. L. Gratch, and S. Loshaek, "Viscosity Relationships for Polymers in Bulk and in Concentrated Solutions," in *Rheology, Theory and Applications*, Vol. 1, F. R. Eirich, Ed., Academic Press, New York, 1956, p. 431.
71. J. J. Benbow, R. V. Charley, and P. Lamb, *Nature*, **192**, 233 (1961).
72. J. J. Benbow, R. N. Brown, and E. R. Howells, *Colloq. Intern. Rheol.*, Paris, 1960.
73. P. L. Clegg, *Trans. Plast. Inst.*, **26**, 151 (1958).
74. J. P. Tordella, *Soc. Plastics Engrs. J.*, **12**, 36 (Feb. 1956).
75. J. P. Tordella, *J. Appl. Phys.*, **27**, 454 (1956).
76. J. P. Tordella, *Rheol. Acta*, **1**, No. 2–3, 216 (1958).
77. R. S. Spencer, *Proc. 2nd Intern. Congr. Rheol.*, Butterworth, London, 1954, p. 20.
78. A. B. Metzner, *Ind. Eng. Chem.*, **50**, 1577 (1958).
79. H. Markovitz, *Trans. Soc. Rheol.*, **1**, 37 (1957).
80. A. Jobling and J. E. Roberts, "Goniometry of Flow and Rupture," in *Rheology, Theory and Applications*, Vol. 2, F. R. Eirich, Ed., Academic Press, New York, 1958, p. 503.
81. D. L. McIntosh, "Elastic Effects in the Extrusion of Polymer Solutions," Ph.D. thesis, Washington University, St. Louis, 1960; *Dissertation Abstr.*, **21**, 1493 (1961).

82. A. B. Metzner, E. K. Carley, and I. K. Park, *Mod. Plastics*, **37**, No. 11, 133 (July 1960).
83. R. S. Spencer and R. E. Dillion, *J. Colloid Sci.*, **3**, 163 (1948).
84. T. Arai and T. Hiroshi, *Trans. Soc. Rheol.*, **7**, 333 (1963).
85. R. A. Wessling and T. Alfrey, Jr., *Trans. Soc. Rheol.*, **8**, 85 (1964).
86. R. A. Wessling, *Trans. Soc. Rheol.*, **9.1**, 95 (1965).
87. M. O. Longstreth and T. Alfrey, Jr., *Am. Soc. Mech. Engrs. Trans.*, **82B**, 167 (1960).
88. T. Alfrey, Jr., and E. F. Gurnee, "Dynamics of Viscoelastic Behavior," in *Rheology, Theory and Applications*, Vol. 1, F. R. Eirich, Ed., Academic Press, New York, 1956, p. 387.
89. T. Alfrey, Jr., *Soc. Plastics Engrs., J.*, **5**, No. 2, 68 (April 1965).

CHAPTER 7

CHARACTERIZATION OF DEFORMATION IN POLYCRYSTALLINE POLYMER FILMS

ROBERT J. SAMUELS

Research Center, Hercules Incorporated, Wilmington, Delaware

I. Introduction

A knowledge of the morphology of deformed polycrystalline polymers is an essential prerequisite for the development of any theoretical interpretations of deformation mechanisms. The purpose of this chapter is to demonstrate how the morphology of deformed polycrystalline polymers can be characterized in a unified, comprehensive fashion. A model is developed to describe the polycrystalline state and is subsequently used to characterize the deformation behavior of the polymer. The theoretical foundations of the various independent experimental techniques (e.g., x-ray diffraction, birefringence, sonic modulus, etc.) used to characterize polycrystalline polymer morphology are then developed. Emphasis is placed on the interdependence of the information obtained from these independent experimental techniques as a consequence of the model used to describe the system. The emphasis on interdependence of derived information leads naturally toward new insights into both the molecular and larger morphological characteristics of deformed polycrystalline polymers.

A polycrystalline polymer film can be characterized according to phenomena observed at different levels of structure in the sample, i.e., the gross film characteristics, the characteristics of the spherulitic matrix of the film, or the molecular characteristics of the spherulite substructure. When the film is deformed, changes occur at all of these morphological levels. As stress is applied to the film, a gross sample deformation occurs. This, in turn, results in spherulite deformation, which consequently leads to a reordering of the crystalline and noncrystalline regions of the polymer. All of these morphological changes must be examined on the same samples if any attempt is to be made to deduce the pattern of the deformation process.

The approach developed in this chapter is similar to stop-action photography. A set of deformed isotactic polypropylene films will be used almost exclusively throughout the chapter as the model deformation system. These samples have undergone the simplest deformation process attainable: a continuous, uniform, uniaxial deformation of the polycrystalline film with a corresponding affine deformation of the spherulites. Each of the samples has been deformed to a different extension, however, so that together they correspond to different still sequences of a continuous drawing process. The films, after deformation, were well aged to insure that no changes would occur in the crystal content of the samples over the period in which the large number of measurements were taken. The assumption has been made that, to gain some insight into the complex interactions that occur as a consequence of polycrystalline film deformation, comprehensive examination of the simplest deformation process is potentially most lucrative.

Thus, this chapter is a review of morphological methods for characterizing deformation mechanisms in polycrystalline polymers arranged in an order which stresses the interdependence of these methods. The examples are designed not only to illustrate the application of each morphological technique but also to demonstrate the diagnostic advantages of examining one material with many different morphological tools. The chapter is not a comprehensive review of work that has been done in the area of polymer morphology but instead emphasizes a *modus operandi* for morphological investigations.

II. Molecular Orientation

A. INTRODUCTION

The term *polycrystalline polymer* indicates that the polymer molecules are assumed to be arranged in crystalline and noncrystalline phases. The number and nature of the phases present in the polymer are subjects of active investigation and speculation at present. Certainly, the molecules will be ordered differently in each phase and the observed properties of the polymer will be a combination of the individual properties of the molecules in each phase. The following survey is a comprehensive interpretation of the nature of the phases in a polycrystalline polyolefin polymer in light of existing evidence.

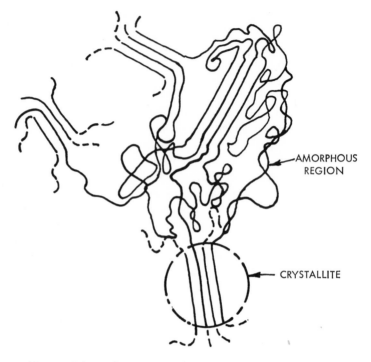

AMORPHOUS
REGION

CRYSTALLITE

Fig. 1.　Schematic representation of the fringed micelle model.

An early two-phase model of the structure of a polycrystalline polymer was the fringed micelle model. This is illustrated schematically in Figure 1. The polymer is assumed to be composed of crystalline and amorphous regions. If the path of a long polymer molecule is followed through this system, some portions of the molecule will be found in the crystalline regions, while other portions wander randomly between the crystallites and contribute to the amorphous region. The crystallites are assumed to form when sections of several polymer molecules become arranged in an ordered array and spontaneously crystallize to form an intermolecular crystal. Since portions of each molecule contribute to several crystallites, the whole polymer can be considered as an interconnected network of crystalline and amorphous regions.

In recent years, it has been observed that many polymers can be obtained from solution in the form of single crystals and that the

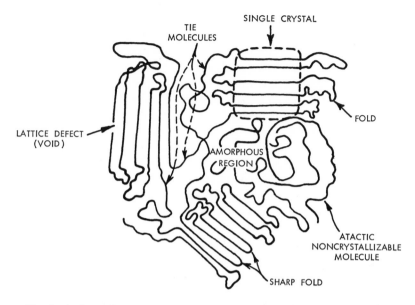

Fig. 2.   Schematic representation of the single crystal–amorphous model.

molecules in these crystals are folded back and forth in an ordered array (1,2,3).   There is growing evidence that the crystallites in the bulk polymer may also be in the form of chain-folded, single crystals (4,5,6).   Thus, the fringed micelle model is slowly evolving into a chain-folded, crystal–amorphous region model (Fig. 2) in which the individual crystallite may be formed from molecules folded back on themselves.   The individual single crystal lamellae are believed to be connected to each other through tie molecules so that they will respond cooperatively to a deformation.   The amorphous region may consist of the following: molecules whose complex tacticity prevents their crystallization, molecules excluded from the crystal due to their molecular weight difference, portions of molecules whose complex tacticity prohibits their inclusion in the folded crystals, the disordered fold regions on the surface of the crystal, and tie molecules which meander randomly before participating in another chain-folded crystal.

The polycrystalline phase model becomes more uncertain when the deformed polymer is considered.   When a polymer sample is deformed, the applied stress is distributed differently to the molecules in the different phases. Some of the crystals will disrupt, some melt, others tilt and reorient, while some of the noncrystalline region will orient and possibly crystallize. It is in this area of oriented polymers that the fringed micelle model is often found to persist in some form. The reason for its persistence is that it is a reasonable model with which to consider the crystallization process in highly oriented polymer systems. In such a system, the molecules are visualized as oriented in the deformation direction with a high degree of intermolecular order. This should lead to an increased probability of an ordered bundle of molecules spontaneously crystallizing into an intermolecular crystal whose chain axis is oriented in the deformation direction. In support of this model, x-ray diffraction studies have shown that the polymer chain axis in the crystallites is oriented in the deformation direction in highly oriented fibers and films.

Recent studies have thrown some doubt into this last stronghold of the fringed micelle or intermolecular crystal model. Keller (7) has obtained electron photomicrographs of highly oriented polyethylene fibers in which he observes lamella-like structures whose chain axis is oriented in the fiber axis direction, whereas the long axis of the lamella is perpendicular to the fiber axis direction. Geil, Peterlin, and Kiho (8) have obtained similar results from annealed, fully drawn fibers which formed across cracks in drawn, chain-folded, single crystals. Kiho, Peterlin, and Geil (9) have demonstrated that twinning and phase changes occur when single crystals of polyethylene are deformed, and suggest that molecular tilting and slip which is parallel to the chain axis contribute to the deformation mechanism. Hansen and Rusnock (10) have proposed that the controlling mechanism in the cold drawing of polymers is one of crystallographic slip. They suggest that slip can occur in an ($hk0$) crystal plane only in steps of unit c-axis spacings. A slip mechanism of this type is analogous to the collapsing of a set of playing cards set on edge and would result in an orientation of the molecular axis in the direction of the applied stress and an increase in the length of the fibril. The number of permitted slips, and hence the natural draw ratio, is determined by the angular position of the slip planes, slip becoming virtually impossible as the planes become nearly parallel to the draw direction. Thus, Hansen and Rusnock calculated that for a (100)[001] slip in nylon 66, a slip of one unit cell spacing in the chain axis direction would cause the nylon to elongate to a draw ratio of 4.0, a value close to the observed natural draw ratio, with the molecules becoming oriented within 14° of the draw direction.

Keith and Passaglia (11) considered the mechanism by which crystal dislocations can contribute to the orientation process of chain-folded single crystals and decided that twinning would be the most likely mechanism. They also concluded that appreciable deformation of chain-folded polymer crystals cannot be attributed solely to dislocation mechanisms but that molecular rearrangements other than slip must be involved. Instead, very large deformations, such as occur in the drawing of fibers, probably involve complex catastrophic processes unless the stress is applied normal to the chain axis of the crystal, in which case, a simple unzipping of the folded chains to give a fine fibrillar texture would be most likely to occur.

The conclusions of Keith and Passaglia (11) are in agreement with the observations of Kiho, Peterlin, and Geil (9), but not necessarily with the proposal of Hansen and Rusnock (10) because their theory requires a fairly large slip mechanism. Certainly, the nature of the morphology of crystallites in drawn polymers is currently an open question. Chain-folded, single crystal lamellae can exist in undeformed bulk polymer and oriented annealed polymer; however, this does not at present eliminate the possibility of intermolecular crystals existing as well. Therefore, it must be concluded that the nature of the structure of the crystallites in deformed polymers is uncertain, but that there is an increasing inclination toward the chain-folded, single crystal lamella model.

The chain-folded, single crystal lamella–noncrystalline region model of a polycrystalline polymer is still a two-phase system of distinct crystalline and amorphous regions. The amorphous region is characterized in an x-ray diffraction pattern by its diffuse scattering halo as differentiated from the discrete Bragg reflections from the crystalline region. Bonart, Hosemann, Motzkus, and Ruck (12) have pointed out that diffuse scattering need not come only from amorphous polymer but can also arise from imperfections in the crystal lattice. This important observation has led to new insights into the nature of the defect structure of crystalline polymers.

All crystal structures when examined by x-ray diffraction at room temperature have nonideal crystal lattices. Because of this nonideal structure, the observed x-ray diffraction exhibits a decrease in intensity of the crystal Bragg reflections and an increase in the amount of diffuse scatter. As long as the long-range order in the crystal lattice is conserved, only lattice imperfections of the first kind will contribute. The diffuse scatter of the first kind will be due to deviations in the lattice caused by thermal motion of the atoms in the crystal, frozen displacement or strains, vacancies, dislocations, and other lattice imperfections which cause only short-range disturbances. The general characteristic of distortions of the first kind is that the integral intensity of the Bragg reflections decreases with increasing scattering angle but the reflection widths do not change with scattering angle.

A two-phase system should consist of a well-ordered crystalline region and gradual transitions from the well-ordered crystal to the noncrystalline region should not exist. Regions of intermediate order, "paracrystal," in which long-range order in the crystal is destroyed are characterized as lattice distortions of the second kind. X-ray diffraction from a paracrystalline lattice is characterized not only by a decrease in the integral intensity of the Bragg reflection with increasing scattering angle but by an increase in the reflection width with increasing scattering angle as well. Lattice distortions of the first and second kinds thus manifest different x-ray diffraction behavior. If a polycrystalline polymer is to be considered as a two-phase system of crystalline and amorphous regions, it must first be established that the polymer does not have an appreciable paracrystalline character.

Ruland (13) has demonstrated how x-ray diffraction analysis can differentiate between polymers which can be considered as two-phase systems of crystalline and noncrystalline regions and those which have distinct paracrystalline character. The effect of lattice distortions of all kinds (thermal motion, first kind, and second kind) can be examined in terms of the behavior of a disorder function, D, expressed as

$$D = \exp\left(-ks^2\right) \qquad (1)$$

where s is $2 \sin \theta / \lambda$, λ is the wavelength of the x-radiation, θ is the scattering angle, and k characterizes the disorder and may be represented as

$$k = 0.5B + 0.7a \tag{2}$$

where B is a spherically averaged temperature factor taking into account lattice imperfections of the first kind, and a is a factor determined by lattice imperfections of the second kind. For a two-phase system of crystalline and noncrystalline regions, only short-range lattice imperfections would be allowed and therefore $k = 0.5B$ for this condition. Thus, Ruland found that isotactic polypropylene must be characterized as a two-phase system (14) since $k = 0.5B$, irrespective of the thermal treatment of the polymer (i.e., the same value of k was found whether the sample was heated to the melting point and then quenched in water at room temperature, or subsequently annealed for 0.5 hr at 160°C, or for 1 hr at 105°C, or if a highly atactic sample was examined). Nylon 6 and 7 on the other hand (13), had varying values of k ($k \neq 0.5B$), depending on the thermal treatment, and these materials had regions of paracrystalline disorder. The boundaries between crystalline and noncrystalline regions in these polymers thus are not well defined. Even this condition does not mean that separate domains do not exist, but that their boundaries are not sharply characterized.

Thus, a two-phase, crystal–amorphous model of a polycrystalline polymer is a realistic one but must be treated with caution, unless a comprehensive analysis of the phase character of the material has been obtained with Ruland's method. Once the two-phase character of the polymer is known with certainty, then the usual methods of x-ray diffraction, density, infrared, etc., can be utilized to determine the fraction of each phase present in other samples of the polymer.

For the purposes of this chapter, the polymer to be studied will be assumed to have a two-phase character. A model such as this seems reasonable even for the extreme case of nylon 6 and 7 because separate domains probably exist even though their boundaries are not sharply characterized. The model polymer that will be used in all of the examples will be isotactic polypropylene which has been shown by Ruland to be a two-phase system.

The reason for stressing the two-phase system is that this model is an excellent one for representing the change in properties of a polycrystalline polymer with deformation. Starting with a two-phase model, the observed density, birefringence, sonic modulus, infrared dichroism, and x-ray diffraction data can all be interrelated and cross correlated. The informational agreement between the results of all of these independent physical measurements on a solid polymer is strong support for the validity of the model. Similarly, molecular information obtained with this model agrees with information obtained from other independent methods such as flow birefringence in dilute solution. Finally, no model has been developed which is practically as useful as the two-phase model for interpretation, on the molecular level, of the deformation behavior of films and fibers.

How is the two-phase model characterized? What are the most important parameters with which to describe the changes in the observed properties of the polycrystalline polymer when it is deformed? These questions were answered by Hermans nearly 20 years ago (15). Each phase of the polymer is assumed to have intrinsic properties of its own which are the same as those the polymer phase would have if it existed as a unique, perfectly uniaxially oriented entity. Thus, the crystals will have an intrinsic Young's modulus in the longitudinal and transverse directions, an intrinsic birefringence, characteristic infrared transition moments, and unique x-ray diffraction behavior. The amorphous region will have its own unique intrinsic properties as well. The observed properties of the polycrystalline polymer will be a result of the mixing of these unique phases. In an unoriented polymer, the phases are mixed at random and, hence, any difference in observed properties will be a result of the relative amounts of the two phases present.

$$P_{\text{unoriented}} = \beta P_c + (1 - \beta)P_{\text{am}} \qquad (3)$$

where P is the observed property of the polycrystalline polymer, P_c is the intrinsic property of the crystalline region, P_{am} is the intrinsic property of the amorphous region, and β and $(1 - \beta)$ are fractions of crystalline and amorphous material, respectively.

Since the intrinsic properties of the phases are defined for ideally oriented states, they are inherently anisotropic. For this reason, orienting these phases in a given direction due to some deformation mechanism will lead to a manifestation of the anisotropic character of the phases. Thus, a quantitative measure of the orientation of each phase will be required, whether the observed property of the oriented polymer is a direct measure of the average anisotropy or is a difference measurement of a given property of the oriented polymer relative to the unoriented state. Any observed anisotropic property of the oriented polycrystalline polymer, $\Delta P_{\text{oriented}}$, will, therefore, be a function of the properties of each of the phases, and may be expressed as

$$\Delta P_{\text{oriented}} = \beta P_c f_c + (1 - \beta) P_{\text{am}} f_{\text{am}} \qquad (4)$$

where f_c and f_{am} are the orientation functions characterizing the average orientation of the crystalline and amorphous phases, respectively. For example, birefringence, which is an anisotropic property (difference between refractive indexes), can be described by the equation

$$\Delta_T = \beta \Delta_c^0 f_c + (1 - \beta) \Delta_{\text{am}}^0 f_{\text{am}}$$

where Δ_T is the measured birefringence of the oriented polymer and Δ_c^0 and Δ_{am}^0 are the intrinsic birefringence of the crystalline and amorphous regions, respectively. Analogously, the difference between the sonic modulus of an oriented and unoriented sample (ΔE^{-1}) can be described by an equation having the same form:

$$\tfrac{3}{2}(\Delta E^{-1}) = \beta f_c / E_{t,c}^0 + (1 - \beta) f_{\text{am}} / E_{t,\text{am}}^0$$

where $E_{t,c}^0$ and $E_{t,\text{am}}^0$ are the intrinsic sonic lateral moduli of the crystalline and amorphous regions, respectively.

Thus, the most important parameters with which to characterize an anisotropic property of an oriented, polycrystalline polymer film are the fraction of each phase present, the intrinsic property of each phase, and the orientation function of each phase. In general, the fraction of each phase present in the polymer is determined by a density or x-ray diffraction measurement. For the density determination of the fraction of each phase present, the density of the crystalline and amorphous phases must be known individually. The crystal density d_c can be calculated from the known unit cell structure and the unit cell monomer molecular weight, while the amorphous region density d_{am} is obtained from measurements on the pure amorphous polymer when it is available, or by extrapolation of the density–temperature curve of the melted polymer. The fraction of crystals, β, present in the polymer sample is then given by the relation:

$$\beta = d_c(d - d_{\text{am}}) / d(d_c - d_{\text{am}}) \qquad (5)$$

where d is the measured density of the sample. The density of the polymer sample may be determined by any of a number of standard experimental methods (16) including pycnometry, hydrostatic weighing, and use of a density gradient tube.

The other common method for determining the fraction of crystals present in a polycrystalline polymer sample is x-ray diffraction. This general method utilizes either the intensity relations between crystalline peaks and the amorphous background or the absolute intensity of one of them to determine the amount of crystalline and amorphous material present in the sample (17,18). The x-ray method of crystallinity determination developed by Ruland (13,14) accounts for the effects of lattice distortions of the first and second kind through the use of the disorder function, as was discussed above, and this is the most reliable x-ray crystallinity method available at the present time.

The infrared absorption of a polymer can be used to obtain a measure of the fraction of crystals in the sample (19). If separate infrared absorption bands due only to the crystalline and amorphous regions can be identified, the infrared method can be used to obtain absolute values of the fraction of crystals. If, as is more common, absorption bands characteristic exclusively of the crystalline region, but none characteristic exclusively of the amorphous region are present, the infrared method can be used as a relative measure of the fraction of crystals which then must be calibrated by an absolute method such as density or x-ray diffraction. Since infrared measurements are fast, simple, and sensitive to small changes in crystallinity, they are useful for determining the fraction of crystals even as a relative method.

Attempts have been made to use nuclear magnetic resonance (NMR) techniques to determine the fraction of crystals in a polymer (20). The method has been found to correlate with density and x-ray diffraction measurements for a number of polymers, but only over limited temperature regions. The method is sensitive to the motion of molecules and, hence, restraints imposed by orientation will also influence the observed results. These complicating features of the method, added to the difficulty of obtaining data, make NMR an impractical method for phase concentration determinations at present.

The major experimental problem in the characterization of the morphology of deformed polycrystalline polymer films has not been the determination of the fraction of each phase present in the two-phase system because some methods for this characterization have been available for many years.   Instead, the main problem has been the determination of the intrinsic properties and the orientation function of each phase of the polymer.   The general two-phase approach to this problem suggested by Hermans (15) lay dormant for some time, but growing interest in recent years has given new impetus to this area of investigation.   Quantitative measurements of the intrinsic properties and orientation functions of the phases of some polyolefin polymers can now be obtained by means of x-ray diffraction, sonic modulus, birefringence, and infrared dichroism. These solid-state measurements not only yield information about the orientation of the molecules in each of the phases, but also provide considerable insight into the structure of the molecules being studied.   The purpose of Sections II-B through II-E is to elucidate how this important morphological information is obtained.   Much of the work discussed is new, some is not fully tested, but as the picture unfolds, a unified view of the interdependence of the various methods should develop.

B. Wide Angle X-Ray Diffraction: Determination of the Crystalline Orientation Function

1. Introduction

In a polycrystalline polymer film, the crystalline region is composed of crystallites.   The exact nature of the crystallites is not known.   Their dimensions can be determined from small angle x-ray diffraction measurements and from analysis of the breadth of the wide angle diffraction arcs.   There is growing evidence that at least some of these crystallites may be in the form of folded single crystals. Whatever the exact nature of the crystallites, they are composed of an ordered array of polymer molecules.   The characteristics of the array are described by the unit cell of the crystal.   The unit cell is the smallest subunit which, on repetition, will yield the crystal structure.   The unit cell quantitatively describes the manner in which the polymer molecules pack into the crystal lattice.   Figure 3 is a picture of the unit cell of polyethylene which is orthorhombic with the dimensions shown (21).   The orthorhombic unit cell consists of three perpendicular axes of dissimilar lengths.

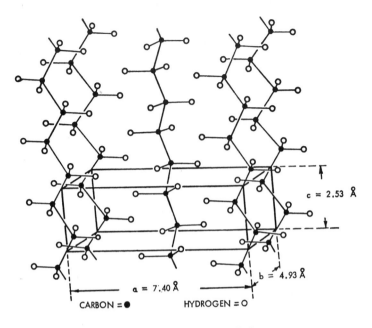

Fig. 3.   Unit cell of polyethylene.

Let us consider the x-ray diffraction behavior of a single crystal which has an orthorhombic unit cell.   If a monochromatic x-ray beam impinges on this crystal, and the crystal is rotated, a diffraction pattern will be obtained which is characteristic of the size and shape of the unit cell.   Bragg (22) has shown that if a plane of atoms is regarded as a mirror, i.e., reflects a portion of the x-rays at an angle of reflection equal to the angle of incidence, the condition for reinforcement of waves from successive planes of the same kind in the crystal is that the difference in the lengths of the paths of the reflected rays be equal to an integral number of wavelengths.   If θ is the angle between the incident ray and the reflecting plane, the condition for reinforcement of all the reflected rays is

$$n\lambda = 2d \sin \theta \tag{6}$$

where $n = 0, 1, 2, 3$, etc., λ is the wavelength, and d is the interplanar spacing.   As the crystal is rotated, the different parallel sets of planes will be brought into a position of reflection and the reinforced

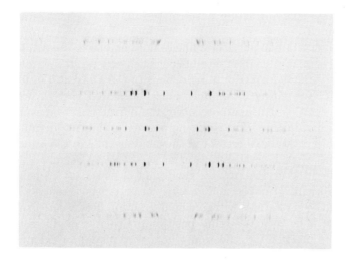

Fig. 4.   Single crystal rotation x-ray diffraction pattern of salicylic acid.

reflected beam will impinge on a photographic film and manifest itself as a sequence of black spots (Fig. 4).   The directions of the reflected beams are governed entirely by the geometry of the lattice (by the orientation and spacings of the planes of atoms).   Thus, the size and shape of the unit cell determine the direction the reflected beams will take.

General equations have been derived which characterize the reflections from various unit cells.   In the orthorhombic unit cell, the general equation takes the form

$$d_{(hkl)} = \frac{1}{\left(\dfrac{h^2}{a^2} + \dfrac{k^2}{b^2} + \dfrac{l^2}{c^2} \right)^{\frac{1}{2}}} \tag{7}$$

or

$$\sin^2 \theta = \frac{\lambda^2}{4} \left(\frac{h^2}{a^2} + \frac{k^2}{b^2} + \frac{l^2}{c^2} \right) \tag{8}$$

Here $d_{(hkl)}$ is the interplanar spacing for the reflecting plane designated by its Miller indices (22) h, k, and l while a, b, and c are the unit cell dimensions.   Other unit cell structures are characterized by a different arrangement of reflections.   Thus, from an analysis of the crystal's diffraction pattern, the unit cell structure may be determined.

A polycrystalline polymeric film is not composed of a single crystal but of many small crystallites. Each crystallite will act as a single crystal and contribute reflections to the x-ray diffraction pattern. The resulting wide angle x-ray diffraction pattern will then be the result of the superposition of the diffraction from each crystallite individually. The resulting pattern is thus a function of the number of crystallites present in the x-ray beam and the orientation (position) of each crystallite relative to the x-ray beam.

If all of the crystallites are randomly oriented in the polymeric film (unoriented sample) and an x-ray beam is sent through the film, some of the crystallites will give no diffraction because none of their planes will be in a reflecting position. Other crystallites will be lying in a position for reflection of a crystal plane; for example, the (110) plane. All of the crystallites which are in the position to reflect the (110) plane will give a reflected beam at the same angle with reference to the incident x-ray beam. The locus of all directions making that angle with the incident beam is a cone having the incident beam as its axis. The resulting cone impinges on the flat x-ray film and appears as a ring on the developed film. This is called the Debye-Scherrer ring and its position on the film characterizes the interplanar spacing which yielded the reflection by means of eq. (6). Other crystallites in the polymer film will be in a position to reflect different crystal planes and the sum of all of these reflections will be the final powder pattern obtained from the sample (Fig. 5).

If the crystallites are not oriented at random but have their crystal axes oriented around a preferred direction, the Debye-Scherrer ring will not be complete but instead will take the form of an arc. In uniaxial orientation, the crystal axes (such as the c axis) will have a preferred direction with respect to a reference direction in the sample (i.e., the axis of a fiber or the stretch direction of a film). At the same time, the crystal axes will have a random orientation in the plane perpendicular to the reference axis in the sample. Thus, if one takes a diffraction photograph of a uniaxial sample with the incident x-ray beam parallel to the stretch direction (reference direction) a Debye-Scherrer powder pattern will be obtained (Fig. 5). Such a pattern demonstrates that the crystallites are randomly ori-

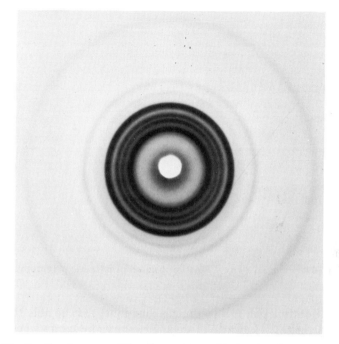

Fig. 5. Powder x-ray diffraction pattern of isotactic polypropylene.

ented in the plane perpendicular to the reference direction. If one now takes a diffraction photograph of the same uniaxial sample with the incident x-ray beam normal to the stretch direction, an oriented rotation pattern will be obtained (Fig. 6). This pattern demonstrates that the crystallites have a preferred orientation with respect to the reference direction. As the crystallites orient more perfectly with respect to the reference direction, the Debye-Scherrer arc will get shorter and shorter. In the limit of perfect orientation, a single crystal rotation pattern made up of sharp spots (similar to Fig. 4) would be obtained.

The foregoing shows that a range of behavior can be expected from the wide angle x-ray diffraction of a uniaxially oriented polymer film, ranging from a single crystal rotation pattern to a powder pattern, depending on the orientation of the crystallites in the polymeric film sample. The average orientation of the crystallites can be formalized on a quantitative numerical basis using Hermans'

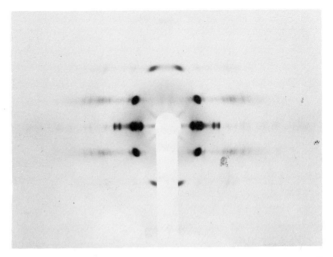

Fig. 6.   Oriented x-ray diffraction pattern of isotactic polypropylene.

orientation function f.   To do this, one must first be able to determine quantitatively the average angle of orientation a given crystal axis makes with the reference direction.   Several models have been proposed as a general solution to this problem and they are reviewed below.

2. A Model for the System

The orientation function f describes the orientation of the crystallite axis relative to some reference direction in the sample.   The orientation function is defined (23) as

$$f_x \equiv (3 \overline{\cos^2 x} - 1)/2 \qquad (9)$$

where $\overline{\cos^2 x}$ designates the average cosine squared value of the angle x between the reference direction in the sample and the x crystallographic direction.

Stein (24) has set up a generalized model for uniaxial crystal orientation.   He represented his coordinate system as shown in Figure 7.

The Z axis of the X, Y, Z Cartesian coordinate system is taken as the stretching direction (reference direction).   The angles α, β, and

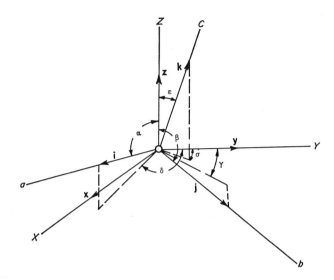

Fig. 7.   Generalized model for crystal orientation (R. S. Stein, ref. 24).

ϵ are measured between the Z axis and the a, b, and c crystallo-graphic axes, respectively.   Since uniaxial orientation is assumed, and there is cylindrical symmetry about the c axis, the three additional angles δ, γ, and σ between the Y axis (taken to be the plane of the polymer film) and the projection of the three crystallographic axes in the XY plane vary randomly.

The three orientation functions which define the degree of orientation of the three crystallographic axes with respect to the stretching direction are then defined [see eq. (9)] as

$$f_\alpha = (3\,\overline{\cos^2 \alpha} - 1)/2$$

$$f_\beta = (3\,\overline{\cos^2 \beta} - 1)/2 \qquad (10)$$

$$f_\epsilon = (3\,\overline{\cos^2 \epsilon} - 1)/2$$

If the orientation is random, the crystallographic axes take all directions with equal probability, the value of $\overline{\cos^2 x}$ obtained by averaging over the surface of a sphere with all directions equally probable is $\frac{1}{3}$, and the orientation function is zero.   Thus, for completely random orientation, $f_\alpha = f_\beta = f_\epsilon = 0$.   If one of the axes is completely oriented with respect to the stretch direction, e.g., $x = 0°$, then $\overline{\cos^2 x} = 1$ and $f_x = 1$. If an axis tends to be perpendicular to the stretching direction, e.g., $x = 90°$, then $\overline{\cos^2 x} = 0$ and $f_x = -\frac{1}{2}$. The orientation function for each axis can thus vary from a value of $+1$ for parallel orientation with respect to the stretching direction, through a value of zero for random orientation, to a value of $-\frac{1}{2}$ if the axis is oriented perpendicular to the stretching direction.

Stein developed this model for the study of polyethylene, which has an orthorhombic unit cell, i.e., the a, b, and c axes are mutually perpendicular. An orthogonality relationship exists among the three mutually perpendicular directions

$$\overline{\cos^2 \alpha} + \overline{\cos^2 \beta} + \overline{\cos^2 \epsilon} = 1 \qquad (11)$$

and therefore

$$f_\alpha + f_\beta + f_\epsilon = 0 \qquad (12)$$

Thus, for this model, the three orientation functions are not independent and only two of them are required to characterize the orientation.

The orthogonality relationship which makes this model so flexible demands that the a, b, and c crystal axes must be perpendicular and, hence, the unit cell of the material being investigated must be of the isometric, tetragonal, or orthorhombic system.

Wilchinsky (25) showed how this model could be extended to a more general treatment in which $\overline{\cos^2 x}$ can be determined, even if there are no reflecting planes normal to the chosen crystallographic direction. Figure 8 illustrates the model in its more general form. Here, the vector $\mathbf{Z}$ represents the stretch direction (reference direction, see Fig. 7). OC, OU, and OV represent a Cartesian coordinate system. The desired crystallographic direction is then fixed to coincide with the OC axis. Within this restriction, the crystal may be fixed at any arbitrary position with respect to the OU and OV axes. The position of the crystal, with respect to the reference direction $\mathbf{Z}$ is then specified by the angles α, β, and ϵ as indicated and

$$\mathbf{Z} = (\cos \alpha)\mathbf{i} + (\cos \beta)\mathbf{j} + (\cos \epsilon)\mathbf{k} \qquad (13a)$$

N is a unit vector along the normal to a set of reflecting planes in the crystal. An (hkl) reflecting plane is represented by triangle abc in

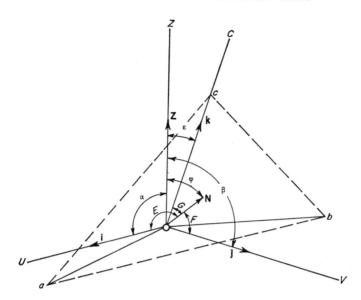

Fig. 8.   Generalized model for crystal orientation (Z. W. Wilchinsky, ref. 25).

Figure 8, where a, b, and c are the crystal axes directions.   The vector $\mathbf{N}$ can be described in terms of its components along the OC, OU, and OV axes, i.e.,

$$\mathbf{N} = (\cos E)\mathbf{i} + (\cos F)\mathbf{j} + (\cos G)\mathbf{k}$$

$$\mathbf{N} = e\mathbf{i} + f\mathbf{j} + g\mathbf{k} \tag{13b}$$

where $e = (\cos E)$, $f = (\cos F)$, and $g = (\cos G)$ are the direction cosines of $\mathbf{N}$ with respect to the OC, OU, and OV axes and $\mathbf{i}$, $\mathbf{j}$, and $\mathbf{k}$ are unit vectors along these axes.

   The angles α, β, and ϵ now represent the angles the OC, OU, and OV axes make with the reference direction.   Angle ϵ also represents the angle between the chosen crystallographic direction and the reference direction in the sample.   The same orthogonality relationship defined in Stein's model [eq. (11)] holds for the newly defined Cartesian coordinate system OC, OU, and OV, e.g.,

$$\overline{\cos^2 \epsilon} + \overline{\cos^2 \alpha} + \overline{\cos^2 \beta} = 1$$

The angle between the plane normal $\mathbf{N}$ and the reference direction $\mathbf{Z}$ is ϕ. From vector algebra $\cos \phi = \mathbf{N} \cdot \mathbf{Z}$ and thus

$$\cos \phi = \mathbf{N} \cdot \mathbf{Z} = e \cos \alpha + f \cos \beta + g \cos \epsilon$$

The desired average is then given by the expression

$$\overline{\cos^2 \phi_{hkl}} = e^2 \overline{\cos^2 \alpha} + f^2 \overline{\cos^2 \beta} + g^2 \overline{\cos^2 \epsilon} + 2ef \overline{(\cos \alpha)(\cos \beta)} + 2fg \overline{(\cos \beta)(\cos \epsilon)} + 2ge \overline{(\cos \epsilon)(\cos \alpha)} \quad (14)$$

The quantity $\overline{\cos^2 \phi}$ for the planes (hkl) is determined experimentally from the distribution of diffracted intensity from these planes, while the coefficients e, f, and g are calculated from a knowledge of the crystal structure.

Since there are six unknowns in eq. (14), the desired quantity $\overline{\cos^2 \epsilon}$ must be evaluated by solving six simultaneous equations involving the six unknowns. Thus, eq. (14) evaluated for five sets of reflecting planes having normals in different directions from each other would contribute five of the required equations while the orthogonality relation, eq. (11), would provide the sixth.

Sack (26), using a model equivalent to Wilchinsky's gave a more detailed and complete theory for the problem of calculating f_x for a prescribed direction in the lattice from the known values of $\overline{\cos^2 \phi}$ in other directions. His solution is equivalent to Wilchinsky's. He does, however, derive some simpler formulas which are valid, if certain symmetry conditions are satisfied. He also considers modifications to the solution which would arise from such experimental limitations as overlap of several Debye-Scherrer rings, limited accuracy in measurements, or, if more data are available than would ideally suffice to calculate the required solution.

From an analysis of symmetry conditions, Sack determined that, for a macroscopically uniaxial aggregate of orthorhombic crystals in which the OC, OU, and OV axes are identified with the crystal axes, the minimum number of independent values of $\overline{\cos^2 \phi}$ which must be determined is 2. This system is equivalent to Stein's model (Fig. 7) and shows it to be a special case of the more general treatment. For tetragonal and hexagonal lattices, Sack finds that the minimum number of independent values of $\overline{\cos^2 \phi}$ is one.

This new formalism makes it possible for the investigator to consider studies of the effects of deformation and processing on the properties of polymers which he could not do quantitatively before. An excellent example is polypropylene. Natta (27) has shown that isotactic polypropylene has a monoclinic unit cell structure. Since the b axis of polypropylene is perpendicular to its (040) planes, its orientation relative to the reference (stretch) direction could be measured directly, i.e., $\overline{\cos^2 \beta} = \overline{\cos^2 \phi_{040}}$ where β is the angle between the stretch direction and the b axis of the crystal.

The c axis of the isotactic polypropylene unit cell is parallel to the helical axis of the isotactic polypropylene molecule in that cell. Determination of the average orientation of the c axis with respect to the reference direction is thus a determination of the average orientation of the helical molecules of the crystalline region with respect to the reference direction. Since the three crystallographic axes in a monoclinic unit cell are not mutually perpendicular (the angle between the a and c axis in isotactic polypropylene is 99° 20′) (27) and no (001) pure c axis reflections are present in the x-ray diffraction pattern, Wilchinsky's generalized approach to the determination of the c axis orientation given above must be used.

Wilchinsky (28) derived from symmetry considerations the following expression relating the variation in intensity with azimuthal angle of the (110), (040), and (130) reflections to the orientation function parameter which characterizes the c axis,

$$\overline{\cos^2 \epsilon} = 1 - \frac{(1 - 2 \sin^2 \rho_2)\overline{(\cos^2 \phi_1)} - (1 - 2 \sin^2 \rho_1)\overline{(\cos^2 \phi_2)}}{\sin^2 \rho_1 - \sin^2 \rho_2}$$

(15)

where the subscripts refer to the respective planes, ρ is the angle between the plane normal and the b axis and is calculated from unit cell dimensions, and ϵ is the angle between the reference direction and the c axis. Thus, the orientation function for the c axis of isotactic polypropylene could be calculated once $\overline{\cos^2 \phi_{hkl}}$ had been determined for two reflecting crystal planes [e.g., the (040) and (110) planes].

3. Experimental Determination of $\overline{\cos^2 \phi_{hkl}}$

Once the unit cell of a polymer crystal is known and the reflections have been indexed (i.e., the hkl Miller index assignments are known for each reflection) $\overline{\cos^2 \phi_{hkl}}$ can be determined experimentally. The discussion that follows will be general but the example used will be data from uniaxially oriented, isotactic polypropylene film. Since a uniaxial system is being used, the azimuthal intensity distribution need only be measured with normal incidence of the x-ray beam or θ_{hkl} incidence of the x-ray beam. If the sample did not have cylindrical symmetry in the xy plane (i.e., were not randomly oriented in the plane perpendicular to the stretching direction), the azimuthal intensity distribution of a given plane would have to be measured at various tilt angles of the sample (i.e., a pole figure analysis of each reflecting plane would be required). In this chapter, we will treat only the uniaxial orientation case. The biaxial case is reviewed by the author elsewhere (29).

Two methods are available for determining the intensity distribution around a Debye-Scherrer arc (i.e., an azimuthal scan). One may photograph the x-ray diffraction pattern of the sample and then examine the desired reflection with a microphotometer, or one may use a diffractometer and measure the intensity distribution directly. The diffractometer technique has the advantage of greater accuracy and flexibility and it is this technique which is used in the experimental examples.

In practice, the diffractometer is set at the radial ($2\theta_{hkl}$) angle of the desired hkl plane to be examined. The sample is then rotated slowly through an angular range large enough to include the azimuthal angles $\rho = 0\text{--}90°$. Here ρ, the azimuthal angle, is defined as the angle between the stretching direction and the plane of measurement of θ_{hkl}. The intensity at each angle ρ, is recorded directly in volts on a strip chart. These intensity values can be converted directly to counts/sec by a suitable scale factor correction. Figure 9 is a typical azimuthal trace of the b axis reflection [(040) plane] from oriented isotactic polypropylene film. The measured intensity I_{meas} in counts/sec at each azimuthal angle ρ must be corrected for absorption, polarization, background, and incoherent scattering before it can be used to evaluate $\overline{\cos^2 \phi_{hkl}}$. The equation used for correction of the measured intensity is

$$I(\rho) = (I_{\text{meas}} - I_{\text{background}})K_{\text{polarization}}K_{\text{absorption}} - I_{\text{incoherent}} \quad (16)$$

The background intensity $I_{\text{background}}$ arises from electronic circuit noise, air scattering, cosmic radiation, etc., and is determined experi-

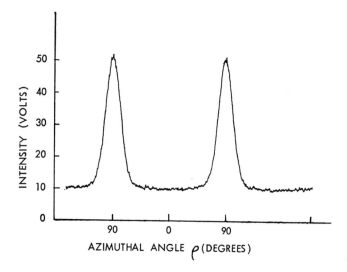

Fig. 9. Typical (040) azimuthal intensity distribution scan from oriented, isotactic polypropylene.

mentally by measuring the scattered intensity with no sample in the beam over the complete azimuthal angular range at the radial $(2\theta_{hkl})$ angle being investigated. The unpolarized incident x-ray beam is partially polarized by the sample and this polarization effect is corrected for by the expression

$$K_{\text{pol}} = 2/[1 + \cos^2 (2\theta_{hkl})] \qquad (17)$$

The absorption correction depends on the angle of incidence of the x-ray beam relative to the polymer film surface. Gingrich (30) has shown that for the incident beam normal to the film surface

$$K_{\text{abs}} = \frac{\mu t (\sec 2\theta_{hkl} - 1)}{1 - \exp [-\mu t (\sec 2\theta_{hkl} - 1)]} \qquad (18)$$

where μ is the linear absorption coefficient and t is the thickness of the sample film. If the incident beam is inclined at the angle, θ_{hkl} to the polymer film surface, the absorption correction has the form

$$K_{\text{abs}} = \frac{\exp (\mu t \sec \theta_{hkl})}{\sec \theta_{hkl}} \qquad (19)$$

The final correction is for incoherent scattering (Compton scattering) which originates from the collision of the x-ray beam with loosely bound or free electrons. This correction is generally small in the low radial angle region of the radiation intensity curve. The incoherent scattering intensity is described by the expression

$$I_{inc} = C_{inc} \left[\sum_i \left(Z_i - \sum_j f_j^2 \right) \right] \tag{20}$$

where Z_i is the atomic number of the atom of type i, f_j^2 is the atomic form factor for this atom at the angles of measurement, where the sum over j is over all of the electrons of the atom, and C_{inc} is an experimentally determined, angularly independent constant. This constant is evaluated by assuming that at a sufficiently large radial angle (i.e., 50°) all of the scattering is incoherent, then at that radial angle (31)

$$I_{inc}(50°) = (I_{meas} - I_{background})K_{pol}K_{abs} \tag{21}$$

C_{inc} is then evaluated by equating eqs. (20) and (21).

Substituting the above corrections into eq. (16), one obtains the following final equation for the corrected intensity $I(\rho)$ in the case of normal incidence of the x-ray beam:

$$I(\rho) = \left\{ (I_{meas} - I_{background}) \left(\frac{2}{1 + \cos^2 2\theta_{hkl}} \right) \right.$$
$$\left. \times \left[\frac{\mu t \ (\sec 2\theta_{hkl} - 1)}{1 - \exp \left[-\mu t \ (\sec 2\theta_{hkl} - 1) \right]} \right] \right\} - I_{inc} \tag{22}$$

Since the radial angle is held constant (at $2\theta_{hkl}$) during an azimuthal scan, the polarization, absorption, and incoherent scattering corrections are constant over the whole azimuthal angular range.

The measured intensities need not be obtained in volt readings from a strip chart, but in more sophisticated, automatic x-ray diffractometers may be obtained directly in counts per second at fixed angular intervals by use of a step scanner. The data are then recorded directly on punched tape which is fed to a computer for analysis. This system of data collecting and analysis has the advantage of increased accuracy, flexibility, safety, and efficiency. Figure 10 is a photograph of an automatic x-ray diffraction apparatus in use in the author's laboratory (32). A is the automatic controller and logic system. Angular intervals, angular range, and type of motions are

Fig. 10.   Automated x-ray diffractometer for orientation studies.

programmed on this panel. B is the diffractometer intensity accu-
mulator system.   The system can accommodate intensity measurements in the form of counts for a fixed time or time for a fixed count
as well as continuous scan. C is the automated goniometer with step
scanning facilities. D is a digital tape punch which records the intensity data from B in machine language, while E is a digital printer
which will type out the intensity data from B simultaneously with
the operation of the tape punch D, in lieu of the tape punch, or
remain off during the data gathering operation.   This diffractometer can obtain the pole figure of three independent reflections (i.e.,
at three $2\theta_{hkl}$ positions) automatically, during one complete operation.   Thus, the azimuthal scans of as many as three reflecting
planes [e.g., the (040), (110), and (130) crystal planes of isotactic
polypropylene] for a uniaxially oriented polymer film can be
obtained automatically in this instrument.

For a system which is oriented uniaxially

$$\overline{\cos^2 \rho_{hkl}} = \frac{\displaystyle\int_0^{\pi/2} I(\rho) \cos^2 \rho \sin \rho d\rho}{\displaystyle\int_0^{\pi/2} I(\rho) \sin \rho d\rho} \tag{23}$$

Equation (23) can be evaluated graphically by multiplying the corrected intensities by the appropriate values of the sine and cosine functions and plotting the results against the azimuthal angle. The areas under the curves can then be measured with a planimeter and used to determine $\overline{\cos^2 \rho_{hkl}}$. As mentioned above, a much more rapid and reliable method of calculating $\overline{\cos^2 \rho}$ is to allow a computer to numerically calculate the value from the input raw intensity data.

Polyani (33) has shown from geometrical considerations that for normal incidence of the x-ray beam $\overline{\cos \phi_{hkl}} = \overline{\cos \theta_{hkl}} \times \overline{\cos \rho_{hkl}}$ and hence

$$\overline{\cos^2 \phi_{hkl}} = \overline{\cos^2 \theta_{hkl}} \times \overline{\cos^2 \rho_{hkl}} \tag{24}$$

If θ_{hkl} incidence of the x-ray beam is used, however, then

$$\overline{\cos^2 \phi_{hkl}} = \overline{\cos^2 \rho_{hkl}} \tag{25}$$

Once $\overline{\cos^2 \phi_{hkl}}$ has been determined experimentally, the orientation function of the desired crystal axis can be calculated. If a pure axial crystal plane reflection is available, $\overline{\cos^2 X}$ for that reflection is equal to $\overline{\cos^2 \phi_x}$, i.e.,

$$\overline{\cos^2 \phi_{h00}} = \overline{\cos^2 \alpha}$$

$$\overline{\cos^2 \phi_{0k0}} = \overline{\cos^2 \beta} \tag{26}$$

$$\overline{\cos^2 \phi_{001}} = \overline{\cos^2 \epsilon}$$

These values can be used directly in eq. (10) to calculate the orientation function. If a pure axial crystal plane reflection is not available in the x-ray diffraction pattern, Wilchinsky's method of analysis [eq. (14)] must be used. This was illustrated earlier for isotactic polypropylene film. Thus, from an analysis of the azimuthal intensity distribution of the (040) plane $\overline{\cos^2 \phi_{040}}$ is obtained. Since this is an (0k0) pure b-axis reflection, $\overline{\cos^2 \phi_{040}} = \overline{\cos^2 \beta}$. However, since no (001) pure c-axis reflections are present in the x-ray diffraction pattern of isotactic polypropylene, $\overline{\cos^2 \epsilon}$ could not be obtained in this simple manner.   Instead, Wilchinsky had to resort to eq. (14) which reduced to eq. (15) when the symmetry characteristics of the monoclinic unit cell were considered.   The angle ρ in eq. (15) was calculated for the isotactic polypropylene crystal from a geometric analysis of the unit cell structure.   The values of ρ for the (040), (110), and (130) diffraction planes were calculated to be (28)

$$\rho_{(040)} = 0°$$

$$\rho_{(130)} = 46.7°$$

$$\rho_{(110)} = 72.5°$$

Then $\overline{\cos^2 \epsilon}$ could be calculated by substituting the appropriate ρ_{hk0} with the experimental $\overline{\cos^2 \phi_{hk0}}$ for two reflections into eq. (15). The f_ϵ value for the isotactic polypropylene sample is then determined by substituting $\overline{\cos^2 \epsilon}$ into eq. (10) and solving.

4. Application of f_x to Crystallite Orientation Processes

Deformation of a polycrystalline polymer film results in an orientation of both crystalline and amorphous regions with respect to the deformation direction.   The change in the orientation of the crystalline region can be followed quantitatively by determining the change in the orientation functions for the crystal axes as a function of the deformation of the sample.   Similarly, if oriented polymer film is heat treated at fixed elongation, or shrunk, or solvent swollen, or undergoes any other physical treatment, resulting changes in the orientation of the crystallite can be followed quantitatively by studying the changes in the crystal orientation function.   Wide angle x-ray diffraction has been used to determine the crystal orientation function of cellulose (15), polyethylene (34,35), isotactic polypropylene (28,34,36), poly-1-butene (34), and polyethylene terephthalate (37).

The changes that occur in the orientation of the crystallites as a sample is deformed uniaxially can be represented graphically by an orientation function triangle diagram (24).   The character of this triangle plot results from the model used to characterize $\overline{\cos^2 \phi_{hkl}}$. The orthogonality of the axes of the model results in the interdependence of the three orientation functions [see eq. (12)]. As a result of the model, only two orientation functions are required to characterize the average orientation of all three axes. Therefore, the state of orientation may be represented as a point on a two-dimensional plot with two of the f's as coordinates.

Figure 11 is an orientation function triangle diagram representing data from samples of isotactic polypropylene which have undergone different physical treatments. Because of restrictions on the range of values imposed by eq. (12), all points must occur within the indicated triangular area. When a sample film is unoriented, all of the crystallites are randomly arranged in the film and, hence, $f_\alpha = f_\beta = f_\epsilon = 0$ (in Fig. 11, $f_c = f_\epsilon$). Thus, the origin in Figure 11 represents the unoriented state of the material. When one of the crystal axes is oriented parallel to the stretch direction, $f_x = 1.0$ and, therefore, each apex of the triangle diagram represents a state of perfect orientation of one of the three axes along the stretch direction. For example, the apex $c_\parallel$ in Figure 11 corresponds to a state where $f_\epsilon = 1$,

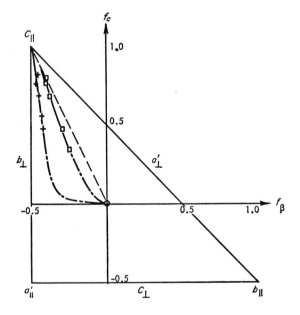

Fig. 11. Orientation function triangle diagram for the b and c axes of isotactic polypropylene (R. J. Samuels, ref. 36).

$f_\beta = f_\alpha = -\frac{1}{2}$. This is a state of orientation where the c axis of the crystal lies in the stretching direction and the b and a' axes* are oriented perpendicular to the stretching direction. One further characteristic of the orientation function triangle is that the center of each side of the triangle diagram corresponds to a state of orientation where the crystal axis lies perpendicular to the stretching direction. Thus, the point where the b axis of the crystal is perpendicular to the stretching direction and the other two axes are random is in the center of the side marked $b_\perp$.

The importance of the orientation function triangle diagram is that one can describe uniaxial orientation processes in terms of the path of a point, which corresponds to a given state of average orientation of the crystallites in the sample, around this triangle. To illustrate the power of this form of representation of uniaxial orientation processes, let us examine the orientation behavior of the isotactic polypropylene samples. For isotactic polypropylene, the c axis of the crystal corresponds to the helical chain axis of the molecule. Generally, when a sample is deformed, the polymer chain axis orients in the direction of the deformation. Thus the c axis of the crystal orients in the direction of the deformation in cold-drawn, isotactic polypropylene. The resulting x-ray diffraction pattern for c-axis orientation is shown in Figure 12a. Both melt-spun, isotactic polypropylene fibers (38) and cold-drawn fibers annealed close to the melting point (36) exhibit two types of crystal orientation simultaneously. Most of the crystals exhibit the expected c-axis orientation but some of the crystals exhibit a'-axis orientation. In a'-axis orientation, the polymer chain axis is nearly perpendicular to the stretching direction and the a' axis of the crystal is oriented toward the stretch direction. This type of orientation appears in the x-ray diffraction pattern as shown in Figure 12b. Thus, a'-axis orientation seems to be the high temperature form in oriented isotactic polypropylene.

* Due to the monoclinic unit cell structure of isotactic polypropylene, the a axis of the crystal does not fall on the OV axis of Wilchinsky's Cartesian coordinate system (Fig. 8). For this reason, f_α does not represent the true a-axis orientation of the crystal, but instead represents the orientation of the OV axis. This axis is designated as the a' axis and is related to the true a axis through the deviation of the β monoclinic angle from 90°. The β angle equals 99° 20' (27) for isotactic polypropylene and, thus, the a'-axis orientation is close to the true a-axis orientation.

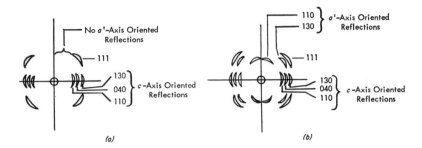

Fig. 12. X-ray diffraction diagrams of isotactic polypropylene fibers: (a) cold-drawn; (b) annealed or spun (R. J. Samuels, ref. 36).

The curves in the orientation function triangle diagram (Fig. 11) represent data from (1) cold drawn, isotactic polypropylene film (34), (2) hot drawn (110°C) polypropylene film (36), and (3) melt-spun, isotactic polypropylene fibers (36).   In this diagram, $f_c = f_e$, the change in the designation indicating that this is the orientation function representing the orientation of the molecular helical chain axis in the crystalline region of the polymer.   The orientation function triangle diagram of the data obtained from cold-drawn, isotactic polypropylene film is a straight line from the origin to the apex $c_\parallel$ of the triangle.   Movement of a point from the origin to the apex of the triangle is movement in the direction of increasing elongation of the sample.   The straight line from the origin to the $c_\parallel$ apex for the cold-drawn, isotactic polypropylene fibers indicates that the a' and b axes of the crystallite rotate randomly about the c axis as the c axis is turned toward the stretching direction.

The curve for the melt-spun fibers in Figure 11 (+ points) is bowed toward the $b_\perp$ side of the orientation function triangle diagram.   The f_c values for these samples were obtained from azimuthal intensity distribution scans of the (040) reflection and the bimodal (110) reflection, and application of the resulting data to eqs. (15) and (10).   The results plotted in the orientation function triangle diagram indicate that the b axis of the crystal orients toward the perpendicular to the spinning direction (fiber axis) faster than the a' axis does, as the c axis orients in the fiber axis (deformation) direction.   Examination of the x-ray diffraction pattern (Fig. 12b) indicates that this a'- and b-axis behavior is observed because a fraction of the crystallites have the a' axis oriented toward the deformation direction, while the rest have the a' axis oriented perpendicular to the stretch direction. Thus, the average a'-axis orientation of all the crystallites is effectively orienting more slowly toward the perpendicular to the deformation direction than is the b axis of the crystallites.

The curve for the hot-drawn (110°C), isotactic polypropylene films in Figure 11 ($\square$ points) is intermediate between the curves for the cold-drawn, isotactic polypropylene film and the melt-spun, isotactic polypropylene fiber. The same rationale as that presented above for the melt-spun fibers can be used to explain the $b_\perp$ bow of this curve. A much smaller fraction of a'-axis oriented crystallites must be present at this temperature than at the melt spinning temperature, however, since the $b_\perp$ bow for the hot-drawn films is much less than that for the melt-spun fibers. One can conclude from these observations that the amount of a'-axis orientation increases with draw temperature and decreases with extension.

The above discussion of the orientation behavior of isotactic polypropylene crystallites has been used to illustrate the power of the crystal orientation function and its representation in an orientation function triangle diagram. This system of representation can clearly illustrate the effects of environment on the orientation behavior of crystallites during deformation. A knowledge of crystallite orientation behavior is vital to understanding the effect of processing and fabrication conditions on the mechanical properties of commercial films.

C. Sonic Modulus: Determination of the Amorphous Orientation Function

1. Introduction

When a polycrystalline polymer film is oriented, the deformation ultimately leads to a rearrangement of the crystalline and amorphous regions of the film. To describe the morphological state of the deformed film completely, one must have a knowledge not only of the final state of orientation of the crystallites but of the final state of orientation of the amorphous region as well. Thus, a knowledge of the value of the amorphous orientation function is essential if the state of orientation of a deformed film is to be described.

Due to the less ordered arrangement of the molecules in the amorphous region, an experimental determination of its orientation function becomes more elusive than that of the crystalline regions. In principle, the distribution of orientations of the amorphous chains could be determined from an analysis of the intensity distribution of the amorphous halo in the wide angle x-ray diffraction pattern (39). However, the separation of the amorphous from the crystalline contributions to the pattern is very difficult experimentally. Similarly, in principle, infrared measurements might be used to determine the amorphous orientation function (40). However, characterization of the amorphous contribution to the infrared spectra is experimentally very difficult and the transition moment angle at the frequency used in the investigation must be known. The use of infrared measurements for the determination of molecular and orientation characteristics of polymers is treated more fully in a later section of the chapter.

A study of the manner in which a sonic wave propagates through a polycrystalline polymeric medium can give considerable insight into the nature of the orientation of the two phases in the medium. To understand the rationale behind this measurement the models used to describe sound propagation must first be examined.

2. A Model for the System

The nature of sound propagation in uniaxially oriented fibers has been examined by Ward (41,42) and Moseley (43). Ward considered a partially oriented fiber as an aggregate of units whose optical and elastic properties are those of a highly oriented fiber. The polymer is thus treated as a single-phase system whose properties are a function of the average distribution of the units which make up the system. The units are assumed to be ideally elastic materials which possess transverse isotropy and whose properties are unchanged by the process of orientation. Using this model, Ward applied a continuum mechanics analysis to define the elastic constants for the system and then further evaluated these in terms of the orientation distribution of the units. In this way he was able to demonstrate, as a special case of the aggregate theory, the following relationship between the sonic modulus E, the intrinsic lateral (transverse) modulus of the perfectly oriented fiber E_t^0, and the average angle θ between the direction of sound propagation and the symmetry axis of the units

$$1/E = (1 - \overline{\cos^2 \theta})/E_t^0 \qquad (27)$$

Moseley examined the sonic modulus of several stiff, oriented polymers whose glass transition temperatures were very far above the room temperature at which they were measured. He found that in this frozen state, the sound propagation of the samples was independent of the degree of crystallinity. For this reason, he developed a theory of sound propagation which considered the fiber as a single-phase system. He then analyzed the molecular processes by which the sound was propagated through the system. Moseley reasoned that if sound is sent across an array of parallel molecules, the sonic energy is transmitted from one molecule to another by the stretching of intermolecular bonds (Fig. 13). On the other hand, if the sound is sent along the length of a bundle of parallel molecules, the sound is propagated principally by the stretching of chemical bonds in the backbone of the polymer chain. For partially oriented polymer molecules the molecular motion due to sound transmission is presumed to have right-angle components along and across the

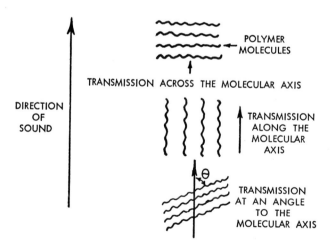

Fig. 13. Possible modes of sound transmission in polymers.

direction of the molecular axis. The magnitude of either of these two components is a function of the angle θ between the molecular axis and the direction of sound propagation. The only physically meaningful assumption about the method of addition of the two components is that any fiber deformation that occurs during sound transmission is the sum of the intramolecular and intermolecular deformations. Thus, he concluded that a series addition of the intra- and intermolecular force constants, weighted by an average orientation parameter, had to be used to represent the propagation of a sound pulse through a fiber. Assuming the force constants to be proportional to the corresponding sonic moduli, Moseley derived the following equation for the sonic modulus E:

$$1/E = [(1 - \overline{\cos^2 \theta})/E_t^0] + (\overline{\cos^2 \theta}/E_1^0) \qquad (28)$$

where E_t^0 is the intrinsic lateral (transverse) modulus of a fully oriented fiber (i.e., where all of the molecules are aligned parallel to each other and perpendicular to the direction of sound propagation) and E_1^0 is the intrinsic longitudinal modulus of a fully oriented fiber (i.e., where all of the molecules are aligned parallel to each other and parallel to the direction of sound propagation). Because of the high longitudinal sound velocity along the chain axis of the molecule, the second term on the right in eq. (28) was considered to be negligible. Thus, eq. (28) reduced to

$$1/E = (1 - \overline{\cos^2 \theta})/E_t^0 \qquad (29)$$

Equation (29) is identical to eq. (27). Thus, both authors, starting from different models, obtained the same equation to describe the mechanism of sound propagation in a uniaxially oriented, single-phase polymer system.

Polyethylene, nylon 6,10, Teflon (44), and polypropylene (36,45) all show changes in sonic modulus with percentage crystallinity when measured at room temperature. These materials do not behave as single-phase systems, but as two-phase systems when responding to a sonic pulse. Polypropylene behaves as a two-phase system even at temperatures far below its glass transition temperature. To be applicable to such polymers, the sonic modulus theory must be extended to a two-phase model, and must take into account different intermolecular force constants for the crystalline and amorphous regions.

To study the orientation behavior of isotactic polypropylene, the sonic modulus equations were extended to a two-phase system (36) by use of a mixing equation involving bulk compressibilities. For a homogeneous ideal mixture, both the density and bulk compressibility are additive properties. If isotactic polypropylene is considered an ideal mixture of amorphous and crystalline phases, then the mixing equation takes the form

$$K = \beta K_c + (1 - \beta)K_{am} \qquad (30)$$

where K is the bulk compressibility of the mixture, K_c is the bulk compressibility of the crystal regions, K_{am} is the bulk compressibility of the amorphous regions, β is the fraction of crystalline material, and $(1 - \beta)$ is the fraction of noncrystalline material.

An equation of this form has been found to be valid for isotropic suspensions of solids in liquids (46) and for polyethylene over a range of crystallinities and temperatures (47). The bulk compressibility K is related to the bulk modulus B, Young's modulus E, and Poisson's ratio ν, by the expression (48)

$$K = 1/B = 3(1 - 2\nu)/E \qquad (31)$$

Waterman (49) found that $\nu = 0.33$ for isotactic polypropylene at room temperature. For this specific case, then

$$K = 1/B = 1/E = 1/\rho C^2 \qquad (32)$$

where the sonic modulus (Young's modulus) $E = \rho C^2$ for a long, thin rodlike sample (50). Here ρ is the density and C the sonic velocity. Equation (29) is assumed to be valid for each phase. Then combination of eqs. (29), (30), and (32) yields the following expression for the measured sonic modulus E_{or} of the oriented sample:

$$1/E_{or} = (\beta/E_{t,c}^0)(1 - \overline{\cos^2 \theta_c}) + [(1 - \beta)/E_{t,am}^0](1 - \overline{\cos^2 \theta_{am}}) \qquad (33)$$

where the subscripts c and am stand for the crystalline and amorphous regions, respectively.

For an unoriented sample, $\overline{\cos^2 \theta} = \frac{1}{3}$ and eq. (33) reduces to

$$3/2E_u = \beta/E_{t,c}^0 + (1 - \beta)/E_{t,am}^0 \qquad (34)$$

where E_u is the measured sonic modulus of the unoriented sample.

Equation (34) predicts that the sonic modulus will vary in a pre-scribed way as a function of the fraction of crystals in the unoriented samples. Thus, by studying the change in the sonic modulus of unoriented samples with crystallinity, the intrinsic lateral modulus of the crystalline and of the amorphous regions of the polymer may be determined.

The orientation function f_x was defined as

$$f_x \equiv (3 \overline{\cos^2 x} - 1)/2 \tag{9}$$

where x represents the angle between the polymer chain axis and a specified reference direction in the sample (here the stretch direction which is also the direction of sound propagation). By applying this expression to eq. (33) and combining it with eq. (34), the following expression for the sonic modulus of oriented, isotactic poly-propylene at room temperature is obtained:

$$\tfrac{3}{2}(\Delta E^{-1}) = \beta f_c/E_{t,c}^0 + (1 - \beta)f_{am}/E_{t,am}^0 \tag{35}$$

where

$$(\Delta E^{-1}) = (E_u^{-1} - E_{or}^{-1})$$

and f_c and f_{am} are defined orientation functions for the crystal and amorphous phases, respectively. Thus, according to eq. (35), once the intrinsic lateral moduli of the crystalline and amorphous regions have been determined from measurements on unoriented samples, the amorphous orientation function f_{am} of any oriented sample can be determined from experimental values of the sonic modulus E_{or}, the fraction of crystals β, and the crystal orientation function f_c.*

3. Determination of the Sonic Modulus

The sonic modulus for a long, thin, rodlike sample is given by the expression (50)

$$E = \rho C^2$$

where E is the sonic modulus (Young's modulus), ρ is the density,

---

* The derivation of eq. (35) is strictly valid only when Poisson's ratio, ν, has a value of 0.333. Thus, eq. (35) is valid for isotactic polypropylene at room temperature. When a temperature other than room temperature is used to obtain sonic moduli of isotactic polypropylene, or when other polymers are studied whose Poisson's ratio is different from 0.333 at the measurement temperature, (31) must be used to correct the theoretical derivation for the new Poisson's ratio.

and C is the sonic velocity.   Obviously, a long, thin, rodlike sample
must be used to determine the sonic modulus.   Fibers make ideal
specimens but long, thin films can also be used.   Film samples
1–5 mils thick, 1 mm wide, and 15–25 cm long are quite satisfactory.
The density of portions of this film can be measured easily and
rapidly in a density gradient column (51).

The sonic velocity (the velocity of a longitudinal wave in the
material) is usually determined by measuring the transit time of a
sound pulse between two tranducers coupled to a test specimen.
The transit time of the pulse over a given length of specimen is
inversely proportional to the propagation velocity in the sample.
Figure 14 represents schematically the general features of an experimental apparatus for measuring the sonic velocity of a polycrystalline polymer specimen.   The sample film strip is clamped at one
end, passed over a pulley, and kept taut by a 10-g weight (for light
tension) attached to the other end.   This arrangement can be varied by replacing the pulley and weight with a clamp attached to a
strain gauge and applying a force of 10 g to the strip in this manner.
An optical bench is a good support for the test equipment.   The
transmitter and receiver (transducers) can then be moved along the
optical bench and the distance between the probes can be read from
a meter stick.   The sound pulse is supplied by a pulse propagation
meter.   The KLH Series Four Pulse Propagation Meter, which produces longitudinal waves at a frequency of 10 kc/sec has been the
standard workhorse in this field.   The operator simply places the
transducers a given distance apart in contact with the specimen and
reads the transit time directly in microseconds (μsec) on the panel
meter or the attached recorder.   This is measured at several dis-

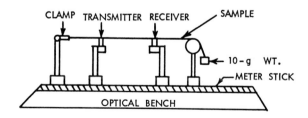

Fig. 14.   Schematic representation of sonic velocity apparatus.

ROBERT J. SAMUELS

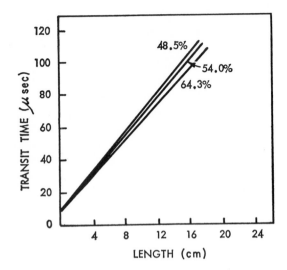

Fig. 15. Sonic velocity determination of unoriented polypropylene films of
different crystallinities.

tances between probes and the results are plotted as in Figure 15.
The slope of the line is equal to $1/C$ (μsec/cm).

The Series Four Pulse Propagation Meter is no longer available
from the KLH R & D Corp. [Dr. Morgan, who designed the meter,
is now the supplier (H. M. Morgan Co., Inc.) of the PPM-5 (Fig.
16)]. This is a solid-state version of the Series Four with some
added conveniences. The apparatus is supplied with a bench,
clamps, and transducer mounts. One of the transducers moves
along the sample at a constant rate and the transit time is monitored
directly as a function of the distance between the probes on an
attached recorder. The moving probe cycles continuously. The
output for two cycles is illustrated in Figure 17. The operator
simply calculates the sonic velocity from the reciprocal of the slope of
the line plotted by the instrument. The reproducibility from cycle
to cycle is about 1%.

Once the sonic velocity has been measured, the sonic modulus is
calculated directly from the known density and velocity values.
The sonic velocity measurement is very rapid, and requires but a few

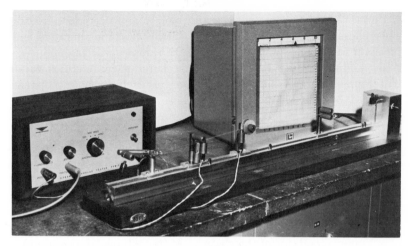

Fig. 16. Dynamic Modulus Tester PPM-5 (H. M. Morgan Co., Inc.).

minutes. This makes it a highly attractive tool for orientation studies.

4. Application of the Sonic Modulus to Orientation Studies

The sonic modulus is a mechanical measurement. It measures the Young's modulus of the specimen. The sonic pulse causes real displacements of the molecules from their equilibrium position as it passes through the sample.

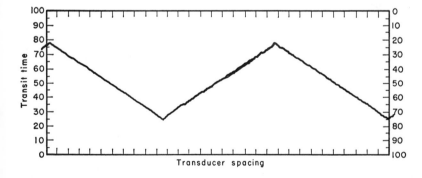

Fig. 17. Recorder trace from the Dynamic Modulus Tester PPM-5.

In the two-phase system, the intermolecular force constants for each phase are different. Since the intrinsic lateral modulus is a mechanical measurement, the unoriented samples will have nonzero sonic modulus values. In fact, the measured sonic modulus value will represent the additive properties of each individual phase weighted by the amount of each phase present. Equation (34) describes the law of addition of the properties of the two phases. Table 1 lists the variation in sonic modulus with β for a series of unoriented, compression molded, isotactic polypropylene films (36) (see Fig. 15). The intrinsic lateral modulus of the crystal $E_{t,c}^0$ and of the amorphous region $E_{t,\mathrm{am}}^0$ can be calculated from these experimental values of the sonic modulus and the fraction of crystals β by using eq. (34).

The intrinsic lateral modulus of the crystal $E_{t,c}^0$ is the intermolecular Young's modulus of the chains. If the polymer chains are in a folded, single crystal with the chain axes all aligned along a vertical axis Z and in a given crystallographic plane [such as the (110) for polyethylene], the intrinsic lateral modulus may be visualized as the force per molecule required to separate the planes of molecules [(110) faces] a given distance by applying the force in the direction perpendicular to the Z axis and normal to and away from the crystal face. The intrinsic lateral modulus of the isotactic polypropylene crystal, calculated from the experimental values of E_u and β determined at room temperature, is:

$$E_{t,c}^0 = 3.96 \pm 0.09 \times 10^{10} \text{ dyne cm}^{-2}$$

Another method of measuring the intrinsic lateral modulus of the

TABLE 1

Sonic Modulus (E_u), Crystal Fraction (β), and Density of Unoriented Compression-Molded Isotactic Polypropylene Films (36)

Sample	$E_u \times 10^{-10}$, dyne cm^{-2}	Density, g cm^{-3}	β
1	2.27	0.8875	0.410
2	2.48	0.8936	0.485
3	2.63	0.8980	0.540
4	3.01	0.9061	0.643

crystal is to measure the force required to cause a given displacement of a known plane in the crystal lattice. The displacement of the crystal lattice is measured as a change in the Bragg angle spacing for the given lattice plane in the wide angle x-ray diffraction pattern. This technique was used by Sakurada, Ito, and Nakamae (52) to determine the intrinsic lateral modulus of the isotactic polypropylene crystal. They obtained a value of $E_{t,c}^0 = 2.8 - 3.1 \times 10^{10}$ dyne cm^{-2} for isotactic polypropylene. This is rather good agreement with the value obtained from the sonic modulus method, especially when one realizes that different investigators (53,54), each using the same stress–x-ray diffraction technique, obtained values of $E_{t,c}^0$ for polyethylene ranging from 2.2×10^{10} to 4.2×10^{10} dyne cm^{-2}.

The intrinsic sonic modulus of the amorphous region is defined as the transverse Young's modulus that the amorphous chains would have in a perfectly oriented fiber. The lateral forces between amorphous chains in this system would be expected to be lower than those in the crystal lattice. The calculated intrinsic lateral modulus of the amorphous region of isotactic polypropylene is

$$E_{t,\mathrm{am}}^0 = 1.06 \pm 0.01 \times 10^{10} \text{ dyne cm}^{-2}$$

which is, as predicted, lower than the value of the intrinsic lateral modulus obtained for the crystal.

Once the material constants for the polymer have been determined from sonic modulus studies on unoriented specimens, the sonic modulus of oriented specimens can be examined. The sonic modulus of an oriented film is a measure of the total orientation of the sample. It is a function of the orientation of both the crystalline and amorphous regions as well as that of the amount of each phase present in the sample. The rule for addition of the properties of each of the phases is defined by eq. (35). The dependence of the sonic modulus E_{or} on elongation for a series of isotactic polypropylene films cast and drawn at 110°C (36) is illustrated in Table 2 and Figure 18. Since the intrinsic lateral moduli are now established, eq. (35) permits calculation of the amorphous orientation function f_{am} for any oriented, isotactic polypropylene sample from experimental values of the sonic modulus (E_{or}), density (β), and the crystal orientation function (f_c) determined by x-ray diffraction. Values of f_{am} determined for the oriented, isotactic polypropylene samples are listed in Table 3.

TABLE 2

Sonic Modulus (E), Crystal Fraction (β), and Density of Cast
and Annealed Polypropylene Films (36)

Sample	Elongation, %	$E \times 10^{-10}$, dyne cm^{-2}	Density, g cm^{-3}	β
1	0	2.86	0.9034	0.608
2	50	3.09	0.9052	0.630
3	100	3.59	0.9056	0.636
4	200	5.07	0.9055	0.636
5	300	6.19	0.9060	0.643
6	400	6.55	0.9052	0.630

The orientation of the crystalline and amorphous phases can be
determined quantitatively by a combination of density, sonic modulus, and x-ray diffraction methods. With the sample thus finger-
printed, a potential exists for relating the observed mechanical
properties of these samples to their morphological characteristics.

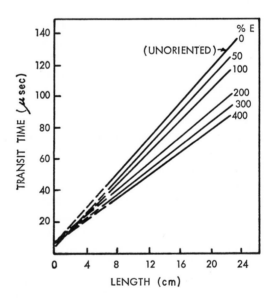

Fig. 18.   Sonic velocity determination of oriented polypropylene films.

TABLE 3
Birefringence (Δ_T) and Orientation Function (f)
Data for Isotactic Polypropylene Films (36)

Sample, cast and annealed	Elongation, %	Birefringence, $\Delta_T \times 10^3$	X-ray		Sonic modulus, f_{am}
			f_c	$-f_\beta$	
2	50	5.759	0.3472	0.2485	−0.0951
3	100	10.19	0.4790	0.2898	+0.0232
4	200	19.64	0.6805	0.3775	+0.2849
5	300	22.25	0.7728	0.4022	+0.3877
6	400	25.30	0.8016	0.3988	+0.4321

Similarly, with this information, the utility of other techniques may be extended and new molecular information about the polymers obtained. For example, by combining this information with birefringence measurements a rapid method of obtaining the orientation functions can be developed and at the same time a method for determining the intrinsic birefringence of each of the phases is obtained.

D. Birefringence: Determination of the Total Molecular Orientation

1. Introduction

Birefringence is a measure of the total molecular orientation of a system. It is defined as the difference in the principal refractive index parallel $n_\parallel$ and perpendicular $n_\perp$ to the stretch direction for a uniaxially oriented specimen. The refractive index, in turn, is a measure of the velocity of light in the medium and is related to the polarizability of the chains. Thus, the final measured birefringence is a function of contributions from the polarizabilities of all of the molecular units in the sample.

The velocity of light in the crystalline region will be different from that in the amorphous region. Since the molecules are optically anisotropic, there is a difference in the principal polarizability parallel $P_\parallel$ and perpendicular $P_\perp$ to the chain axis. If all of the anisotropic molecules are randomly distributed in the specimen (unoriented sample), the average contribution to the refractive index will be the same in all directions and the birefringence, $\Delta_T = (n_{\parallel} - n_{\perp})$, will be zero (i.e., the specimen is optically isotropic). If all of the molecules are perfectly aligned parallel to the stretch direction, the birefringence will be a combination of the ideal polarizability anisotropy of each region; that is, each region will have the birefringence of a fully oriented fiber of that phase alone. The measured birefringence of oriented sample films will have intermediate values between these extremes, depending on the orientation and amount of each phase present in the specimen.

For a uniaxially oriented, polycrystalline polymer (assuming a two-phase model of crystalline and amorphous regions) the measured birefringence Δ_T may be defined as (15)

$$\Delta_T = \beta\Delta_c^0 f_c + (1 - \beta)\Delta_{am}^0 f_{am} \tag{36}$$

Here Δ_c^0 is the intrinsic birefringence of the perfectly oriented crystal and Δ_{am}^0 is the intrinsic birefringence of a perfectly oriented amorphous region (i.e., the birefringence the amorphous chains would have in a perfectly oriented amorphous fiber). Equation (36) neglects an added term, the form birefringence (55), which takes into account the effect of the shape of the crystallite in the medium. The effect has been found to contribute 5–10% of the total birefringence, depending on orientation, to the measured birefringence of polyethylene (56). The form birefringence, if any, of isotactic polypropylene has not been measured and therefore the contribution of this term to eq. (36) will be considered negligible for the purpose of this discussion.

Birefringence should be an excellent property to use for the study of orientation in a polycrystalline polymer. Once the intrinsic birefringences have been obtained, the amorphous orientation function could be determined, for instance, by a combination of density, x-ray diffraction, and birefringence measurements. The problem in applying this method to a wide spectrum of polymers has been the difficulty in obtaining intrinsic birefringence values of the polymer. Unlike the sonic modulus, for which the intrinsic lateral moduli can be obtained from measurements on unoriented samples of different crystallinities, the birefringence of an unoriented specimen is zero. Thus, one cannot measure directly the intrinsic birefringences of a polymer by examining the birefringence of unoriented specimens.

This has been a major handicap in the use of birefringence measurements for orientation studies on polycrystalline polymers. To date, birefringence has been used to determine the amorphous orientation function of only two polymers, polyethylene (57) and polypropylene (36). For the polyethylene studies, the birefringence of the crystalline region was assumed to be the same as that obtained from single crystals of low molecular weight (C_{36}) paraffin homologs (58), whereas the intrinsic birefringence of the amorphous region had to be calculated theoretically from bond polarizability values. By use of these values, in combination with x-ray diffraction and density measurements, the birefringence contribution of the crystalline and amorphous regions was calculated. The amorphous orientation function of the oriented samples was then estimated from these results.

A study of the birefringence of oriented, isotactic polypropylene films was carried out to test the validity of amorphous orientation functions obtained from sonic modulus measurements. In the process of doing this, a method was demonstrated for obtaining the intrinsic birefringence of the crystalline and amorphous regions from experimental data, and hence the practical utility of the birefringence measurement was extended. Before considering this work in more detail, a description of the experimental determination of birefringence will be given.

2. Determination of Birefringence

Light propagates by wave motion through a medium. The standard relationship between frequency ν, wavelength λ, and velocity c for wave motion is

$$c = \nu\lambda \tag{37}$$

The frequency of a beam of monochromatic light never changes, even if the light enters an entirely different material. The wavelength and velocity of the same light beam do change if it enters a different material. The index of refraction n of a particular material is defined as

$$n = c_0/c_m = \lambda_0/\lambda_m \tag{38}$$

The subscripts 0 and m stand for vacuum and the medium, respectively. The velocity in air is almost the same as that in vacuum and hence the refractive index of air may be assumed equal to 1.0

in most cases (the actual value is $n = 1.0003$). A light wave is slowed down as it goes from vacuum to any material and hence the refractive index is greater than 1.0 for all materials.

The mechanism of light propagation through a medium is related to the polarizability of the molecules in the medium.

$$P = N\alpha \qquad (39)$$

where P is the polarizability per unit volume, N is the number of molecules, and α is the polarizability of a molecule. The polarizability α in turn, is a consequence of the electric field of the incoming light wave E inducing a dipole moment μ in the electrons of the molecule. The polarizability α is the ratio of the induced dipole

$$\alpha = \mu/E \qquad (40)$$

moment to the applied field and therefore is essentially a measure of the mobility of the electrons in the molecules.

The origin of the refractive index of a material is the polarizability of the molecules. The refractive index can be related to the principal polarizability of the molecule by the Lorenz-Lorentz equation

$$(n^2 - 1)/(n^2 - 2) = \tfrac{4}{3}\pi P \qquad (41)$$

Most molecules are optically anisotropic (have different polarizabilities along and across the molecule). When the anisotropic molecules are randomly distributed throughout the medium, the average polarizability in any direction is the same and a single refractive index will characterize the system.

When a polymer film is uniaxially oriented, the chain axis of the molecules orients with respect to the deformation axis and the molecules have cylindrical symmetry around the deformation axis. Since the optically anisotropic molecules now have a preferred direction with respect to the deformation axis of the film, the film now manifests anisotropic optical properties. For this system, the principal refractive index for the light vibrating in the deformation (stretch) direction is different from that for light having its electrical vibration in the plane perpendicular to the deformation direction. The birefringence is used to characterize such an optically anisotropic system. For a uniaxial system, the birefringence Δ_T is defined as the difference between the principal refractive index in the stretch direction $n_\parallel$ and the principal refractive index perpendicular to the stretch direction $n_\perp$:

$$\Delta_T = n_\parallel - n_\perp \qquad (42)$$

To determine Δ_T for a sample, the individual principal refractive indices of the sample parallel and perpendicular to the deformation axis must be determined and then subtracted, or else the difference in the two principal refractive indices must be measured directly.

The direct measurement of the individual principal refractive indices of the sample is a tedious and difficult procedure, fraught with possible errors, and very time consuming. Monochromatic light is used because different frequencies of light will have different velocities in the sample. The sample to be measured is split into several subsamples which must be placed in a series of immersion liquids of different refractive index. They are then examined under a polarizing microscope. The plane of the incident light is adjusted parallel to the stretch direction of the subsamples and the refractive index $n_\parallel$ is determined by the Becke method from examination of the series. The plane of the incident light is then adjusted perpendicular to the stretch direction of the subsamples and the refractive index $n_\perp$ is determined by another examination of the series. The observer, when using the Becke method, looks for a thin band of light at the edge of the sample. This band of light is called the Becke line. Under the microscope, the narrow bright band moves toward the medium of higher refractive index if the focus of the microscope is raised, and toward the medium of lower refractive index if the focus is lowered. With the use of monochromatic light and controlled temperature an accuracy of ± 0.001 in the refractive index may be reached by successive approximation when proper selection of immersion media is made (59).

A more rapid and popular method for determining birefringence is the use of a compensator to determine the phase difference (retardation) between two mutually perpendicular, plane-polarized wave motions emerging from the sample. Since the sample is anisotropic, the velocity of the wave passing through the sample parallel to the stretch direction will be different from the velocity of the wave oriented perpendicular to the stretch direction. This velocity difference between the two waves causes a phase difference in the emerging rays. The phase difference, called the retardation, is proportional to the birefringence. By measuring the retardation, the birefringence can be determined from the relation

$$\Delta_T = \lambda R/t = \Gamma_\lambda/t \qquad (43)$$

where λ is the wavelength of light (mμ), R is the retardation (phase difference in wave numbers—dimensionless), Γ_λ is the retardation at the specified λ (mμ), and t is the thickness of the sample (mμ).

Experimentally, the retardation of the sample is determined by matching the sample's retardation with an equal and opposite retardation using a standard calibrated crystal. The standard compensating crystal with its vernier and holder is called a compensator. Several different compensator designs are available. They all work on the same physical principle but differ in their mechanical arrangement. The most common are the Babinet compensator, which is made up of a quartz wedge whose thickness can be varied by sliding different regions of the wedge into the light path; the Berek compensator, which varies the thickness and hence the retardation by a rotation of the calibrated crystal; and the Ehringhaus rotatable compensator with a crystal combination plate. The author uses a Zeiss polarizing microscope with Ehringhaus compensators. The advantage of the Ehringhaus compensator is that the usual zero reading with the sample removed from the beam is eliminated. Instead, due to the properties of the combination plate, the zero-order, black extinction line is obtained by rotating the crystal plate first in one direction and taking a reading and then in the opposite direction for a second reading. The average value of the two vernier readings gives the correct retardation of the sample in degrees of inclination of the plate. The result can be converted to the desired phase difference in millimicrons by calculation. Actually, the Ehringhaus compensator comes with a table of phase differences for given inclination angles so that the retardation in millimicrons can be obtained rapidly.

The experimental arrangement of the sample and optical components in the microscope is shown in Figure 19. The light beam passes through a polarizer which has its axis at 45° to the stretch direction of the sample film. The incident beam E can be resolved into a component $E_\parallel$, parallel to the stretch direction, and a component $E_\perp$, perpendicular to the stretch direction. The phase difference between these two components in the emerging wave depends on the birefringence of the sample. The emerging ray

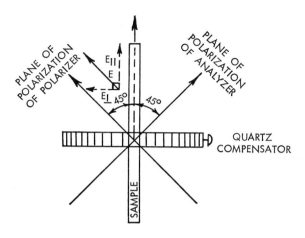

Fig. 19. Arrangement of the optical system for a retardation measurement. The incident light beam is directed perpendicular to the surface of the page.

then enters the compensator whose direction of retardation is opposite to that of the sample film. Finally, an analyzer is oriented perpendicular to the plane of polarization of the incident light and 45° to the stretch direction of the sample film. Initially, white light is used to find the zero-order, black extinction fringe in the compensator. This fringe is set on the cross hairs of the microscope and then a monochromatic interference filter is placed in the beam. This sharpens the extinction line under the cross hairs which are then recentered. White light must be used initially because the fringes of all orders are black under monochromatic radiation and hence white light must be used to identify the zero-order fringe.

The author uses a green line ($\lambda = 546$ mμ) interference filter and has used both quartz and calcite Ehringhaus compensators. With the quartz compensator, one has an available measuring range of about seven orders, while with the calcite compensator the available measuring range is about 133 orders. The time required to obtain a retardation measurement on a transparent film is only a few minutes. Thus, the birefringence measurement is very rapid.

To calculate the birefringence of the sample, the measured retardation Γ_λ (mμ) and the measured thickness (mμ) (1 mm = 10^6 mμ) are substituted into eq. (43).

A method has been developed for determining the retardation

from the light intensity transmitted through a system similar to Figure 19 but without the compensator (60). This method is most suitable for dynamic birefringence measurements and will not be considered in detail here.

3. Application of Birefringence to Orientation

Birefringence is used routinely in many industrial film and fiber plants as a measure of the average orientation of the sample. This reasoning ignores the complex nature of birefringence and treats the sample as if it were a single, average-phase system. This single, average-phase, molecular orientation concept pervades much of the literature and has led some investigators (42,61,62) to attempt, on this basis, to correlate birefringence with other properties of the polymer. Other investigators have realized the complex nature of birefringence, but due to the lack of availability of one or both of the intrinsic birefringence values, they could not make full use of bire-fringence data (34,63). A knowledge of the intrinsic birefringence of the crystal and amorphous regions of a polycrystalline polymer is essential if maximum information is to be obtained from the meas-ured birefringence of a sample.

The intrinsic birefringence of the crystal and amorphous regions of a polymer can be obtained if the fraction of crystals and the crystal and amorphous orientation functions are known for a series of oriented samples. Equation (36) can be rewritten in the follow-ing form:

$$\Delta_T/\beta f_c = \Delta_c^0 + \Delta_{am}^0[(1 - \beta)/\beta](f_{am}/f_c) \tag{44}$$

Equation (44) predicts a straight line relationship between $\Delta_T/\beta f_c$ and $[(1 - \beta)f_{am}/\beta f_c]$. Experimental values of Δ_c^0 and Δ_{am}^0 can then be obtained directly from the intercept and slope, respectively, of the resulting line.

A birefringence study was made on isotactic polypropylene films (36) to test the validity of amorphous orientation functions obtained from sonic modulus measurements. Data from Section II-C, Table 3 were plotted according to eq. (44) which is illustrated in Figure 20. As predicted by the equation, a straight line is obtained. Each point on this plot is a combination of a birefringence, a density, an x-ray diffraction, and a sonic modulus measurement. The ratio f_{am}/f_c should be very sensitive to errors in f_{am}. This, then, was a

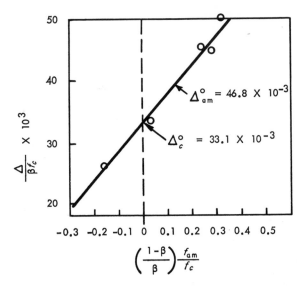

Fig. 20.   Determination of intrinsic birefringences (R. J. Samuels, ref. 36).

good check on the validity of the two-phase sonic modulus theory. At the same time, it demonstrated an experimental method for determining the intrinsic birefringence of the crystal and amorphous regions of a polymer.   The intrinsic birefringence of the crystalline ($\Delta_c^0 = 33.1 \times 10^{-3}$) and amorphous ($\Delta_{am}^0 = 46.8 \times 10^{-3}$) regions of isotactic polypropylene were calculated from the intercept and slope of the line in Figure 20.   This was the first time intrinsic birefringences of a polycrystalline polymer had been obtained from measurements made directly on the polymer.

The above results indicate the potential importance of being able to fingerprint the morphological features of a polycrystalline polymer.   By characterizing the morphological features of the isotactic polypropylene samples, the intrinsic birefringences of the polymer could be determined.   With this new information, other characteristics of the polymer can be elucidated.   For example, the contribution of the separate phases to the total measured birefringence can be calculated.   These are plotted for isotactic polypropylene film samples in Figure 21 where $\Delta_c = \Delta_c^0 f_c$ and $\Delta_{am} = \Delta_{am}^0 f_{am}$.   The negative amorphous contribution at low elongations in Figure 21 indi-

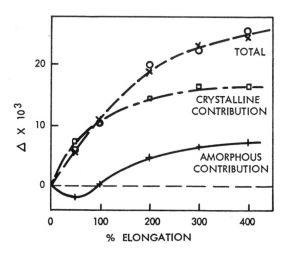

Fig. 21. Crystalline and amorphous contributions to the birefringence of deformed, isotactic polypropylene films: (O) $\Delta = \Delta_T$ (experimental); (X) $\Delta = \Delta_T$ (calculated); (□) $\Delta = \beta\Delta_c$ (calculated); (+) $\Delta = (1 - \beta)\Delta_{am}$ (calculated) (R. J. Samuels, ref. 36).

cates that the amorphous chains are oriented toward the perpendicular to the direction of stretch. As elongation of the film continues, the amorphous chains turn into the stretch direction and the amorphous birefringence becomes positive. The crystalline orientation contribution to birefringence is fairly high even at low elongations. The low measured birefringence at low elongations is obtained even with the high crystalline orientation because the negative effect of the amorphous birefringence decreases the experimental value. Similar results were observed from orientation studies on low density polyethylene (57). Low density polyethylene is about 50% crystalline and structurally similar to polypropylene; therefore, the similar behavior of their crystalline and amorphous regions is not surprising.

These results illustrate the danger of treating birefringence as a measure of the average orientation of the sample. The measured birefringence of the low elongation samples of isotactic polypropylene films indicates little orientation of the system when, in fact, there is considerable orientation of the crystalline and amorphous regions of the samples. The birefringence value may lead to confusion when it is compared with other properties of the film, such as its mechanical properties, if the investigator assumes the films have little molecular orientation. On the other hand, a comprehensive morphological interpretation of the structure of the film could make apparently confusing data—obtained from samples with the same measured birefringence but different mechanical properties—seem obvious.

Once the crystalline and amorphous intrinsic birefringences have been obtained, a very rapid method can be developed for determining the crystalline and amorphous orientation functions of the polymer. The equation for birefringence is

$$\Delta_T = \beta \Delta_c^0 f_c + (1 - \beta)\Delta_{am}^0 f_{am} \tag{36}$$

where Δ_c^0 and Δ_{am}^0 are now known constants and Δ_T can be determined in a matter of minutes. The sonic modulus equation for an oriented polymer film is:

$$\tfrac{3}{2}(\Delta E^{-1}) = \beta f_c/E_{t,c}^0 + (1 - \beta)f_{am}/E_{t,am}^0 \tag{35}$$

where $E_{t,c}^0$ and $E_{t,am}^0$ are now known constants and (ΔE^{-1}) can be determined in a matter of minutes. The fraction of crystals β can be determined from a density measurement, which takes a few hours to equilibrate in a density gradient column, but requires not more than a few minutes of operator time. Substituting these measured results from the sonic modulus, birefringence, and density determinations into eqs. (36) and (35) and solving them simultaneously leads to the desired values for the orientation functions f_c and f_{am}.

The intrinsic birefringence of the crystal was found to be smaller than that of the amorphous region of isotactic polypropylene. The smaller value of Δ_c^0 is due to the ordered lattice arrangement of neighboring molecules in the crystal. A bond within the molecule experiences not only the electromagnetic field of the incident light but also the polarization field of the surrounding molecules. This internal field is more anisotropic in the ordered lattice of the crystal than in the disordered amorphous phase. Consequently, there are different effective polarizabilities in the two phases. Both Stein (64) and Vuks (65) have derived theories which satisfactorily account for internal field effects in the polyethylene crystal. They each concluded that, due to internal field effects, the intrinsic birefringence of the crystal will be lower than that of the amorphous material. Experimental results with isotactic polypropylene are consistent with their theoretical conclusion.

As discussed earlier, birefringence is a result of the polarizability of the anisotropic units which make up the polymer chain. It should therefore be possible to obtain some information about the polarizability of the monomer units within the chain from birefringence studies on polymers. Since, in general, due to internal field effects, it is impossible to calculate the anisotropic polarizability of bonds by using, as point of departure, the optical properties of the substance in the crystalline state (66), Δ_{am}^0 and not Δ_c^0 should be used in any attempt to calculate monomer polarizability differences.

The principal polarizability difference per monomer unit $(a_\parallel - a_\perp)_{mer}$ can be calculated for isotactic polypropylene from the expression (57)

$$(a_\parallel - a_\perp)_{mer} = \frac{\Delta_{am}^0}{\frac{2}{9}\pi[(\bar{n}^2 + 2)^2/\bar{n}]N_{mer}} = +4.49 \times 10^{-25} \text{ cm}^3 \quad (45)$$

where

$\Delta_{am}^0 = 46.8 \times 10^{-3}$,

$\bar{n}$ is the average refractive index of the polymer $= 1.51$, and

$$N_{mer} = \left(\frac{\text{amorphous density}}{\text{molecular weight mer unit}}\right) \times \text{Avogadro's number}.$$

The measured value of $(a_\parallel - a_\perp)_{mer}$ is a characteristic property of the polymer molecule and hence should be the same whether obtained from measurement of birefringence in the bulk polymer or from flow birefringence measurements of polymer solutions. Tsvetkov (67) has tabulated values for the anisotropy per monomer unit calculated from flow birefringence data for numerous polymers. The value of $(a_\parallel - a_\perp)$ reported for polypropylene is $+3.5 \times 10^{-25}$ cm^3. The value calculated from Δ_{am}^0 was $+4.5 \times 10^{-25}$ cm^3. The agreement between measurements in the solid state and in dilute solution is quite striking, especially since the $(a_\parallel - a_\perp)$ values for polymers are known to range from -35×10^{-25} cm^3 for poly-p-chlorostyrene to $+21 \times 10^{-25}$ cm^3 for ethylcellulose.

The birefringence measurement is thus a rapid and powerful tool for the study of morphological characteristics of deformed polycrystalline polymers. Since birefringence is a measure of the total molecular orientation of the two-phase system, its examination in conjunction with other physical measurements (sonic modulus, x-ray diffraction, density, etc.) yields considerable insight into the molecular characteristics of the bulk polymer. The major problem in its use is a knowledge of the intrinsic birefringences of the polymer system to be investigated. One possible route to the experimental evaluation of these morphological constants has been described above. The principle for their evaluation is a general one, however, and any combination of physical measurements which yields values for β, f_{am}, and f_c can be used in conjunction with birefringence to determine Δ_c^0 and Δ_{am}^0. Once these have been evaluated for a given polymer, the birefringence measurement can be used as an important method for the comprehensive, quantitative evaluation of the morphology of the deformed polymer.

E. Infrared Dichroism: Molecular Structure and Orientation Parameters

1. Introduction

So far, the primary emphasis in this chapter has been on techniques for the quantitative determination of molecular orientation in polycrystalline polymers. As a secondary consequence of these methods, certain molecular structure parameters such as the intrinsic birefringences and the intrinsic lateral moduli of the system have become available. In this section, there will be a turnabout, and the quantitative determination of a molecular structure parameter—the transition moment angle—will be the principal quantity of the measurement, while the molecular orientation results will be a secondary consequence of the overall technique. For this reason, a general knowledge of the infrared absorption process and of the theoretical approach to structural infrared studies is necessary for comprehension of the importance of the transition moment angle determination. A number of excellent books (68,69) and reviews (70,71) are available on this subject, and it is not within the scope of this chapter to give a detailed account of infrared structure analysis. An attempt will be made, however, to outline the general theory of infrared spectroscopy to orient the reader for the discussions on molecular structure and orientation parameters which follow.

Mention should be made at the outset, however, that infrared dichroism is an important quantitative technique for the determination of crystalline and amorphous orientation functions. The nature of the measurement holds out the promise of being able to measure rapidly and quantitatively the crystalline and amorphous orientation functions separately. Since this method, like the previous techniques, is nondestructive, it has a potentially important role in the *"in vivo"* study of deformation processes in polymers. The interdependence of the two important facets of infrared dichroism, molecular structure and orientation parameters, is the subject of this section.

2. General Infrared Theory

Absorption in the infrared region of the spectrum is a function of the internal energy of the molecule. Molecules absorb energy in this region according to quantum rules, and hence a continuous spectrum of energy absorption is not observed; instead, energy is absorbed only at discrete frequencies. When absorption occurs in this radiant energy region, it changes the ground state rotational and vibrational energy levels of the molecule and this produces rotational and vibrational motion of the atoms constituting the molecule. Since the rotational energy changes are smaller than the vibrational energy changes, the effect of the rotational modes of motion on the observed spectra is to act as a sort of background which influences the shape of the vibrational energy bands. Hence, unless one is examining the spectra under high resolution to resolve the rotational energy absorption, the observed spectra can be considered as a measure of the vibrational modes of energy absorption of the molecules, i.e., a vibrational spectrum.

The simplest model for representing the vibratory motion of the atoms resulting from the process of energy absorption in a molecule is the motion of a group of balls (atoms) of different masses held together by springs (chemical bonds). Application of a force to displace and then release a ball would correspond to the absorption of infrared energy. Upon being released the ball is set into vibratory motion. The frequency of the oscillation and the amount of displacement of the ball is a function of the spring (force) constant k, the mass of the ball m, and the original applied force F. For the same initial displacement, the motion of the ball will be changed by substituting another spring for the first or by changing the mass of the ball. The force required to stretch a spring a short distance d is given by Hooke's law

$$F = kd \qquad (46)$$

Thus, for a given applied force the displacement of the ball will be a function of the spring (force) constant.

In a simple diatomic molecule (I—II), the only vibration which can occur is a periodic stretching along the I—II bond. Stretching vibrations resemble the oscillations of two bodies connected by a spring and for this case Hooke's law, eq. (46), is applicable as a first approximation. The vibrational frequency ν (cm^{-1}) for stretching the I—II bond is given by the expression

$$\nu = (1/2\pi c)(f/\mu)^{\frac{1}{2}} \qquad (47)$$

where c is the velocity of light, f is the force constant of the bond, and μ is the reduced mass of the system defined as

$$\mu = \frac{m_I m_{II}}{m_I + m_{II}} \qquad (48)$$

where m_I and m_{II} are the individual masses of atoms I and II.

In actual practice, one is generally interested in molecules composed of numerous atoms having bonds of differing strength. Thus, the motion of not two balls connected by a spring, but, instead, many balls and springs, is involved when energy is absorbed. The motion of a particular group of balls may predominate in a given applied force (absorption frequency) region, however, depending on the masses and spring constants involved. For instance, if a given ball is displaced and the adjoining balls have large masses and spring constants, the motion will be essentially isolated to the originally displaced ball because the displacement of the adjoining balls will be negligible.

Thus, infrared absorption by the molecule changes the vibrational energy level of the system. The problem is to correlate this energy change with the structure and force constants of the molecule. For a harmonic oscillator approximation, this problem can be treated as a classical analysis of a vibrating system. The problem is then reduced to a determination of the classical vibration frequencies associated with the normal modes of oscillation of the system. The characteristic of a normal mode is that all nuclei in the molecule move with simple harmonic motion of the same frequency. Determination of the normal frequencies and the forms of the normal vibrations is then the primary problem in correlating the structure and internal forces of the molecules with the observed vibrational spectrum.

The problem is rather complex. A nonlinear molecule of n atoms has $3n$ degrees of freedom which are distributed as 3 rotational, 3 translational, and $3n - 6$ vibrational motions, each with a characteristic fundamental band frequency (linear molecules have $3n - 5$ vibrational modes). The fundamental band frequency may be defined as the absorption frequency at which a normal mode of the molecule is singly excited. Obviously, for a high polymer molecule, where n is large, the number of fundamental vibrations of the molecule can become inordinately large. Not all of these modes will be infrared-active, however. No absorption will take place in the infrared, unless there is a change in the dipole moment of the molecule during the normal vibration. Thus infrared absorption will occur only for unsymmetrical vibrations.

Often, there is no change in the dipole moment of the molecule during a particular normal vibration. For example, the symmetrical stretching frequency ν_1 of the linear CO_2 molecule is inactive in the infrared because the dipole moment changes produced by the two C=O bonds are equal and opposite and therefore cancel.

$$\overset{\leftarrow}{O}=C=\vec{O}$$

The asymmetric stretching vibration ν_3 is active in the infrared, however, with a transition moment lying along the major axis of the molecule

$$\vec{O}=\overset{\leftarrow}{C}=\vec{O}$$

because the relative motion of the atoms brings the carbon atom alternately closer to one and then to the other oxygen atom. This relative motion necessarily causes a corresponding relative motion of the charge centers. Since a change in dipole moment occurs for a molecule whenever a change in position of the centers of positive and negative charge resulting from atomic motion occurs, ν_3 is infrared-active.

Stretching of bonds is not the only vibrational mode of deformation in a molecule. In fact, if there is a total of $3n - 6$ vibrational modes in a molecule, $n - 1$ of these are stretching modes and $2n - 5$ are due to other deformations such as bending, wagging, twisting, and rocking. The bending mode of the CO_2 molecule, for instance,

$$\overset{\uparrow}{O}=\underset{\downarrow}{C}=\overset{\uparrow}{O}$$

has its transition moment perpendicular to the major axis of symmetry of the CO_2 molecule. All of these deformation modes contribute to the vibrational spectra. Depending on the nature of the vibration, they may have transition moments (direction of change in the dipole moment), parallel, perpendicular, or at some angle to the major axis of symmetry of the molecule. This characteristic of directionality can be utilized in structure determinations.

Thus, not all of the normal modes of vibration are infrared-active. To further complicate the problem of structure analysis from the observed infrared spectra, not all of the absorption frequencies observed experimentally are the result of normal modes, since the harmonic oscillator model is an oversimplification. In reality, overtone and combination bands appear due to the effect of anharmonic terms in the potential function. An already complicated theoretical solution of the observed spectrum is thus further confused because the general potential function contains more independent force constants than can be determined from the experimental data. The number of theoretical constants can be reduced, however, by making assumptions about the nature of the force field in the molecule. One such assumption is that contributions to the potential energy arise only from the stretching of chemical bonds and from the bending of angles between chemical bonds. This is known as the valence force–field (VFF) assumption (72). Another assumption is to include in the potential function interaction terms which represent the existence of weak central forces between nonbonded atoms (73). This is known as the Urey-Bradley force field (UBFF). Finally, a central force field (CFF) which takes into account the interactions between every pair of atoms has also been applied in normal coordinate calculations of the vibrational spectrum to reduce the number of force constants.

The determination of the normal modes and their frequencies depends on the solution of a $3n \times 3n$ determinant. Methods exist for simplifying this computational problem which require some molecular structure to be assumed as a model. From the symmetry characteristics of the assumed molecular structure, the general determinant can be resolved into more easily computed subdeterminants of lower order (by application of group theory), each of which involves only normal frequencies of a given symmetry class. Thus, by assuming the molecular structure and the nature of the interactions (force field), the complexity of the determination of the vibrational modes of the molecule can be reduced.

Any experimental method which can help to identify the modes of motion of an observed absorption band can be of great help in resolving the validity of a normal coordinate analysis of the observed spectra. The normal coordinate analysis always involves more force constants than observed frequencies. While, in general, one can calculate a set of force constants that are reasonable and will reproduce the observed frequencies, the assignment of a specific frequency to a particular vibration of the molecule is not always possible (68). Similarly, since the molecular structure and force field are assumed in the calculation, independent methods for corroborating assignments are necessary in order to determine which assumed molecular structure is correct. Simply matching the observed frequencies is not a sufficient solution. Some form of independent experimental check on the calculated band assignments is necessary.

One method of qualitatively checking the assignment of a vibrational frequency as characteristic of a particular mode of motion of a group of atoms is by relating the frequency assignment for that vibrational mode to the characteristic absorption frequency of a similar group of atoms in another, known molecule. It has been demonstrated that, although a normal mode involves, to some extent, all the atoms in a molecule, a large number of these vibrations are localized to a high degree of approximation in particular bands or in small groups of atoms. These *group frequencies* persist, irrespective of the type of molecule in which the bond or group of atoms occurs. Thus, there is an invariance of the bond-stretching force constants and, therefore, the frequency of a normal vibration which is mainly confined to the stretching and contraction of one bond in the molecule may be estimated from eq. (47), for instance, which was derived for a diatomic molecule. Stretching frequencies calculated in this manner should be regarded as only approximate, however. Angle-bending force constants, which are generally about an order of magnitude smaller than stretching constants, and interaction constants are much more sensitive to the environment. In general, vibrational frequencies of particular groups of atoms will remain nearly constant from molecule to molecule so that some theoretical vibrational mode assignments for groups of atoms in large molecules can be checked by comparing the frequencies with those of small molecules in which the vibrations of the same groups have been determined in detail.

Another helpful method for checking spectral band assignments is isotopic substitution (71). The most useful form of isotopic substitution in a molecule is that of deuterium for hydrogen. In this case, the force constants do not change and only the masses of the hydrogen atoms are changed in magnitude. For this reason, the observed frequency of the hydrogen modes in the molecule will be reduced [see, for example, eq. (47)] and the method therefore serves to identify hydrogen modes in the spectra. Similarly, bands of complex origin (where there is interaction between hydrogen modes and other modes) will usually become evident, since the interaction will generally disappear on deuteration. Obviously, if different kinds of hydrogen atom groupings are present in the molecule and they can be selectively deuterated, specific information about the assignments of these groups could be ascertained.

It was mentioned earlier that no absorption will take place in the infrared region unless there is a change in the dipole moment of the molecule during the normal vibration and, therefore, absorption will only occur for unsymmetrical vibrations. Thus, each of the infrared-active vibrational modes, be they stretching, bending, twisting, rocking, or a combination of these, will have a transition moment parallel, perpendicular, or at some angle α_ν to the major axis of symmetry of the molecule (Fig. 22). The transition moment angle expected for each observed frequency can be calculated theoretically on the basis of the assumed molecular structure and the normal coordinate vibrational mode assignment. If the vibrational mode assignment is correct, the theoretically calculated transition moment angle for a given observed frequency and the experimentally determined value at that frequency should be the same. The meas-

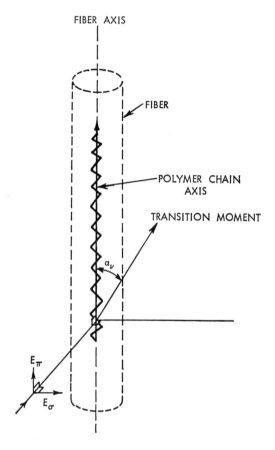

Fig. 22.   Schematic representation of perfect orientation parameters.

ured transition moment angle α_ν is thus a desirable experimental parameter.

The transition moment is a vector quantity and thus has both magnitude and direction.   The intensity of the infrared absorption band depends upon the angle the electric vector in the incident radiation makes with the transition moment.   In particular, the intensity is proportional to the square of the scalar product of the transition moment and the electric field vectors.   The absorption coefficient k for a particular direction of the incident electric field vector E_0 may be written

$$k = P^2 \sum_i \cos^2 \Psi_i \qquad (49)$$

where Ψ_i is the angle between the transition moment direction of the ith absorbing center and the direction of the electric vector, P is the magnitude of a vector $\mathbf{P}$ proportional to the transition moment, and the summation extends throughout unit volume of the specimen. Equation (49) forms the basis for the utilization of polarized infrared radiation as a powerful tool in the study of the spectra and structure of oriented polymers.

3. Infrared Measurement of the Transition Moment Angle

The directional character of the transition moment leads one logically to the study of oriented systems for its experimental determination. The problem is similar to that of birefringence with an added complication. The orientation of the transition moments must be considered as well as the orientation of the molecules. The two-phase character of birefringence data can be avoided in infrared studies, however, at least in the initial investigation. When the polymer crystallizes into an ordered lattice, the increase in the symmetry of the molecules results in fewer and sharper bands when compared with those in the amorphous region, as in Figure 23. Simi-

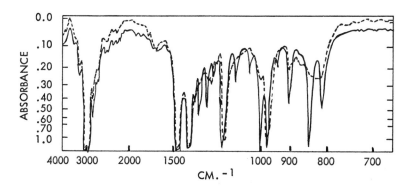

Fig. 23.   Infrared spectra of polypropylene: (——) crystalline; (— — —) amorphous.

larly, external lattice vibrations may even produce new bands. As a result, bands which are exclusively crystalline, bands which are amorphous, and bands which are a mixture of the contributions from both regions (mixed bands) are identifiable. Thus, by a judicious choice of absorption frequencies, the characteristics of each phase can be examined separately. Since bands characteristic of the individual phases in the film can be identified, some measure of anisotropy that can be correlated with the orientation of the phase would be desirable. The anisotropy parameter used in infrared studies is the infrared dichroic ratio.

When a polymer is uniaxially oriented, the principal axes of the refractive index ellipsoid are directed parallel and perpendicular to the fiber axis (stretch direction). When the incident electric vector is parallel to a principal axis, the radiation will traverse the specimen with its plane of polarization unaltered. Thus, only optical density measurements made with the electric vector parallel and perpendicular to the orientation axis will be meaningful. The significant factor in interpreting the optical anisotropy of the sample in terms of molecular orientation is the dichroic ratio

$$D = k_\pi/k_\sigma \qquad (50)$$

where k_π and k_σ are the principal absorption coefficients for radiation vibrating parallel and perpendicular to the direction of the orientation axis, respectively. The quantity actually measured experimentally is the dichroic ratio

$$D' = \epsilon_\pi/\epsilon_\sigma \qquad (51)$$

where ϵ_π and ϵ_σ are the optical density, log I_0/I, parallel and perpendicular to the orientation axis, respectively (see Fig. 24) and I_0 and I are the incident and transmitted intensities, respectively. The quantity D' equals D, if there are no scattering or reflection losses. Throughout the following discussion, both ϵ_π and ϵ_σ are assumed to have been corrected for nonabsorptive energy losses and equal k_π and k_σ, respectively.

The relationship between the infrared dichroism D, the orientation of the molecules, and the direction of the transition moments was solved by Fraser (74). Consider a fiber where all of the polymer chain axes are oriented parallel to the fiber axis. Optical den-

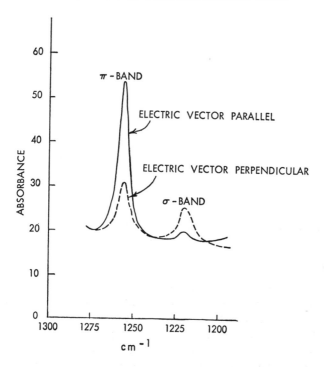

Fig. 24.   Infrared dichroism of cast, uniaxially oriented polypropylene film.

sity measurements are made with the electric vector parallel and perpendicular to the fiber axis, respectively.   For this ideal system, the incident electric vector, ϵ_π (see Fig. 22), will be parallel to the polymer chain axis as well as to the fiber axis; hence the angle Ψ between the transition moment direction and the electric vector direction will equal α_ν.   Thus, from eq. (49), the expression for the parallel principal absorption coefficient is

$$k_\pi = NP^2 \cos^2 \alpha_\nu \qquad (52)$$

where N is the number of absorbing centers per unit volume and α_ν is the transition moment angle for the centers absorbing radiation of frequency ν.   From similar arguments for the perpendicular electric vector, ϵ_σ, the expression for the perpendicular principal absorption coefficient is

$$k_\sigma = \tfrac{1}{2}NP^2 \sin^2 \alpha_\nu \tag{53}$$

Thus, for perfect alignment of the molecules parallel to the fiber axis, the dichroic ratio is related to the transition moment angle by the expression

$$D_0 = k_\pi/k_\sigma = 2 \cot^2 \alpha_\nu \tag{54}$$

where D_0 is the dichroic ratio of an ideally oriented polymer.

When all of the molecules are randomly oriented in a film or fiber (optically isotropic), then

$$k_\pi = k_\sigma = \tfrac{1}{3}NP^2 \tag{55}$$

And, thus, the dichroic ratio for an unoriented specimen is

$$D = k_\pi/k_\sigma = 1.0 \tag{56}$$

In a real polymer film or fiber, the molecules are only partially oriented and there must be some measure of the degree of orientation of the molecules relative to the orientation axis. Fraser considered this problem from two different points of view (75). One approach, generally utilized by spectroscopists, will be discussed here. The other approach, more suitable for quantitative studies, will be discussed in Section II-E-4.

An oriented polymer system can be considered as composed of a fraction f of perfectly oriented chains and a fraction $(1 - f)$ of perfectly random chains. From eqs. (52), (53), and (55) therefore

$$k_\pi = NP^2[f \cos^2 \alpha_\nu + \tfrac{1}{3}(1 - f)]$$

and $\tag{57}$

$$k_\sigma = NP^2[\tfrac{1}{2}f \sin^2 \alpha_\nu + \tfrac{1}{3}(1 - f)]$$

The dichroic ratio for this model, in terms of D_0 [eq. (54)] and f, thus takes the form

$$D = \frac{1 + \tfrac{1}{3}(D_0 - 1)(1 + 2f)}{1 + \tfrac{1}{3}(D_0 - 1)(1 - f)} \tag{58}$$

The following expression can then be obtained by solving eq. (58)

in terms of the fraction of perfectly oriented chains

$$f = \frac{(D - 1)(D_0 + 2)}{(D + 2)(D_0 - 1)} \tag{59}$$

For any given uniaxially oriented sample, the fraction of perfectly oriented chains is a constant. The value of the constant is unknown, however. The spectroscopist is interested in determining the value of D_0 and, hence, of α_ν for each absorption frequency from the measured dichroic ratios of the sample. This value of α_ν then can be compared with the calculated values from the theoretical molecular models proposed. To accomplish this, the value of the unknown constant f must be determined.

One method of estimating the value of f is to use a reference band in the observed infrared spectrum. The investigator who uses this method assumes that the transition moment angle for a particular group frequency observed in the spectrum (for example, a CH_2 stretching mode) is the same as either the one calculated from the molecular structure of the polymer (76) or the one determined for a similar group vibration in a smaller model compound. By use of this assumed value of α_ν and the measured value of D, the value of the constant f in eq. (59) is calculated. This calculated value of f then is used as a constant in eq. (59) for the calculation of α_ν from the measured D values at the other observed absorption frequencies. A reference band calculation of this type must be treated with some caution, since in it either the transition moment angles are assumed to be invariant from compound to compound, which assumption is questionable, or the possible effects of lattice interactions on the transition moment are ignored.

Another method for estimating f, developed by Fraser (77), permits determination of a minimum value for the fraction of perfectly oriented molecules f_m. To use this method, D_0 in eq. (59) is allowed to tend to infinity for $D > 1$ or zero for $D < 1$, giving

$$\begin{aligned} f_m &= (D - 1)/(D + 2) &&\text{when } D > 1 \\ f_m &= 2(1 - D)/(D + 2) &&\text{when } D < 1 \end{aligned} \tag{60}$$

In practice, the most dichroic parallel and perpendicular bands in the observed spectrum are used to calculate the values of f_m. The greatest calculated value of f_m is then chosen for use in estimating the transition moment angle for other bands in the spectrum. By substituting the calculated f_m and the measured dichroic ratio D into eq. (59) a value of D_0 is obtained. The estimated value of α_ν [from eq. (54)] is then limited to the following range:

$$\cot^{-1} \left[\frac{f_m(D + 2) + 2(D - 1)}{2f_m(D + 2) - 2(D - 1)} \right]^{\frac{1}{2}} \leq \alpha \leq \cot^{-1} (D/2)^{\frac{1}{2}}$$

$$\text{when } D > 1$$

and

$$\cot^{-1} \left[\frac{f_m(D + 2) - 2(1 - D)}{2f_m(D + 2) + 2(1 - D)} \right]^{\frac{1}{2}} \geq \alpha \geq \cot^{-1} (D/2)^{\frac{1}{2}}$$

$$\text{when } D < 1$$

(61)

This method for the determination of f has the advantage that neither the correctness of the molecular model nor the invariance of the transition moment angle from compound to compound must be assumed. The utility of the method in molecular conformation studies has been clearly demonstrated by Tsuboi in his study of the molecular conformation of the α form of poly-γ-benzyl-L-glutamate (78). The method has the disadvantage, however, of determining only an estimated value of f (i.e., the minimum value f_m).

The primary difficulty in the experimental determination of α_ν by the infrared technique is that operationally there is one equation with two unknowns [eq. (59)]. What is needed to solve this dilemma is an independent, quantitative, experimental method for determining the unknown constant f.

4. X-ray Diffraction and Determination of the Transition Moment Angle

In a uniaxially oriented polymer, the molecules are not ideally oriented and it is necessary to have some measure of the degree of orientation of the molecules relative to the orientation axis. Fraser (75) considered this problem from two different points of view. One of these was discussed above in Section II-E-3. The other approach has generally been ignored until recently (36). Instead of considering the oriented sample as containing a certain fraction f of perfectly oriented molecules and a fraction $(1 - f)$ of randomly oriented mole-

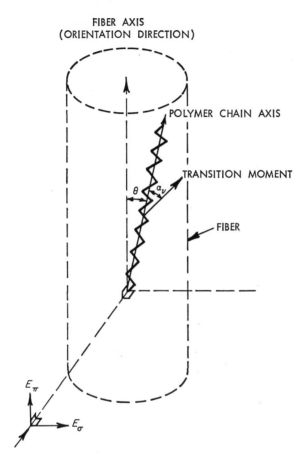

Fig. 25. Schematic representation of uniaxial orientation parameters.

cules, all of the molecules can be considered to be oriented at some average angle θ, relative to the orientation direction, as in Figure 25. The dichroic ratio for this model, in terms of D_0 and the average orientation angle θ, is given by the expression

$$D = \frac{1 + (D_0 + 1) \overline{\cos^2 \theta}}{1 + \frac{1}{2}(D_0 - 1) \overline{\sin^2 \theta}} \qquad (62)$$

which is analogous to eq. (58). The expression for θ is then

$$\overline{\sin^2 \theta} = \frac{2(D_0 - D)}{(D_0 - 1)(D + 2)} \tag{63}$$

A further analysis of these expressions, to find the relation between the fraction of oriented chains f of the first model and the average angle of orientation θ of this model, leads to the expression

$$\left[\frac{3 \overline{\cos^2 \theta} - 1}{2}\right] = \frac{(D - 1)(D_0 + 2)}{(D + 2)(D_0 - 1)} = f \tag{64}$$

Thus, Fraser's fraction f is the same as Hermans' orientation function [eq. (9)] when it is evaluated in terms of the average orientation of the polymer molecules. This deduction is very useful. Sections II-B, II-C, and II-D discussed various methods for determining quantitatively the crystalline and amorphous orientation functions of a uniaxially oriented, polycrystalline polymer film or fiber. Any of these methods, coupled with the infrared dichroic measurements of absorption bands from the appropriate phase will yield quantitative values for the transition moment angles.

For example, the infrared absorption bands from the crystalline region of a uniaxially oriented polymer can be measured and the dichroic ratios determined. The orientation function for the crystalline region of the polymer film f_c can be determined quantitatively from azimuthal, wide angle x-ray diffraction measurements of the appropriate reflecting planes of the crystal and application of Wilchinsky's method (Section II-B) for its calculation. Substitution of the measured f_c and D values into eq. (64) yields directly the value of D_0 and, hence, that of the transition moment angle for that absorption frequency. No assumptions are necessary in this calculation as to the correctness of a molecular model, the invariance of the transition moment angle from compound to compound, or of any approximate character for f.

A higher degree of reliability in the experimentally determined transition moment angle can be achieved if samples with different degrees of molecular orientation are examined by both the infrared dichroic and x-ray diffraction methods. Equation (64) predicts that a plot of f_c, the crystal orientation function determined from x-ray diffraction measurements, versus $(D - 1)/(D + 2)$, where D is

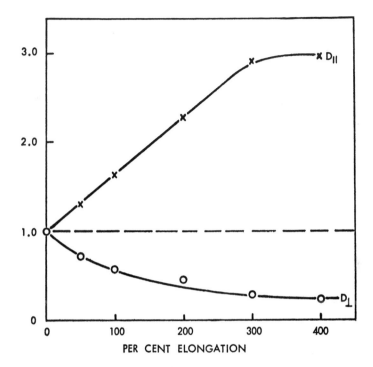

Fig. 26.   Relation between the elongation of isotactic polypropylene films, $D_\parallel$ and $D_\perp$: (O) 1220 cm^{-1} band; (X) 1256 cm^{-1} band (R. J. Samuels, ref. 36).

determined from infrared measurements, will be linear with a zero intercept.   A least-square evaluation of the data from this line should yield a value of the slope, $(D_0 + 2)/(D_0 - 1)$, corrected for random errors and sample differences.   A quantitative value of the transition moment angle can then be calculated from the resultant D_0 value by use of eq. (54).

Some results of this approach with data from uniaxially oriented, isotactic polypropylene and polyethylene films will now be presented to illustrate the general utility of the method.   Figure 26 and Table 4 give experimental infrared data from the same cast, isotactic polypropylene films for which the x-ray diffraction, birefringence, and sonic modulus data are listed in Table 3.   The 1220 cm^{-1} σ band is generally agreed to be due to absorption by the crystalline phase

TABLE 4
Infrared Data for Cast Isotactic Polypropylene Films (36)

Cast sample	Elongation, %	$D_{\parallel}$, 1256 cm^{-1}	$D_{\perp}$, 1220 cm^{-1}
1	0	1.00	1.02
2	50	1.30	1.39
3	100	1.63	1.77
4	200	2.28	2.19
5	300	2.92	3.58
6	400	2.96	4.21

only (see Fig. 23), although there is no general agreement as to the assignment of the vibrational modes of the chain characterized by this band, as shown in Table 5. There is agreement in the vibrational mode assignments only to the extent that the various authors (70,71,79–82) characterize the band as represented by the motion of one, two, or a mixture of three of the following bonds: CH_2, CH, C—CH_3 or an equatorial C—C. Isotactic polypropylene has a hel-

TABLE 5
Various Assignments for the 1220 cm^{-1} and 1256 cm^{-1} Infrared
Absorption Bands of Isotactic Polypropylene[a]

σ(1220 cm^{-1}) E	π(1256 cm^{-1}) A	Ref.
γ_w(C—H)	γ_w(C—H)	70
γ_w(C—H)	γ_w(C—H)	79
γ_r(CH) mixed with γ_r(CH$_2$), γ_r(CH$_3$)	γ_r(CH) mixed with γ_r(CH$_2$), γ_r(CH$_3$)	71
γ_t(CH$_2$)(35), δ_{ax}(CH)(15), ν_{eq}(C—C)(15)	γ_t(CH$_2$)(35), δ_{ax}(CH)(15), δ_{eq}(CH)(15)	80
γ_t(CH$_2$)(25), δ(CH)(15), ν_{eq}(C—C)(20)	γ_t(CH$_2$)(35), δ_{ax}(CH)(20), δ_{eq}(CH)(10)	81
γ_t(CH$_2$)(26), δ(CH)(20), ν(C—CH$_3$)(19)	δ(CH)(37), γ_t(CH$_2$)(19), γ_r(CH$_3$)(15)	82

[a] Symbols: σ, perpendicular band; π, parallel band; E, phase difference = $\pm 2\pi/3$; A, phase difference = 0; γ_w, wagging; γ_r, rocking; γ_t, twisting; δ, bending; ν, stretching; ax, axial; eq, equatorial; numbers in parentheses indicate approximate potential energy distribution, %.

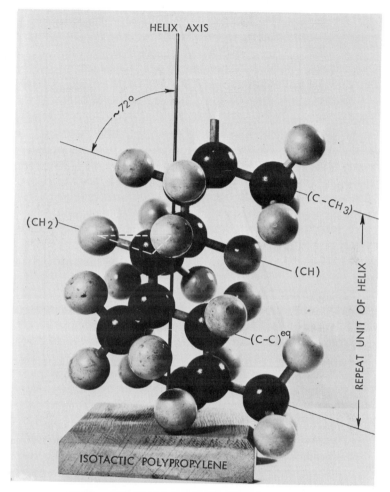

Fig. 27.   Conformation of isotactic polypropylene.

ical conformation in the crystalline phase (27) and each of these bonds makes an angle of about 72° with the helical axis as shown in Figure 27.

In Figure 28, the crystal orientation function f_c (from x-ray diffraction, Table 3) is plotted against the infrared 1220 cm^{-1} σ-band dichroic ratio according to eq. (64).   The experimental points fit

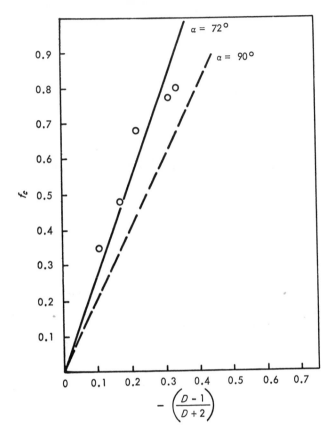

Fig. 28. Determination of the transition moment angle α_ν for the 1220 cm^{-1} band in isotactic polypropylene (R. J. Samuels, ref. 36).

a calculated line drawn for $\alpha_\nu = 72°$ quite reasonably. This indicates that the transition moment is directed along one of the bonds mentioned above. The dashed line in Figure 28 has been calculated for $\alpha_\nu = 90°$ (i.e., where the transition moment of the σ band is perpendicular to the helical axis of the chain). An assignment of $\alpha_\nu = 90°$ for the 1220 cm^{-1} band would not be reasonable in light of the experimental data.

A further example of the general utility of this method is illustrated in Figure 29, which represents the data of Norris and Stein

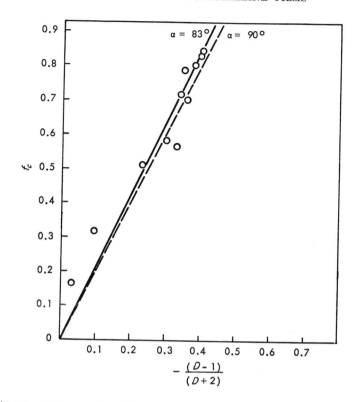

Fig. 29.   Determination of the transition moment angle α_ν for the 730 cm^{-1} band in low-density polyethylene based on the data of R. S. Stein and F. H. Norris (57) (R. J. Samuels, ref. 36).

(57) for low-density polyethylene.   Polyethylene has a planar zig-zag conformation in the crystal lattice.   The 730 cm^{-1} σ band absorbs only in the crystalline region and has been assigned as a CH$_2$ rocking mode.   The α_ν for this motion would be expected to be 90° to the backbone axis.   From Figure 29, $\alpha_\nu \cong 83°$ seems to be more reasonable.   The observed difference in α_ν between theory and experiment may be due to experimental difficulties in separating the 730–720 cm^{-1} doublet into its components, orientation effects of the amorphous region in the 720 cm^{-1} band affecting the resolution of the 730 cm^{-1} band, or possible coupling effects from bending and stretching modes.

The above method of relating orientation functions to infrared dichroism is a general one. If absorption takes place only in the crystalline phase, the α_ν values can be determined quantitatively for any crystal by using Wilchinsky's method for determining f_c. This method can thus have considerable utility in structure analysis.

Absorption, however, does not always take place in only one phase of the polymer. There are some infrared absorption bands which are characterized by absorption in both the crystalline and amorphous phases. The 1256 cm^{-1} π band of isotactic polypropylene is just such a mixed band. Since this band results from absorption in both the crystalline and amorphous regions, its infrared dichroism would be expected to correlate with some average orientation function. Such an average can be expressed as the average orientation of each phase weighted by the amount of each phase present

$$f_{\mathrm{av}} \equiv \beta f_c + (1 - \beta)f_{\mathrm{am}} \tag{65}$$

where β, the fraction of crystals, is determined from density, f_c is derived from x-ray diffraction, and f_{am} is determined from sonic modulus. Figure 30 is a plot of f_{av} against $(D - 1)/(D + 2)$ for the 1256 cm^{-1} π band of the oriented, isotactic polypropylene samples. A good linear plot with a zero intercept is obtained as predicted.

If, in the spectra of the polymer film, absorption bands are present which are characteristic of the amorphous region only, then the infrared dichroism of these bands should correlate linearly with the amorphous orientation function according to eq. (64). Structural interpretation of the absorption spectra of amorphous polymers is quite complex (83) and this technique would be very helpful in characterizing assignments. To the author's knowledge, no experimental work has been done along these lines.

The above discussion has emphasized how quantitative structural information about polymer chains can be obtained by means of x-ray diffraction, birefringence, and sonic modulus measurements along with infrared dichroism studies on the uniaxially oriented polymer films. A primary purpose of this chapter is to elucidate methods for the determination of the orientation characteristics of deformed polymer films. Obviously, once the value of the transition moment

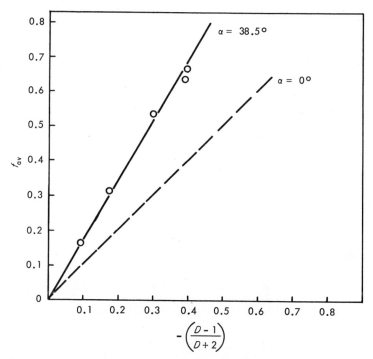

Fig. 30. Relation between the average orientation function f_{av} and the infrared dichroism of the 1256 cm^{-1} band in isotactic polypropylene (R. J. Samuels, ref. 36).

angle for a given absorption frequency has been quantitatively determined by the above method, eq. (64) may be used to determine the orientation function from infrared dichroism measurements only. Thus, infrared dichroism may be used to obtain quantitative values for f_c and f_{am}, provided absorption bands characteristic of each individual phase can be obtained. If, as has been observed with isotactic polypropylene, only bands uniquely characteristic of the crystalline region are available, the infrared dichroic measurement can be combined with sonic modulus or birefringence measurements to determine f_c and f_{am} quantitatively. Infrared dichroism is thus a powerful tool for quantitatively characterizing the molecular morphology of uniaxially deformed polymer systems.

III. Spherulite Deformation

A. Introduction

In the previous section, the polycrystalline polymer was considered as a two-phase system composed of crystallites and noncrystalline regions whose individual orientations could be followed by a number of independent physical methods. In many polymers, under appropriate processing conditions, the crystallites and noncrystalline regions will arrange themselves in an ordered superstructure called a spherulite. The polymer film will then be composed of many spherulites and any deformation process will involve the deformation of the spherulites as well as the orientation of the crystallite and noncrystalline substructure.

Polyethylene, polypropylene, nylon, Teflon, polyethylene terephthalate, polyhexamethylene sebacamide, and poly-γ-methyl-L-gluta-

Fig. 31. Phase microscope photograph of an isotactic polypropylene spherulite.

mate are some of the polymers that form one or more types of spherulites. A spherulite may be described as a spherically symmetrical aggregate of crystalline and amorphous (noncrystalline) polymer. The crystallites are arranged in large part along radial fibrils which have a common center at the site of primary nucleation. Figure 31 is a phase-contrast photograph of an isotactic polypropylene spherulite. Notice the radial fibrils all emanating from the site of primary nucleation. The crystallites are aligned mainly along these radial fibrils, although some secondary crystallization can occur in the interradial region. The noncrystalline polymer can also be inter- or intraradial but is found predominantly in the interradial regions.

The intraradial crystallites generally have a preferred orientation with respect to the radial direction in the spherulite. This orientation has been determined by micro x-ray diffraction (84,85) and electron diffraction studies (86,87) of selected regions of spherulites. The polymer chain axis in the crystallites has generally been found to be oriented perpendicular (or at some angle toward the perpendicular) to the radial direction in the spherulite and hence is tangential with respect to the spherulite. This observation has led to recent speculation that the crystallites arranged along the radii of the spherulite may be in the form of a folded-chain lamella with the chain axis oriented perpendicular to the spherulite radius.

Depending on the processing conditions, the same polymer may form spherulites having positive, negative, or mixed birefringence. The birefringence of the spherulite Δ_s is defined as $(n_r - n_t)$, where n_r is the refractive index parallel to the radial fibril and n_t is the refractive index perpendicular (tangential) to the radial fibril. If the noncrystalline polymer is assumed to be isotropic (by no means true), then the birefringence of the spherulite will be directly related to the birefringence of the crystallites along the fibril. In polymers, the largest refractive index γ is along the chain axis while the refractive indices α and β are perpendicular to the chain axis ($\beta > \alpha$).

If the crystallite is biaxial (i.e., $\beta \neq \alpha \neq \gamma$), then Δ_s can be either negative or positive. For example, if the chain axis is assumed to be perpendicular to the radial axis, i.e., γ is perpendicular, two possibilities can be considered. (1) If α is parallel to the radius, the spherulite birefringence Δ_s is negative because α is less than both β and γ (see above) and hence $n_t > n_r$. (2) If β is parallel to the radius, n_r can be greater or smaller than n_t. Since the chain axis is rotationally symmetrical around the radial axis, n_t will equal the average refractive index of the two (γ and α) refractive indices. Depending on the magnitude of α, then, n_t can be either greater or smaller than n, (which is dependent only on β). If n_t is greater than n_r, the birefringence will be negative [as in (1)], but if n_t is smaller than n_r, the birefringence will be positive. Hence, the sign of the spherulite birefringence is a function of crystallite orientation.

Polyhexamethylene sebacamide provides a practical illustration. This polymer forms either negative or positive spherulites depending on the processing conditions. Micro x-ray diffraction (84) studies have shown that the chain axis of this polymer is perpendicular to the radial axis of the spherulite for both positive and negative birefringences. The a axis of the crystallite was found to be parallel to the radial direction in the positively birefringent spherulite, while the b axis of the crystallite was found to be oriented parallel to the radial direction in the negatively birefringent spherulite. The refractive index of the a axis is therefore greater than the refractive index of the b axis, and the average refractive index of the b and c axes is less than the refractive index of the a axis.

Positive and negative spherulites of the polypeptide poly-γ-methyl-L-glutamate can be obtained by careful precipitation from solution (88). The type of spherulite obtained will be a function of the time the solution is aged before precipitation occurs. Numerous types of positive, negative, mixed, or ringed spherulites can be obtained from isotactic polypropylene, depending on temperature and time of annealing (89). Similarly, many other polymers yield different spherulitic forms, depending on the conditions used in their preparation (90). Clearly, the morphology one obtains in a spherulite will be some function of the processing conditions used to make the film.

Just as spherulite morphology is a function of processing conditions, so too is the resulting size of the spherulite. The spherulites grow from primary nuclei and, hence, the size of the spherulite is governed by the number of primary nuclei present. The number of primary nuclei present is a function of the growth temperature, maximum temperature, time of melting, and the concentration of impurities present. Generally, large spherulites are formed from melts which have been heated to a high enough temperature so that any trace of previous crystallinity is lost, and from samples crystallized at low degrees of supercooling where spontaneous nucleation rates are low. The size of spherulites can range from as large as centimeters to less than a micron in diameter, depending on the processing conditions used.

The size of the spherulite dictates the method available for its measurement. Spherulites centimeters in diameter can be measured directly (with a ruler). The size of spherulites larger than a few microns in diameter can be determined with a polarizing microscope, or, within a limited size region, by means of light scattering measurements. For spherulites smaller than a few microns in diameter, the only methods of size measurement available are the electron microscope and light scattering techniques.

An orientation step is often included during the production of a polymer film to impart added mechanical strength to the product. If the undeformed film was spherulitic, the spherulites will be deformed during the orientation step. Depending on the speed and temperature of the orientation step, spherulite deformation may or may not be affine. An affine deformation of the spherulite is one in which a given deformation of the film will lead to an equivalent deformation of the spherulite. Takayanagi, Minami, and Nagatoshi (91) have shown that this type of deformation occurs in polypropylene when the temperature of the deformation process is at the crystal absorption temperature of the dynamic mechanical spectra (110°C) and the speed of orientation is not so rapid as to exceed the energy limits at that temperature for mobility of the crystallites. The film stretched under these conditions does not manifest necking but deforms uniformly. Films produced under these conditions had the best combination of high orientation, breaking strength, and modulus of elasticity that Takayanagi could produce. Clearly, a knowledge of spherulite size, extension ratio (i.e., amount of deformation), and the mechanism of deformation of the spherulites is desirable.

One method that can be used to obtain information about the mechanism of spherulite deformation is wide angle x-ray diffraction. Wilchinsky (92) derived a theory relating the crystal orientation function (Section II-B) of uniaxially oriented samples to the deformation of the spherulites. Examination under the electron microscope of surface replicas from deformed spherulites had revealed (93) that the spherulitic shape becomes ellipsoidal with elongation. Starting with this observation as his model, Wilchinsky made the following assumptions about the deformation mechanisms in the spherulite:

(1) The crystallites remain undeformed during spherulite deformation.

(2) The crystallites are connected by tie molecules.

(3) The chain axis of the crystallites tends to orient in the direction of the deforming strain.

He then calculated values of the crystal orientation function at various extension ratios λ for two kinds of spherulite deformation in a fiber. One (expressed as f_λ), he assumed to have uniform deformation within the spherulite; in the other $(f_{\lambda'})$, he included the enhanced orientation of the inner regions of the spherulite when it is deformed. Both were derived for simple elongation at constant volume. In both, the spherulite was assumed to increase in length by an extension ratio λ while it decreased $\lambda^{-\frac{1}{2}}$ in the other two orthogonal directions. Equating λ of the spherulite with λ of the deformed sample implies an affine deformation. The two models differ in the nature of the crystalline displacement within the affinely deformed spherulite.

Wilchinsky found that data from uniformly deformed polyethylene films agreed with his theoretical curve for enhanced orientation of the regions of the spherulites. The experimental values of the crystal orientation function f_c for the uniaxially deformed isotactic polypropylene films reported in Table 3 fall between Wilchinsky's two theoretically derived curves (Fig. 32). This indicates that an affine deformation of the spherulites occurred when these isotactic polypropylene films were uniformly drawn at 110°C. The exact nature of the crystallite displacement mechanism (uniform or enhanced orientation) within the spherulites, however, is uncertain.

A much more direct method for examining the morphology of deformed spherulites is the combination of electron microscopy and electron diffraction. Generally, both of these measurements can be made with the same instrument (electron microscope), often on the same portion of the specimen. In theory, one can obtain microscopic resolution on the order of a few angstroms and, hence, small regions of deformed spherulites could be seen by means of this tech-

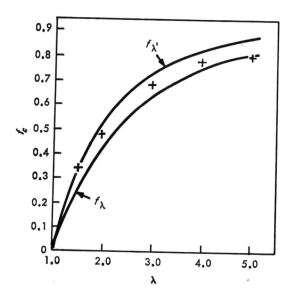

Fig. 32. Relation between the predicted and experimental crystal orientation functions and the extension ratio of the sample: (——) predicted; (+) experimental (R. J. Samuels, ref. 36).

nique. Electron diffraction patterns then could be obtained from the same specimen region and the crystallite orientation analyzed. Theoretically, this tool should be useful in the elucidation of spherulite deformation mechanisms; practically, however, little use has been made of the technique for this purpose. The primary reasons for this lack of use are the short exposure times available and the difficulty in sample preparation.

Electron scattering intensities are very high with respect to those of x-rays, the ratio of x-ray to electron scattering intensities being $1:10^6$. This means that several seconds of exposure to an electron beam will yield an electron scattering pattern equivalent in intensity to a diffraction pattern obtained from several hours of exposure to x-rays. An added complication is that the polymer structure is quickly destroyed by the electron beam; therefore, little time is available experimentally to align a desired region of the sample in the microscope and obtain both a micrograph and an electron diffraction pattern.

Similarly, the size of the specimen that can be observed must be varied according to the type of radiation used. While the linear dimension of the specimen is about 1 mm for x-rays, it is only about 10^{-5} to 10^{-4} mm for electrons. Polymer spherulites of general interest are of linear dimensions considerably greater than several hundred angstroms, and consequently sample preparation is a major problem in electron microscopy.

Most investigators have had to resort to either surface replication techniques (90,93–95), which allows only observation of the surface deformation of the spherulites, or fracture surface studies (96) to obtain specimens for observation. Ingram and Peterlin (86) have shown how this limitation could in part be overcome by preparing solution-cast polyethylene spherulites on Mylar film. The Mylar film with the attached spherulite (two-dimensional) was then deformed 30%, backed with carbon, shadowed, and examined by electron microscopy and diffraction techniques. In this manner, the authors were able to show evidence for the drawing out of oriented fibers from interspherulitic boundaries, tilting and/or rotation of the crystallographic axes of lamella regions, lamella slip, and possible phase changes during deformation. Similarly, Selikhova, Markova, and Kargin (87) have been able to obtain electron diffraction patterns from the radial fibrils of polypropylene spherulites which argee with micro x-ray diffraction results (85).

These techniques are limited by the fact that bulk samples of three-dimensional spherulites, which are the specimens of most practical interest, could not be examined. Bassett (97) has developed a selective dissolution technique which allows thin surface layers from bulk samples to be detached for study in the electron microscope. The resulting detachment replica diffracts electrons and, hence, the same region can be examined by electron diffraction and microscopy techniques. The problem with a technique of this type is that it is impossible to know whether the detachment replica under observation actually represents the structure present in the bulk sample or if the dissolution procedure produced morphological changes in the specimen.

Another approach to the problem of bulk sample preparation is that of ultramicrotoming sections for examination. Rusnock and Hanson (98) have developed a sample embedding technique which allowed them to ultramicrotome 500-Å sections of nylon 66 spherulites from molding pellets and filaments and examine them under the electron microscope. No attempt was made to obtain electron diffraction patterns from the samples, although there seems no reason why this should not have been possible. Attempts should certainly be made to extend this technique to other polymers. Clearly, sample preparation is a major problem in the application of electron microscopy and diffraction techniques to the study of spherulite deformation mechanisms.

An attractive extension of the electron diffraction technique has recently been developed by Bassett and Keller (99). From detachment replicas of bulk polyethylene films, they have obtained small angle electron diffraction patterns, which are similar in form to small angle x-ray diffraction patterns. Spacings as large as 1400 Å were observed from replicas of sections of drawn, linear polyethylene films which had originally contained ringed spherulites in the undrawn film. These spacings were obtained from replica regions which were observed to contain a quasi-periodic arrangement of dark lines having separations which varied between 500 and 5000 Å in different regions in the field. The small angle spacing was therefore interpreted as arising from the quasi-periodic structure. Thus, wide angle electron diffraction, small angle electron scattering, and visible observations of small regions of bulk samples can be obtained with the electron microscope.

The electron microscope will certainly contribute significant information on deformation mechanisms in spherulitic polymers. The technique, however, is a destructive one, i.e., the sample to be examined must be destroyed during its examination. To be able to observe the deformation behavior of spherulites in bulk polymer samples under controlled conditions of temperature and stress without destruction of the sample is highly desirable. Small angle light scattering (SALS) is the only system now available for quantitatively satisfying this condition. Bulk samples up to 10 mils thick and spherulites ranging from 0.1 to 50 μ in diameter have been examined nondestructively by means of this technique. By examining changes in the light scattering patterns at small angles from the incident beam, as will be described in the following section, information can be obtained about the size and deformation of the spherulites in the sample. For this reason, the SALS technique is an important tool for elucidation of the morphology of deformed spherulitic films. It is a new technique, however, and almost no advantage has been taken of it for the solution of practical film problems.

B. Small Angle Light Scattering (SALS)

1. Undeformed Spherulites

A spherulite is a three-dimensional, spherically symmetrical aggregate of crystalline and noncrystalline polymer. Any theory which purports to describe small angle light scattering (SALS) behavior of spherulites must use as its model the known three-dimensional structure of the spherulite as observed by light and electron microscopy. The amplitude method (100) for calculating the intensity envelope of scattered radiation is most useful for the theoretical small angle light scattering approach since it requires a model for its derivation. By choosing a model which realistically represents the known morphological characteristics of the spherulites, the observed small angle light scattering behavior of spherulitic film should be predictable.

Typical SALS patterns obtained from undeformed, isotactic polypropylene spherulites are shown in Figure 33. These patterns were observed photographically by means of the system illustrated schematically in Figure 34 (101). A continuous wave He–Ne gas laser is used as the polarized, monochromatic ($\lambda = 6328$ Å) light source. The vertically polarized incident light from the laser, E_0 impinges on the sample and is scattered by the spherulites. For uniaxially deformed samples, the stretch direction (SD) is aligned parallel to the plane of the incident light (Z). The scattered ray E passes through the analyzer and impinges on the flat-plate photographic film. The position of the scattered reflection is defined by two angles θ and μ. The radial angle θ is defined as the angle between the unit vectors S_0 and S which specify the direction of the incident and scattered beams, respectively. The azimuthal (tilt) angle μ is defined as the angle between the Z axis (stretch direction) and the projection of the difference $S - S_0$ between the unit vectors on the yz plane.

An analyzer is placed between the sample and the photographic film in Fig. 34. When the plane of the analyzer is horizontal (i.e., the plane of the analyzer and the plane of the incident light are

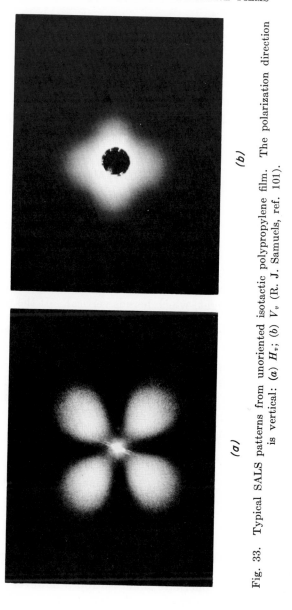

(a)

(b)

Fig. 33. Typical SALS patterns from unoriented isotactic polypropylene film. The polarization direction is vertical: (a) H_v; (b) V_v (R. J. Samuels, ref. 101).

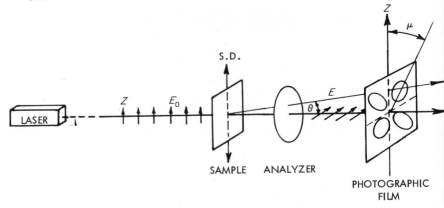

Fig. 34. Diagram of the photographic SALS system (R. J. Samuels, ref. 101).

perpendicular), an H_v SALS pattern is obtained (Fig. 33a). By convention, the subscript in the term H_v defines the plane of the incident polarized light, which here is vertical. If the plane of the analyzer is vertical (i.e., the plane of the incident polarized light and the plane of the analyzer are parallel) a V_v pattern is obtained (Fig. 33b).

Since the undeformed spherulite is spherical, the simplest model to apply to the amplitude method to describe the observed scattering behavior would be that of a uniform isotropic sphere of polarizability α_0 and radius R_0. Starting with this model, the following expression is obtained for the scattered intensity (100):

$$I = AV_0^2\alpha_0^2[(3/U^3)(\sin U - U \cos U)]^2 \qquad (66)$$

where I is the intensity, A is a proportionality constant, V_0 is the volume of the isotropic sphere, the shape factor

$$U = (4\pi R_0/\lambda') \sin (\theta/2)$$

λ' is the wavelength of light in the medium, and θ is the polar scattering angle.

The undeformed spherulite is not isotropic, however (see Section III-A); instead, it has different radial and tangential refractive indices due to the ordered arrangement of anisotropic crystallites along its radii. Thus, a more reasonable model with which to represent the undeformed spherulite is that of an anisotropic sphere in an isotropic medium.

Stein and Rhodes (102) considered the problem of the SALS patterns to be expected from a homogeneous anisotropic sphere in an isotropic medium and derived the following general equations for the intensity of the scattered light from a V_v or H_v measurement:

$$I_{V_v} = A V_0^2 (3/U^3)^2 [(\alpha_t - \alpha_s)(2 \sin U - U \cos U - \text{Si } U)$$
$$+ (\alpha_r - \alpha_s)(\text{Si } U - \sin U)$$
$$+ (\alpha_t - \alpha_r) \cos^2 (\theta/2) \cos^2 \mu (4 \sin U - U \cos U - 3 \text{ Si } U)]^2 \quad (67)$$

$$I_{H_v} = A V_0^2 (3/U^3)^2$$
$$[(\alpha_t - \alpha_r) \cos^2 (\theta/2) \sin \mu \cos \mu (4 \sin U - U \cos U - 3 \text{ Si } U)]^2$$
$$(68)$$

where V_0 is the volume of the anisotropic sphere, α_t and α_r are the tangential and radial polarizabilities of the sphere, α_s is the polarizability of the surroundings, θ and μ are the radial and azimuthal scattering angles, respectively (Fig. 34), and A is a proportionality constant. Si U is the sine integral defined by

$$\text{Si } U = \int_0^U \frac{\sin x}{x} \, dx \quad (69)$$

and is solved as a series expansion sum for computational purposes; U has the same definition as in the isotropic sphere model

$$U = \frac{4\pi R_0}{\lambda'} \sin (\theta/2) \quad (70)$$

except that R_0 is now the radius of the anisotropic sphere. The quantity U has the same definition for both models since they both are spherical and U depends only on the shape of the model. In the limit of isotropic spheres, eqs. (67) and (68) reduce to eq. (66).

Figure 35 contains the predicted H_v and V_v SALS patterns for undeformed, isotactic polypropylene spherulites (101), derived by using eqs. (67) and (68) with the shape factor defined by eq. (70). These theoretical SALS patterns are plotted with contour lines of constant intensity. The inner contour line of a lobe is the line of highest intensity. The contour lines then decrease in intensity going out from the center of a lobe, the outermost contour line being of lowest intensity. The spherulite birefringence $\Delta_s = (n_r - n_t)$ is

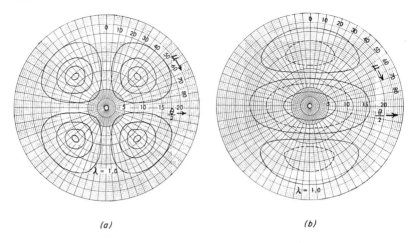

(a)   (b)

Fig. 35.   Theoretical SALS patterns for unoriented isotactic polypropylene film.   The polarization direction is vertical: (a) H_v; (b) V_v (R. J. Samuels, ref. 101).

known to equal -0.003 for isotactic polypropylene (89) and, hence, an $(\alpha_t - \alpha_r)$ of $+0.003$ was used in eqs. (67) and (68).   In order to derive these equations, α_t, α_r, and α_s had to be assumed to be sufficiently close in magnitude that the incident light wave is not affected appreciably on passing through the sphere boundary, while the large dependency of the observed V_v SALS intensity on the azimuthal angle μ indicated that the coefficient $(\alpha_t - \alpha_r)$ in the third term of eq. (67) must be as large as or larger than the coefficients $(\alpha_t - \alpha_s)$ and $(\alpha_r - \alpha_s)$ of the first and second terms (102).   Thus, the difference between the polarizability of the spherulite and its surroundings must be equal to or less than the anisotropy of the spherulite.   Since the surroundings of a polypropylene spherulite in a film are other polypropylene spherulites, these conditions seem reasonable.   In calculating the theoretical V_v SALS pattern in Figure 35, the polarizability of the surroundings α_s was assumed to be equal to the tangential polarizability of the sphere α_t while the difference between the radial polarizability α_r and that of the surroundings is assumed to be equal to $(\alpha_t - \alpha_r)$.   The agreement between the theoretical SALS patterns in Figure 35 and the experimental SALS patterns in Figure 33 demonstrates that the model of the spherulite as an anisotropic sphere in an isotropic medium represents validly the observed SALS behavior of undeformed, isotactic polypropylene spherulites. Stein and Rhodes (102) reached a similar conclusion about the validity of the model in representing the observed SALS patterns from undeformed polyethylene spherulites.

The average size of the undeformed spherulites can be determined from an analysis of the H_v SALS pattern. The intensity of the H_v SALS from undeformed spherulites will go through a maximum with increasing radial scattering angle $\theta/2$ independent of the azimuthal angle examined, as illustrated in a later discussion of photometric measurements of H_v SALS patterns (see Fig. 41, $\lambda = 1.0$). The value of U in eq. (70) at this $(\theta/2)_{max}$ equals 4.09. Thus, the average radius R_0 of the anisotropic spheres (undeformed spherulites) can be obtained from a determination of $(\theta/2)_{max}$ from the H_v SALS pattern and subsequent solution of eq. (70).

2. Deformed Spherulites

a. Theory. The unique characteristic of the SALS technique is that with it the deformation behavior of spherulites can be followed nondestructively under controlled conditions of temperature and stress. The theoretical interpretation of the SALS behavior of undeformed spherulites extended the isotropic scattering equation [eq. (66)] specifically to include the anisotropy of polarizability. To interpret the SALS behavior of deformed spherulites, the anisotropy of the shape of the spherulite imposed by deformation must also be considered. The small angle light scattering theory has been extended to uniaxially deformed spherulites (101). As a result, the observed changes in the SALS patterns with deformation can be interpreted theoretically in terms of the size of the original undeformed spherulite and the subsequent deformation the spherulites have undergone.

When a spherulite undergoes a uniform uniaxial deformation, it changes its shape from a sphere to a prolate spheroid. The axial ratio of the resulting spheroid is a function of the extension of the spherulite. Thus, the shape of the model used to represent the small angle light scattering from a deformed spherulite must be changed from a sphere to a prolate spheroid.

A change in the model from a sphere to a spheroid results in a change in the definition of the shape factor U in eqs. (66)–(68).

The model used is an ellipsoid of revolution with semiaxes R and vR. The shape factor for this model is defined by the expression (103–105)

$$U = \frac{4\pi R}{\lambda'} \sin\left(\frac{\theta}{2}\right)(\sin^2 \Psi + v^2 \cos^2 \Psi)^{\frac{1}{2}} \qquad (71)$$

where λ' is the wavelength of light in the medium, θ is the radial angle defined previously, and Ψ is the angle between the Z axis (stretch direction) and the vector $\mathbf{S}$ (the difference $\mathbf{S} - \mathbf{S}_0$ between the unit vectors). Since the relation between the angle Ψ and the azimuthal angle μ (106) is $\cos \Psi = \cos(\theta/2) \cos \mu$, the expression for U can be written in the form

$$U = \frac{4\pi R}{\lambda'} \sin\left(\frac{\theta}{2}\right)\left[1 + (v^2 - 1)\cos^2\left(\frac{\theta}{2}\right)\cos^2 \mu\right]^{\frac{1}{2}} \qquad (72)$$

For eq. (72) to be useful, R and v must be defined in terms of the extension ratio λ_s of a uniaxially deformed spherulite at constant volume.

When a spherulite of radius R_0 undergoes a simple extension at

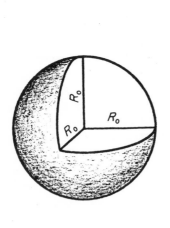
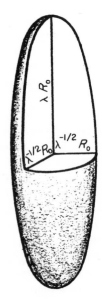

Fig. 36. Schematic representation of spherulite deformation.

constant volume, the radius is extended $\lambda_s R_0$ in the direction of extension while it contracts $\lambda_s^{-\frac{1}{2}} R_0$ in the two perpendicular directions (Fig. 36). The resulting deformed spherulite is a prolate spheroid with semimajor axis $\lambda_s R_0$ and semiminor axes $\lambda_s^{-\frac{1}{2}} R_0$. The model used to derive eq. (72) was an ellipsoid of revolution with semimajor axis vR and semiminor axes R. By identifying the dimensions of the prolate spheroid (spherulite) with those of the model used to derive eq. (72), the following expressions are obtained:

$$R \equiv \lambda_s^{-\frac{1}{2}} R_0$$
$$vR \equiv \lambda_s R_0$$
$$v \equiv \lambda_s^{\frac{3}{2}} \qquad (73)$$

Substituting these definitions for R and v into eq. (72), the following final form for the shape factor is obtained:

$$U = \frac{4\pi R_0 \lambda_s^{-\frac{1}{2}}}{\lambda'} \sin\left(\frac{\theta}{2}\right) \left[1 + (\lambda_s^3 - 1)\cos^2\left(\frac{\theta}{2}\right)\cos^2 \mu \right]^{\frac{1}{2}} \qquad (74)$$

The deformed spherulite is an anisotropic spheroid and eqs. (67) and (68) represent the behavior of an anisotropic spheroid in an isotropic medium when U is defined by the general shape factor given in eq. (74). Thus, eqs. (67) and (68), U being defined by eq. (74), are general equations describing the scattering from anisotropic spheroids in an isotropic medium. The models of an anisotropic sphere in an isotropic medium, and an isotropic sphere, represent limiting cases of the general equations.

b. Photographic SALS Measurements. Flat-plate photographs of the H_v SALS patterns from the same isotactic polypropylene films from which all of the data in Tables 2–4 were obtained are shown in Figures 37 and 38. The photographs were obtained with the experimental arrangement shown schematically in Figure 34. Each of the H_v SALS patterns for the affinely deformed films has a matching theoretical pattern alongside. The general form of eq. (68) [i.e., U defined by eq. (74)] was used to calculate the theoretical H_v SALS patterns to be expected for a spherulite of a given initial radius R_0 as it is extended uniformly to the different extension ratios, λ_s. The theoretical H_v SALS patterns were calculated for a spherulite having an initial radius $R_0 = 0.66$ μ and a value of $(\alpha_t - \alpha_r) = +0.003$. The agreement between the experi-

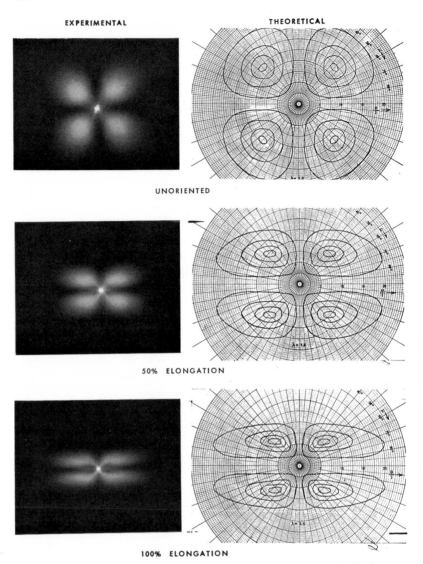

Fig. 37. Theoretically predicted and experimentally determined changes in the H_v SALS pattern of isotactic polypropylene film with elongation. The polarization direction and the film stretch direction are vertical (R. J. Samuels, ref. 101).

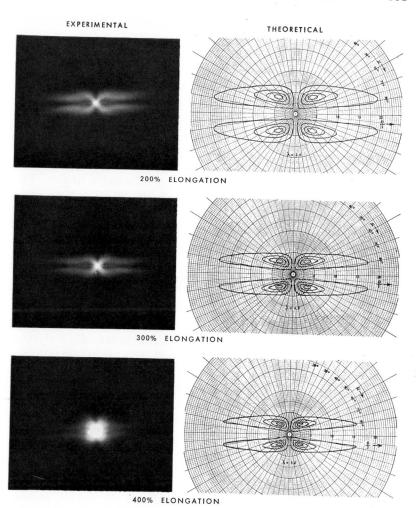

EXPERIMENTAL THEORETICAL

200% ELONGATION

300% ELONGATION

400% ELONGATION

Fig. 38. Theoretically predicted and experimentally determined changes in the H_v SALS pattern of isotactic polypropylene film with elongation. The polarization direction and the film stretch direction are vertical (R. J. Samuels, ref. 101).

mentally determined and theoretically calculated patterns is evident, particularly the following features:

(1) The change in shape of the clover leaf patterns with elongation of the experimental and theoretical figures is the same.

(2) The theoretical lobes are drawn with contour lines of constant intensity. The center contour line of the lobe is the one of highest intensity, while the outermost line of the lobe is that of lowest intensity. The corresponding contour lines from pattern to pattern are of identical intensities. As the elongation increases, the high intensity regions move to lower radial angles while the lower intensity contour line stays elongated to a high radial angle. The experimental flat-plate photographs of the H_v SALS pattern show this theoretically predicted intensity behavior. There is an increase in the concentration of intensity at low radial angles of the lobe and a decrease in the intensity of the extended wings of the lobe at higher radial angles with increasing elongation of the samples.

Similarly, by using the general form of eq. (67) [with U defined by eq. (74)] one may calculate the theoretical V_v SALS patterns to be expected from a spherulite of a given radius R_0 as it is extended uniformly to different extension ratios λ_s. The calculated V_v SALS patterns for a spherulite having an initial radius $R_0 = 0.66$ μ, $\alpha_t - \alpha_r = +0.003$, $\alpha_t - \alpha_s = 0.000$, and $\alpha_r - \alpha_s = +0.003$ are illustrated for several elongations in Figure 39.

Here, again, the theoretical lobes are drawn with contour lines of constant intensity. The outermost continuous line in each lobe is the one of weakest intensity, the dashed line is next strongest, and the inner continuous line is of highest intensity. This highest-intensity lobe is found only near the center of the pattern and is not found in the vertical lobes. The corresponding contour lines from pattern to pattern are of identical intensities. At the side of each calculated pattern is an experimental V_v SALS pattern obtained from the same isotactic polypropylene cast films ($R_0 = 0.5$–0.7 μ) used to obtain the H_v SALS patterns.

The black spot in the center of each pattern is caused by a beam stop which was found necessary to prevent overexposure due to the high intensity at the center of the pattern. The agreement between the theoretically predicted and experimentally observed V_v SALS patterns is striking. The vertical lobes of the pattern obtained from the undeformed spherulites ($\lambda = 1.0$) extends to higher radial

EXPERIMENTAL   THEORETICAL

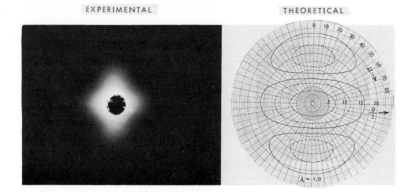

UNORIENTED

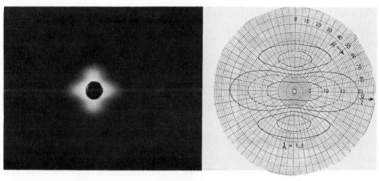

50% ELONGATION

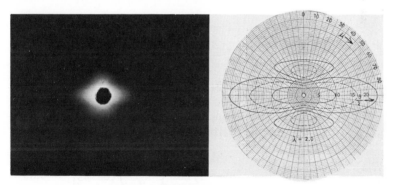

100% ELONGATION

Fig. 39. Theoretically predicted and experimentally determined changes in the V_v SALS pattern of isotactic polypropylene film with elongation. The polarization direction and the film stretch direction are vertical (R. J. Samuels, ref. 101).

($\theta/2$) angles than does the horizontal lobe. On deformation of the spherulite by as little as 50% elongation, the situation is reversed and the horizontal lobe now extends to a higher radial angle than that of the vertical lobes. Continued extension of the spherulites to 100% elongation serves to further increase the disparity in radial angle between the horizontal and vertical lobes.

Thus, by using a model which is consistent with the known structure of the spherulites (i.e., an anisotropic spheroid in an isotropic medium), the photographically observed SALS behavior of uniaxially deformed spherulites can be described theoretically.

c. Photometric SALS Measurements. A theoretical explanation of the observed SALS patterns from deformed films gives the experimenter considerable insight into the effect of processing conditions on the behavior of the spherulites. A quantitative numerical measurement of the deformation the spherulite has undergone would be even more useful to him. With the equations for the anisotropic spheroid, a method can be developed for determining both the initial radius R_0 and the extension ratio λ_s from an analysis of the small angle light scattering pattern from deformed spherulites.

Due to the dependency of the terms $(3/U^3)(4 \sin U - U \cos U - 3 \operatorname{Si} U)$ in eq. (68) on U, a maximum intensity will always be observed at a value of $U = 4.09$, provided $[\cos^2 (\theta/2)]^2$ in eq. (68) has a value of unity. Thus, if the radial intensity distribution at a fixed azimuthal angle μ_1 is measured, a maximum will be found in the plot of $I_{H_v}/[\cos^2 (\theta/2)]^2$ versus $\theta/2$. The value of U at this ($\theta_{\max,1}/2$) at the fixed μ_1 will be 4.09. Under these conditions, eq. (74) can be written in the form

$$\frac{4\pi R_0 \lambda_s^{-\frac{1}{2}}}{\lambda'} \sin\left(\frac{\theta_{\max,1}}{2}\right) \left[1 + (\lambda_s^3 - 1) \cos^2\left(\frac{\theta_{\max,1}}{2}\right) \cos^2 \mu_1\right]^{\frac{1}{2}} = 4.09$$

(75)

If this is repeated at a second azimuthal angle μ_2 to determine ($\theta_{\max,2}/2$), eq. (74) may be written as

$$\frac{4\pi R_0 \lambda_s^{-\frac{1}{2}}}{\lambda'} \sin\left(\frac{\theta_{\max,2}}{2}\right) \left[1 + (\lambda_s^3 - 1) \cos^2\left(\frac{\theta_{\max,2}}{2}\right) \cos^2 \mu_2\right]^{\frac{1}{2}} = 4.09$$

(76)

By solving eqs. (75) and (76) simultaneously, the following expression for λ_s is obtained:

$$\lambda_s = \left[1 + \left\{ \frac{4[\sin^2(\theta_{\max,1}/2) - \sin^2(\theta_{\max,2}/2)]}{\sin^2\theta_{\max,2}\cos^2\mu_2 - \sin^2\theta_{\max,1}\cos^2\mu_1} \right\} \right]^{\frac{1}{2}} \quad (77)$$

The original undeformed radius of the spherulite R_0 may then be calculated by substitution of the value of the extension ratio of the spherulite λ_s obtained from eq. (77) into eq. (75) or (76). Thus, in theory, both the extension ratio and the original radius of the undeformed spherulite can be determined from an analysis of the radial intensity distribution at two azimuthal angles of the H_v SALS pattern from deformed spherulites.

An experimental arrangement that has been used by the author to obtain the radial intensity distribution at different azimuthal angles is illustrated schematically in Figure 40 (101). The plane of polarization of the laser beam is fixed parallel to the stretch direction of the sample (for an oriented film) and the plane of polarization of the analyzer is perpendicular to both the plane of the laser beam and the stretch direction of the sample. The orientations of the laser beam, the sample, and the analyzer are fixed in this relationship. When the H_v SALS pattern is tilted at an angle μ all three components (the laser, sample, and analyzer) are tilted together. The photomultiplier measures I_{H_v} (the intensity of the H_v scattered radiation) as a function of the radial angle θ at each tilt angle μ. A

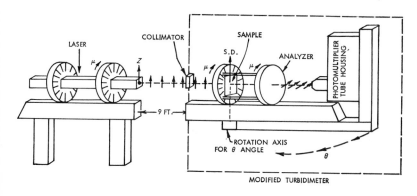

Fig. 40.   Diagram of the photometric SALS system (R. J. Samuels, ref. 101).

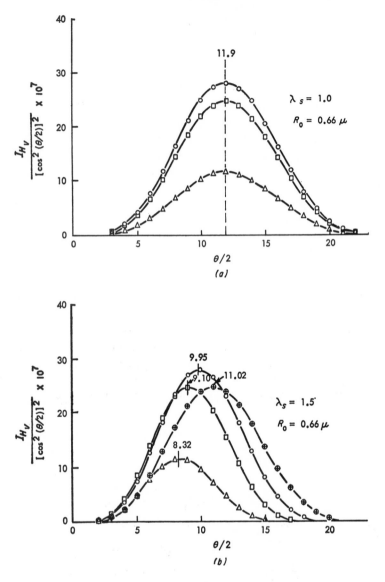

Fig. 41. Theoretical H_v SALS intensity distribution curves: ($\triangle$) $\mu = 20$; ($\square$) $\mu = 35°$; ($\bigcirc$) $\mu = 45°$; ($\oplus$) $\mu = 55°$ (R. J. Samuels, ref. 101).

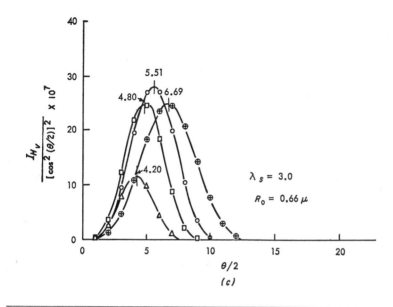

(c)

background measurement is made for each sample at $\mu = 0°$. The background is then subtracted from the intensity, I_{H_v} (measured), obtained at each tilt angle μ to give the final I_{H_v} used for interpretation.

The above theory predicts that both the extension ratio of the spherulite λ_s and the original radius of the undeformed spherulite R_0 can be determined from an analysis of the radial intensity distribution at two azimuthal angles of the H_v SALS pattern and substitution of the results into eqs. (77) and (75) or (76). The radial intensity distribution at different azimuthal angles μ can be calculated theoretically from eqs. (68) and (74) for isotactic polypropylene samples with given values of R_0 and λ_s. Figure 41 contains the theoretically calculated radial intensity distributions for spherulites with an initial radius of the undeformed spherulite of $R_0 = 0.66 \mu$, $\alpha_t - \alpha_r = +0.003$, and $\lambda_s = 1, 1.5,$ and 3 at azimuthal angles of $\mu = 20, 35, 45,$ and $55°$. For the undeformed spherulite, the extension ratio λ_s equals unity and $\theta_{max}/2$ for all azimuthal angles μ has the same value [follows eq. (70)]. The calculated $\theta_{max}/2$ for the given λ_s and R_0 values is designated by the numbered line at the

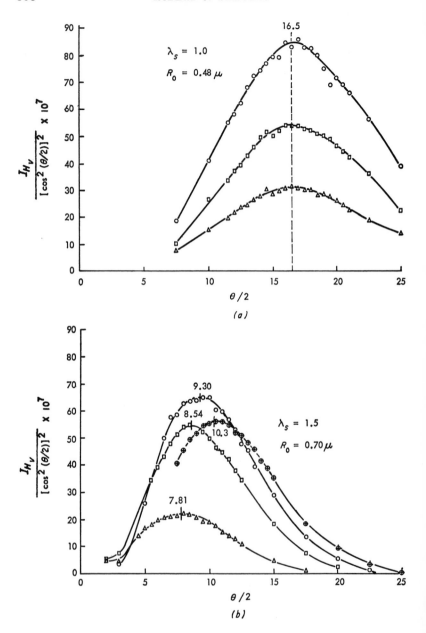

Fig. 42. Experimental H_v SALS intensity distribution curves; ($\triangle$) $\mu = 20°$; ($\square$) $\mu = 35°$; ($\bigcirc$) $\mu = 45°$; ($\oplus$) $\mu = 55°$ (R. J. Samuels, ref. 101).

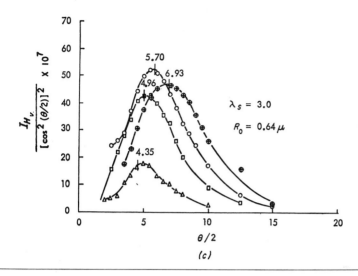

$\lambda_s = 3.0$

$R_0 = 0.64\,\mu$

$\theta/2$

(c)

top of each peak. At 50% elongation of the spherulite ($\lambda_s = 1.5$), the radial curves have a different $\theta_{max}/2$ for each azimuthal angle [follows eq. (74)]. The peak intensity is still at $\mu = 45°$ but the position of the peak has shifted to a lower $\theta_{max}/2$ than was calculated for the undeformed spherulite. As expected, the position of $\theta_{max}/2$ for $\mu = 45°$ has shifted to an even lower radial angle when the spherulite has been elongated 200% (λ_s 3.0), and the $\theta_{max}/2$ values for the other azimuthal angles are all different from each other.

Figure 42 contains the experimentally determined radial intensity distribution curves obtained from isotactic polypropylene cast spherulitic films that were affinely deformed to extension ratios of $\lambda_s = 1, 1.5$, and 3. These are the same films from which the SALS photographs were obtained. The similarity between the theoretically predicted curves (Fig. 41) and the experimental curves is obvious.

The intensity magnitudes on the ordinate of Figure 41 have not been scaled to those in Figure 42 [i.e., the proportionality constant A in eq. (68) has not been used]. The $\theta_{max}/2$ value for the undeformed sample in Figure 42 is the same for all values of the azimuthal angle μ as expected. Similarly, the $\theta_{max}/2$ values for the $\lambda_s = 1.5$ and $\lambda_s = 3.0$ samples have different values for different azimuthal angles. Since the extension ratio λ_s was known for the isotactic polypropylene films examined, $\theta_{max}/2$ could be calculated for the experimentally determined intensity distribution curves at each extension ratio. The numbered line at the top of each peak is the calculated $\theta_{max}/2$ value for the given λ_s and R_0 values. The agreement between the experimental and calculated maxima is excellent. In all of these respects, the experimentally observed SALS intensity distributions from the isotactic polypropylene films agree with those predicted by theory.

Thus, a quantitative numerical determination of the deformation the spherulite has undergone during mechanical or thermal treatment of the film can be obtained from H_v SALS photometric measurements. To do so, these measurements are treated theoretically by assuming that the deformed spherulite is an anisotropic spheroid in an isotropic medium, a model consistent with the known morphological characteristics of the spherulite. Therefore, the deformation of spherulites relative to the imposed uniaxial deformation of the film under different conditions of temperature and stress can be studied quantitatively. Similarly, deviations from the predicted form of the SALS pattern could yield valuable information about complex changes occurring in the spherulites. Such changes would be expected at large uniaxial deformations where disruptive processes are known to occur within the spherulites. Certainly, small angle light scattering is an important new technique for nondestructively evaluating spherulite deformation processes in polymer films (and fibers) under controlled conditions of temperature and stress.

IV. Conclusions

In this chapter, an attempt has been made to demonstrate a unified, comprehensive approach to the characterization of the deformation of polycrystalline polymer films. To do so, the different morphological levels of the film were defined as gross sample deformation (characterized by the sample extension ratio), spherulite deformation, and molecular orientation. The objective was to be able to relate morphological information obtained from numerous independent experimental techniques at different morphological levels in a simple fashion.

To simplify the problem of attaining this objective, a single set of isotactic polypropylene samples which had all undergone a uniform, uniaxial deformation was used as the model system. The characteristics of these samples were then examined at the different morphological levels. The molecular level was considered first. A two-phase model of a polycrystalline polymer was proposed and the important morphological parameters were shown to be the fraction of crystals present in the sample, the intrinsic properties of the crystalline and noncrystalline regions of the polymer, and the orientation of each of these regions. This information alone acted as a unifying thread with which to relate morphological data obtained from x-ray diffraction, sonic modulus, birefringence, density, and infrared dichroism measurements, and led to new combinations of these techniques for the rapid acquisition of morphological data. It further developed insight into the significance of the observed information from these different methods so that fundamental molecular parameters such as the intrinsic lateral moduli and the intrinsic birefringences of the crystalline and noncrystalline regions, the principal polarizability difference per monomer unit, and the infrared transition moment angles could be obtained for the isotactic polypropylene samples.

Spherulite deformation and its relation to gross sample deformation was then considered. By relating the molecular orientation of the crystalline region (the crystal orientation function f_c) to the gross sample deformation (λ), the spherulites were shown to be deformed affinely. In addition, the observed small angle light scattering (SALS) from the deformed spherulites could be used to give quantitative information about the size of the original, undeformed spherulites and the deformation the spherulite had undergone. Obviously, for a nonaffine deformation, the SALS technique will similarly indicate how much deformation the spherulite has undergone in comparison with the different deformation of the gross sample.

Thus, by using a unified, comprehensive approach to the characterization of the deformation of polycrystalline polymer films, molecular, spherulitic, and gross sample deformation characteristics have been related. With the insights and information that have been made available, this approach can now be extended to more complicated deformation processes and different experimental techniques. In this manner, a comprehensive theoretical interpretation of the numerous deformation processes occurring in polycrystalline polymer films should eventually be developed.

362   ROBERT J. SAMUELS

References

1. A. Keller, *Phil. Mag.*, **2**, 1171 (1957).
2. E. W. Fischer, *Z. Naturforsch.*, **12a**, 753 (1957).
3. P. H. Till, *J. Polymer Sci.*, **17**, 447 (1957).
4. P. H. Geil, *J. Appl. Phys.*, **33**, 642 (1962).
5. R. P. Palmer and A. J. Cobbold, *Makromol. Chem.*, **74**, 174 (1964).
6. A. Keller and S. Sawada, *Makromol. Chem.*, **74**, 190 (1964).
7. I. L. Hay and A. Keller, *Nature*, **204**, 862 (1964).
8. A. Peterlin, H. Kiho, and P. H. Geil, *Polymer Letters*, **3**, 151 (1965).
9. H. Kiho, A. Peterlin, and P. H. Geil, *J. Appl. Phys.*, **35**, 1599 (1964).
10. D. Hansen and J. A. Rusnock, *J. Appl. Phys.*, **36**, 332 (1965).
11. H. D. Keith and E. Passaglia, *J. Res. Natl. Bur. Std.*, **68A**, 513 (1964).
12. R. Bonart, R. Hosemann, F. Motzkus, and H. Ruck, *Norelco Reptr.*, **7**, 81 (1960).
13. W. Ruland, *Polymer*, **5**, 89 (1964).
14. W. Ruland, *Acta Cryst.*, **14**, 1180 (1961).
15. P. H. Hermans, *Contribution to the Physics of Cellulose Fibres*, Elsevier, New York, 1946.
16. G. M. Brauer and E. Horowitz, in *Analytical Chemistry of Polymers III*, G. M. Kline, Ed., Interscience, New York, 1962, Chap. I.
17. W. O. Statton, in *Handbook of X-rays in Research and Analysis*, E. F. Kaelble, Ed., McGraw-Hill, New York, to be published, Chap. 21.
18. V. D. Gupta and R. B. Beevers, in *The Encyclopedia of X-rays and Gamma Rays*, G. L. Clark, Ed., Reinhold, New York, 1963, pp. 783–789.
19. R. Zbinden, *Infrared Spectroscopy of High Polymers*, Academic, New York, 1964, Chap. I.
20. I. Y. Slonim, *Russ. Chem. Rev. English Transl.*, **31**, 308 (1962).
21. C. W. Bunn, *Trans. Faraday Soc.*, **35**, 482 (1939).
22. H. P. Klug and L. E. Alexander, *X-Ray Diffraction Procedures*, Wiley, New York, 1954.
23. J. J. Hermans, P. H. Hermans, D. Vermaas, and A. Weidinger, *Rec. Trav. Chim.*, **65**, 427 (1946).
24. R. S. Stein, *J. Polymer Sci.*, **31**, 327 (1958).
25. Z. W. Wilchinsky, *J. Appl. Phys.*, **30**, 792 (1959).
26. R. A. Sack, *J. Polymer Sci.*, **54**, 543 (1961).
27. G. Natta, *Nuovo Cimento*, **15**, Series 10, Suppl., 40 (1960).
28. Z. W. Wilchinsky, *J. Appl. Phys.*, **31**, 1969 (1960).
29. R. J. Samuels, *Norelco Reptr.*, **10**, 101 (1963).
30. N. S. Gingrich, *Rev. Mod. Phys.*, **15**, 90 (1943).
31. R. S. Stein, J. Powers, and S. Hoshino, *Office Naval Res. Tech. Rept.*, No. **33**, July 7, 1961.
32. The automatic system was obtained from General Electric Company for control of their XRD-5 diffractometer.
33. M. Polanyi, *Z. Physik*, **7**, 149 (1921).
34. S. Hoshino, J. Powers, D. G. Legrand, H. Kawai, and R. S. Stein, *J. Polymer Sci.*, **58**, 185 (1962).

35. R. S. Schotland, Ph.D. thesis, Polytechnic Institute of Brooklyn, Brooklyn, N.Y., 1962.
36. R. J. Samuels, *J. Polymer Sci.*, **A3**, 1741 (1965).
37. G. Farrow and J. Bagley, *Textile Res. J.*, **32**, 587 (1962).
38. M. Compostella, A. Coen, and F. Bertinotti, *Angew. Chem.*, **74**, 618 (1962).
39. M. E. Millberg, *J. Appl. Phys.*, **33**, 1766 (1962).
40. R. S. Stein and D. G. Legrand, *Office Naval Res. Tech. Rept.*, **No. 25**, Sept. 1, 1960.
41. I. M. Ward, *Proc. Phys. Soc. London*, **80**, 1176 (1962).
42. I. M. Ward, *Textile Res. J.*, **34**, 806 (1964).
43. W. W. Moseley, Jr., *J. Appl. Polymer Sci.*, **3**, 266 (1960).
44. W. H. Church and W. W. Moseley, Jr., *Textile Res. J.*, **29**, 525 (1959).
45. S. E. Ross, *Textile Res. J.*, **34**, 565 (1964).
46. R. J. Urick, *J. Appl. Phys.*, **18**, 983 (1947).
47. H. A. Waterman, *Kolloid Z. Z. Polymere*, **192**, 9 (1963).
48. *Encyclopaedic Dictionary of Physics*, Vol. 2, Pergamon Press, New York, 1961, p. 21.
49. H. A. Waterman, *Kolloid Z. Z. Polymere*, **192**, 1 (1963).
50. J. W. Ballou and S. Silverman, *Textile Res. J.*, **14**, 282 (1944).
51. G. Oster and M. Yamamoto, *Chem. Rev.*, **63**, 257 (1963).
52. I. Sakurada, T. Ito, and K. Nakamae, *Kobunshi Kagaku*, **21**, 197 (1964).
53. I. Sakurada, T. Ito, and K. Nakamae, *J. Japan Soc. Testing Mat.*, **11**, 683 (1962).
54. M. Horio, *Symposium at Polytechnic Institute of Brooklyn*, Sept. 7, 1963.
55. O. Wiener, *Abhandl. Saechs. Akad. Wiss. Leipzig, Math. Physik. Kl.*, **32**, 507 (1912).
56. F. A. Bettelheim and R. S. Stein, *J. Polymer Sci.*, **27**, 567 (1958).
57. R. S. Stein and F. H. Norris, *J. Polymer Sci.*, **21**, 381 (1956).
58. C. W. Bunn and R. de P. Daubeny, *Trans. Faraday Soc.*, **50**, 1173 (1954).
59. E. M. Chamot and C. W. Mason, *Handbook of Chemical Microscopy*, Vol. 1, Wiley, New York, 1951.
60. R. S. Stein, S. Onogi, and D. A. Keedy, *J. Polymer Sci.*, **57**, 801 (1962).
61. H. De Vries, *On the Elastic and Optical Properties of Cellulose Fibers*, Sehotanus and Jeus, Utrecht, 1953.
62. H. M. Morgan, *Textile Res. J.*, **32**, 866 (1962).
63. G. Farrow and J. Bagley, *Textile Res. J.*, **32**, 587 (1962).
64. R. S. Stein, private communication.
65. M. F. Vuks, *Opt. i Spektroskopiya*, **2**, 494 (1957).
66. M. V. Volkenstein, *Configurational Statistics of Polymeric Chains*, Interscience, New York, 1963, Chap. 7.
67. V. N. Tsvetkov, in *Newer Methods of Polymer Characterization*, B. Ke, Ed., Interscience, New York, 1964, Chap. 14.
68. H. A. Szymanski, *Infrared Theory and Practice of Infrared Spectroscopy*, Plenum Press, New York, 1964.
69. G. W. King, *Spectroscopy and Molecular Structure*, Holt, Rinehart and Winston, New York, 1964.
70. S. Krimm, *Fortschr. Hochpolymer.-Forsch.*, **2**, 51 (1960).

364 ROBERT J. SAMUELS

71. C. Y. Liang, in *Newer Methods of Polymer Characterization*, B. Ke, Ed., Interscience, New York, 1964, Chapt. 2.
72. G. Herzberg, *Infrared and Raman Spectra of Polyatomic Molecules*, D. Van Nostrand, New York, 1945.
73. H. C. Urey and C. A. Bradley, *Phys. Rev.*, **38**, 1969 (1931).
74. R. D. B. Fraser, in *Analytical Methods of Protein Chemistry*, Vol. 2, P. Alexander and R. J. Block, Eds., Pergamon, New York, 1960, Chap. 9.
75. R. D. B. Fraser, *J. Chem. Phys.*, **21**, 1511 (1953).
76. E. M. Bradbury, A. Elliot, and R. D. B. Fraser, *Trans. Faraday Soc.* **56**, 1117 (1960).
77. R. D. B. Fraser, *J. Chem. Phys.*, **29**, 1428 (1958).
78. M. Tsuboi, *J. Polymer Sci.*, **59**, 139 (1962).
79. M. P. McDonald and I. M. Ward, *Polymer*, **2**, 341 (1961).
80. T. Miyazawa and Y. Ideguchi, *Bull. Chem. Soc. Japan*, **36**, 1125 (1963).
81. T. Miyazawa, *J. Polymer Sci.*, **C7**, 59 (1964).
82. R. G. Snyder and J. H. Schachtschneider, *Spectrochim. Acta*, **20**, 853 (1964); *J. Polymer Sci.*, **C7**, 85 (1964).
83. S. Krimm, *J. Polymer Sci.*, **C7**, 59 (1964).
84. A. Keller, *J. Polymer Sci.*, **17**, 351 (1955).
85. H. D. Keith, F. J. Padden, Jr., N. M. Walter, and H. W. Wyckoff, *J. Appl. Phys.*, **30**, 1485 (1959).
86. P. Ingram and A. Peterlin, *Polymer Letters*, **2**, 739 (1964).
87. V. I. Selikhova, G. S. Markova, and V. A. Kargin, *Vysokomolekul. Soedin.*, **6**, 1136 (1964).
88. S. Ishikawa, T. Kurita, and E. Suzuki, *J. Polymer Sci.*, **A2**, 2349 (1964).
89. F. J. Padden, Jr. and H. D. Keith, *J. Appl. Phys.*, **30**, 1479 (1959).
90. P. H. Geil, *Polymer Single Crystals*, Interscience, New York, 1963, Chap. 4.
91. M. Takayanagi, S. Minami, and H. Nagatoshi, *Asahi Garasii Kogyo Gijutsu Shoreikai*, **7**, 127 (1961).
92. Z. W. Wilchinsky, *Polymer*, **5**, 271 (1964).
93. R. S. Stein, M. B. Rhodes, P. R. Wilson, and S. N. Stidham, *Pure Appl. Chem.*, **4**, 219 (1962).
94. V. I. Selikhova, G. S. Markova, and V. A. Kargin, *Vysokomolekul. Soedin.*, **6**, 1132 (1964).
95. V. A. Kargin, T. D. Sogolova, and L. A. Nadareishvili, *Vysokomolekul. Soedin.*, **6**, 1407 (1964).
96. C. J. Speerschneider and C. H. Li, *J. Appl. Phys.*, **33**, 1871 (1962).
97. D. C. Bassett, *Phil. Mag.*, **6**, 1053 (1961).
98. J. A. Rusnock and D. Hansen, *J. Polymer Sci.*, **A3**, 647 (1965).
99. G. A. Bassett and A. Keller, *Phil. Mag.*, **9**, 817 (1964).
100. A. Guinier, G. Fournet, C. B. Walker, and K. L. Yudowitch, *Small Angle Scattering of X-Rays*, Wiley, New York, 1955.
101. R. J. Samuels, *J. Polymer Sci.*, **C13**, 37 (1966).
102. R. S. Stein and M. B. Rhodes, *J. Appl. Phys.*, **31**, 1873 (1960).
103. L. C. Roess and C. G. Shull, *J. Appl. Phys.*, **18**, 308 (1947).
104. A. L. Patterson, *Phys. Rev.*, **56**, 972 (1939).
105. A. Guinier, *Ann. Phys. Paris*, **12**, 161 (1939).
106. R. S. Stein and J. J. van Aartsen, private communication.

CHAPTER 8

THE TECHNOLOGY OF MELT CASTING

JAMES L. HECHT

E. I. du Pont de Nemours & Co., Inc., Buffalo, New York

I. The Melt-Casting Process

Melt casting is the process by which a thermoplastic melt is forced through a slit die which shapes it in the form of a film. Melt casting is used to make most films which are commercially available, and for many of these melt casting is essentially the only commercial process.

Polyolefins, polyesters, polystyrene, polyamides, and polyvinylidene chloride are examples of polymers from which films are usually made by melt casting. Also, a substantial amount of the film made from polyvinyl chloride and fluorocarbon polymers is made by melt casting.

The melt-casting process consists of six parts:

(1) Pumping the melt.
(2) Improving the melt quality.
(3) Passing the melt through a die to form a film.
(4) Quenching the film.
(5) Conveying the film toward the windup or the next process step.
(6) Measuring film thickness.

Usually, the polymer is melted and pumped by a screw extruder. Because of its importance in commercial operations, the technology of extrusion has been studied extensively during the past 15 years and what once was an art is now a science. The resulting theory and practice which have evolved are well summarized in detail in two recent books (1,2). Consequently, in this review of the melt-casting process we shall not cover the pumping process which precedes it except when reference to extrusion is necessary in the discussion of specific casting problems.

Between the extruder (or other melt-pumping device) and the casting die is a section which improves the quality of the melt. The most important function here is melt filtration. This usually is done by passing the melt through a series of screens which are set against a breaker plate for support. Besides filtration, this section can contain a valve for controlling the extrusion pressure, and usually will contain instrumentation for measuring the pressure and temperature of the melt. Finally, the holdup provided by this section decreases temperature inhomogeneities in the melt.

There are two basic types of film dies: flat and tubular. A flat-film die consists of a heated body to which are connected two parallel lips. These form a slot which gives the film its shape. As a rule one of these lips is rigid, while the other can be differentially adjusted along its length, as described in Section II-D. At either end of the lips are end blocks which confine the melt to flowing through the forming slot. If the casting process requires formation of a bead (i.e., a narrow section at the edge of the web which is considerably thicker than the base sheet, as is often required if the film is to be stretched), this is accomplished by inserting bead plates between the lips and the end blocks. A diagram of a flat-film die is given in Figure 1.

The basic features of dies used to make tubular films are described in Section V-A.

The hot thermoplastic melt emerging from the die must be quenched (i.e., cooled in a controlled manner to a solid plastic film). In the case of flat films, quenching usually is done either by casting on a cooled drum, or by casting into a water bath. These processes are discussed in detail in Section IV-B. Quenching processes for tubular films are somewhat more complicated and are covered in Section V-B.

The film is usually conveyed by nip rolls (i.e., two counter-rotating rolls, one of which is driven, that are in contact with each other under pressure). Proper placement of additional rolls, which can be either driven or idler rolls (i.e., driven by the film itself), is important in conveying the film, particularly for feeding the film to nip rolls. The technology of film handling is similar to that used for a wide variety of webs.

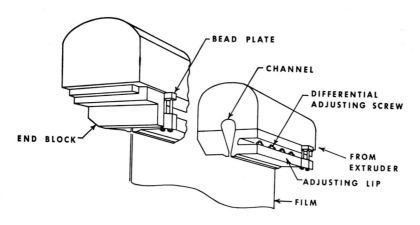

Fig. 1. Flat film die.

The melt-casting operation usually also includes continuous in-line measurement of film thickness. The instrumentation used for this purpose is discussed in some detail in Section II-E.

Following casting, the film is wound onto a mill roll. However, often there are one or more postcasting operations before windup. These include orientation, surface treatment to promote adhesion, and coating.

In this chapter the emphasis is on broad principles which apply to many materials. The subject matter is organized on the basis of the principal problems which exist in melt casting commercially acceptable film. These are: (1) to obtain a film with minimum variations in thickness; (2) to obtain a film with the required appearance and properties; (3) to obtain as high a productivity as possible from a given investment.

The science and technology associated with each of these three problems will be discussed in the next three sections for the case of flat films. In the final section of this chapter, the application of some of these principles to tubular casting is discussed, along with certain special characteristics of circular-die casting.

II. Minimizing Variations in Thickness

Gauge variations and gauge-related defects (poor roll formation, coating defects caused by gauge variations, etc.) usually constitute the greatest cause of film rejections and customer complaints. Gauge control is emphasized in this chapter because of the commercial importance of this property, because the origin and correction of gauge variations lie mainly within the melt-casting process, and because the same principles with respect to minimizing gauge variations can be applied to all film-forming materials.

A. Gauge Variations in the Longitudinal Direction

Gauge variations in the longitudinal direction arise in two ways: (1) pulsations in the rate at which the melt is pumped; (2) uneven draw of the film. The latter may be caused either by pulsations in the takeoff drive or by vibrations in the film before it is quenched.

Variations in thickness in the longitudinal direction usually are not as serious as variations in the transverse direction. They do not, for example, contribute to poor roll formation. However, longitudinal gauge variations, as will be shown later, are an important cause of thickness variations in the transverse direction. Also, for many uses, films must have a minimum point thickness, and thus a high gauge variation in the longitudinal direction increases the average film thickness at which this point specification can be met (and therefore increases the cost of the film to the user). Finally, there are some uses for film where the point thickness must be in a relatively narrow range, and for these uses large variations in any direction become intolerable.

If longitudinal gauge variations are being caused by pulsations in the melt-pumping rate and the melt is being pumped by an extruder, this usually indicates that the metering section of the extruder has a pumping capacity greater than either the rate at which the screw is being fed or the rate at which the polymer is being melted. The preferred solution, therefore, is to increase the rate-limiting flow so that it exceeds by a considerable amount the pumping capacity of the metering zone. When melting is the problem, the use of higher temperatures at the back end of the extruder (i.e., the end where the feed enters) may offer a quick solution, with preheating the feed a second step, if necessary. If feeding is the problem, changes in the physical form of the feed, changes in the feeding mechanism, and changes in the design of the feed section of the screw (e.g., deepening it) should be considered.

When the limiting rate cannot be increased sufficiently, then it may be necessary to reduce the capacity of the metering zone of the extruder. Unfortunately, this reduces manufacturing capacity. It can be accomplished without equipment changes by two separate methods: slowing down the extruder or increasing the back pressure (e.g., by adjusting the extrusion valve). Often a combination of these is desirable. When action of this type is necessary, however, the best long-range solution may be to build another screw with a somewhat shallower metering zone.

Sometimes, throughput fluctuations arise after a period of satisfactory operation. Obviously, this results from a change, and the proper corrective action involves eliminating the change. Often the cause is wear of the screw flights which can be corrected by building up the screw flights to the original dimensions.

When an effort is being made to reduce pulsations in pumping, attention should be paid to factors which might cause pulsations in the rate of feeding and melting. Such variations can produce fluctuations in flow even when the metering zone limits the flow. Needless to say, there should be no fluctuations in temperature throughout the extruder.

The uniformity of the takeoff drive is best determined by a stroboscope if the drive is rotating at 100 rpm or more. Otherwise, the best procedure is to attach a tack generator to the drive and measure either the voltage or the current with a high-speed recorder. Possible pulsation in the extruder drive should also be checked. The latter often can be spotted as a variation in melt pressure.

The key to reducing longitudinal gauge variations is to correctly analyze the cause. This may prove difficult. A good first step is to note if there is a regular pattern in gauge variation. Often it is necessary to hypothesize a cause and test the hypothesis by trying a solution to see if it results in substantial improvement.

B. The Origin of Gauge Variations in the Transverse Direction

Gauge variations in the transverse direction arise from four distinct mechanisms: (1) uneven flow of melt across the length of the die; (2) edge thickening as a result of necking down; (3) changes caused by uneven shrinkage and expansion during cooling; (4) gauge variations in the longitudinal direction.

Nonuniform melt flow across the die is the most important cause of gauge variations in the transverse direction. This uneven flow arises from a variety of causes. There is a significant pressure drop along the length of the die, and thus the pressure which causes the flow of melt through the die land is not constant. Another major source of flow variation arises from temperature variations which give nonuniformities in the viscosity of the melt. Finally, it is not possible to set the lips of a casting die in such a way that the die opening does not change as a result of distortion of the lips under pressure.

Because of the importance of the uneven flow of melt across the length of the die, this is discussed in detail in the next section.

When a film is melt cast, it necks down so that the width of the web is less than the length of the casting die. The degree of necking down depends principally on the polymer melt, the draw ratio, and the distance between the die lip and the point where the film is quenched. Necking down is accompanied by a thickening of the film, particularly at the edges, but significant thickening often extends for quite a few inches. Prevention of severe necking down usually can be accomplished by keeping the distance small between the die and the point where the film is quenched. When this is done, a sheet of good gauge can be made by adjusting the lips of the casting die to compensate for the necking down at all points except the edges, and these are usually trimmed. An additional precaution which should be taken to produce film of good gauge is not to change the distance between the die lips and the point of quenching. Otherwise the gauge profile may be altered as a result of an increase or decrease in necking down.

If during quenching, a section or lane of the film is cooled more rapidly than the remainder of the sheet, this can result in large variations in thickness because a lane which is cooled more rapidly thickens at the expense of the neighboring melt. Fortunately, differential cooling of this type can usually be relatively well controlled.

Finally, gauge variations in the longitudinal direction cause non-uniformities in the transverse direction. This occurs because the pulsations in flow and takeoff by which these are produced are not always uniform across the sheet. Gauge variations in the longitudinal direction also cause problems, in the control of gauge variation (see Section II-F).

C. Analysis of Melt Flows

In this section, we shall discuss in detail the major cause for gauge variations in the transverse direction: the uneven flow of melt across the length of the die.

The melt flow at various points along the length of a film die can be quantitatively related in terms of flow variables. Most thermoplastic melts obey the so-called "power law" (3).

$$\frac{dv}{dx} = f_0 \left(\frac{S}{S_0}\right)^n \qquad (1)$$

where v is the linear velocity of flow at any point in the melt, in in./sec.

x is the direction, perpendicular to the wall of the die, in in.

S is the shear stress of the melt, in psi.

f_0 is the "generalized fluidity" of the material, in $\sec^{-1}$. It is equal to the shear rate when the shear stress is at a standard state, S_0.

S_0 is the standard-state shear stress (usually 1 psi).

n is the stress exponent for the material and can vary between 1 and 4.5.

The constants f_0 and n depend on the polymer involved. The constant f_0 also is strongly dependent on temperature. For a Newtonian polymer, the stress exponent, n, is equal to unity.

Applying eq. (1) to an infinitesimal length of a narrow slit gives the following relationship for the case of S_0 equal to 1 psi (4).

$$dQ = \frac{f_0 h^{n+2}}{2^{n+1}(n+2)} \left(\frac{p}{t}\right)^n dL \tag{2}$$

where L is the length of the die, in in.

Q is the flow, in in.^{3}/sec.

h is the die opening, in in.

p is the pressure of the melt at the beginning of the land, in psi.

t is the corrected length of the land,* in in.

Equation (2) delineates the variables which govern the flow at any point: the pressure of the melt at the beginning of the land, the viscosity (i.e., the fluidity) of the melt, the die opening, and the length of the land. Of these four, only the length of the land can be precisely controlled.

Materials with high-stress exponents will have a large pressure drop across the length of the die. This is because the shear stress is low in the channel which runs from the feed point to the end of the die, and therefore the viscosity is much higher than in the die land where the material is subjected to a high shear stress. Thus, the pressure required for polymer to flow down the channel (so that melt can be fed to the far end of the die) is significant compared to the pressure drop in the die land itself. On the other hand, for a polymer melt whose rheology is close to Newtonian it is easy to design the die so that the pressure drop in the channel is small compared to that in the narrow land.

---

* The corrected length is the actual length plus a correction factor for the entrance effect.

The magnitude of the problem and its dependence on the rheology of the melt can best be shown by computation of an index of uniformity (uI), which is the ratio of the flow rate at the far end of the die to that at the feed end. Carley (4) showed that for a die with a circular cross section, or one which approaches a circular cross section, the uI is given by

$$uI = \left[1 - \frac{[(1 + 0.05n)\, n\alpha L]^{(n+1)/n}}{(1 + 0.05n)(n + 1)} \right]^{n} \tag{3}$$

where the constant α is defined by

$$\alpha^{n+1} = \frac{(n + 3)h^{n+2}}{2\pi n(n + 2)t^{n}\, R^{n+3}} \tag{4}$$

where R is the radius of the channel, in in.

Using these equations, the following values were calculated by Carley (4) for a center-fed die 48 in. long (i.e., L equal to 24 in.), with a channel radius, R, of 0.75 in., $h = 0.01$, and $t = 0.50$ in. The importance of the stress exponent is readily apparent.

n	uI
1	1.000
2	0.970
3	0.852
4	0.665

With this die, the adjustment devices which are described in the next section would be satisfactory for most, if not all, polymers. However, for a thicker sheet (requiring a greater value of h) it would be impossible to compensate by mechanical adjustment if the stress exponent of the melt were high. Film of relatively uniform gauge might be made by initially setting the lips in a wedge (i.e., with a smaller lip opening at the feed end than at the dead end), but this technique is limited and, even where it is sufficient to give a film with uniform thickness, it results in film which has not been uniformly drawn across the sheet. This, in turn, can result in nonuniform properties (e.g., gloss). Another technique is to increase the radius of the channel, but this increases the holdup time of the melt. A preferred approach is to vary the length of the land to compensate for the pressure drop in the channel. This has the advantage that the entire sheet receives the same shear stress and is cast with the same draw ratio.

The design of a hopper lip with a variable land length is based on the relationship (ref. 5)

$$t = t_f - \left[\frac{L^{n+1}h^{n+2}(n+3)}{2\pi(n+2)R^{n+3}}\right]^{1/n}\left(\frac{n}{n+1}\right)\left[1 - \left(\frac{Z}{L}\right)^{\frac{n+1}{n}}\right] \quad (5)$$

where t is the length of the land (corrected) at point Z along the die, in in.; Z is the distance from the far end of the die, in in. (and equals L at the feed end); and t_f is the length of the land at the feed end, in in.

As a matter of practicality, it is best to calculate the land length at the far end [i.e., let Z equal zero in eq. (5)], and then let the land length vary linearly between the two ends. This closely approximates the exact computation and greatly facilitates machining the lip.

In the design of a film-casting unit, the decision must be made whether to use an end-fed die or a center-fed die. The latter decreases problems associated with pressure drop in the channel, but the end-fed die has the advantage that it is more streamlined. In view of the ease with which the pressure-drop problem can be solved, the end-fed die usually is preferred.

Change in the melt viscosity caused by temperature differences constitutes an important source of gauge variations. The fluidity of the polymer usually can be correlated empirically by the equation

$$f_0 = ae^{bT} \quad (6)$$

where T is the temperature, in °K, and a and b are constants.

Differentiating eq. (6),

$$\frac{df_0}{dT} = bf_0$$

and, for small changes in temperature

$$\frac{\Delta f_0}{\Delta T} = bf_0 \quad (7)$$

Since from eq. (2) the flow rate, Q, is proportional to f_0

$$\frac{\Delta Q}{Q}(100) = \frac{\Delta f_0}{f_0}(100) \quad (8)$$

and, substituting eq. (7) into eq. (8), the variation in flow rate equals

$$\frac{\Delta Q}{Q}\,(100) = 100\,(b)(\Delta T) \qquad (9)$$

For polyethylene, the constant b is equal to about $0.025/°K$ (6). Thus, a gauge variation of about 2.5% results from a temperature variation of only 1°C.

Temperature variations in the melt arise in three distinct ways.

1. The melt from the extruder is not uniform in temperature. These temperature variations can be decreased by increasing the flow time between the extruder and the casting die provided the additional residence time is not harmful. If this is purposely done, it usually also is possible to obtain some mixing without creating holdup spots.

2. The melt is heated or cooled in the casting die if the temperature of the die is not the same as that of the melt. Isothermal flow would prevent variations of this type, but often this can not be done because of process considerations. Experimental data for a case where the melt enters the die at a temperature 20°C higher than the die temperature are given in Figure 2. The melt tempera-

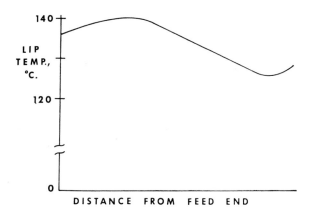

Fig. 2. Variation of the temperature of the lips of a casting die.

ture is less at the far end of the die because the melt is cooled substantially while flowing in the channel, and this is reflected in the lip temperatures. The temperature of the lip is less than that of the die because the lip is cooled by the surroundings.

3. The temperature of the die is also influenced by thermal effects at both ends. This is shown in Figure 2. The decrease in lip temperature which takes place from the feed end to the dead end is reversed at both ends.

In designing a casting die, it is important to design both for uniformity and for constancy of temperature. Heating with a low-viscosity liquid (e.g., water, oil, or Dowtherm, depending on the temperature), provides better control than electrical heating, provided the temperature of the liquid is adequately controlled and a uniform flow rate along the die is maintained. When good process control is needed and the added cost is justified, the best practice is to heat the die and the lips separately with heat-transfer fluids whose temperatures are independently controlled.

A final source of gauge variation comes from changes in the lip opening. These are unavoidable because the lips are deflected by the melt pressure and this movement is not uniform along the length. The effect of a change in lip opening on the flow rate is readily obtained from eq. (2):

$$\frac{\Delta Q}{Q}(100) = (n + 2)\frac{\Delta h}{h}(100) \tag{10}$$

Equation (10) shows that even with Newtonian fluids the effect of a change in lip opening is large. With non-Newtonian materials, it is even greater. If a section of the lip is flexed 1 mil during the casting of polyethylene ($n = 2.8$) and the opening is 0.020 in., the resulting gauge variation would be 24%.

Several unique designs have been patented to prevent deflection of the lips and the gauge variations which result (7–9). This problem is discussed further in Section III-A with respect to the problem of producing thin films.

D. MEANS OF GAUGE ADJUSTMENT

From the previous section it is apparent that in order to compensate for flow variations which cannot be controlled, there must be some means of adjusting flows across the width of the casting die. In addition, it is not always desirable to have a flat sheet as the product of the casting operation. When the film is oriented, it frequently is necessary to contour the cast sheet to minimize the gauge variation in the stretched film.

The flow is regulated across the length of the casting die by adjusters which warp the die lip. This is a very effective method of adjustment. As shown by eq. (10), a relatively small change in lip opening produces a large change in flow rate. Because this method is very satisfactory, an alternative means—that of purposely varying the temperature across the lip—does not appear to be practiced by commercial film manufacturers.

Figure 3 is a schematic diagram of an adjusting device. There are a number of requirements which such an adjuster should meet:

1. The response should be linear. That is, the deflection of the lip, δA, should as nearly as possible be directly proportional to the adjustment, δB. If this condition is met, the person making the

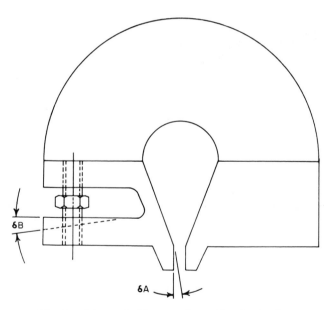

Fig. 3.   Schematic diagram of an adjusting device.

adjustment can make a good estimate of the correction required. For example, he will know that a 5% change in gauge requires turning the adjusting bolt about 45°. As will become apparent (Section II-F), linear response of adjusters is critical for the success of an automatic gauge control system.

2. The interaction should be relatively small. When an adjustment is made, there not only is a deflection of the lip δA at the point of adjustment, but the lip also is deflected a smaller amount at other points along the length. The interaction is defined as the percentage deflection at an adjacent bolt, or $100 \times (\delta A_{n+1})/(\delta A_n)$. For good adjustment, the interaction should be less than 50%. When this condition is met, the measured interaction at bolt $(n + 2)$ is only 10–20%.

3. There should be a balance between adjustability and flexibility. The adjustment must be sufficient to give the required changes in flow. However, as a rule the more adjustable the lip, the more easily it is distorted by the pressure of the melt. Thus, a rigid design would be more appropriate for making a film requiring a nar-

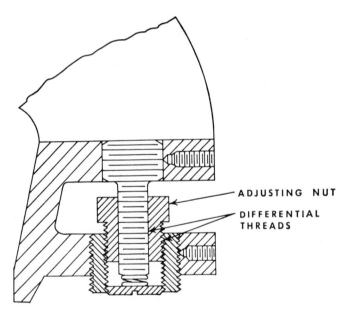

Fig. 4. Adjustment mechanism (12).

row lip opening than for the production of film requiring a wide opening. A related design problem is whether to make both lips flexible, thereby doubling the adjustability, or to have one lip rigid. As a rule, sufficient adjustment can be obtained with a single adjustable lip, and the use of a stationary lip has advantages with respect to both lower cost of equipment and stability of operation.

4. The adjusting device should be of such a nature that it remains functional over a long period of time. Most adjusting devices depend on the proper functioning of threaded components. If proper design and maintenance procedures are not followed, the combination of high temperature and long periods of continual use will result in expensive shutdowns as a result of frozen bolts.

The performance of the adjusting device varies greatly with the design of the adjuster and the geometry of the lip. A large number of different mechanical adjusting devices have been described in detail in the patent literature (e.g., 10–17). Figures 4 and 5 demon-

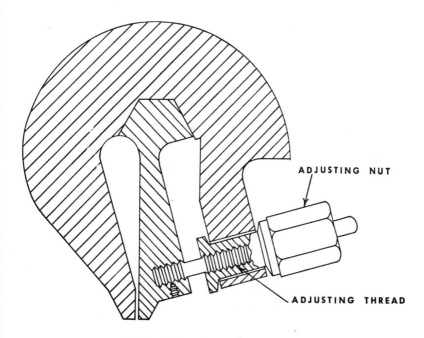

Fig. 5. Adjustment mechanism (14).

strate two of these. Some adjusting devices are quite novel. In one design, the adjustment is controlled by the use of electric heaters to expand the bolts (11). Another design utilizes an adjusting bar (17).

The performance of a die lip can be evaluated by measuring deflections as a function of adjustment, using an instrument such as the Model 230 P-2 Electro-Probe manufactured by Federal Products Corp., Providence, Rhode Island. Measurements with an instrument of this type show that performance often can be greatly improved by machining slots between the adjusters to reduce the stiffness of the lip. Data in Figure 6 show the large influence that the design of such slots can have on the performance of the lip.

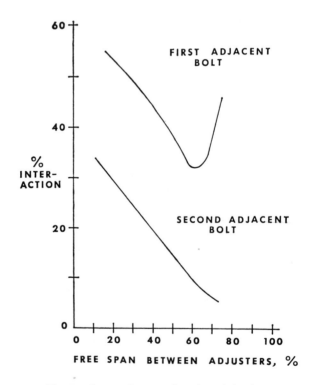

Fig. 6. Interaction as a function of slotting.

Fig. 7. Test section of a die lip.

If a die lip is to be built from a new design, it may be desirable to build a short section to test various slotting arrangements. Figure 7 illustrates the evaluation of such a section.

E. Means of Gauge Measurement

A means of continuously measuring film thickness is mandatory for good gauge control unless frequent breaks can be made in the roll. Fortunately, a variety of instruments is available. The choice of the best instrument varies with the requirements of the application.

The most frequently used type of gauging instrument is the β-ray gauge. In this method, the film whose thickness is being measured functions as a barrier to β-ray particles. Other techniques for film measurement utilize x-rays, ultraviolet light, and interferometry. Pneumatic devices also have been used.

A schematic diagram of a β-ray gauge is shown in Figure 8. The instrument consists of a radioisotope source which emits β particles, a radiation detector such as an ionization chamber, and a recorder.

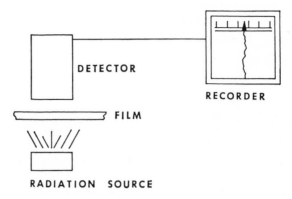

Fig. 8.   Schematic diagram of a β-ray gauge.

The film being gauged is passed between the source and the detector and absorbs radiation to an extent which is related to the thickness of the film.   By Lambert's law:

$$I = I_0 e^{-\mu x} \tag{11}$$

or

$$X = \frac{1}{\mu} \ln \frac{I_0}{I} \tag{12}$$

where I is the radiation which can be detected, I_0 is the incident radiation, μ is the linear absorption coefficient of the film, x is the thickness of the film.

The utility of the β-ray gauge derives from the fact that there are a number of radioactive materials with different absorption characteristics which meet the requirements of a satisfactory source material.   These requirements include long half-life, availability at reasonable cost, and the ability to be handled safely under the conditions necessary for use in a commercial instrument.   Because of the variety in radioactive sources which can be used, measurements can be obtained with accuracy over a wide range of thicknesses.

The variation in the absorption of β-rays as a function of the unit weight of the absorber (which, of course, is proportional to the thickness of the absorber) is shown in Figure 9, for three common radioactive sources.   From these data, the ranges given in Table 1

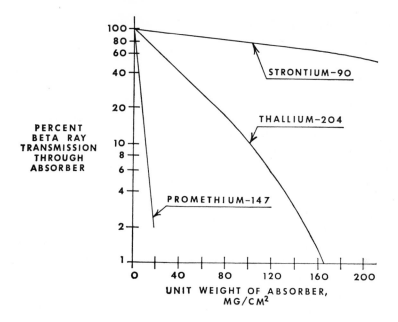

Fig. 9. Transmission curves for β-ray sources (18).

TABLE 1
Comparison of Radioactive Sources

Radioactive source	Approximate useful range of thickness measurement, mils	Years of useful life (19,20)	Comments
Strontium-90	15–200	20	Inexpensive
Thallium-204	3–30	4	
Promethium-147	0.8–8	2	Very weak source; cannot be used in standard source containers
Krypton-85	0.3–5	12	Relatively inexpensive; most widely used source for plastic films

can be calculated. Strontium-90 is an excellent source for the measurement of thick films, while thallium-204 is useful for films of intermediate thickness. For thin films, krypton-85 is the best radioactive source.

A properly designed β-ray gauge measures thickness with an accuracy of 1% for thicknesses greater than 0.5 mil. With thinner films there is a loss in accuracy because of significant absorption of radiation in the air between the source and the detector. The design of the throat—the distance between the source and the detector—is particularly important when gauging thin films because the error in measurement depends on the length of the air gap and and on how well the temperature of the air is controlled.

β-Ray gauges are expensive. An installed unit costs between $15,000 and $50,000.

X-ray gauges are based on the absorption characteristics shown in Figure 10. At a characteristic wavelength of the material there is a peak in the absorption coefficient. Consequently, a gauging device which gives accurate thickness measurements can be constructed by choosing a wavelength just to the left of a critical wavelength, and comparing the absorption at this wavelength with a reference point just to the right of the peak.

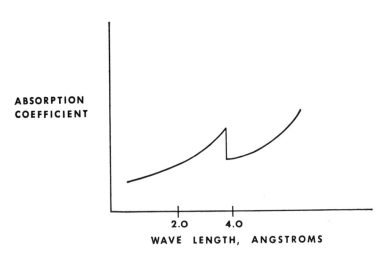

Fig. 10.   X-ray absorption spectra.

As a rule, x-ray gauges are not as accurate as β-ray gauges. Their use is further restricted by difficulties resulting from alignment problems which cause significant errors because of the small cross-sectional area of the crystal detectors. However, x-ray gauging does have a place. For example, these gauges are less expensive than β-ray gauges, the installed cost ranging from $8,000 to $20,000.

Gauges based on the absorption of ultraviolet rays are the least expensive of the instruments based on absorption techniques. Installed units cost from $3,000 to $10,000. The principle on which UV gauges are based is similar to that used with x-ray gauges. Absorption at a wavelength where the UV absorption peaks is compared to the absorption at a reference wavelength where only a small amount of absorption takes place. In addition to low cost, ultraviolet gauging has the advantage of giving relatively accurate readings with films as thin as 0.1 mil. However, at best, ultraviolet gauges have an accuracy of only 1–5%, and additional errors of considerable magnitude can occur if there are temperature

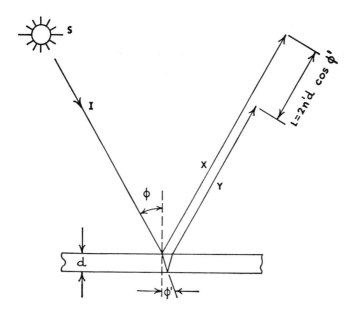

Fig. 11. Optical principle of an interferometric gauge.

variations in the film.   Changes in composition may also cause serious errors.

The interferometric device measures film thickness by equating it to the optical path difference, or time delay, between the reflections of an incident light wave from the two surfaces of the film (21).   The basic optical principle is diagrammed in Figure 11.   A beam of radiation, I, is emitted from a light source, S, and is directed toward the film surface.   A portion of this light, X, is reflected from the first surface of the film.   The remainder of the beam, $I - X$, continues into the film until it reaches the second surface, at which point a second reflection occurs, Y.   Beam Y travels an additional optical path in propagating through the film, and the observed length of this path, L, is directly proportional to the film thickness, d, because

$$L = 2n'd \cos \phi' \qquad (13)$$

where n' is the refractive index of the film and ϕ' is the angle of refraction.

The optical path difference L, and therefore the thickness of the film is measured by a Michelson interferometer.

A pneumatic gauge-measuring device is shown schematically in Figure 12 (22).   The thickness-sensing element is a precision air nozzle N, which is held in a fixed position with respect to a rigid platen, P, over which the film strip, F, is drawn.   A spring-loaded drag level, D, serves to keep the film sample snug against the platen. The inner diameters of the orifice, O, and the nozzle, N, are calculated to produce a nozzle pressure, PN, which is linearly proportional to the distance between the tip of the nozzle and the film surface exposed to the air stream.

Devices of this type are less expensive than those previously described.   However, they are not usable with thin films (0.5 mil or less), they are not accurate, and the lack of accuracy becomes particularly bad when the film is moving at high speeds, because the air layer which rides the film interferes with the measurement. Also, the instrument shown in Figure 12 has the film in contact with the platen, and this cannot be allowed if a scratch-free film is required.   Noncontacting pneumatic gauges have been built, but these are even less accurate.   Because of these debits, pneumatic gauges find very little application.

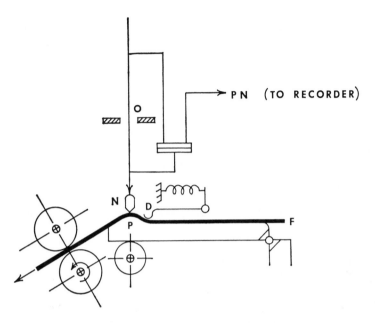

PN (TO RECORDER)

Fig. 12. Schematic diagram of a pneumatic instrument to measure film thickness.

F. Means of Gauge Control

The elements for a system of gauge control are a means of gauge adjustment and the instrumentation to measure continuously the gauge variation in the transverse direction. However, even when both are available, this is not enough. They must be combined into an effective system.

This system can be either manual or automatic. Until recently, only manual systems were used, but the literature suggests that automatic systems are beginning to play a role in commercial film manufacturing operations (23–28). However, whether the system is manual or automatic, the same basic principles apply. These are:

1. Account must be taken of the response of the adjusters. It was pointed out (Section II-D) that even with a well-designed adjusting device, the interaction at the adjacent bolt may be as great as 50%. If gauge adjustment is to be rapid, the complex

388 JAMES L. HECHT

response of the lip to adjustment must be known and taken into account.

2. A gauge profile continuously measured in the transverse direction is not a true gauge profile. Since the sheet is moving, the measurement is made on a diagonal across the sheet. Thus, the variation at any point consists of the "true" variation plus the variation in the longitudinal direction. The "true" variation, which is what adjustments should be based upon, is best approximated by minimizing gauge variation in the longitudinal direction and then averaging the results of several measurements in the transverse direction.

3. Often it is necessary to cast the film so that the gauge profile is not flat. A frequent cause for contour casting is that the cast film is to be stretched subsequently. For example, in the production of film from polyethylene terephthalate it is necessary to cast the film thicker at the center than at the edges in order that the stretched film have a minimum variation in thickness (29). This can be carried out in two ways. The most desirable method is to gauge the stretched film and then adjust the cast film on the basis of this information. However, there are situations when it is necessary that the film be controlled to a contoured profile. One example is when there is a relatively long period between the time when the film is cast and the time when the gauge profile of the stretched film is measured.

4. Some system of lane identification must be developed. In other words, there must be a way to relate to a specific adjusting bolt on the casting die the section of the film which is too thick or too thin. This is not always easy. For example, changes in width take place during casting as a result of necking down.

In the case of a manual control system, the application of these principles is part of the skill of the technician who has the responsibility for making adjustments of the gauge. With an automatic system, readings from the instrument which measures gauge variation are fed to a computer which interprets what should be done and signals this information to mechanical devices which operate the adjusters.

Automatic control systems have many advantages. Use as a labor-saving device is, of course, one. Of great importance is that a properly designed automatic control system will result in production quality being processed more rapidly after startup, thereby decreasing waste. Similarly, the automatic system will produce film with better uniformity of thickness, which results in better film quality and less rejects for quality. The automatic system also is particularly advantageous for a film casting operation where physical access to the adjusters is difficult.

These advantages for automatic control are more than balanced, in many cases, by the high cost of an automatic control system and the need for some technical development for each new use. However, as automatic systems become better developed, their use will expand.

III. Obtaining the Desired Film Properties

A. Thickness

The film thicknesses which can be produced during casting are a function of the lip opening and the viscoelastic properties of the melt. It is desirable that a range of thicknesses be producible with a given lip opening because often it is desired to make a variety of film thicknesses without a shutdown to adjust the casting die.

The thinnest film which can be made with a given lip opening is restricted by the maximum draw ratio. Draw ratio is defined as the lip opening divided by the thickness of the film. The maximum draw ratio usually occurs when the cast film begins to tear, and it is primarily a function of the melt (i.e., the polymer, any additives present, and the melt temperature). Physical factors, such as lip opening and thickness of the bead or edges, can have a significant effect, however. The wide variation in drawability which exists between materials is demonstrated by the fact that it is not uncommon to use a draw ratio of 20:1 in the commercial production of polyethylene (30), yet films have been made of materials which could not be drawn 2:1.

The upper limit of film thickness will be approximately the thickness of the lip opening. Since many thermoplastic melts have a tendency to "bulge" because of viscoelasticity—i.e., to expand after they emerge from an extrusion orifice—it is possible with some materials to produce a film which is somewhat thicker than the die opening, particularly if the lip opening is small. On the other hand, with other materials it is necessary to have a draw ratio of at least 1.2:1 in order to cast a sheet which does not have folds in it.

Casting very thick and very thin films can result in special problems. In the case of a thick sheet, difficulty can be experienced with gauge control because the deflection of the lip is a smaller fraction of the slot when the lip opening is wide. This difficulty can be circumvented by using a lip with a diverging land, which controls thickness by the pressure drop in the throat.

Casting a thin film becomes a problem if the melt is of such a nature that a high draw ratio cannot be used. In this case, it is necessary to go to a narrow lip opening. Three problems occur under these circumstances, however.

1. Melt fracture is more likely to occur and this may seriously limit production rates. This is discussed in more detail in Section IV-A, and in Chapter 6.

2. If the die land is not perfectly machined, gauge bands will occur which cannot be eliminated by adjustment.

3. It may be difficult to maintain the narrow lip opening because of the high melt pressure which results from the small opening. In order to minimize the opening of the lips, the casting die should be designed to prevent sliding of the lips, and the adjusting bolts should be designed to give maximum compressive forces to counteract the flexing of the lips. In addition, the pressure in the die should be minimized. This can be done by using a short land length and by casting at the maximum melt temperature consistent with other process requirements. Increasing the melt temperature has the additional advantage of increasing the drawability of the melt.

B. WIDTH

The width of the cast film is the width of the slot in the casting die minus the necking down which occurs during casting. This necking down can be quite large. It is decreased by decreasing the distance between the die lips and the point at which the film is quenched, or by decreasing the draw ratio.

Production of the desired width is important. The film must be wide enough to provide the width required for finished film, but not much wider or else waste is increased.

If a variety of film widths is required, it may be desirable to build several pairs of lips with different widths or, with thermally stable polymers, to use a rod which plugs part of the lip.

C. Beads and Edges

The edge of the film can cause problems, particularly if the edge consists of a bead to permit stretching.

A bead is best formed by an independently adjustable orifice placed between the die lips and the end block (31). A device of this type is shown in Figure 13. Simpler but less effective methods also can be used. These include a fixed-slot orifice, the opening of which corresponds to the shape of the bead desired; drastically decreasing the length of the land at the edges of the lips to provide greater flow at the edges; thickening the edge by permitting "neck-in"; or directing an air jet at the edge of the film.

Leakage of polymer at the edges occurs if the end blocks are not constructed to provide an effective seal between themselves and

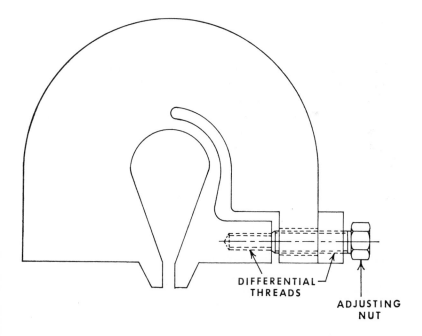

Fig. 13. Adjustable bead plate.

the lips. Although more elaborate methods have been used, satisfactory results are usually obtained by lapping the surfaces involved.

Independent control of the melt temperature at the edges is often desirable. Consequently, it is good practice to place an electric heater against the end block. This can be used to compensate for the cooling which takes place at the edge, and to provide additional heat should it be desirable to have smoother flow or more flow.

It is difficult to cast film so that the edge is at the desired thickness. Usually the edge is too thick as a result of necking in. When this occurs, the waste can be cut down by modifying the casting die to decrease the flow at the edges, and thereby compensate for the thickening caused by necking in. Tapering the land length at the edge is one way by which this can be done.

D. OPTICAL PROPERTIES

Optical properties such as gloss, haze, and clarity are influenced significantly by the melt-casting process. Since internal haze is much less than surface haze for films which do not contain opaque additives, optical properties depend primarily on surface irregularities which scatter light. Both casting and quenching variables are important with respect to the formation of these irregularities.

Films with high gloss have low haze and high clarity, if the film does not contain opaque additives. But the relationship between gloss, haze, and clarity is not exact. The nature of the irregularity on the film surface makes a significant difference in the manner in which light is scattered. Thus two films may have the same gloss when the incident light is at a 60° angle, but the gloss will differ when the incident light is at an angle of 85°. A detailed discussion of the nature of optical properties is given by Clegg and Huck (32).

The formation of surface irregularities which give haze, low clarity, and low gloss usually results from either melt–flow phenomena or crystallization. Surface haze can be produced also in other ways. When film is quenched in water, liquid vaporization may mar the film surface. Haze also can be produced by draw fracture, caused by excessive drawing of the melt.

The effect of melt–flow phenomena is complicated by the fact that at least two different mechanisms work to produce surface irregularities. These are (1) "extrusion haze" or "land fracture" which manifests itself as a high-frequency irregularity on an otherwise uniform extrudate, and (2) "orange peel" or "melt fracture" which shows as a distorted extrudate, described as wavy.

Surface haze produced by melt–flow phenomena is affected by design of the film die, shear history of the melt, melt temperature, and throughput rate. Details of these effects on fracture are given in Chapter 6.

Increasing the extrusion rate increases surface roughness. However, since low extrusion rates increase film costs, it is not desirable to control optical properties by decreasing the extrusion rate. Sometimes, of course, this must be done.

Probably the most effective way to achieve high gloss is to increase the melt temperature which, in turn, decreases the viscosity of the melt and the elasticity of the melt. Unfortunately, this approach is usually limited by degradation of the polymer. When degradation is a problem it may be desirable to heat the melt in the die lips (33,34).

Increasing the shear of the melt has been reported as another means of improving the optical properties of film. This can be done either by placing a throttling device between the extruder and the die, or else by a high-pressure die (32,35).

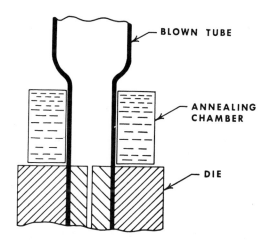

Fig. 14. Annealing process.

Surface irregularities arising from crystallization are minimized by rapid cooling, once the film is below the crystalline melting point. Quenching processes are described in detail in Sections IV-C and V-B.

The optical properties of film have been improved by annealing the film prior to quenching, as shown in Figure 14 (32,36,37). Holding the extrudate for 2–3 sec at a temperature above the melting point relaxes stresses formed in the die.

Drawing the hot melt decreases surface haze by spreading out irregularities. Stretching (i.e., cold-drawing) has the same effect, but also changes physical properties.

Thus, a large number of process variables significantly affect optical properties. When clarity is a problem, however, particular attention should be devoted to what usually is the most important variable of all: the polymer.

E. Surface Properties

The ease with which the film is handled in subsequent operations, such as packaging, depends on a variety of factors including roll formation, thickness uniformity, modulus, and the surface properties of the film. The latter is probably best described quantitatively by the coefficient of friction. A low coefficient is desirable to give good slip and to reduce blocking (Chapter 9, p. 425).

Surface irregularities of the type which give haze in film are very effective in reducing the coefficient of friction. A smooth, glossy surface, on the other hand, has a tendency to block. Thus, changes in process conditions which improve gloss also increase the coefficient of friction. As a result, film products often are a compromise and are made to give sufficient clarity for the required use and still maintain enough slip. When product requirements demand a very high gloss and the resulting product does not have sufficient slip, the problem often is best solved by techniques other than those related to melt casting, such as the addition of a slip agent, or sizing.

F. Physical Properties

The physical properties of films are not greatly affected by variations in the melt-casting process, provided certain secondary effects do not occur. These secondary effects include:

1. Cold-drawing. This occurs whenever stretching takes place below the crystalline melting point. It increases strength properties such as tenacity, decreases elongation, and increases shrinkage. When film is quenched in water or made by certain tubular processes it is impossible to eliminate cold-drawing, and therefore film properties appear to be affected by the melt-casting process.

2. Polymer degradation. This may decrease molecular weight and thereby decrease both strength properties and elongation.

3. Fracture. Severe fracture decreases both strength properties and elongation.

G. Appearance Defects

The appearance of the film is an important property—how important depends on its use. For most film applications there are some appearance defects which can be tolerated, but other defects make the film unusable.

The nature of appearance defects varies with the film and the casting process. In this section, discussion will be limited to the types of defects which occur in a variety of films.

Before proceeding, something should be said about recognizing and analyzing defects. An inexperienced person often will fail to see defects which are serious enough to make the film unusable. To spot defects, the film must be viewed under proper conditions of illumination, both at a low angle and at the perpendicular. Also, many defects cannot be seen at film speeds greater than 50–100 ft/min, and some defects cannot be detected even at lower speeds. Thus, it may be desirable to make provision for additional inspection at low speeds.

When a visual defect has been observed, it is desirable to establish whether the defect is on only one surface. This gives an important clue as to the origin of the problem. If the defect is on only one surface, it will no longer be visible if covered with a drop of a liquid which has a refractive index equal (or nearly equal) to that of the film.

Streaks or die lines are defects which often appear on one surface. As their name indicates, they appear as small longitudinal lines. In addition to being an appearance defect, they often constitute weak points in the film since they represent localized thin spots.

A nick in the die lip gives a streak which appears as a single continuous line on one surface. The only way this streak can be eliminated is to suspend operations either until the lip is repaired, or a substitute lip can be mounted on the die, or an entirely new casting die has been put into operation. Streaks on one side also can arise from the buildup of degraded material on the die lip. These streaks can be eliminated by cleaning the lips. This is done by mounting a flat piece of soft metal (e.g., brass) in a holder and rubbing this against the inside of the lip in such a way as to force material in the lip through the slot. Unfortunately, this involves a significant loss of material and time.

Streaks are not necessarily confined to one side. A disturbance of melt flow—such as is caused by the use of a "spider" in a circular die —can cause memory effects which give streaks. A similar source of streaks may be a particle that has become lodged in or ahead of the die lip. Screen packs are used to prevent this and, if it occurs, a leak in the screen is indicated. Often such a particle can be dislodged by cleaning the lips, but sometimes this cannot be done with a large particle, in which case the run must be shut down and the casting die cleaned.

Scratches are another defect which usually appears on only one surface. These occur when film is dragged over a surface which is not moving, such as an idler roll which is not rotating freely.

One of the most common defects found in films is fisheyes or gels, also known as blobs. They consist of unmelted or partially melted particles which appear as small, nonhomogeneous areas in the film. The nature of the resin can be important, since the cause often is highly crosslinked material which does not completely melt but is elastic enough to go through the filter. Polymers have been made which were so prone to give gel that acceptable film could not be made from them. Usually, the most effective method of minimizing gel is to use a finer filter. If the problem is only one of melting, an increase in back pressure during extrusion results in considerable improvement of film quality.

The use of a finer filter also removes more of the contamination which often is present. Specks of contamination in film usually are obvious but, if transparent, will appear similar to gel.

If the melt contains bubbles, the appearance may be similar to gel particles, but usually the film contains visible craters on the surface. In extreme cases, bubbles cause holes in the film. Bubbles can form in a number of ways: if the polymer is wet, the water becomes steam; air can be trapped in the melt; or volatile materials can be present or be caused to form by decomposition.

Small dimples can look very much like bubbles. Dimples are formed by small particles (such as dirt) adhering to surfaces which touch the film, particularly where the film is hot.

Holdup spots can lead to thermal degradation which shows up as gel, bubbles, discolored particles, or discolored streaks. Holdup also can cause other visual defects because of the difference in viscoelastic properties between the bulk of the melt and the material which has been held up. Because of these problems, melt-casting systems should be designed to avoid dead spots. The effectiveness with which flow channels have been streamlined can be measured by adding a pigmented melt a short time before a shutdown. Holdup spots are then determined by disassembling the system and observing where the original melt remains.

Holdup can also occur on a pitted surface. Consequently, all surfaces in contact with the melt should have a good finish and be of material which will not react with the hot melt. For example, it has been recommended that in the casting of polyvinyl chloride film, all metal surfaces in contact with the melt be polished, given a hard chromium plating, and then be repolished (38).

Elimination of holdup spots has an additional advantage in that it allows going from one melt composition to another without a large loss of transition material. Back-to-back casting of different batches increases machine efficiency, but this advantage is lost if a long transition period is required to flush out the previous melt. Minimizing the transition period is also aided by scheduling in such a way that each successive batch is more viscous than the previous one.

Appearance defects often are associated with the quenching operation. When drum casting is used, pucker marks can occur (see Section IV-C). If water quenching is used, craters may be formed as a result of water boiling at the film surface, or because air is released from the water at the film surface. The latter problem can be solved either by deaerating the water, or by using cooled water. The problem of water boiling at the film surface can be alleviated by proper circulation of water in the quench tank.

Simonds, Weith, and Schack list a large number of appearance defects and their causes (39).

IV. Maximizing Productive Capacity

A. FACTORS INFLUENCING PRODUCTIVITY

The formation of films by melt casting is a manufacturing operation and—as in any manufacturing operation—an important measure of success is the productivity for a given investment. Productivity, in turn, is the product of throughput rate and efficiency. The latter is the fraction of time that good film is produced.

The throughput rate of a melt-casting operation can be limited by any of the following: (1) melt-pumping capacity, which usually is extruder capacity; (2) melt fracture in the die lips; (3) quenching capacity; (4) the ability to convey the film.

As a rule, the ability to convey the film will not prove limiting. Normal web handling equipment can operate at speeds well over 1000 ft/min.

Melt extruders are available as standard equipment in sizes up to 10 in. in diameter and capacities of 2000–5000 lb/hr. Special extruders for film lines have been built as large as 15 in. in diameter. Figure 15 shows extruder capacities listed by manufacturers as a function of screw diameter. The capacity is, of course, dependent on many factors other than the diameter of the screw. These include the length of the extruder, the geometry of the screw, and factors associated with the polymer such as density, softening point, melt viscosity, bulk density of the feed, and the maximum temperature which can be permitted during extrusion. The capacity also depends on the melt quality required (i.e., uniformity of temperature and throughput). Usually, better quality is achieved by decreasing throughput. Since extrusion of film demands better uniformity than most other extrusions, the capacity at which a film extrusion line is operated is even less, at times, than the lower limit suggested by manufacturers.

Data in Figure 15 indicate that a single melt extruder gives throughput rates well in excess of what normally is desired for a film line. However, nonuniform melt temperatures are produced when large extruders are used with viscous melts, and this has been known to act as a limit in sizing a film line.

The phenomenon known as melt fracture was mentioned in Section III-G and is discussed further in Chapter 6. For polymers which experience melt fracture there is an obvious upper limit on

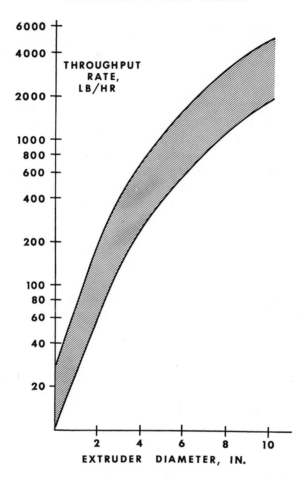

Fig. 15. Extruder capacity as a function of screw diameter. The upper limit is approached as: (*a*) the extruder has a high ratio of length to diameter; (*b*) the polymer has high density; (*c*) the polymer has low melt viscosity; (*d*) variations in temperature and throughput can be tolerated.

the throughput per inch of width. However, this throughput often can be made nonlimiting by taking those steps which are obvious from rheological studies, such as maximizing the draw ratio, maximizing the melt temperature, and optimizing the geometry of the casting die.

The rate of quenching—i.e., cooling the melt until it has solidified—often limits throughput. Because of the importance of quenching with respect to both productivity and film quality, it is discussed in detail in Sections IV-C and V-B.

If any of the above four criteria limits the production rate, the design of the entire unit should be based on this throughput to minimize investment. Many times, of course, the design capacity is not limited by any of the above, but by the requirements for film. In this case, a business judgment, based on sales forecasts, must be made to determine the design basis. A high-capacity unit has the advantage that less investment is required per unit of productive capacity and the operating expenses are less per unit of production. However, these advantages are lost if excess capacity remains idle for a long period of time.

Now let us turn to those factors which influence efficiency. Production of film can be lost in a variety of ways, the most important of which are:

1. Time necessary to start up and make film with an acceptable gauge variation. This problem has been discussed in detail (Section II).

2. Time lost as a result of the filter becoming plugged. Because of its importance this is discussed in detail in Section IV-B.

3. Loss which occurs because film does not meet specifications.

4. Loss which occurs because of equipment failures (e.g., a power failure, or a malfunctioning tube in a control circuit).

5. Loss which occurs in going from one film type to another, particularly when there is a change in composition.

A frequent cause for the production of film which must be scrapped is the use of raw material which does not meet specifications. The other common causes of off-specification film are inadequate process control, because of either instrument or equipment failure, and incorrect operation caused by human error.

Maintenance of an adequate level of in-process quality control is the most effective means of minimizing losses caused by the production of off-specification material. The specific properties to be monitored and the frequency with which these properties are checked depend on the manufacturing operation and the quality standards for the product. Other means which are effective in reducing losses resulting from off-specification film include (*1*) monitoring the performance of key equipment, *(2)* using alarms to alert operating personnel when nonstandard conditions exist, *(3)* employing check instruments to measure very critical variables, and *(4)* setting specifications for all raw materials.

In conclusion, it should be mentioned that there is a double stake in operating at a high efficiency because a loss in efficiency hurts not only productivity but also yield. It is good practice in a commercial film operation to analyze quantitatively the causes of losses in productivity. This, in turn, will indicate the areas where action should be taken.

B. Filtration

The need for filtration of the melt has been mentioned before. Filtration serves a wide variety of purposes, from removing contamination to filtering out some of the crosslinked polymer which otherwise would show up as gel or fisheyes. The degree of filtration required is a function of the polymer and how the film is used.

Usually, melt filtration is carried out by placing a screen pack (a series of screens) against a breaker plate which supports the screens. The breaker plate can be as thin as $\frac{1}{8}$ in. (depending on the melt pressure). It contains a large number of holes, usually from $\frac{1}{8}$ to $\frac{3}{8}$ in. in diameter, which permit passage of the melt. How-

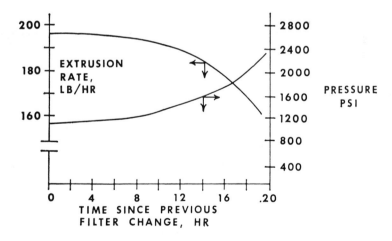

Fig. 16. Effect of filter plugging on extrusion rate.

ever, a variety of other filtration devices can be used which provide finer filtration, more surface area, or both.

Filtration often causes a loss in throughput because a pressure drop occurs as the melt flows through the filter. When the filter is properly designed, this pressure drop is small initially, but as the filter begins to plug, the pressure drop increases and the rate of extrusion decreases. The decrease in the throughput rate during an extrusion run is shown in Figure 16.

Plugging of the screen pack is more than a capacity problem. It frequently causes gauge variations as a result of a decrease in throughput or an increase in melt temperature (caused by the higher back pressure), or a combination of both. In addition, the increase in extrusion temperature can adversely affect the film in such ways as discoloring the polymer or increasing gel formation. Any of these factors can dictate when it is necessary to shut down the process and change the filter.

There are three different ways of operating when a filter is becoming plugged.

1. Run the extruder at constant speeds and, as the throughput decreases, slow down the takeoff to maintain the desired thickness. The throughput rate would then follow the curve shown in Figure 16.

2. Increase the speed of the extruder, as the filter plugs, to keep the throughput constant. This has the serious disadvantage, however, that it further increases the melt temperature, and consequently this method of operation usually is not desirable to use.

3. Start with a partially closed extrusion valve between the extruder and the film die. Then, as the filter starts to plug, compensate for the increased pressure drop across the filter by decreasing the pressure drop across the valve. Thus, the throughput and the extrusion temperature remain constant. This method has obvious advantages with respect to control, and usually is the most satisfactory method of operation. There are exceptions, however. For example, this method results in some decrease in productivity, if the extrusion rate is limiting the rate of film production.

When it becomes necessary to change a filter there is a loss in productivity during the time the process is shut down. However, this is usually much less costly than the time it takes to start again and resume making film which meets specifications. Even during a brief shutdown, temperatures in the casting die and the extruder are changed, and it then takes a considerable time for these temperatures to return to equilibrium, during which time the gauge is constantly shifting.

Loss in productivity and yield caused by a filter change can be minimized by designing the filter in such a way that replacement takes a minimum time, thereby minimizing the upset caused by the change. A number of designs have been developed to do this, some of which claim to give almost no interruption in flow. Also, self-cleaning screen packs are beginning to be available.

When significant losses result from filter changes, efforts should be made to increase the time between which filter changes are necessary. This can be done in a number of ways:

1. Make certain that the degree of filtration used is necessary. If product quality can be maintained with a coarser filter, the time between filter changes will be increased.

2. Decrease the amount of filterable material in the melt. The material plugging the filter may be dust which gets into the polymer or reworked material because of careless handling procedures. This could be ascertained by obtaining a chemical analysis of the material.

3. Increase the area available for filtration.

4. Provide several levels of filtration. In other words, allow several screens to become plugged slowly rather than one screen to become plugged rapidly. If a screen pack is used, this can be accomplished by having the melt first strike a coarse screen, followed by a screen of intermediate size, and then a fine screen, with each of these screens physically separated by a coarse screen.

Film quality considerations are important in the design of a filtration system. In particular, care must be taken to make certain that the melt cannot circumvent the filter. Consequently, when good filtration is critical the melt should pass through two filters. Then, should the melt leak around one filter, or go through a hole which develops in the filter, there still is adequate filtration.

Another important design point is that the filter should not be used for anything but a filter. While it often is necessary to increase back pressure in order to improve the extrusion, the use of a fine screen pack to increase the pressure is foolish because it leads to rapid plugging. As a rule, the use of an extrusion valve to control back pressure is the preferred method.

A final design consideration is that the temperature of the polymer increases when there is a large pressure drop. The method of calculating this increase in temperature is outlined by Bernhardt (1). With polyethylene, the average temperature increases 3.5°C/100 atm. Since the temperature of some of the polymer will increase much more, this effect may be significant when making film from a temperature-sensitive material.

In summary, good filtration is necessary for good film quality. It often is necessary to pay for this quality improvement in terms of productivity and yield, but usually the price is small when the filtration system is properly engineered.

C. QUENCHING

The process by which the hot molten sheet cools is critical in a number of ways. As previously mentioned, quenching capacity often limits the total capacity of a film line. Even more important in the selection of the quench process are the following product requirements:

1. A flat sheet must be obtained. This is not very difficult with some materials, such as polyethylene, but it imposes definite restrictions with materials where the melt has high shrinkage forces. Melts such as these generally require a quenching process in which the film is supported.

2. A prescribed time–temperature history for the film is often necessary. Many thermoplastic polymers crystallize. The degree of crystallinity, which depends on the time–temperature history of the polymer, affects a wide range of properties, from surface properties (such as clarity) to the ability to cold-draw the film.

3. Excessive necking down (i.e., decrease in the width of the sheet) must be prevented.

4. Surface defects must be avoided. Specific problems with respect to surface defects are discussed in Section III-G as well as later in this section.

5. Uniform cooling of the sheet is necessary to insure uniformity of thickness and optical properties.

There are two types of processes by which flat films are quenched: drum casting and water quenching. Schematic diagrams of these are shown in Figures 17 and 18. It is also possible to have the

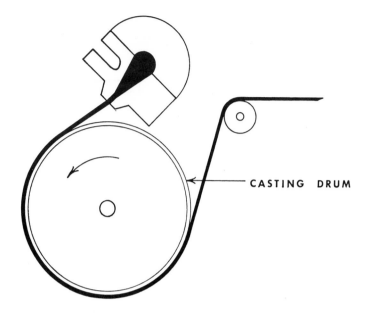

Fig. 17.   Drum casting process.

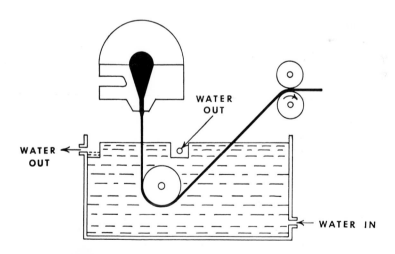

Fig. 18.   Water-quench process.

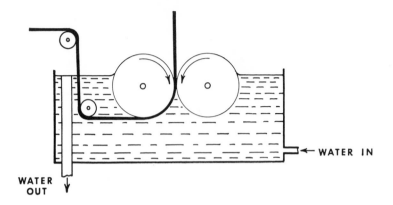

Fig. 19.   Combination of water and drum casting.

essential features of the two methods combined, as shown in Figure 19 (40).

In order to calculate the rate of film cooling, drum casting can be considered as a case of heat transfer by conduction in which

$$\frac{1}{h} = R_W + R_D + R_0 \tag{14}$$

where h is the heat transfer coefficient between the cooling water and the film, in Btu/hr-ft^2-°F, R_W is the resistance to heat transfer between the water which cools the drum and the drum itself, R_D is the resistance to heat transfer of the drum wall, and R_0 is the resistance to heat transfer between the film and the surface of the drum.

For good heat transfer, the best design is a drum which is double-walled to give a baffled annular space through which cooling water can be pumped at a high linear velocity (41).   This gives a low value of R_W.   If the diameter is 4 ft or less, this construction can be made conveniently and inexpensively from centrifugally cast tubing.

A well-designed drum will also have a relatively thin outer wall to minimize R_D.   This is particularly important if the drum is made of stainless steel because it has a low thermal conductivity.

In general, the heat transfer coefficient is determined by the

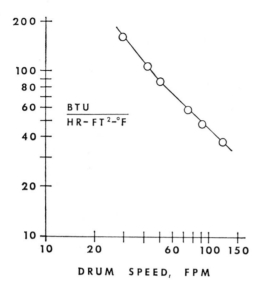

Fig. 20. Heat transfer coefficient as a function of drum speed.

resistance between the film and the surface of the drum. This is illustrated by the data in Figure 20 which correlate measured heat-transfer coefficients (i.e., $1/R_0$) with drum speed. Since $1/(R_W + R_D)$ usually is greater than 300 Btu/hr-ft²-°F, the resistance between film and drum is completely controlling for film cast in this manner.

The relatively poor heat transfer at high speeds results from entrapment of an air layer between the film and the drum. As the drum speed increases, so does the thickness of the air layer.

The air layer causes important film quality problems also. Ripples and puckers are formed when the film does not adhere to the drum and differential cooling and lateral shrinkage of the film occur. Excessive neck-in and uneven neck-in (i.e., scalloping) can also occur.

These quality problems are eliminated if the film adheres satisfactorily to the drum. The most effective step which can be taken to give satisfactory adhesion is to heat the drum (42). Pinning the edges with air jets has been used to decrease neck-in and avoid scalloping (43). The use of a contact roll or an air knife to squeeze out the air layer is effective, but often is accompanied by other effects on film quality which preclude its use (42). Recently, the use of vacuum suction has been disclosed (44).

Heating the drum to adhere the cast film has several disadvantages. The rate of cooling is slower and a larger drum is required. In addition, the film leaves the drum hot and, if wound up shortly thereafter, will block (i.e., the film will stick to itself). To circumvent this difficulty, the film can be further cooled on a cold drum which follows the quench drum.

Returning to the calculation of the film cooling rate, in the case of a thin film (1–2 mils thick) the resistance to heat transfer through the film is small compared to the other resistances. Thus, the film temperature can be assumed to be uniform throughout the thickness of the film. If the additional assumption can be made that the temperature of the surface of the drum does not change significantly while in contact with the film, heat-transfer predictions can be calculated by the standard heat-transfer equation:

$$Q = hZL \frac{T_{F_1} - T_{F_2}}{\ln\left[(T_{F_1} - T_W)/(T_{F_2} - T_W)\right]} = WC_p(T_{F_1} - T_{F_2}) \quad (15)$$

where Q is the heat removed from the film, in Btu/hr

h is the heat-transfer coefficient, as defined in eq. (14)

Z is the width of the cast film, in ft

L is the length of contact of the film on the drum, in ft

T_{F_1} is the temperature of the unquenched film, in °F

T_{F_2} is the temperature of the quenched film, in °F

T_W is the average temperature of the cooling water, in °F

W is the weight of the film, in lb/hr

C_p is the specific heat of the film, in Btu/lb-°F

With thicker films it is necessary to consider the problem as one of unsteady-state conduction of heat, particularly, if the heat transfer coefficient is relatively high. Fortunately, unsteady-state heat-transfer calculations can be carried out without difficulty using Gurney-Laurie charts (45).

The same basic heat-transfer relationships hold for water quenching except that the heat-transfer coefficient consists only of the resistance between the water layer and the film. Moreover, in water quenching, the film is cooled from both surfaces, thereby at least doubling the rate of cooling. Nevertheless, it still is usually necessary to use unsteady-state heat transfer calculations for thicker films.

There is a difference between drum casting and water quenching in the effect of film speed on the heat transfer coefficient. In drum casting, there is a decrease in the heat transfer coefficient at high film speeds. On the other hand, with water quenching just the opposite appears to occur. Although no quantitative data are available, improved heat transfer at high film speeds has been qualitatively observed.

The opposing effects of film speed on the heat-transfer coefficient may explain the contradictory observations reported in the literature for the most effective means of rapidly quenching to obtain high clarity. Several writers have commented that optimum clarity can be obtained only with water quenching (42,46). On the other hand, Rose implies that along with other advantages, the quench drum gives as good or better clarity (47). Both views can be correct. A thin film traveling at a low linear speed can be cooled more rapidly on a drum, but at high speeds, and particularly with thick films, water quenching has an advantage as a means of rapid cooling.

There is a wide variety of additional advantages and disadvantages to both processes. The use of a quench drum has the following disadvantages in addition to those which are caused by the air layer (e.g., slower cooling rates, surface defects such as ripples):
(1) Drum casting requires a larger investment. Also, the drum is a maintenance problem, because it must be kept free of impressions.
(2) With some melts, a deposit builds up on the drum. This necessitates periodic cleaning in order to maintain film quality.

The list of disadvantages for the water-quench process is much longer:
1. The film is not supported during quenching. As a result, this process is not suitable for many melts, particularly those with low melt strength.
2. The hot film can cause water to boil at the film surface which, in turn, mars the surface of the film. This can often be prevented by properly designing the water circulation system. Also, the use of mixtures of water and glycerol, or water and ethylene glycol, have been reported to help in this respect (40).
3. It is difficult to avoid vibrations in the water level which, in turn, produce variations in gauge and clarity. A good deal can be

done to minimize this effect by proper design of the film-handling equipment and positioning of weirs. For example, a baffle or weir which is placed in the center of the quench tank prevents the surface of the water near the die from being disturbed by water running back into the bath from the film leaving the tank (48).

4. "Bellying"—a distortion of the film due to the weight of the water—can occur.

5. If the melt contains an organic plasticizer which is not soluble in water, this often forms a scum on the surface of the water which in turn leads to film quality problems. Similarly, many of the materials added as slip agents are surface-active, and these can cause problems in a quench system.

6. Water-quenched film tends to have more imbalance in physical properties than film made by drum casting because there is usually a small amount of cold-drawing in the machine direction when the partially cooled film is at a temperature just below the crystalline melting point.

7. With the water-quench process the removal of water from the film can become a problem.

Future research will undoubtedly be directed at further reducing the effect of the air layer so that high-speed processes can be developed which take full advantage of the many inherent advantages of drum casting. Greater utilization of combinations of drum casting and water quenching may also take place in the future.

V. Tubular Processes

Tubular processes are widely used in film manufacturing operations. Although most of the basic principles which guide the design and operation of these processes are the same as for flat films, very different equipment is used for casting and quenching. Consequently, tubular equipment and its technology are discussed in the next two sections. We shall then conclude with a comparison of the relative advantages of tubular and flat-film processing.

A. CIRCULAR DIES

A circular die can be viewed as a center-fed, flat-film die which has been twisted so that the two ends come together and are joined. Thus the analysis of melt flows given before (Section II-C) can be applied to circular die casting. But the equipment is very different. Whereas, the basic components of a flat die are the two parallel lips, a circular die consists of a mandrel, which is the inner lip, and a concentric ring, which forms the outer lip.

Tubular dies usually have less adjustability than flat dies. Although a large number of different designs for flexible lips have been built for circular dies (49–54), none of these has proved completely satisfactory, usually because when the lip is adjusted at the point of a gauge band the lip tends to distort at some other point. The gauge also can be adjusted by mechanically shifting the two concentric rings which form the lips, but only a limited improvement can be obtained by this technique.

Controlling thickness by varying the temperature around the die has been reported (55).

Because of the difficulty in correcting thickness differences it is particularly desirable that circular dies have temperature uniformity. Usually, circular dies are heated electrically to keep the installation compact and permit room for the many auxiliaries which are associated with a circular die. A good practice for heating is to use resistance heaters cast in aluminum. These give

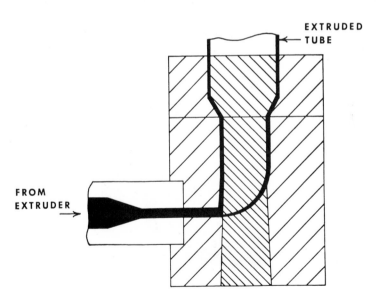

Fig. 21. Offset circular die.

rapid, uniform transfer of heat (42).   The use of induction heating has been reported to give good uniformity, but this method is expensive (42,56).

A major design decision which has to be made for circular dies is the choice of method to support the mandrel.   The most widely followed practice is shown in Figure 21.   In this offset die, the mandrel is anchored at the rear of the die and the polymer melt is introduced from one side.   Since this can lead to nonuniform flow, or a weld line where the two streams meet, designs have been based on a center feed, as shown in Figure 22.   However, the center-fed die gives weld lines created by the "spiders" holding the mandrel in place.   Which design is best depends on how pronounced the weld lines are and what the eventual use of the product is.   A single weld line can be eliminated by slitting the film at the point of the weld line.   On the other hand, the welds caused by spiders will often be less pronounced than those produced in an offset die. In these cases, the center-fed die may be preferred if the tube is to be used as tubing.   In all cases, the design of the die is critical.

Good flow distribution can usually be obtained for the die shown in Figure 21 by proper design of the channel leading to the land. A wide variety of "pressure distributors" have been used.   For example, dies have been designed with a breaker plate around the mandrel in the form of a ring (55).   Factors influencing a particu-

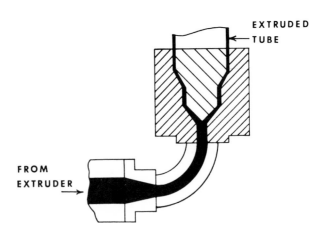

Fig. 22.   Center-fed circular die.

lar design include size of the die, rheology of the melt, and thermal stability of the melt. Holdup time in the die should be minimized for materials which have poor thermal stability.

An advantage of circular die casting is that thickness variations which originate in the die can be distributed by rotating the entire die slowly on its axis. This improves roll formation by eliminating bands which arise when layers of a thick section are stacked on top of each other. A similar result can be obtained using a stationary die, but rotating either the extruder (57) or the windup (58). The latter has the additional advantage of randomizing all gauge variations, not alone those caused by the die.

B. Tubular Quenching Processes

There is much greater variety in quenching methods for tubular processes than for flat films.

Tubular processing usually is carried out vertically. The only advantage of casting horizontally is that a tall building is not required, but this is offset by several processing disadvantages, including distortion of the tube by gravity. The choice between extruding upward or downward varies with circumstances. Some quenching processes require downward extrusion. Also, downward extrusion is necessary for operability if the melt has low strength. When there is no disadvantage to extruding vertically upward, this method usually is preferred because the heavy extruder and its auxiliaries can be located at ground level.

In all tubular processing, air is injected into the tube to prevent it from collapsing. In some processes air is added to expand the diameter of the tube immediately after it emerges from the circular die. This process of "blowing" the hot tube can be considered as melt casting a film which is then drawn in both the transverse and longitudinal directions.

The tubular quenching process is frequently closely related to a subsequent orientation step (i.e., stretching the film below its crystalline melting point). In this chapter, only the quenching part of the process is discussed.

There are three ways that tubular films have been cooled: by air, by direct contact with liquids (usually water), and by conductive cooling from a metal surface.

Figure 23 shows a tubular process using air cooling (59). The

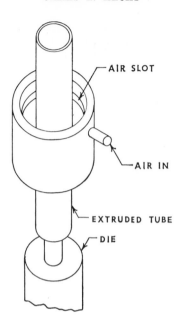

Fig. 23.   Tubular casting process using air cooling.

outside of the tube is cooled by air from a cooling ring.   Since a high-velocity air stream would mark the film, large volumes at low velocity are used.   Although the cooling rate in such a system is not high, the heat-transfer coefficient being about 5 Btu/hr-ft^2-°F, the method is simple and cheap, and appears very satisfactory for certain types of films.   A 20–50% increase in cooling rate can be obtained by adding fresh cool air to the inside of the bubble.

It was previously pointed out that different cooling rates during quenching can have a marked influence on thickness variations. Consequently, cooling rings have to be carefully designed for uniform air flow.   In addition, it is common practice to rotate the cooling ring to randomize those variations which do exist (42).

Several different types of water-quenching processes have been developed.   Figure 24 shows the use of a wetted-wall column (60). The cast tube, which is supported by an air bubble trapped between the die and pinch roll, is cooled by direct contact with the water which flows down the inside wall of the column.

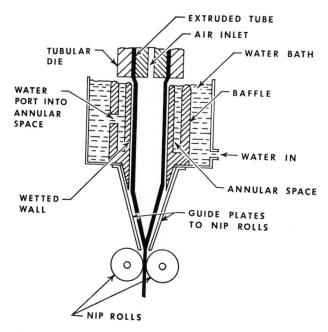

Fig. 24. Tubular casting process using water quenching.

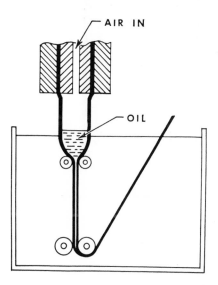

Fig. 25. Tubular casting process using water quenching.

Direct extrusion into a water bath can also be used (61,62).   As shown in Figure 25, the tube is prevented from collapsing under the water pressure by a constant head of liquid within the tube.   Usually a hydrocarbon oil is used.   Above the liquid level, the tubing is supported by air pressure.

Conductive cooling can be considered the tubular equivalent of drum casting because heat is transferred by conduction between the film and a cooled metal surface.   Whereas in drum casting the air layer serves no useful purpose and is detrimental because it causes

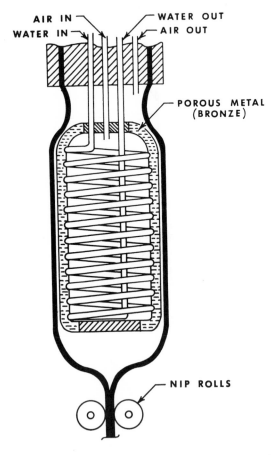

Fig. 26.   Tubular casting process using a cooling mandrel.

a decrease in the heat-transfer coefficient, in tubular processes an air layer is necessary as a lubricant between the film and the stationary metal surface.

Several different processes have been developed where the tube is drawn over a cooled mandrel (63–66). Basically they are all similar to that shown in Figure 26. In some cases, the film is expanded in diameter and allowed to travel some distance before touching the mandrel; in others, the mandrel is very close to the die and has a diameter slightly less than the diameter of the die.

The mandrel can be either a solid or porous metal. The use of porous bronze, or other similar porous structures, permits passage

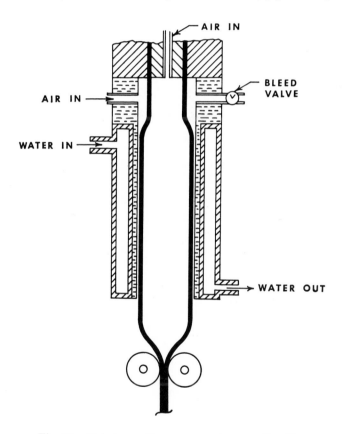

Fig. 27. Tubular casting process using a cooling die.

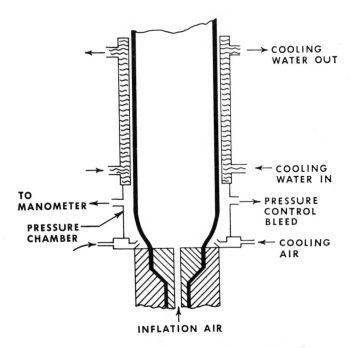

Fig. 28.   Tubular casting process using a cooling die.

of air through the mandrel to give continuous lubrication along
its length.   When a mandrel made of solid metal is used, lubrication
is provided by the air accompanying the advancing film.   Both
types of mandrels are cooled by water passing through coils embedded in the surface.

A properly designed mandrel is tapered to accommodate shrinkage
of the tube during cooling.   The degree of the taper varies with such
factors as type of polymer and throughput rate, but typically it
runs 1–4 mils per inch of diameter per inch of length.   Another
design feature is that the mandrel usually contains pressure-relief
outlets along its length to permit withdrawal of gas along the length
of the mandrel.   A proper surface on the mandrel can be important
with respect to eliminating sticking and scratching of the film (67).

Conductive heat transfer can be used to cool the outer surface
of the film.   An example is the cooling die shown in Figure 27
(68).   In this process, air expands the tube and keeps it in contact

with the die mold, while a second airstream lubricates the wall of the mold. A more elegant cooling die is shown in Figure 28 (67). Uniform contact of the film with the die is obtained by a high pressure in the bubble. This is made possible by having the space between the die and the cooling jacket enclosed and pressurized.

In all conductive cooling applications, the air layer used for lubrication should be optimized. If it is too thick, heat transfer is poor. But if it is not maintained thick enough, the lubrication is inadequate and operating problems develop. Usually, good performance is obtained when the air layer is 1–10 mils thick. A well-designed conductive cooling system gives heat-transfer coefficients greater than 50 Btu/hr-ft^2-°F.

These are the three basic types of quenching methods. There are times when it is advantageous to combine two of these, such as water quenching and conductive cooling (69,70).

C. Comparison of Tubular and Flat-Film Processes

The decision whether to use a tubular process or a flat die involves complex questions. The answer depends on a variety of technological considerations dealing with the process, on market considerations such as the quality of the slit roll needed, and on business factors such as the nature and volume of markets.

If the film is to be oriented, a major technological consideration is choosing between tubular and flat-die processing is how stretching can best be done.

Another major technological consideration is whether the film must be rapidly cooled to meet product requirements and, if so, how this can best be accomplished. The flat-film process has been described as best for rapid quenching (42), but this comment probably does not take into account recently developed techniques. The fact is that there is no simple answer as to which method is best suited for rapid quenching, because the question cannot be considered apart from the answer to other questions. Can water quenching be used? If drum casting is to be used, how hot does the drum have to be in order that the film stick to the drum? What is the desired line speed? The answers to these and other questions have a critical bearing on whether tubular or flat-film processing will give the most rapid quenching.

When product requirements can be met by either process, the

choice is based on economic considerations. The required investment is of great importance in such a case. As a rule, a tubular
process requires considerably less investment provided it can be run
at an equivalent speed. This is particularly true for most tubular
orientation processes.

Another important economic consideration is waste. An advantage of the tubular process is that there is no edge trim. Some
tubular processes require large changes to vary the width, however,
and these often would be inferior, in this respect, to a flat-film
process for a product which was produced in a number of different
widths.

Tubular film sometimes is not slit, but is used in tubular form
(e.g., to make bags). For these markets, tubular processes have
an advantage over flat-film processes with respect to both the
appearance of the bags and the cost of manufacturing them.

But flat-film processes also have advantages. Usually they give
film of more uniform thickness. The ability of the tubular process
to randomize gauge variations is a counterbalancing factor and, as
a result, tubular processes may at times give mill rolls of better
appearance. But often it is important that the point-to-point
variation in the film be small, and when this is the case, the advantage currently lies with the flat-film process.

There are other advantages to flat-film processing. Thick films
usually cannot be made by tubular processes. Tubular processing
gives problems which are not encountered in flat-film processes with
respect to startup and wrinkling. Also, when film is cast with a
volatile plasticizer, a flat-film process permits removal of solvent
from both surfaces.

This is the picture with today's technology. Film-making processes will be improved in the future, and with these improvements
may come shifts with respect to flat-film versus tubular processing.
Nevertheless, it is probable that both processes will continue to
find wide application.

Acknowledgments

Many of the author's associates with E. I. du Pont de Nemours and Company
assisted in the preparation of this chapter. In particular, grateful acknowledgment should be made to G. H. Hockeborn for his assistance in searching the
patent literature; Miss Jean P. Ouderkirk for library searches; D. T. Bottorf
and R. W. McClure for their contributions to the sections on gauge adjustment

and gauge measurement; and R. A. Radecki for his assistance with the drawings. The following reviewed parts of the manuscript and contributed many valuable suggestions: R. J. Albert, R. L. Burton, G. I. Deak, C. E. Dengler, M. Goldman, V. C. Haskell, C. E. Holcomb, R. A. Hovermale, S. W. Lasoski, Jr., J. B. Lyon, W. P. Weisenberger, and H. H. Yang. Finally, grateful acknowledgment is made to the management of E. I. du Pont de Nemours and Company for encouragement and support, as well as permission to publish this work.

References

1. E. C. Bernhardt, *Processing of Thermoplastic Materials*, Reinhold, New York, 1959, pp. 154–248.
2. J. M. McKelvey, *Polymer Processing*, Wiley, New York, 1962, pp. 228–298.
3. M. Reiner, *Deformation and Flow*, H. K. Lewis, London, 1949, p. 102.
4. J. F. Carley, *J. Appl. Phys.*, **25**, 1118 (1954).
5. J. F. Carley, *Mod. Plastics*, **33**, 127, 130, 132, 134, 136 (Aug. 1956).
6. J. M. McKelvey, *Ind. Eng. Chem.*, **46**, 660 (1954).
7. E. E. Heston (to National Rubber Machinery Co.), U.S. Pat. 2,975,475 (Mar. 21, 1961).
8. W. Seifreid and W. Ott, U.S. Pat. 3,000,054 (Sept. 19, 1961) (to Kalle & Co. Akt.-Ges.).
9. H. Reifenhäuser, U.S. Pat. 3,057,010 (Oct. 9, 1962) (to Reifenhäuser A. G.).
10. W. E. Velvel, U.S. Pat. 2,765,492 (Oct. 9, 1956) (to E. I. du Pont de Nemours).
11. R. E. Lowey, Jr., U.S. Pat. 2,938,231 (May 31, 1960) (to Blaw-Knox Co.).
12. E. J. Moore, U.S. Pat. 3,035,305 (May 22, 1962) (to E. I. du Pont de Nemours).
13. F. R. Nicholson, U.S. Pat. 3,039,043 (June 19, 1962) (to E. I. du Pont de Nemours).
14. F. R. Nicholson, U.S. Pat. 3,067,464 (Dec. 11, 1962) (to E. I. du Pont de Nemours).
15. H. A. Konopacke and J. C. Cunningham, U.S. Pat. 3,096,543 (July 9, 1963) (to Olin Mathieson).
16. E. J. Moore, U.S. Pat. 3,102,302 (Sept. 3, 1963) (to E. I. du Pont de Nemours).
17. Brit. Pat. 789,105 (Jan. 15, 1958) (to Monsanto Chemical Co.).
18. C. G. Dell, E. I. du Pont de Nemours & Co., private communication.
19. G. A. Gerhard, *Mod. Plastics*, **40**, 134, 137, 141, 200 (1963).
20. Oak Ridge Isotopes Catalog, 1966.
21. Product Bulletin, *DuPont 500 Optical Thickness Gage*, Instrument Products Div., E. I. du Pont de Nemours, Wilmington, Del.
22. F. A. Obstfeld and C. G. Heisig, *Mod. Plastics*, **39**, 119, 122, 124, 174 (Jan. 1962).
23. F. M. Alexander, U.S. Pat. 2,909'660 (Oct. 20' 1959) (to Industrial Nucleonics).

24. F. M. Alexander, U.S. Pat. 3,000,438 (Sept. 19, 1961) (to Industrial Nucleonics).
25. E. J. Moore, U.S. Pat. 3,122,782 (Mar. 3, 1964) (to E. I. du Pont de Nemours).
26. C. N. Jolliffe and J. A. Parkins, U.S. Pat. 3,122,783 (Mar. 3, 1964) (to E. I. du Pont de Nemours).
27. C. N. Jolliffe, U.S. Pat. 3,122,784 (Mar. 3, 1964) (to E.I. du Pont de Nemours).
28. H. E. Sponaugle, *Soc. Plastics Engrs. J.*, **19**, 561 (1963).
29. M. Miller and D. R. McGregor, U.S. Pat. 2,920,352 (Jan. 12, 1960) (to E. I. du Pont de Nemours).
30. L. E. Dowd, *Mod. Plastics*, **40**, 142, 147, 204 (1962).
31. K. L. Knox, U.S. Pat. 2,686,931 (Aug. 24, 1954) (to E. I. du Pont de Nemours).
32. P. L. Clegg and N. D. Huck, *Plastics (London)*, **26**, 114 (April 1961); **26**, 107 (May 1961).
33. D. Hartland, U.S. Pat. 2,702,408 (Feb. 22, 1955) (to Union Carbide Corp.).
34. P. Alexander, U.S. Pat. 2,876,497 (Mar. 10, 1959) (to Union Carbide Corp.).
35. S. J. Skinner, U.S. Pat. 3,125,620 (Mar. 17, 1964) (to Monsanto Chemical Co.).
36. J. Pilaro and R. Kremer, *Mod. Plastics*, **37**, 115, 206 (1960).
37. W. L. Peticolas, Can. Pat. 645,071 (July 17, 1962) (to E. I. du Pont de Nemours).
38. *Plastics (London)*, **26**, 119 (Sept. 1961).
39. H. R. Simonds, A. J. Weith, and W. Schack, *Extrusion of Plastics, Rubber, and Metals*, Reinhold, New York, 1952, pp. 264–266.
40. A. F. O'Hanlon and J. W. Cornforth, U.S. Pat. 2,728,951 (Jan. 3, 1956) (to Imperial Chemical Industries).
41. R. L. Wibbens, *Plastics Technol.*, **7**, 35 (Apr. 1961).
42. D. Grant, in *Plastics Progress 1961*, P. Morgan, Ed., Macmillan, New York, 1962, pp. 151–178.
43. W. A. Chren and C. H. Hofrichter, Jr., U.S. Pat. 2,736,066 (Feb. 28, 1956) (to E. I. du Pont de Nemours).
44. A. N. Aronsen, U.S. Pat. 3,154,608 (Oct. 27, 1964) (to Crown Zellerbach Corp.).
45. R. H. Perry, Ed., *Chemical Engineers' Handbook*, 4th ed., McGraw-Hill, New York, 1963, pp. 10–16.
46. E. G. Fisher, in *Polythene*, A. Renfrew and P. Morgan, Eds., 2nd ed., Interscience, New York, 1960, pp. 519–548.
47. R. A. Rose, *Plastics Inst. (London), Trans.*, **30**, 118 (1962).
48. E. Overgage and G. John, *Brit. Plastics*, **32**, 512 (1959).
49. M. O. Longstreth and J. E. Tollar, U.S. Pat. 2,769,200 (Nov. 6, 1956) (to Dow Chemical).
50. D. F. Bartoo, U.S. Pat. 2,805,446 (Sept. 10, 1957) (to Toegepast Natuurwetenschappelijk Onderzoek ten Behoeve van de Volksgezondheid).
51. R. H. B. Buteux, J. R. Cann, and J. W. Cornforth, U.S. Pat. 2,952,872 (Sept. 20, 1960) (to Imperial Chemical Industries).

52. M. O. Longstreth and J. E. Tollar, U.S. Pat. 2,963,741 (Dec. 13, 1960) (to Dow Chemical Co.).
53. T. M. McCauley and G. Svoboda, U.S. Pat. 2,978,748 (April 11, 1961) (to Koppers Co.).
54. H. O. Corbett, U.S. Pat. 3,088,167 (May 7, 1963) (to National Distillers and Chemical Co.).
55. R. A. Wallis, *Intern. Plastic Engr.* **4**, 37 (1964).
56. D. W. Stephenson and R. J. Schroyer, U.S. Pat. 3,123,699 (March 3, 1964) (to Blaw-Knox Co.).
57. H. Domininghaus, *Soc. Plastics Engrs. J.*, **20**, 593 (1964).
58. C. A. Kronholm, U.S. Pat. 2,844,846 (July 29, 1958) (to Chester Packaging Co.).
59. E. D. Fuller, U.S. Pat. 2,461,975 (Feb. 15, 1949) (to Visking Corp.).
60. R. H. B. Buteux and J. W. Cornforth, U.S. Pat. 2,863,172 (Dec. 9, 1958) (to Imperial Chemical Industries).
61. W. T. Stephenson, U.S. Pat. 2,452,080 (Oct. 26, 1948) (to Dow Chemical Co.).
62. K. G. Francis and J. W. McIntire, U.S. Pat. 2,717,424 (Sept. 13, 1955) (to Dow Chemical Co.).
63. M. T. Cichelli, U.S. Pat. 2,987,765 (June 13, 1961) (to E. I. du Pont de Nemours).
64. C. E. Berry and M. T. Cichelli, U.S. Pat. 2,987,767 (June 13, 1961) (to E. I. du Pont de Nemours).
65. E. L. Fallwell, U.S. Pat. 3,092,874 (June 11, 1963) (to E. I. du Pont de Nemours).
66. M. Goldman and M. Wallenfels, U.S. Pat. 3,141,912 (July 21, 1964) (to E. I. du Pont de Nemours).
67. P. Harrison, *Plastics (London)*, **29**, 50 (Feb. 1964).
68. G. W. Jargstorff and C. A. Joslin, U.S. Pat. 2,519,375 (Aug. 22, 1950) (to Union Carbide Corp.).
69. H. W. Heisterkamp and W. S. Ladegaard, U.S. Pat. 3,090,998 (May 28, 1963) (to Dow Chemical Co.).
70. R. H. Ralston, U.S. Pat. 3,142,092 (July 28, 1964) (to Hercules Powder Co.).

CHAPTER 9

THE FRICTION AND LUBRICATION
OF POLYMER FILMS

D. K. OWENS

*E. I. du Pont de Nemours & Co., Spruance Film Research
and Development Laboratory, Richmond, Virginia*

I. The General Nature of Friction

A. History

From ancient paintings and writings it appears that the Egyptians, Greeks, and Romans were fully aware of the effects of friction and the use of lubricants. Paintings dating ca. 2000 B.C. depicted men pouring lubricating oil into the path of a heavy sledge, and a chariot dating ca. 1400 B.C. found in a tomb still bore traces of the original lubricant on its axle. Later, writers such as Aristotle showed that they had a practical knowledge of friction and lubricants and of the use of hard sheet metal facings on bearings to reduce wear. Although the ancients were well acquainted with friction and had considerable insight into its control, almost 2000 years elapsed before any quantitative study of friction was made. The first quantitative work was done by Leonardo da Vinci who showed by experimentation that friction was proportional to load, and that friction was apparently independent of the area of the sliding bodies. Da Vinci's work on friction, as was the case with so much of his research, was considerably ahead of its time, and it was not until 200 years later that interest again arose in the area of friction. In the 17th century the industrial revolution began, and engineers became interested in new forms of motive power. Naturally, friction and its control assumed considerable practical importance. In this atmosphere of engineering development, the modern period of friction research began. The first workers of this period were Frenchmen, and in 1699 the French engineer Amontons described his first experiments on the nature of friction (1a), which amounted actually to a rediscovery of the principles enunciated earlier by da Vinci. His first paper described what have become known as the two main "laws" of friction or Amontons' laws: (1) frictional force is proportional to the load, and (2) friction is independent of the size of the sliding bodies. These observations were confirmed by all subsequent workers. The French school of friction research, beginning with Amontons and culminating in the research of Coulomb, regarded friction as arising from surface roughness. Coulomb (1b) believed that the friction of sliding bodies represented the force required to raise the points of roughness or asperities over one another, i.e., the roughnesses of the sliding bodies interlocked and the force of friction was that force necessary to overcome this interlocking. In contrast to the French point of view, the position was quite different in England. Desagu-liers in England had observed strong adhesion between lead spheres when they were pressed together and he considered that similar forces might be involved in friction.

The English philosopher Leslie pointed out the most serious defect in the French view. He postulated that sliding bodies are not continually ascending on the asperities, but must alternately rise and fall. Consequently, energy consumed in lifting the body up an asperity would be recovered when the body came down, and there would be no net loss of energy in such a process. He also had a negative attitude toward the influence of adhesion on the friction of sliding bodies and considered that friction arose from deformation losses in the sliding bodies; energy was lost by one body plowing through the other. For the next 250 years almost nothing was contributed to the resolution of this question of the mechanism of friction until the investigation of Holm in 1946 and Bowden and Tabor in 1950 gave a clear picture of the mechanism of friction between solid bodies.

B. Metallic Friction

1. Introduction

Because of its technological importance, the friction of metals has received the bulk of study, and only in recent years has polymer friction been given attention. Since most of the knowledge of polymer friction has arisen from an extension of the theories of metallic friction, and because similar mechanisms apply, it is useful to consider first the problem of metallic friction before undertaking a discussion of polymer friction.

Most of the phenomena of metallic friction can be explained in terms of an adhesion mechanism which embodies the following physical principles:

(1) All surfaces, however carefully prepared, are covered with roughnesses or asperities which are large in comparison with molecular dimensions.

(2) When two such surfaces are placed in contact, they do not touch over the whole of the apparent area of the specimens, but

only on the asperities. The true area of contact is then governed by the deformation properties of the asperities.

(3) Because of the high unit pressure and intimate contact over the real area of asperity contact, strong adhesion or junction formation occurs, which is analogous to cold welding.

(4) In order to slide one surface over the other, these junctions must be sheared and the observed frictional force is essentially the shearing force of the junctions.

Each of these principles will be examined in detail.

2. Area of Contact

It has been shown that even the most carefully polished metal surface is covered with irregularities which are quite large on a molecular scale (2,3,4). Since surface forces due to molecular interaction operate only over a range of 2–3 Å and decrease as the inverse seventh power of the distance of separation, it is obvious that the total interaction of one surface with another in the form of frictional effects must occur only where asperities touch. It is then necessary to examine the factors affecting the nature and extent of asperity contact.

Consider a single asperity of spherical shape and radius r pressing into a flat surface of softer metal under a load W. As W is increased, the elastic limit of the softer metal is exceeded and it begins to deform plastically. With further increases in W, the deformed region increases in size until a steady value of mean pressure is reached. This value, p, is the yield pressure of the metal and can be considered equivalent to the hardness of the metal. This behavior forms the basis for the familiar Brinell and Rockwell hardness tests (5a). The process has been studied on a larger scale, and it has been found that the onset of plastic deformation is dependent upon the radius of the asperity (6). Thus it can be shown that even in a material such as hard tool steel, for an asperity radius of 10^{-4} cm, full plasticity is reached at loads of the order of 10^{-3} g.

In this analysis it can be seen that the real area A of asperity contact of metals will be influenced only by the load and the yield pressure or hardness of the metal and is given by the relation

$$A = W/p \tag{1}$$

In the case of multiple asperity contact, each asperity will support a load $W_1 = A_1 p$, and the total load supported will be equal to the summation of the individual asperity loads over the summed asperity areas so that the total area of contact will be independent of the number of asperities. It can be concluded from the foregoing discussion that: (1) the real area of contact of metals is proportional to the load, and (2) the real area of contact of metals is independent of the number of asperities or surface roughness. It then follows that the real area of contact in metals is independent of the apparent area of the specimens. Measurements of the electrical resistance of metallic contacts by Holm (7) and by Bowden and Tabor (8) have confirmed the validity of these conclusions.

3. Adhesion of Metals

The third principle of metallic friction postulates strong adhesion between surfaces at points of asperity contact. As long ago as 1724, Desaguliers demonstrated strong adhesion between freshly cut lead surfaces which were wrung together by hand (9). It is easy to show also, in the case of clean metals sliding over one another, that strong adhesion has occurred because there is considerable plucking and tearing of the surfaces and transfer of metal between the sliding parts. If, however, two clean metals are pressed together without sliding, it is difficult to demonstrate any normal adhesion between them. There are two reasons why adhesion between metals in static contact is not ordinarily observed.

Measurements of adhesion are conducted after removal of the load. Although the asperities are deformed plastically under load, the regions around them are not so heavily loaded and are deformed elastically. When the load is removed, the movement of the specimens caused by the release of the elastic stresses breaks the junctions and little or no adhesion is observed. In the case of ductile metals such as indium, the normal adhesion between two indium spheres is of the same order of magnitude as the force used to press them together (10). It is believed that the ductility of the metal enables the junctions to accommodate the release of elastic stresses without rupture.

The second reason for the apparent lack of normal adhesion between most supposedly clean metal surfaces is simply that the surfaces are not really clean. In the atmosphere, metal surfaces are covered with films of oxide or adsorbed gases and water vapor so that intimate metallic contact is prevented. During sliding, these surface films are broken and significant adhesion is observed. Cocks (11) has shown that metals with normal loading frequently exhibit very high values of electrical resistance which fall drastically when sliding begins.

4. Sliding of Metals

The fourth principle of metallic friction, the shearing of adhesive junctions during sliding, will now be considered. It has been shown that when two metal surfaces are placed in contact, the asperities will deform plastically until their total cross-sectional area is sufficient to support the load. For metals this area is proportional to the load and the yield pressure according to eq. (1). If over this area A, adhesive junctions of shear strength s are formed, the force to shear the junctions will be

$$F = As \tag{2}$$

and by substitution from eq. (1)

$$F = Ws/p \tag{3}$$

This equation has been found to hold over a wide range of conditions (12).

The coefficient of friction is defined by the following expression

$$\mu = F/W \tag{4}$$

and, since s and p are material constants, the coefficient of friction can also be expressed by

$$\mu = s/p \tag{5}$$

This last expression shows that the coefficient of friction of metals is determined solely by the ratio of two material constants and not by external conditions of load, size, shape, or surface finish of the metals. Thus a fairly quantitative explanation of Amontons' laws is available.

Although these equations provide a simple explanation of friction, the quantitative values obtained from applying these equations to actual frictional data are not satisfying. The yield pressure p and shear strength s of a metal may be determined on bulk material in a variety of ways. These values may be inserted in eq. (5) to give a predicted value for the coefficient of friction. Values of μ predicted in this way are, however, always found to be lower than the observed values, and values of s determined from friction measurements are always greater than those obtained by other methods. McFarlane and Tabor (10) have shown that the chief deficiency in eq. (5) is its treatment of p and s as independent properties. They have shown that the applied tangential force prior to sliding causes further plastic flow at the asperity contacts, resulting in a growth of the area of real contact or junctions to values greater than predicted for static loading from eq. (1). This junction growth gives an apparent value of s higher than expected from measurements on bulk metals.

5. Static and Kinetic Friction

From the beginning of organized research in friction, it has been recognized that there are two different friction coefficients. The static coefficient (μ_s) represents the peak force required to initiate sliding, and the kinetic coefficient (μ_k) represents the forces of steady state or continuous sliding. With most metals, there is normally very little difference between the two, but polymers often do exhibit sizable differences, μ_s being much higher than μ_k. For polymers, the magnitude of μ_s is dependent on the time in which the surfaces have been in stationary contact before sliding begins. This time dependence will be discussed in the section on polymer friction.

A frequent phenomenon of friction is that of stick-slip motion between sliding bodies. Such motion may occur at a very slow rate or at a high enough rate to give an audio-frequency squeak. Bowden and Tabor (13) have given an interesting analysis of stick-slip motion. For it to occur, there appear to be two requisites: the sliding system must possess sufficient elasticity to permit one member of the sliding pair to move with the other, and μ_s must be time-dependent. If a stationary, upper surface held in an elastic mounting is placed under load on a lower, movable surface, adhesive junctions will form. When the lower surface is driven forward, the upper surface will travel with it until the elastic restoring force of the system exceeds the strength of the junctions. The junctions then break and the upper surface moves in a direction opposite to that of the lower surface. This is the slip phase. When the upper surface comes to rest, adhesion occurs, and the stick phase is repeated.

An oscillographic recording of this type of motion gives a sawtooth trace. A deficiency of measuring friction under stick-slip conditions is that the relative velocity of the bodies during slip is not easily measured. The upper body accelerates as it breaks away and decelerates as it comes to rest. It is clear that this phenomenon is more a characteristic of the measuring system than of the surfaces themselves, although it is greatly reduced or absent in lubricated systems where adhesion is low. Stick-slip behavior is useful for determining μ_s from the peaks of the stick cycles. When the natural frequency of the measuring system is allowed to establish itself, loading time between cycles is very reproducible. Sometimes a spring is deliberately introduced into the measuring system to enhance stick-slip motion (see Section II).

6. Types of Lubricated Friction

It is generally recognized that there are two main types of lubricated friction. The first is the hydrodynamic type favored by high sliding speeds and low pressures. It is characterized by the presence of a layer of lubricant between the sliding surfaces of such thickness that the surfaces do not touch. The forces of friction are essentially those due to shearing of the lubricant layer and are dependent primarily on speed of sliding and lubricant viscosity. This type of lubrication is found usually in machinery as typified by the journal bearing whose geometry is purposely designed to favor hydrodynamic conditions. Hydrodynamic friction may be important in high-speed polymer-film friction and will be discussed later.

The second type of friction is called the boundary type and is favored by high pressures and slow sliding speeds. It is characterized by considerable contact between surfaces. The previously outlined theories were developed for this type of friction. Unlubricated sliding is almost always of the boundary type, except where sliding speeds are so great that frictional heating causes surface melting and this layer of molten material acts as a hydrodynamic film. It should be recognized that the type of friction experienced is not primarily determined by the properties of the lubricant or surfaces, but by the conditions under which sliding occurs. Thus as speed is varied, a sliding system may pass from boundary to hydrodynamic friction. There is frequently a region of "mixed" lubrication where there may be some asperity contact and boundary friction while there is still an appreciable hydrodynamic component. It may be generally said that friction will increase with speed in hydrodynamic sliding and remain essentially independent of speed in boundary sliding.

7. Boundary Lubrication

Sir William Hardy in his classical researches on boundary films (14) showed that large reductions in the friction of metals could be achieved by the application of very thin layers of organic fatty compounds. He found fatty acids more effective on metals than fatty alcohols which were, in turn, more effective than paraffins. He also found that friction decreased with increasing chain length up to about C_{12} and was constant thereafter. Gregory (15) showed that cadmium was effectively lubricated by a monomolecular layer of lauric acid. Other workers using the Blodgett technique (16) of applying monomolecular layers of fatty polar compounds have confirmed these results (17–19).

Friction measurements have shown for layers of paraffins and fatty alcohols on heated metals, that lubrication breaks down at a temperature corresponding to the crystalline melting point of the fatty compound (20). Fatty acids present a somewhat different behavior. It was found that the lubrication of heated metals covered with a fatty acid film did not break down until a temperature considerably higher than the melting point of the acid was reached (15). This temperature corresponded to the melting point of the metal soap of the fatty acid. It was also observed that fatty acids on nonreactive metals such as platinum did not exhibit this behavior, but behaved in the same manner as paraffins or fatty alcohols. Fatty amines and amides would also be expected to be good boundary lubricants.

Electron diffraction studies have revealed that polar fatty compounds are oriented on the metal surface with the terminal methyl group outermost and in close-packed array. Breakdown of the lateral orientation of the molecules was found to occur at the same temperatures as the breakdown of lubricating ability (21).

Investigations by Bowden and Leben (22), Gregory and Spink (23), and Greenhill (24), using the Blodgett technique for applying successive monolayers of stearic acid to metals, showed that in most cases 1–3 monolayers gave very low friction on the first traverse.

For materials other than fatty acids, a much greater thickness was necessary to give low friction. Although 1 or 2 monolayers of fatty acid gave low friction on metals, they broke down rapidly upon repeated sliding. As many as 50 monolayers were necessary for good durability of the lubricating film on repeated sliding. An excellent review of the boundary lubrication of metals has been published by Bowden, Gregory, and Tabor (25).

It appears that a good boundary lubricant is one which is capable of adsorption on the metal surface to give a close-packed oriented monolayer. Such a layer is very strong because of the van der Waals forces of lateral cohesion between molecules. The greatest effect is achieved when the lubricant is also firmly attached to the substrate by chemical bonding as in the case of fatty acids on metals. The importance of close packing is illustrated by the work of Cottington, Shafrin, and Zisman (26) who measured the friction and contact angles of long-chain monolayers on steel and glass prepared by the retraction technique (27a,b,c). They showed that friction reached a minimum and the contact angle a maximum at a chain length of C_{14}. The lower contact angles and higher frictions for shorter chain lengths indicated that these materials were not in a state of closest packing and lacked the necessary lateral cohesion of a good lubricant.

Under conditions of boundary lubrication Amontons' laws are valid. The function of the lubricant is to reduce the extent of metal contact by interposing between the surfaces a strong layer of low shear strength. This layer reduces the s term of eqs. (2) and (5) and thereby reduces the force of friction. Neither hardness nor area of contact is affected because these quantities are still determined by the physical properties of the underlying metal substrate. In the case of surfaces covered with adsorbed monolayers, shearing at asperity contacts is believed to occur mostly in the plane of the terminal methyl groups. Bailey and Courtney-Pratt (28) measured the shear strength of calcium stearate monolayers on mica and found it to be 250 g/mm². Area of contact was measured interferometrically and could be determined accurately by virtue of the extreme smoothness of the mica. The load was very small. Using their value of s in eq. (5) for lubricated copper, a value for $\mu = 0.005$ is predicted which is far lower than actually observed. This discrepancy can be explained by the observations of Bridgman (29), who found that the shear strength increased almost proportionally with pressure. Under the high unit loadings at asperity contacts, the shear strength of a fatty acid monolayer would be much higher than the figure found by Bailey and Courtney-Pratt.

C. Polymer Friction

Most of the basic work on polymer friction has been done with fibers or with large blocks of polymer; very little fundamental work on polymer films is to be found in the literature. There is no reason, however, why the general mechanisms of fiber and bulk polymer friction cannot be applied to film as well, and in the following discussion, these materials will be considered as interchangeable.

Work by Shooter and Tabor (12), Shooter (30), and King and Tabor (31) has indicated that the adhesion mechanism of friction originally developed for metals can be applied to polymers over a limited load range. Thus, Shooter and Thomas (32) found that the friction of polymers at high loads (1–4 kg) was directly proportional to the load, i.e., Amontons' laws were obeyed. At smaller loads, many workers have shown several important differences between the frictional behavior of polymers and metals. Unlike metals, the coefficient of friction of polymers is not independent of load at lower loadings but increases as the load is reduced (33–37). The friction of polymers has also been found to decrease with increasing surface roughness in contrast to the friction of metals which has been shown to be independent of surface roughness (38,39). It will now be shown how a single general law of friction which reconciles these differences may be derived.

1. Mechanisms

Schallamach (40), investigating the friction of rubber on glass, found that the coefficient of friction was given by

$$\mu = cW^{-\frac{1}{3}} \tag{6}$$

where W is the load and c is a proportionality constant. The frictional force was given by

$$F = cW^{\frac{2}{3}} \tag{7}$$

He found for a rubber sphere pressed onto a glass plate that the area of contact was also proportional to $W^{\frac{2}{3}}$. Gralén (37) studying fibers found that friction was described by the relationship

$$F = aW + bW^n \qquad (8)$$

where a and b are constants.

It was subsequently found by Howell (41), Lincoln (38), Huffington (42), and Howell and Mazur (43) that the friction of polymers was expressed more simply by

$$F = kW^n \qquad (9)$$

Except for variation of the exponent, this last expression is identical with eq. (7) and implies that the adhesion theory of friction for metals is applicable to polymers in that friction results from adhesion at asperity junctions, which have a constant shear strength s. If s is constant, it follows that the area of real contact will also vary as W^n and most workers have explained the difference between polymer and metal friction on the basis of the variation of contact area with load.

2. Deformation and Area of Contact of Polymers

The work of Schallamach on rubber and that of Lincoln on nylon using small loads gave results indicating that the area of contact of these materials varied as $W^{\frac{2}{3}}$. This value is in accord with Hertz's equations for elastic deformation (44). Howell and Mazur found that the force of friction and area of contact of nylon fibers on glass varied as $W^{0.72}$ and Finch (45) found that the area of contact of nylon fibers varied as $W^{0.89}$. In the case of complete plastic deformation, the Hertz equations yield a value of 1 for n, making the area of contact directly proportional to the load. It appears that the deformation of polymers is intermediate between plastic and elastic, and that their frictional behavior is strongly influenced by their deformation properties.

Pascoe and Tabor (46) have shown that the deformation properties of polymers can be conveniently measured by means of spherical indenters pressed into a block of the polymer under a known load. Using a steel ball, they found that the load W and the indentation diameter d were related by the expression

$$W = k_1 d^m \qquad (10)$$

The exponent m is a material constant of the substance under test and varies from 2 for plastic deformation to 3 for elastic deformation. Pascoe and Tabor have further shown that the projected area of the indentation A_I is given by

$$A_I = (\pi/4)(1/K)^{2/m}W^{2/m}D^{[(2/m)-4]/m} \qquad (11)$$

where D is the indenter diameter. Assuming that the area of real contact of the indenter is a constant fraction of A_I, they concluded that the frictional force should vary as $W^{2/m}$ or the coefficient of friction, μ, as $W^{(2/m)-1}$. From the projected area of the indentation and the load causing it, the yield pressure or hardness p of the polymer can be calculated. Table 1 shows some typical values of m, p, $2/m$, and $(2/m) - 1$ taken from Pascoe and Tabor.

The well-known low friction of polytetrafluoroethylene cannot be anticipated from these data, for on the basis of hardness, one would expect its friction to be greater than that of nylon assuming comparable shear strengths. The variation of A with W for nylon is in good agreement with that found by Howell and Mazur. Assuming the shear strength of the junctions to be constant and equating fiber diameter with indenter diameter [eq. (10)] Pascoe and Tabor predicted that, for nylon, $\mu = cW^{-0.26}D^{0.52}$ where constant c includes the shear strength of the polymer. Experimental data showed that this relationship was true not only of fibers, but of bulk materials over the enormous load range of 10^{-6}–10^4 g. Some recent work by Adams (47) indicates that the assumption of

TABLE 1
Values of Several Material Constants

Polymer	Deformation index, m	Yield pressure, p, kg/mm^2	Variation of A with W, $2/m$	Variation of μ with W, $(2/m) - 1$
Nylon 66	2.7	7.32	0.74	−0.26
Polyethylene terephthalate	2.7	16.1	0.74	−0.26
Polyethylene	2.7	1.27	0.74	−0.26
Polyvinylidene chloride	2.4	13.0	0.83	−0.17
Polymethyl methacrylate	2.6	14.4	0.77	−0.23
Polytetrafluoroethylene	2.5	2.7	0.80	−0.20

constant junction shear strength in polymers may not be justified.
A correction for pressure appears necessary as in the case of lubricated metals.   Both Pascoe and Tabor and Adams have shown that
junction growth does not occur in sliding polymers as it does with
metals.   In their paper, Pascoe and Tabor also showed that the
projected area of the indentation formed in a polymer by a hard
indenter was dependent on the time of application of the load and
increased proportionally to the logarithm of time.   This behavior
was attributed to the viscoelastic nature of polymers and their well-
known tendency to flow or creep under load.   The variation of μ_s
with time of loading in polymers has been noted earlier and seems to
be explainable on the basis that the area of real contact increases
with stationary contact time.

3. Effect of Roughness

Numerous workers have investigated the friction–roughness relationship for polymers.   Archard (48), Lodge and Howell (49), and
Rubenstein (34) have published theoretical treatments of idealized
cases of multiple asperity contact.   Archard considered the case
of an elastic sphere of radius R pressing into a flat surface under a
load W.   If the sphere is covered with smaller spherical asperities of
radius r and there are n such asperities per cm² of actual contact
area, it can be shown on theoretical grounds that the area of real
contact A will be given by

$$A = kr^{\frac{2}{3}}n^{\frac{1}{3}}R^{\frac{2}{9}}W^{\frac{6}{9}} \tag{12}$$

where k is a constant.

Kaliski (50) has treated the cases of a polymer fiber sliding over a
smooth versus a matte finished chrome pin and has found that the
force of friction is inversely proportional to the square root of the
number of asperities on the pin where plowing of the fiber by the
harder asperities of the matte chrome is assumed.   This treatment
does not explain the observations of many workers that friction is
also reduced when the fiber is roughened and slid on a smooth, hard
surface where a significant plowing effect is unlikely.

Owens (51) has considered the case of a rough polymer film sliding
on a smooth metal surface.   He assumed that Kaliski's relation-
ship, $\mu \propto 1/N^{\frac{1}{2}}$, where N is the number of asperities per unit area
of apparent contact was correct, and that elastic deformation

occurred. Further assuming that the exponent of N is a function of Pascoe and Tabor's deformation index m [eq. (10)], he showed that the following relations must be satisfied: $f(m) = 0$ when $m = 2$ (plastic deformation) and $f(m) = \frac{1}{2}$ when $m = 3$ (elastic deformation). The expression $(m/2) - 1$ satisfied both conditions. This exponent was in fair agreement with experimental data.. Better agreement was obtained when it was assumed that the term $(1/K)^{2/m}$ in eq. (11) contains a factor related to asperity number and that K is proportional to the $N^{(m/2)-1}$ factor derived above. By substituting for K, the expression $(k_2/N)^{[(m/2)-1]2/m}$ is obtained, which simplifies to $k_3 N^{(2/m)-1}$, where N is the number of asperities per unit of surface area. It then follows that the coefficient of friction of a polymer should be proportional to $W^{(2/m)-1}$ at constant asperity number and to $N^{(2/m)-1}$ at constant load. The relationships of load, friction, roughness, and static indentation diameter may then be formulated into the following equations:

$$W = k_1 d^m \tag{13}$$

$$\mu = k_4 W^{(2/m)-1} \tag{14}$$

$$\mu = k_5 N^{(2/m)-1} \tag{15}$$

The polymer film chosen to test these relationships was regenerated cellulose coated with a copolymer of vinylidene chloride (91%) and acrylonitrile (9%). The surface of the copolymer coating was roughened in a controlled manner by adding suitable amounts of powdered clay of average particle diameter 7 microns to the copolymer solution used to coat the regenerated cellulose film. These particles were considered to be the asperites, and their number per unit of area was determined microscopically. The indentation measurements were made by pressing a glass bead 237 microns in diameter into the films and measuring the indentation diameter microscopically as a function of load. The coefficient of static friction of the films was measured with an apparatus to be described in Section II.

A log-log plot of indentation diameter versus load gave a straight line of slope $m = 2.35$. Similar plots of the coefficient of static friction μ_s against load and asperity numbers gave the results shown in Table 2. These results are in close agreement with Pascoe and Tabor's results for bulk polyvinylidene chloride from Table 1.

TABLE 2
Values of m from Graphical Data

Relationship	Experimental value of exponent	Calculated value of m	$2/m$	Method
$W = k_1 d^m$	2.35	2.35	0.85	Static indentation
$\mu_s = k_4 W^{(2/m)-1}$	−0.14	2.33	0.86	μ_s vs. load
$\mu_s = k_5 N^{(2/m)-1}$	−0.15	2.35	0.85	μ_s vs. asperity no.

In summary, the friction of polymers in bulk, fiber, or film form can be expressed over a wide range of conditions by a single general relationship, $F = kW^n$. The value of n is determined by the deformation properties of the materials and varies from $\frac{2}{3}$ for elastic deformation to 1 for plastic deformation. From this point of view, the friction of metals is simply a special case of the general relation where $n = 1$. For polymers, $n < 1$ because these substances deform in a manner intermediate between plastic and elastic. This deformation can be expressed by an easily measured material constant, the deformation index m. It is found that $n = 2/m$. The load and roughness dependence of polymer friction is then a consequence of the fact that the area of real contact does not vary linearly with load but with load to the $2/m$ power. The force of friction results from shearing of adhesive junctions formed over the area of real contact. The intrinsic adhesion of a polymer can be strongly affected by its chemical constitution.

4. Effect of Chemical Constitution

The chemical structure of a polymer has an important effect on its frictional properties, not attributable entirely to changes in hardness or other bulk properties with changes in structure. Zisman (27e) has shown that chlorination increases the friction and surface energy of a polymer while fluorination reduces both variables. Table 3 shows friction data for polyethylene and other polymers which may be considered structurally similar to polyethylene, but with some of the hydrogens replaced by Cl or F. Samples were clean and unlubricated.

The changes in friction appear to be explainable on the basis of

ANC

TABLE 3
Friction Data for Polymers

Polymer	μ_s
Polyvinylidene chloride	0.90
Polyvinyl chloride	0.50
Polyethylene	0.33
Polyvinyl fluoride	0.30
Polyvinylidene fluoride	0.30
Polytetrafluoroethylene	0.04

changes in the intrinsic adhesion of the polymers. Adhesion is due to strong, localized fields of attractive force emanating from the surfaces of all solids and liquids. The well-known low friction and adhesion of polytetrafluoroethylene cannot be explained on the basis of yield pressure, but has been explained on the basis of a large attenuation of the attractive forces due to the screening effect of the fluorine atoms. It seems reasonable, therefore, to speculate that chlorine atoms might enhance the force fields causing higher intrinsic adhesion for the chlorinated polymers. Zisman's work has indeed shown that the chlorinated polymers have more energetic surfaces.

D. LUBRICATION OF POLYMER FILMS

The demonstrated applicability of the adhesion mechanism of friction to polymers implies that polymers can be lubricated in much the same way as metals. There is ample evidence that hydrodynamic lubrication is important in the high-speed processing of polymers, but there is some controversy over whether true boundary lubrication can occur with polymers.

1. Hydrodynamic Lubrication in Polymer Film Friction

Lyne (52) has studied the friction of cellulose acetate yarn running over a smooth chrome pin. Using a white mineral oil lubricant he showed that friction increased with increasing speed. Again, keeping the speed constant at 100 m/min and varying the viscosity of the oil, he found a similar relationship between viscosity and friction. Hansen and Tabor (53) have given Lyne's data an interesting treatment in terms of a conventional journal bearing as found in machinery. For fibers lubricated with an oil of viscosity Z travelling over a pin at velocity V and nominal pressure P, they showed that the coefficient of friction was a function of ZV/P as is the case with a journal bearing. The agreement is surprising in view of the poor hydrodynamic design of a yarn running over a pin where side leakage of lubricant from under the fiber must surely be significant. With films, whose width is considerable compared with a yarn, side leakage will be far less and conditions for hydrodynamic sliding more favorable. Although there are no data in the literature, analogy suggests that hydrodynamic factors may assume major significance in high-speed polymer film friction.

2. Boundary Lubrication of Polymers

There is some controversy in the literature concerning whether effective boundary lubrication of polymers is possible or not. Effective boundary lubrication will be defined as a substantial reduction in the coefficient of friction of sliding bodies resulting from the presence on the surfaces of a material from the class of compounds generally recognized as boundary lubricants (see section on Boundary Lubrication of Metals). It will be assumed that conditions of sliding are such that no appreciable hydrodynamic component exists.

Polar fatty compounds on metals frequently reduce the coefficient of friction from 1 or greater to values of 0.10 and often as low as 0.05. It has been observed by numerous workers, however, that these compounds are much less effective on polymers. Thus, Bowers, Clinton, and Zisman (54a), using solutions of a number of classical boundary lubricants, were able to achieve a coefficient of static friction of only 0.50 for nylon sliding on nylon as compared with the unlubricated value of 0.75. They attributed the relative ineffectiveness of these materials on nylon to the lack of adsorption sites in the polymer precluding oriented monolayer formation. Pascoe (55) has explained similar observations by postulating that the shear strengths of the lubricant and the polymer are not appreciably different and, hence, only small decreases in friction are observed. Rubenstein (56) has theorized that the lubricant may act as a plasticizer for the polymer. While the lubricant may reduce the shear strength s at the sliding interface, the plasticizing action reduces the hardness p so that the s/p ratio remains constant or may even increase. Thus, any decrease in friction due to reduced shear strength may be offset by the increase in friction due to softening.

These theories were developed for liquid lubricants or solutions of solid lubricants in inert solvents. Fort and Olsen (57), working with a variety of fibers, reported poor lubrication in the case of liquid lubricants. With oleic acid, the coefficient of friction of rayon was reduced from 0.72 to only 0.60. Quite different results were obtained using solid lubricants which were applied from solution and tested after evaporation of the solvent. In the case of rayon lubricated with stearic acid and octadecylamine, coefficients of friction were reduced to 0.28 and 0.20, respectively, from an unlubricated value of 0.72. Fort (58) examined the friction of numerous polymer films under a 1 kg load against hemispherical sliders of the same material. Films studied included nylon, polyethylene terephthalate, polyacrylonitrile, cellulose acetate (unplasticized), and polytetrafluoroethylene, although the bulk of the data were for polyethylene terphthalate. In a systematic study of homologous series of fatty alcohols, acids, and alkanes on polyethylene terephthalate film, he found that all of the solid members of each series reduced the coefficient of friction of the film to the same low value of 0.09 from an unlubricated value of 0.29. With liquid members, no values lower than 0.20 were found, although friction did decrease linearly with increasing chain length. For liquid compounds, he found acids more effective than alcohols, and the latter more effective than alkanes for a given chain length. These results closely parallel those of Hardy and Zisman for metals.

Fort also applied solutions of stearic acid in hexadecane to polyethylene terephthalate film and observed low values of the friction coefficient. He proposed that stearic acid was weakly adsorbed at specific sites on the film surface and that hexadecane was able to adlineate to form a condensed, oriented, mixed monolayer over the remainder of the surface and give effective lubrication. He also investigated a monolayer of octadecanol applied by the Blodgett technique to polyethylene terephthalate film and found a low friction coefficient. He also found this monolayer capable of withstanding 100 repeat traverses of the slider at 1 kg load with no evidence of breakdown. This is a startling result in view of the poor durability of a monolayer on metal, despite strong adsorption and high lateral cohesion. Fort has suggested that this behavior is a consequence of the low yield pressure of the polymer (compared with metals) keeping the shearing force per unit of contact area small. This effect would result in a greatly reduced abrasive action of the polymer as compared with metals. Surface migration of lubricant to repair the damage done by sliding was considered unlikely. A final and highly significant observation by Fort was that none of the liquid or solid materials tested showed evidence of plasticizing the film.

Owens (59) has studied the lubrication of vinylidene chloride–acrylonitrile copolymer coatings on cellulose film by a homologous series of solid fatty amides. The amides were applied by dissolving them in the copolymer solution used for coating the cellulose. When the solution was applied to the cellulose and dried with heat, the lubricant diffused to the surface and appeared as solid crystals. As chain length increased, the friction was found to decrease over the range C_{12}–C_{22}. Results are shown in Table 4 for coatings containing the designated lubricants.

Using the critical surface tension of wetting (γ_c) techniques of Shafrin and Zisman and Bernett and Zisman (27c,d), it was shown that the lubricated film surface consisted of a composite of lubricant crystals and polymer. The completeness of surface coverage by the lubricant was a function of chain length; there was no detectable amide on the polymer surface at C_{12} chain length, and coverage was complete at C_{22} chain length. It was concluded that the variations in friction with amide chain length were a consequence of the variation in surface coverage. Wetting measurements showed that the lubricant was present on the film surface in the

TABLE 4
Effect of Amides on Coefficient of Friction

Lubricant	μ_s
None	0.76
Lauramide (C_{12})	0.71
Myristamide (C_{14})	0.61
Palmitamide (C_{16})	0.52
Stearamide (C_{18})	0.35
Behenamide (C_{22})	0.15

bulk, crystalline state. It was demonstrated that the areas of polymer between lubricant crystals were not covered with a mono-layer; none of the films showed evidence of a monolayer. Although the lubricant was not oriented on the film surface as an adsorbed monolayer, Owens proposed that the close packing and high lateral cohesion of the bulk, crystalline configuration conferred structural strength enabling the lubricant to withstand large loads without disruption. Again, there was no evidence of plasticization of the copolymer, even by the C_{12} amide which was completely retained within the polymer.

Allan (60a) has studied the effects of small amounts of oleamide added to polyethylene prior to its extrusion into film. After film formation, he found that friction decreased as the film was aged. At constant additive concentration, the duration of the aging process was shorter for thicker films, which reflected the greater amount of additive available for the surface in thicker films. The contact angle of water on the film scarcely changed during the aging process. Allan proposed that oleamide was surface active in polyethylene and diffused to the air–film interface. There it formed an oriented monolayer with the polar group attached to the medium of higher dielectric constant (the polyethylene). This lubricant layer was weakly adsorbed and could be disrupted by immersion in water or exposure to high humidity with an accompanying increase in fric-tion. The lubricant layer was found to be more stable to moisture when the film surface was flame treated to increase its polarity (60b).

Bowers, Jarvis, and Zisman (54b) in an important paper have recently published several unequivocal examples of surface activ-ity in certain fluorinated lubricants added to polymers. Using 0.5% of tris(1H,1H-pentadecafluorooctyl) tricarballylate in poly-methyl methacrylate, they demonstrated a reduction in γ_c from 39 to 19 dynes/cm. This last value is characteristic of a fully fluorinated surface. The coefficient of static friction was reduced from 0.60 to 0.19. Using bis(1H,1H-pentadecafluorooctyl) tetrachlorophthalate at a concentration of 1% in a film of vinyl-idene chloride (85%)–acrylonitrile (15%) copolymer, friction was reduced from 0.80 to 0.10, and γ_c was reduced to 18 dynes/cm from 38–44 dynes/cm for pure copolymer. These lubricant layers were very durable, showing only a slight rise in friction after 600 repeat traverses at 1 kg load. Evidence indicated that the lubricant layers were self-healing; material removed by sliding was replaced by lubricant diffusing from within the bulk of the polymer. These workers also investigated several other vinyl-type polymers and other additives with similar results.

They observed that the effectiveness of a given additive was dependent upon its organophilic–organophobic balance with respect to the polymer. The intriguing possibility of tailoring the chemical structure of the lubricant for optimum results with a given polymer suggests itself. The reader is referred to the original paper for details.

Despite certain evidence to the contrary, the bulk of research indicates that boundary lubrication of polymers is a reality. Lubrication of films is common practice in technological applications and some polymer films owe their commercial success to the fact that lubrication is possible to overcome undesirable frictional characteristics. The exact mechanism of polymer lubrication is open to question, whether it be either by oriented monolayers, as with metals, or by an unoriented crystalline bulk phase of lubricant on the polymer surface. In all probability, the mechanism is dependent on the nature of the polymer and lubricant and the kinds of interactions which may occur between them.

II. The Measurement of Film Friction

A. Instruments and Methods

The selection or design of instrumentation to measure polymer film friction is a difficult task. Almost every worker in the field has used a different apparatus designed to suit his particular needs. Probably the oldest, simplest, and least satisfactory apparatus for measuring friction is the inclined plane. A horizontal plane surface which can be inclined through a measurable angle is covered with the film to be tested and a block of known weight also covered with film is placed on it. The plane is then slowly inclined until the block begins to slide down. By a simple analysis of the components of force, it can be shown that the coefficient of friction between the film-covered block and plane is equal to the tangent of the angle of inclination. The value measured is the coefficient of static friction.

There are several deficiencies in measuring friction by this method. The time of loading which is the time from the placing of the block on the plane until it begins to slip cannot be controlled. It has already been shown that the coefficient of static friction of polymers varies with the stationary contact time. A second serious deficiency lies in the inability to measure kinetic friction with any accuracy. It can be approximated by measuring the time required for the block to slide a known distance, but any control of sliding speed is impossible. The inclined plane method may be satisfactory for rough measurements or for comparison where two blocks and two films are placed on the same plane, but for work requiring even moderate precision, other methods must be used.

The most satisfactory instruments for measuring friction are those in which the sliding members are hydraulically or mechanically driven. Only in this way can speeds be closely controlled or adjusted. Variable speed is a highly desirable feature of any tester so that films may be studied over the entire range of possible operating conditions. The American Society for Testing and Materials has adopted a tentative method for friction testing of plastic film (5b). It is the only method which might be considered "standard." It embodies the "sled and plane" concept wherein a film-covered sled of fixed weight is slid over a stationary film-covered plane. Alternatively, the sled may be fixed and the plane movable. Both arrangements are shown in the method. Forces are measured either with recorders and strain gauges or with spring-operated dial gauges. Drive mechanisms are self-contained. Another approved arrangement given by the method uses the crosshead and load cell of a universal tensile tester to furnish both the drive and measuring facilities. The friction tester is connected to the tensile tester with a system of cables and pulleys. Coefficients of friction are calculated from the measured force and the sled weight by eq. (4). This method appears to be a flexible one, eminently suited for films. The effects of stationary loading time are taken into consideration, and the sled is covered with sponge rubber to equalize the load over the test area. A number of precautions applicable to any method of friction testing are noted. Samples should be kept free of dust and lint since these particles act to increase surface roughness and prevent proper specimen contact. Fingerprints should also be avoided because the fats they contain may act as lubricants and give false results.

A device operating on the fixed sled–movable plane principle is sold by the Martin Sweets Co.* (61). On this instrument the load can be varied by adding weights to the sled and frictional force is measured by a spring and dial gauge. Another instrument operating on the fixed plane–movable sled principle is the IBM paper friction tester (62). Although developed for tabulating cards, there appears to be no reason why it cannot easily be adapted for film testing. Allan (60c) has described a film friction tester where the film is driven between a stationary sled and plane by a pair of rollers. The sled is attached to a strain gauge to measure force.

A friction tester for films operating on an entirely different principle is sold by the Bell Telephone Laboratories and has been described in the literature (63). In this instrument a small, fixed metal disc covered with film is placed in contact with a larger rotating disc also covered with film. The small disc is attached to a pendulum system. When the small disk is placed in contact with the rotating disc, the force of friction causes the entire pendulum system to rotate through an angle, the sine of which is proportional to the coefficient of friction. Both the force pressing the discs together and the speed of rotation are variable.

A laboratory instrument of great flexibility developed by Fort (58) has wide applicability to film friction measurements. It incorporates a number of features not found in other testers. It consists of a rotating variable-speed turntable to which the film is clamped with a ring. Mounted normal to the turntable circumference is a pivoted, counterbalanced arm having free motion in the vertical and horizontal planes. In Fort's apparatus, the arm carried a hemispherical slider of polymer; for film testing a film-covered block could be substituted. The sample loading is accomplished by placing weights on the arm. Friction between the arm and turntable induces a torque in the arm, and this force is measured with a strain gauge. The turntable was provided with a heater and thermostat to permit measurement of film friction at elevated temperature. The use of a ring mounting which forms a well with the film at the

---

* No attempt has been made to give a complete list of all friction testers which may be commercially available. Failure to mention a particular tester does not imply unsuitability for film testing, nor does the description of certain commercial instruments indicate a special merit not possessed by other similar testers.

bottom was found helpful by Fort for confining liquid lubricants whose properties were to be studied. Most film friction tests are performed in the long or machine direction of the film. A possible shortcoming of the turntable device is that it measures in all directions as it rotates. Possible confusion could arise if the film were to exhibit a directional frictional effect resulting from anisotropy or extrusion effects.

The friction of films can also be measured by an entirely different method which was originally developed for fibers (64) and is almost universally used in the textile industry. It is based on the principle that when a film or fiber is slid over a cylindrical surface, the coefficient of friction μ between film and cylinder is related to the tension T_1 in the film passing onto the cylinder and the tension T_2 in the film leaving the cylinder by the equation

$$T_2/T_1 = e^{\mu\theta} \tag{16}$$

where θ is the contact angle of the film around the cylinder, in radians. The cylinder may be stationary against a moving film or the film may be stationary against a rotating cylinder. This method is known as the capstan method and eq. (16) is referred to in the literature as the belt or capstan equation.

A device operating on this principle for measuring film friction has been described by Owens (59). A rotating variable-speed cylinder is covered with film. A strip of film is hung over the cylinder with one end attached to a strain gauge and a weight is attached to the free end; T_1 is equal to the weight on the free end of the film and T_2 is measured by the strain gauge and a high speed recorder. The contact angle in this case is 180°. If a spring is inserted in the system between the film and strain gauge, stick-slip motion is enhanced and coefficients of static friction are easily measured. This method has the obvious advantage of simulating the geometry of film passing around rolls as commonly found in film processing equipment. With some modification of the measuring devices it could be adapted to measure a continuous film moving over a stationary cylinder. Measuring heads and auxiliary equipment for sensing the tensions in fibers passing over a cylinder are commercially available (65). These heads will, however, only accommodate a film width of about $\frac{1}{2}$ to $\frac{3}{4}$ in. There are a number of variables affecting the results obtained by this method; they have been investigated extensively (33,52,66).

In summary, the sled and plane instruments are useful for routine quality control, but do not appear to possess the necessary flexibility and wide range of operating conditions required for a research tool. The turntable or capstan instruments are more suited for research purposes and the use of electronic force-measuring systems increases the sensitivity of these devices. The capstan method has the further advantage of simulating film processing geometry and of being potentially adaptable to the continuous monitoring of film friction. All of the instruments discussed have one feature in common: they may be used for film-to-film friction or for measuring the friction of films against other materials by making one of the sliding members from the desired test material.

Further discussion of friction testing and measurement is found in Chapter 14.

III. Technological Control of Film Friction

A. INTRODUCTION

The control of polymer film friction is, in many cases, essential to the commercial success of a film. Vinyl and vinylidene chloride films have already been shown to have high friction and adhesion. It is doubtful if films and coatings of these polymers would enjoy the wide usage they do, if the film surfaces could not be modified to reduce friction and adhesion. As processing and packaging machine speeds are increased, new methods must be sought to regulate film surface properties for optimum machine performance.

There are some terms peculiar to the film industry. "Blocking" or "matting" refers to the tendency for adjacent film surfaces to stick together when left under pressure for a long time, as in roll form or in stacks of sheets. This sticking is often so severe that sheets cannot be separated without damage. Obviously, the sticking tendency and friction of a film are manifestations of the same phenomenon, but differ in the time interval over which they occur. The relation between sticking and friction may be complicated by the "creep" properties of the polymer. Friction, in film parlance, is called "slip." This term denotes the slipperiness of a film against itself or other materials and is, in a sense, the opposite of friction in that "poor slip" denotes high friction, and vice versa.

From a consideration of the factors affecting polymer adhesion and friction, there appear to be two practical ways of controlling these properties. The simplest way is to roughen the film surface to reduce friction and adhesion and, similarly, to dust the film with particles to reduce or prevent contact. This latter method is widely used in the rubber industry where powdered magnesium stearate (67), magnesium oxide (68), and talc (69) are commonly used. Machines have been developed to apply talc uniformly to moving films to improve slip and prevent sticking (70).

The second way of reducing friction is that of applying a lubricant to the film surfaces. In commercial practice, these lubricants are mostly derivatives of long-chain fatty compounds. This class of compounds will be recognized as the classical boundary lubricants discussed earlier. The mechanism of their action on polymer friction and adhesion has also been discussed. Many structural modifications have been disclosed in the literature and these modifications generally represent the need for meeting certain lubricant requirements such as enhanced exudation to the film surface, solubility in solvents, high-temperature stability, etc. Subsequent discussions are arranged by film type, although it should be recognized that many compounds are not specific for any given film.

B. Regenerated Cellulose Film

Historically, cellulose was the first polymer film commercially available in large quantities as a general packaging medium. The frictional properties of plain cellulose film are poor and get worse at high humidities. While today very little plain cellulose film is used in the packaging industry, its poor surface properties continue to generate manufacturing difficulties as evidenced by the wealth of antifriction, antisticking compositions disclosed in the patent literature. Some of the compositions which have been applied to plain cellulose film to reduce sticking and improve slip are the following: soap (71a), wax emulsions (72a), stearylamine (71b), cetylamine (71b), N-cyclopentyllauramide (73), N-cyclopentylstearamide (73), stearamide (72b,74), lauryllauramide (72b), myristamide (72b), ethanolstearamide (72b), stearamide combined with stearic acid and sodium silicate (75), stearic acyloin (76a), aluminum tannate (71c), carnauba wax–stearic acid emulsion (77), sodium silicate (74,83a), octadecyltrimethylammonium chloride (78a), dimethyldioctadecylammonium chloride with silica (78b), colloidal silica or kaolin with glyceryl monostearate (79), polymethyl siloxanes (80,81), sodium methanesiliconate (82), and fatty amide emulsions (83b).

C. Vinyl Coatings and Films

Copolymers of vinylidene chloride with various acrylic monomers have found extensive use as coatings for cellulose film because of their superior grease resistance, appearance, and moisture barrier properties. Due to the chlorine content, these copolymers have very poor frictional and blocking characteristics. They exhibit the additional disadvantage of sticking to the hot metal sealers of packaging machines. Numerous antisticking and lubricating compositions have been disclosed. The antisticking compositions fall generally into the category of roughening agents, which are suspended in the copolymer coating solution. Hauser (84) has described the use of small particles of talc, kaolin, bentonite, zinc oxide, and polyethylene to improve the surface properties of vinylidene chloride–acrylonitrile coated cellulose film. Others have used insoluble polyvinyl chloride particles (85,86), chalk (87), silica (88a,b,89a), barium sulfate, magnesium carbonate, powdered glass and mica (89a), and insoluble particles of 1,2-dihydronaphthalene polymer (90). Clay particles also improved the slip of vinylidene chloride copolymer coatings (51).

Many lubricants for vinyl coatings have been described. These are usually long-chain fatty compounds; some are natural products. Commercially useful lubricants include di(n-alkyl) ethers of 12–18 carbons (91a), fatty acid anhydrides (91b), paraffin (89a), di(n-alkyl) ketones of 8 or more carbons (86,88a,b), and miscellaneous waxes such as spermaceti, candelilla, palm, castor, and pentaerythrytol tetrastearate (84). Saturated fatty amides have also been shown to reduce the friction of vinyl coatings (59). A composition containing carnauba wax and a 16–18 carbon fatty acid dissolved in the coating solution is said to produce copolymer coatings which do not adhere to hot metal (87).

A novel approach by Koch (89b) has been to copolymerize vinylidene chloride with octadecyl methacrylate to produce a coating polymer which is said to be slippery and needs no antisticking agents when the octadecyl methacrylate content exceeds 26%. Presumably the long alkyl side chains give the polymer surface "built-in" lubrication. Self-supporting films of polyvinyl chloride have been lubricated with the stearic acid amide of ethylenediamine (92), and N-alkyl substituted fatty amides have been used to prevent sticking of polyvinyl acetal sheets (93).

D. POLYOLEFIN COATINGS AND FILMS

Polyolefin films and coatings represent a rapidly growing segment of the film market. Since these products are made by melt extrusion, some unusual requirements are imposed upon lubricating and antiblocking compositions. They must be stable at the high temperatures, pressures, and shear conditions in the extruder and such compositions should be capable, preferably, of being incorporated into the polymer prior to extrusion for simplicity and economy. Early workers, however, applied materials to the finished film. Olefin polymers were rendered nonadherent by coating the surfaces with dimethyl phthalate (94) and slip was improved by the topical application of an alkylbenzene ethylene oxide adduct from water solution (95). Hull has reported an interesting approach where the lubricants were dissolved in the water of the quench tank for the extruded polyethylene film (96).

Numerous compounds for reducing friction have been developed for addition to the polyolefin prior to extrusion. Polyhydric alcohol esters and metal soaps (97), glyceryl monoesters (98), fatty esters such as dodecyl stearate and lauryl palmitate (99), and N-hydroxyalkyl amides such as N-ethanololeamide (100) have all proved to be effective lubricants for polyethylene film. Probably the most effective lubricants in terms of amount required for good slip are the fatty amides, especially the unsaturated ones. Compounds found in the literature include oleamide (60a,101), erucamide (102), mixtures of erucamide and oleamide (103), and various saturated amides such as myristamide, palmitamide, and lauramide (104). These substances are reported to provide good slip in concentrations of 0.001–0.1% of the polymer. They also appear to be stable at extrusion temperatures.

Roughening of the surface has also been employed as a means of reducing the sticking of films or coatings, as with vinyl products.

Dusting of polyethylene-coated surfaces with 1–10 micron particles of polyethylene powder is claimed to improve slip and reduce blocking (105). It is claimed that 0.2–0.5% silica or diatomaceous earth of particle size 0.5–7.0 μ, added to the polyethylene prior to extrusion, reduces blocking (103,106). The size range is reported to be critical and should be between 2 and 6 μ. The blocking tendency of low-density polyethylene film has been improved by incorporating 0.5–1% high-density polyethylene into the polymer prior to extrusion (107).

A promising approach to roughening the surface of polyethylene coatings on cellulose film is to form hemispherical indentations in the chrome-plated quench roll of the extrusion coater. These indentations are reproduced in the polyethylene coating as hemispherical asperities. Coated films made by this process with erucamide in the polyethylene are said to have excellent clarity and gloss, good slip, and to be nonblocking (102).

E. Miscellaneous Films

A general agent for reducing sticking of a variety of films has been described (108). Cyanuric acid applied to the film from organic solvent forms microcrystals on the surface which are claimed to reduce the sticking of cellulose acetate, vinylidene chloride, vinyl chloride, polyethylene terephthalate, and nitrocellulose-coated cellophane films. For lubricating cellulose triacetate or acetate–butyrate photographic films, poly(dialkyl siloxanes) (109), straight-chain fatty amides (110), and phenyl–alkyl silicone mixtures (111) have been used at the time of manufacture. Photographic films have also been lubricated by adding silicones to the final rinse water prior to drying the film (112).

Tacky wax coatings on regenerated cellulose film have been rendered nonblocking by coating them with palmitic acid and ethylcellulose in 2-propanol (113). Nitrocellulose coatings on cellophane contain paraffin wax for moisture proofness. The wax appears to perform the additional function of imparting good slip (76b,c).

A process for making low-friction nylon castings which may have application in film manufacture consists of uniformly dispersing finely divided polytetrafluoroethylene powder in nylon prior to casting (114).

References

1a. G. Amontons, *Mem. Acad. Roy. Sci.*, **1699**, 206; **1704**, 96.

1b. C. A. Coulomb, *Mem. Sçavans étrangers*, **1785**, 161.

2. A. J. W. Moore, *Proc. Roy. Soc. London Ser. A*, **195**, 231 (1948).

3. S. Tolansky, *Multiple Beam Interferometry of Surfaces and Films*, Oxford University Press, Oxford, England, 1948.

4. J. W. Menter, *J. Inst. Metals*, **81**, 163 (1952).

5a. *ASTM Methods E-10-58T and E18-57T*. American Society for Testing Materials, Philadelphia, Pa., 1958.

5b. *ASTM Method D 1894-61T*. American Society for Testing and Materials, Philadelphia, Pa., 1961.

6. D. Tabor, *The Hardness of Metals*, Clarendon Press, Oxford, England, 1951.

7. R. Holm, *Electrical Contacts*, Almqvist and Wiksells, Uppsala, Sweden, 1946.

8. F. P. Bowden and D. Tabor, *Proc. Roy. Soc. London, Ser. A*, **169**, 391 (1939).

9. J. T. Desaguliers, *Phil. Trans.*, **33**, 345 (1724).

10. J. S. McFarlane and D. Tabor, *Proc. Roy. Soc. London, Ser. A*, **202**, 244 (1950).

11. M. H. Cocks, *Nature*, **170**, 203 (1952).

12. K. V. Shooter and D. Tabor, *Proc. Phys. Soc. London*, **B65**, 661 (1952).

13. F. P. Bowden and D. Tabor, *The Friction and Lubrication of Solids*, Vol. II, Clarendon Press, Oxford, England, 1964, pp. 78–9.

14. Sir W. B. Hardy, *Collected Works*, Cambridge University Press, Cambridge, England, 1936.

15. J. N. Gregory, *C.S.I.R.O. (Australia) Tribophysics Div. Rept.* No. A74, 1943.

16. K. B. Blodgett, *J. Am. Chem. Soc.*, **56**, 495 (1934); **57**, 1007 (1935).

17. T. Isemura, *Bull. Chem. Soc. Japan*, **15**, 467 (1940).

18. T. P. Hughes and G. Whittingham, *Trans. Faraday Soc.*, **38**, 9 (1942).

19. J. J. Frewing, *Proc. Roy. Soc. London, Ser. A*, **181**, 23 (1942).

20. D. Tabor, *Nature*, **145**, 308 (1940); **147**, 609 (1941).

21. K. Tanaka, *Mem. Coll. Sci., Univ. Kyoto*, **21**, 85 (1938); **22**, 377 (1939).

22. F. P. Bowden and L. Leben, *Phil. Trans. Roy. Soc. London, Ser. A*, **239**, 1 (1939).

23. J. N. Gregory and J. A. Spink, *Nature*, **159**, 403 (1947).

24. E. B. Greenhill, *Trans. Faraday Soc.*, **45**, 631 (1949).

25. F. P. Bowden, J. N. Gregory, and D. Tabor, *Nature*, **156**, 97 (1945).

26. R. L. Cottington, E. G. Shafrin, and W. A. Zisman, *J. Phys. Chem.*, **62**, 513 (1958).

27a. W. C. Bigelow, D. L. Pickett, and W. A. Zisman, *J. Colloid Sci.*, **1**, 513 (1946).

27b. W. C. Bigelow, E. Glass, and W. A. Zisman, *J. Colloid Sci.*, **2**, 563 (1947).

27c. E. G. Shafrin and W. A. Zisman, *J. Phys. Chem.*, **64**, 519 (1960).

27d. M. K. Bernett and W. A. Zisman, *J. Phys. Chem.*, **63**, 1241 (1959).

27e. W. A. Zisman, *Ind. Eng. Chem.*, **55**, 19 (1963).

28. A. I. Bailey and J. S. Courtney-Pratt, *Proc. Roy. Soc. London, Ser. A*, **227**, 500 (1955).
29. P. W. Bridgman, *Rev. Mod. Phys.*, **18**, 1 (1946).
30. K. V. Shooter, *Proc. Roy. Soc. London, Ser. A*, **212**, 488 (1952).
31. R. F. King and D. Tabor, *Proc. Phys. Soc. London*, **B66**, 728 (1953).
32. K. V. Shooter and D. Thomas, *Research London* **2**, 533 (1949).
33. H. G. Howell, *J. Textile Inst.*, **42**, T521 (1951).
34. C. Rubenstein, *Proc. Phys. Soc. London*, **B69**, 921 (1956).
35. G. King, *J. Textile Inst.*, **41**, T153 (1950).
36. B. Olofsson, *Textile Res. J.*, **20**, 476 (1950).
37. N. Gralén, *Proc. Roy. Soc. London*, **221**, 491 (1952).
38. B. Lincoln, *Brit. J. Appl. Phys.*, **3**, 260 (1952).
39. D. F. Denny, *Proc. Phys. Soc. London*, **B66**, 721 (1953).
40. A. Schallamach, *Proc. Phys. Soc. London*, **B65**, 657 (1952).
41. H. G. Howell, *J. Textile Inst.*, **44**, T359 (1953).
42. J. D. Huffington, *Research Lab.*, **10**, 163 (1957).
43. H. G. Howell and J. Mazur, *J. Textile Inst.*, **44**, T59 (1953).
44. H. Hertz, *Z. Reine Angew. Math.*, **92**, 156 (1886).
45. B. B. Finch, *Textile Res. J.*, **21**, 383 (1951).
46. M. W. Pascoe and D. Tabor, *Proc. Roy. Soc. London, Ser. A*, **235**, 210 (1956).
47. N. Adams, *J. Appl. Polymer Sci.*, **7**, 2075, 2105 (1963).
48. J. F. Archard, *Nature*, **172**, 918 (1951).
49. A. S. Lodge and H. G. Howell, *Proc. Phys. Soc. London*, **B67**, 89 (1954).
50. E. J. Kaliski, *Textile Res. J.*, **28**, 325 (1958).
51. D. K. Owens, *J. Appl. Polymer Sci.*, **8**, 1477 (1964).
52. D. G. Lyne, *J. Textile Inst.*, **46**, P112 (1955).
53a. W. W. Hansen and D. Tabor, *J. Appl. Phys.*, **27**, 1558 (1956).
53b. W. W. Hansen and D. Tabor, *Textile Res. J.*, **27**, 300 (1957).
54a. R. C. Bowers, W. C. Clinton, and W. A. Zisman, *Ind. Eng. Chem.*, **46**, 2416 (1954).
54b. R. C. Bowers, N. L. Jarvis, and W. A. Zisman, *Ind. Eng. Chem. Prod. Res. Develop.*, **4**. 86 (1965).
55. M. W. Pascoe, Ph.D. dissertation, Cambridge University, Cambridge, England, 1954.
56. C. Rubenstein, *J. Appl. Phys.*, **32**, 1445 (1961).
57. T. Fort, Jr., and J. S. Olsen, *Textile Res. J.*, **31**, 1007 (1961).
58. T. Fort, Jr., *J. Phys. Chem.*, **66**, 1136 (1962).
59. D. K. Owens, *J. Appl. Polymer Sci.*, **8**, 1465 (1964).
60a. A. J. G. Allan, *J. Colloid Sci.*, **14**, 206 (1959).
60b. A. J. G. Allan, *J. Polymer Sci.*, **38**, 297 (1959).
60c. A. J. G. Allan, *J. Polymer Sci.* **24**, 461 (1957).
61. The Martin Sweets Co., 3131 W. Market St., Louisville, Ky.
62. H. O. George and J. E. Arnoult, *Tappi*, **40**, 972 (1957).
63. R. F. Westover and W. I. Vroom, *Soc. Plastics Engrs. J.*, **19**, 1093 (1963).
64. I. J. Saxl, *J. Franklin Inst.*, **221**, 789 (1936).

65. Rothschild Elektronische Mess- und Steuergeräte, Traubenstrasse 3, Zürich 2, Switzerland.
66. W. Wegener and B. Schuler, *Z. Ges. Textil-Ind.*, **64**, 250, 362, 458 (1962).
67. P. A. Davis, Brit. Pat. 346,489 (Apr. 8, 1929).
68. G. Uhl, Brit. Pat. 351,643 (Apr. 30, 1929).
69. E. L. McDowell and L. M. Freeman, U.S. Pat. 2,262,689 (Nov. 11, 1941).
70. R. B. Randall, U.S. Pat. 2,758, 564 (Aug. 14, 1956).
71a. D. E. Drew, U.S. Pat. 2,095,129 (Oct. 5, 1937).
71b. D. E. Drew, U.S. Pat. 2,137,274 (Nov. 22, 1939).
71c. D. E. Drew, U.S. Pat. 2,226,554 (Dec. 31, 1940).
72a. British Cellophane, Ltd., Brit. Pat. 472,264 (Sept. 20, 1937).
72b. British Cellophane, Ltd., Brit. Pat. 523,720 (July 22, 1940).
73. H. R. Arnold and P. R. Austin, U.S. Pat. 2,151,369 (March 21, 1939).
74. L. Veyret, U.S. Pat. 2,623,244 (Dec. 30, 1952).
75. J. D. Pollard, U.S. Pat. 2,206,046 (July 2, 1940).
76a. J. A. Mitchell, U.S. Pat. 2,251,752 (Aug. 5, 1941).
76b. J. A. Mitchell, U.S. Pat. 2,079,379 (May 4, 1937).
76c. J. A. Mitchell, U.S. Pat. 2,193,831 (March 19, 1940).
77. W. D. White, U.S. Pat. 2,316,496 (Apr. 13, 1943).
78a. W. O. Brillhart, U.S. Pat. 2,621,135 (Dec. 9, 1952).
78b. W. O. Brillhart, U.S. Pat. 2,658,843 (Nov. 10, 1953).
79. R. P. Wymbs, U.S. Pat. 2,658,835 (Nov. 10, 1953).
80. W. S. Kather, A. A. Lister, and E. D. Brown, U.S. Pat. 2,803,613 (Aug. 20, 1957).
81. A. L. James, Ger. Pat. 1,026,671 (March 20, 1958).
82. E. J. Dunn, U.S. Pat. 2,930,717 (March 29, 1960).
83a. Société la Cellophane, S. A., French Pat. 990,808 (Sept. 26, 1951).
83b. Société la Cellophane, S. A., French Pat. 849,720 (Nov. 30, 1940).
84. P. M. Hauser, U.S. Pat. 2,525,671 (Oct. 10, 1950).
85. W. Berry, J. Rose, and J. S. Pearce, Brit. Pat. 714,275 (Aug. 25, 1954).
86. E. G. Sanderson and G. I. Evans, Brit. Pat. 766,708 (Jan. 23, 1957).
87. C. H. Hofrichter, U.S. Pat. 2,711,996 (June 28, 1955).
88a. British Cellophane, Ltd., Australian Pat. 57,567 (Feb. 27, 1960).
88b. British Cellophane, Ltd., Brit. Pat. 871,840 (July 5, 1962).
89a. W. T. Koch, U.S. Pat. 3,034,929 (May 15, 1962).
89b. W. T. Koch, Belg. Pat. 567,211 (May 14, 1958).
90. W. L. Hyden and J. A. Mitchell, U.S. Pat. 2,388,326 (Nov. 6, 1945).
91a. British Sidac, Ltd., Brit. Pat. 913,044 (Dec. 12, 1945).
91b. British Sidac, Ltd., Brit. Pat. 913,045 (Dec. 12, 1962).
92. F. W. Duggan and R. P. Stambaugh, U.S. Pat. 2,464,855 (March 22, 1949).
93. D. S. Kaltreider, U.S. Pat. 2,423,981 (July 15, 1947).
94. P. B. Gerhardt and J. V. Sommer, U.S. Pat. 2,257,167 (Sept. 30, 1941).
95. T. F. Banigan, U.S. Pat. 2,519,013 (Aug. 15, 1950).
96. D. R. Hull, U.S. Pat. 2,324,397 (July 13, 1943).
97. C. S. Myers, U.S. Pat. 2,462,331 (Feb. 22, 1949).
98. J. E. Guillet, R. I. Combs, and C. E. Tholstrup, U.S. Pat. 3,057,810 (Oct. 9, 1962).

 99. E. I. du Pont de Nemours & Co., Brit. Pat. 590,055 (July 7, 1942).
100. National Distillers and Chemical Corp., French Pat. 1,313,105 (Dec. 28, 1962).
101. A. E. Symonds, U.S. Pat. 2,770,609 (Nov. 13, 1956).
102. L. F. Gronholz, D. K. Owens, and C. R. Price, Belg. Pat. 636,349 (Feb. 19, 1964).
103. Union Carbide Chemical Co., Australian Pat. 40,011 (July 29, 1958).
104. H. C. Barker, E. E. Lewis, and W. B. Happoldt, U.S. Pat. 2,770,608 (Nov. 13, 1956).
105. W. Schoch and H. G. Deutscher, Ger. Pat. 1,101,751 (March 9, 1961).
106. G. R. Toy and M. E. Krasnow, U.S. Pat. 3,028,355 (Apr. 3, 1962).
107. Badische Anilin und Soda Fabrik, A.G., Ger. Pat. 1,121,328 (Jan. 4, 1962).
108. F. L. Andrew and E. A. Vitalis, U.S. Pat. 2,856,317 (Oct. 14, 1958).
109. W. R. Holm, Belg. Pat. 618,682 (Dec. 7, 1962).
110. J. E. Campbell and J. J. Dennis, Belg. Pat. 617,284 (Nov. 5, 1962).
111. J. Tallet and A. Ben-Ezra, U.S. Pat. 3,080,317 (March 5, 1963).
112. H. Brueggemann, *J. Soc. Motion Picture Television Engrs.*, **66,** 106 (1957).
113. M. C. Funk and J. F. Helms, U.S. Pat. 2,833,671 (May 6, 1958).
114. L. L. Stott, U.S. Pat. 3,126,339 (March 24, 1964).

CHAPTER 10

FILM ORIENTATION

JEAN B. MAURO

U.S. Industrial Chemicals Co., Bridgeport, Connecticut

and

JOSEPH J. LEVITZKY

Packaging Division, Olin Mathieson Chemical Corp., New Haven, Connecticut

I. Introduction

Orientation of polymer films is a means of improving their strength and durability in order to broaden the scope of application in packaging and other uses, as well as making them serviceable in thinner gauges. Among the physical properties of films that may be significantly improved by orientation are tensile and impact strengths, stiffness, clarity, and resistance to crazing. Gas and water-vapor permeability may also be reduced. Among the more common commercially oriented films are polyethylene terephthalate, polyvinylidene chloride, polypropylene, and some crosslinked polyethylenes.

The subject of orientation is here reviewed in terms of the various processes and equipment available, its effect on the properties of polymer films, and the techniques which may be applied to secure orientation.

Orientation can result from any process which causes shear gradients to be developed within a material, e.g., by milling, spinning, injection molding, and extrusion. In these cases, the molecules of a formed plastic specimen are displaced from an essentially random distribution into some more orderly arrangement. More specifically, the molecules tend to become aligned in the direction of the orienting force. For orientation of films, stretching after extrusion is the most feasible technique. The stretching may be either in one direction, i.e., in the transverse direction, resulting in a film that is commonly designated as uniaxially oriented, or the stretching may be in both directions, i.e., in the transverse and machine directions, giving a product which is called biaxially oriented. If the resulting properties are approximately equal in both directions, the film is referred to as *balanced*. At the present time, the formal definition by the American Society for Testing and Materials for oriented polypropylene, one of the commercially significant films, is as follows:

1. An *oriented* film is one which has a tensile strength of at least 15,000 psi in at least one principal direction.

2. A *balanced oriented* film is an oriented film in which the machine direction and transverse direction tensile strengths are within 8000 psi of each other.

II. Theoretical Considerations

Essentially, molecular orientation of thermoplastic film—whether it be unidirectional or biaxial—is a process of stretching the material in such a manner as to line up the molecular chains in a predetermined direction (1). Once lined up, the ordered arrangement is frozen in a strained condition.

In orienting a thermoplastic, three forces must be overcome: (*1*) directed covalent forces within the molecular chain itself, hooking carbon to carbon on the backbone, and to any side chains that are present; (*2*) the frictional constraining force caused by the intertwining of the spring-shaped molecules; and (*3*) the attraction existing between the molecules.

Heat is used to help overcome these internal forces. At room temperature, the translational and vibrational energies of the molecules themselves are less than the sum of the frictional constraining forces and the intermolecular forces. Thus, the molecules are frozen in place. As the temperature of the material is raised, the vibrational energy of the molecules increases, increasing slightly the distance between carbon atoms within the molecules and to a far greater extent the distance between molecules. Increasing the interchain distances lowers the constraining van der Waals forces. With further increases in temperature a point is reached where vibrational energy is just sufficient to overcome the intermolecular attractions. At this point, a pure solid substance would be said to melt. Within a polymeric material, however, there are wide differences among the molecular weights, and therefore the melting points, of the mixtures of various molecules. The term *softening point* is used to designate that temperature at which a sufficient number of molecular segments have loosened sufficiently to result in a softening of the bulk material as determined by various standard tests. Actually, the so-called glass-transition temperature is the common designation used for the temperature at which molecular freezing is just overcome. This temperature often differs substantially from the observable softening point by common tests. The glass transition, though a second-order phenomenon, is determinable with considerable precision (2).

An unoriented thermoplastic may be pictured as a mass of randomly arranged, intricately intertwined, spring-like molecules frozen in place (cf. Chap. 2). Orientation is the ordering of these randomly arranged molecules. The most common method used to orient a thermoplastic is to stretch it after it has been heated to a temperature at which it is soft. This temperature is below the flow temperature at which the molecules would glide readily past one another when the material is stressed, with little intramolecular stretching, but is above the glass-transition point (3).

When the material is stretched at orientation temperature, two important changes occur.

1. The direction of the molecules in the material is changed. A rotating force is exerted which pulls or turns the entangled molecules in a direction more nearly approaching the direction in which the material is stressed (4). The force required for this unraveling of entwined molecules is that needed to overcome the sum of the

frictionlike forces exerted by side-chain entanglement and intermolecular attraction.   This force is nonrecoverable.

2. The molecules are extended like springs.   The force needed
for this extension is that required to overcome the intrachain forces
(that result in coiling) and shape-forming forces (interatomic forces
at fixed bonding angles) within the backbone of the molecules.   As
in a coiled spring, the resulting strain is an elastic deformation, and,
as such, the force is stored as potential energy, recoverable when
molecular strain is relieved.

While the molecules are held in this extended-spring configuration,
the temperature is dropped below the softening point of the material.
This cooling reduces the vibrational energy of the molecules to below
the sum of the intermolecular attractive forces and the mechanical
constraining forces, and freezes the molecules in that strained
position.

The potential energy stored in the extended molecules is the
so-called elastic memory, characteristic of oriented, non-heat-set
thermoplastics; that is, when the material is reheated to the orientation temperature, the molecules tend to return to their original
size and spatial arrangement.   At elevated temperatures, but below
the orientation temperature, recovery manifested as shrinkage will
occur but to a less extent.   The temperature at which recovery
deformation begins to occur can be controlled by the thermal conditions of the process.   The film can be annealed by application of
heat to relax the forces partially while maintaining the film in a
highly stretched condition.   This procedure does result in some
reduction in dimension in the stretched direction or directions.
Such a film is referred to as *heat set*.

The direction of alignment and stretch of molecules depends, of
course, on the direction in which the material is stretched.   Orienting or aligning the long chainlike molecules in one direction puts
them in the best load-bearing position if the material is to be
stressed in tension in that direction.   The tensile load-bearing ability of such a uniaxially oriented material in the direction perpendicular to that of orientation is much lower, however, since an
applied stress tends to pull adjacent molecules apart, rather than
pulling along the molecular chain axis.

Biaxial orientation aligns molecules in both the transverse and
longitudinal directions.   It provides a primarily planar molecular

structure, with the axes of the molecules nearly parallel to the sheet surfaces. Such an arrangement does not provide the maximum strength, since the molecules are not in the optimum load-bearing position, but it provides a balance of improved strength character-istics—both longitudinal and transverse.

Virtually all thermoplastics can be oriented to some extent, but one can orient amorphous materials more readily than crystalline ones. The stretching process frequently induces crystallization (as with natural rubber) or increases crystallization (as with poly-ethylene terephthalate), particularly where the repeating unit has a high degree of symmetry and where there is geometric regularity, as in isotactic and syndiotactic molecules (5). Until the pres-ent decade, attention has been directed to the effects of orientation on non-crystallite-forming polymers. However, with the recent commercialization of the polyolefinic crystallite-forming resins, much research effort has been directed toward the orientation of these materials.

Crystallization is not a problem with polystyrene and related copolymers, acrylics, polyesters, and polycarbonates, since under normal conditions of polymerization and processing no appreciable amount of crystallization occurs. Crystalline materials may exhibit a differential elongation or "necking" during the stretching process, causing sharp differences in thickness of the sheet. Successful orien-tation of such materials can usually be performed only if the sheet is quenched immediately after extrusion, thus limiting the degree of crystallinity. In the case of polypropylene the problem is com-pounded by the tendency of the material to crystallize spontane-ously upon standing for a period of time at room temperature. In practice, polypropylene is invariably oriented in line immediately after the postextrusion quench.

It has been pointed out that orientation is performed by stretch-ing, but stretching does not necessarily mean deformation with strengthening. In the case of low-pressure polyethylene, for instance, the temperature range in which the material is stretched and thus strengthened is limited in the upper temperature regions by the crystalline melting point. The characteristics are most eas-ily explained in relation to a uniaxially stretched test piece. If such a test piece is stretched below the crystalline melting point, the stress increases to a maximum value. If further stretching is carried out, some local constriction will appear and the stress will decrease until such time as the constricted region has taken up a new cross-sectional area, i.e., has necked down. If stretching is continued still further, stress as well as cross-sectional area remain constant but the constriction propagates until the whole of the specimen has taken on the new cross-sectional area.

If stretching is carried out exactly at the crystalline melting point or above, then deformation will take place uniformly along the whole length of the specimen and there will be no constriction. The crystalline melting point thus represents a limiting temperature as regards the type of deformation. Below this temperature, stretching produces deformation with local constriction and a high degree of strengthening; above it, the material stretches without local constriction and without appreciable strengthening.

The behavior just described caused investigators to attempt to define the deformation phenomena by studying the stretching process as typified by bubble formation (6). A sample of low-pressure polyethylene film, clamped like a membrane and subjected to pressure from one side, first buckles until the two main stresses are equal. If the stress maximum is exceeded, a thin-walled bubble suddenly appears at the center; this bubble bursts almost immediately. The bubble consists of uniaxially stretched material.

The behavior is different if the temperature is high but still below the crystalline melting point. The bubble does not burst, but expands until it encompasses the whole of the membrane. X-ray photographs show that in such a bubble stretching has occurred evenly in all directions. The mechanical strength is therefore theoretically uniform in all directions. If, however, the above experiments are carried out at temperatures above the crystalline melting point, there is no appreciable increase in mechanical strength.

Table 1 includes the most important references to patents which claim improvement in film properties as a result of biaxial orientation.

III. Orientation Techniques

Orientation of thermoplastic films in the transverse direction may be carried out by one of three general techniques: (1) drawing of tubularly extruded film over an internal-spreading mandrel; (2)

TABLE 1
Patents on Orientation for Product Improvement

Film	Patent	Date of issue	Inventors
Pyroxylin	U.S. 1,979,762	Nov. 6, 1934	R. G. O'Kane and E. R. Derby
Cellulose ester	U.S. 2,002,711	May 28, 1935	F. A. Parkhurst and G. J. Esselen
Cellulose acetate	U.S. 2,034,716	March 24, 1936	H. Dreyfus
Polyvinyl chloride	U.S. 2,067,025	Jan. 5, 1937	F. Schmidt
	U.S. 2,239,780	Apr. 29, 1941	H. Fikentscher and H. Jacqué
Polystyrene	U.S. 2,047,554	July 14, 1936	E. Fischer
	U.S. 2,074,285	March 16, 1937	E. Studt and U. Meyer
Polyamide	U.S. 2,244,208	June 3, 1941	J. B. Miles, Jr. (to du Pont)
	U.S. 2,307,817	Jan. 12, 1943	P. R. Austin (to du Pont)
	U.S. 2,321,635	June 15, 1943	G. B. Taylor (to du Pont)
Polyvinylidene chloride	U.S. 2,183,602	Dec. 19, 1939	R. M. Wiley (to Dow)
	U.S. 2,344,511	Mar. 21, 1944	E. H. Harder (to Dow)
	U.S. 2,409,521	Oct. 15, 1946	R. M. Wiley (to Dow)
	U.S. 2,448,433	Aug. 31, 1948	C. R. Irons and C. E. Sanford (to Dow)
	U.S. 2,452,080	Oct. 26, 1948	W. T. Stephenson (to Dow)
Polyethylene	U.S. 2,219,700	Oct. 29, 1940	M. W. Perrin, J. G. Paton, and E. G. Williams (to I.C.I.)
	U.S. 2,461,975	Feb. 15, 1949	E. D. Fuller (to Visking)
	U.S. 2,586,820	Feb. 26, 1952	W. F. Hemperly and N. R. Smith (to Carbide)

(*continued*)

466   JEAN B. MAURO AND JOSEPH J. LEVITZKY

TABLE 1 (*continued*)

Film	Patent	Date of issue	Inventors
Rubber hydro-chloride (Pliofilm)	U.S. 2,274,150	Feb. 24, 1942	G. C. Mack (to Wingfoot)
	U.S. 2,301,222	Nov. 10, 1942	H. D. Minnich
	U.S. 2,328,827	Sept. 7, 1943	R. C. Martin (to Wingfoot)
	U.S. 2,328,843	Sept. 7, 1943	H. J. Osterhof (to Wingfoot)
	U.S. 2,334,022	Nov. 8, 1943	H. D. Minnich
	U.S. 2,429,177	Oct. 14, 1947	F. J. Young (to Wingfoot)
Polyethylene tere-phthalate	U.S. 2,497,376	Feb. 14, 1950	J. C. Swallow, D. K. Baird, and B. P. Ridge (to I.C.I.)
	U.S. 2,578,899	Dec. 18, 1951	A. Pace, Jr. (to du Pont)
	U.S. 2,627,088	Feb. 3, 1953	F. P. Alles and W. R. Saner (to du Pont)
	U.S. 2,718,666	Sept. 27, 1955	K. L. Knox (to du Pont)
	U.S. 2,823,421	Feb. 18, 1958	A. C. Scarlett (to du Pont)
	U.S. 2,884,663	May 5, 1959	F. P. Alles (to du Pont)
	U.S. 2,904,841	Sept. 22, 1959	E. F. Haugh (to du Pont)
	U.S. 2,920,352	Jan. 12, 1960	M. Miller and D. R. McGregor (to du Pont)
Polarizing film	U.S. 2,041,138	May 19, 1936	E. H. Land
	U.S. 2,011,553	Aug. 13, 1935	E. H. Land
	U.S. 2,547,763	Apr. 3, 1951	E. H. Land and W. H. Ryan
Polypropylene	U.S. 2,918,394	Dec. 22, 1959	A. F. Smith (to Ethane)
	U.S. 2,931,740	Apr. 5, 1960	U. Riboni (to Montecatini)
	U.S. 2,947,598	Aug. 2, 1960	D. Maragliano and F. Denti (to Montecatini)
	U.S. 2,952,867	Sept. 20, 1960	G. Diedrich, E. Gaube, and K. Richard (to Hoechst)

(*continued*)

TABLE 1 (*continued*)

Film	Patent	Date of issue	Inventors
Polypropylene (*continued*)	Australian 221,446	June 4, 1959	Montecatini, E. DiGiulio, and P. Perrini
	Belgian 549,259	July 7, 1956	Farbwerke Hoechst A. G.
	Belgian 549,523	July 13, 1956	Farbwerke Hoechst A. G.
	Belgian 553,174	Dec. 5, 1956	Montecatini, D. Maragliano, and G. DiGiulio
	Belgian 557,717	May 22, 1957	Montecatini, A. Camerini, and G. Guzzetta
	Belgian 561,515	Oct. 10, 1957	Montecatini, P. Masso, and G. Guzzetta
	Belgian 571,135	Sept. 9, 1958	Imperial Chemical Industries
	Belgian 571,136	Sept. 9, 1958	Imperial Chemical Industries
	British 812,972	May 6, 1959	E. I. du Pont de Nemours & Co.
	British 817,125	July 22, 1959	Montecatini
	British 838,954	June 22, 1960	Montecatini

inflating a tubularly extruded film to a larger diameter, effecting radial stretching; (*3*) passing a flat film through a tenter frame or other device incorporating diverging film-gripping members.

Machine-direction orientation is usually accomplished by drawing the extruded film from the die lips at a faster rate than it is extruded or by passing the film through two or more pairs of nip rolls, subsequent pairs operating at a faster circumferential speed than preceding pairs. Biaxial orientation is achieved by combining the longitudinal and lateral steps sequentially in the process line or by performing both operations simultaneously.

Table 2 includes the more significant apparatus patents for film orientation.

TABLE 2
Patents on Orientation Equipment

Patent	Date of issue	Inventors
U.S. 2,339,451	Jan. 18, 1944	J. Bailey and R. S. Jesionowski (to Plax)
U.S. 2,373,215	Apr. 10, 1945	H. L. Young (to B. F. Goodrich)
U.S. 2,412,187	Dec. 3, 1946	F. E. Wiley, R. W. Canfield, R. S. Jesionowski, and J. Bailey (to Plax)
U.S. 2,473,404	June 14, 1949	F. J. Young (to Wingfoot)
U.S. 2,483,339	Sept. 27, 1949	C. E. Gardner and E. B. Gardner (to Gardner Ind. Associates)
U.S. 2,505,146	Apr. 25, 1950	W. H. Ryan (to Polaroid)
U.S. 2,545,868	March 20, 1951	J. Bailey (to Plax)
U.S. 2,547,736	Apr. 3, 1951	R. P. Blake (to Polaroid)
U.S. 2,571,355	Oct. 15, 1951	C. E. Gardner (to Wingfoot)
U.S. 2,582,165	Jan. 8, 1952	H. Rosenfeld (to Milprint)
U.S. 2,618,012	Nov. 18, 1952	D. T. Milne (to American Viscose)
U.S. 2,878,546	March 24, 1959	P. T. Kastner (to Olin)
U.S. 2,923,966	Feb. 9, 1960	W. R. Tooke and E. G. Lodge, Jr. (to American Viscose)
U.S. 2,978,740	Apr. 11, 1961	K. W. Maier (to Olin)
U.S. 2,988,772	June 20, 1961	H. S. Horn (to Celanese)
U.S. 3,004,284	Oct. 17, 1961	A. P. Limbach (to Carbide)
U.S. 3,014,234	Dec. 26, 1961	H. P. Koppehele (to American Viscose)
U.S. 3,063,090	Nov. 13, 1962	H. P. Koppehele (to American Viscose)
U.S. 3,068,528	Dec. 18, 1962	J. E. Owens (to du Pont)
Canadian 627,801	Sept. 19, 1961	A. Fior, A. Camerini, and V. Rusignuolo (to Montecatini and K. Ziegler)
Canadian 630,530	Nov. 7, 1961	E. F. Haugh (to du Pont)

A. BIAXIAL ORIENTATION

1. Internal Forming by Mandrel

The first method to be developed for biaxial orientation of a polymer film originated in Germany prior to 1932 (7,8). It is obsolete and of academic interest only. This method, shown in Figure 1, was applied to polystyrene and involves extrusion of a tube and subsequent pulling of the tube over a parabolic spreader mandrel. The film take-off rolls were operated at a peripheral speed sufficiently higher than the linear rate of extrusion to provide longi-

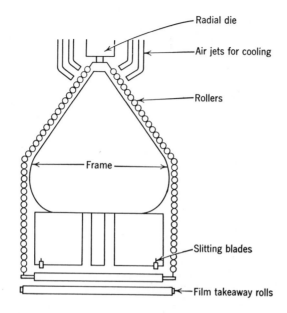

Fig. 1. Device for orientation of film by use of an internal-forming mandrel.

tudinal stretching and thus orientation in the machine direction. As the tube was drawn over the mandrel, it was stretched in the transverse direction. The mandrel was equipped with a number of rollers along its periphery to prevent the film from sticking to it. In practice, the edges of the flattened film tube were slit off as scrap and the resulting two webs of film were wound on separate rolls (9–11).

2. Bubble Process

Another method for orienting tubular film was devised several years ago (12–16). It is commonly called the bubble process and is similar in effect to the internal-forming mandrel technique. The process differs essentially in that the mandrel is replaced by a volume of compressed gas, usually air, locked within the tube. The gas is entrapped between a set of nip rolls and the extrusion die as shown in Figure 2. As a result, the gas remains in a constant position and the extruded tube is actually forced over it. Supplemental forms or

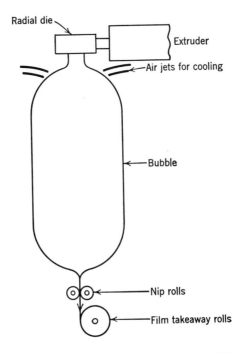

Fig. 2.   Tubular process for biaxial orientation of film.

spreading devices are sometimes used to aid in cooling and shaping
the tube.   The nip rolls, in addition to containing the trapped gas,
are used to draw off the tube at a rate of speed necessary to impart
the desired orientation in the machine direction.   Orientation in
the transverse direction is controlled by the quantity and pressure
of the gas confined within the tube.

This process is a difficult one to control in terms of gauge and cooling, particularly when used with essentially amorphous materials,
since no convenient end point of stretching is present as it is with
crystalline materials.   Modifications of this technique are used in
producing saran film and some polyethylene films.   There are, however, certain advantages in using the bubble process over the flat-
film methods to be discussed later (17).

(a) A simultaneous biaxial stretch is automatically obtained on
blowing and this is readily variable over a wide range.   Even so,

it should be noted that the longitudinal and transverse stretches are not wholly independent variables and these, together with a complex stretching-force distribution and a tendency to instability of the bubble at high transverse stretches, can cause some difficulty when aiming at a fully balanced film.

(b) The radial symmetry tends to produce even heating and is an important asset in assisting uniform orientation in both directions.

(c) Edge scrap, which is a problem in the flat-film processes, can be negligible in the bubble process.

(d) Streamlining of the extrusion die is readily achieved and is important for heat-sensitive materials.

(e) Linear running speeds are half those of comparable flat film processes for a given throughput of polymer, permitting greater ease of control.

B. Longitudinal Orientation

Longitudinal stretching of film has already been mentioned as an integral part of the biaxial orientation process. The most common method employed is the use of two heated pairs of nip rolls, the second pair operating at a greater peripheral speed than the first pair. Conditions of controlled film temperature are maintained. Thus, as the film passes through these rolls, it is heated to the temperature at which satisfactory stretching can occur, and is subjected to sufficient longitudinal tension to accomplish the desired degree of stretching. A second means of performing orientation in the machine direction is with a series of rolls, driven and idler, operating at differential speeds.

Since the longitudinal stretching of polymeric materials is carried out at elevated temperatures, the use of highly polished stretching rolls is necessary to maintain a smooth surface on the film being stretched. This, in turn, preserves transparency and prevents weakening the film through scratching. Unfortunately, inherent in the use of smooth rolls is the tendency of the film to slip transversely, encouraging the phenomenon of necking in, i.e., an uncontrolled reduction in transverse dimension—an abrupt reduction of web thickness upon stretching of crystalline materials.

Normally the higher the rate of longitudinal stretching, the greater is the degree of necking in. One means of minimizing necking in is through the use of idler rolls between two pairs of nip rolls rotating at different rates. The idler rolls should be as close together as possible in order to maintain the effect of always having the film in contact with a roll surface. The greater the distance between the idler rolls, the greater is the opportunity for the film to neck in during transfer from one roll to another. Alternatively, the pairs of nip rolls may be placed as close together as possible.

Another method of reducing neck in, when operating with either differential speed rolls or pairs of fast and slow nip rolls, involves the use of film having beaded edges (18). It has been found that the longitudinal tension applied to the film apparently sets up greater friction between the stretching rolls and the beaded edges of the film than between the rolls and the main body surface of the film. It is this additional friction which prevents excessive necking in. Futhermore, when beaded edges are not provided, the reduction in film thickness may not only be irregular but the reduction in width of the film may be both irregular and excessive. This irregular reduction in width is referred to as *scalloping*, and such a condition results in excessive waste necessitated by trimming and also in process difficulties and film breakage when the longitudinally stretched film is to be oriented in the transverse direction by means of a tenter-frame apparatus (Sect. III-C).

To complete the picture of longitudinal stretching, three other techniques are worth brief mention because of their novelty. The first of these involves pulling the film over a metallic, heated edge. The edge heats the film quickly to the appropriate stretching temperature, the edge serving to localize the heat and therefore the stretching zone. Since the heat is localized, necking in is reputed to be very low (19).

In the second method, the moving film comes into contact with a system of rolls in such a way that the longitudinal axis of the film forms an acute angle with the axis of the rolls. The film leaves the rolls with its longitudinal axis perpendicular to the axis of the rolls. The width of the stretched film is equal to the width of the wraparound on the rolls, this wraparound being greater than the original width of the film since the film strikes the rolls obliquely. The direction of orientation is neither longitudinal nor transverse but rather at a 45° angle to these directions (20).

The third method consists of passing the film through at least three successive stages, each stage consisting of a pair of pressure rolls.   During passage through each stage, the film is compressed to a thinner gauge, thereby effecting machine direction orientation (21).

C. Transverse Orientation

The most commonly employed apparatus for lateral or transverse orientation is the tenter frame, a device long used in the textile industry and more recently adapted to the stretching of plastic films (22,23).   The orientation of thermoplastic films, however, is

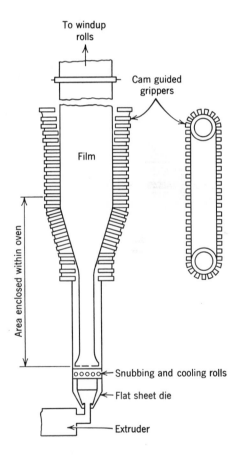

Fig. 3.   Tenter clip device for transverse orientation of film.

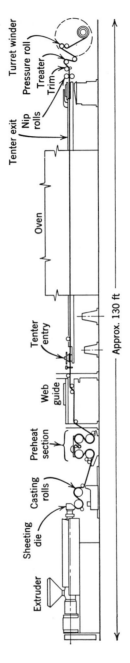

Fig. 4.   Process line for biaxial orientation of polymer film.

more complex than textile stretching because of the nature of the films and their sensitivity to processing conditions.

The apparatus, shown in Figures 3 and 4, includes several major equipment pieces in addition to the extruder. The process train usually includes a series of internally cooled rolls, a stretching device, and an oven enclosing the stretcher and containing a number of carefully controlled temperature zones, followed by slitting and windup equipment.

Tenter frames are designed to be widely adjustable with respect to the width of the web they can accommodate. Such design permits flexibility in the rate and degree of transverse stretching accomplished by the frame. This built-in flexibility is essential in order to provide optimum stretching conditions of various thermoplastics, each having its own particular mechanical and physical properties and treatment requirements. The length of the tenter frame is determined by both the required rate of stretching and the desired rate of production. Some materials require gradual transverse stretching, which must be reflected in increased tenter frame length. At high production rates, the tenter must also be quite long in order to maintain the rapidly moving stretched film under appropriate conditions of tension throughout its frequently complicated thermal treatment. This, of course, necessitates the use of long ovens.

In general terms, the tenter-frame, transverse-direction orientation process may be described as follows. After extrusion, quenching and possibly longitudinal orientation, the temperature of the surface of the sheet is lowered below that desired for transverse orientation (while the interior of the sheet may still remain above that temperature). The sheet is then moved forward through the tenter oven and is reheated to the proper temperature for transverse orientation. The cooling and subsequent reheating of the film aid in bringing it to a uniform temperature throughout its cross section. The sheet is now ready for orientation by the stretching device.

Orientation is accomplished by the use of a number of clamps or grippers mounted on conveyors situated on each side of the sheet. The clamps are guided along channel cams, forming a diverging angle with the center line of the web. This angle determines the rate of transverse pull and is adjustable. Adjustability of this angle is important since some thermoplastic films may require different degrees of stretching than others. A shallower angle will necessitate a longer tenter unit and thus the dimensions of the tenter frame must be selected with a particular thermoplastic in mind or with sufficient flexibility to adapt to any requirements.

A major problem in the construction of the stretching apparatus is the design of the clamps or tenter clips (5). The forces acting on the edges of the sheet during stretching are large, and sufficient to overcome the initial frictional forces between the clamp jaws and the sheet. It is essential, therefore, to translate some of the horizontal force into pressure exerted by the clamp jaws. It will suffice here to call attention to the problem and point out that it has been adequately solved by various clamp designs and by several tenter-frame manufacturers.

It is frequently necessary to cool the clamps in order to increase the strength of the material in the clamped region; however, the temperature of the sheet must be maintained within a certain range. If it is too high or too low, the sheet may melt, tear, or crack. Cooling of the clamp jaws may be done by blowing relatively cool air onto the clamps.

While passing through the oven and stretching equipment, the film, still under tension, may be subjected to various heat-setting and annealing treatments which serve to relieve locked-in stresses and improve thermal stability of the finished film. Upon leaving the oven, the film is quenched to room temperature by passage through cooling cabinets, over chilled rolls, or simply by exposure to the atmosphere. At this time the grippers are released from the film, edge trim is slit away, and the finished film is wound into rolls.

Accurate control of film temperature is absolutely essential since the amount of orientation effected by stretching is related inversely to the temperature, provided the temperature is above the second-order transition point. Stretching at high temperature has little beneficial effect upon orientation; stretching at temperatures just above the transition point is most effective. However, because of difficulties in working with most materials at these lower temperatures, intermediate temperatures are usually used in practice.

During processing it is also important that the thickness of the extruded sheet be maintained accurately. Thick and thin sections cool and stretch at unequal rates, tending to produce areas of anisotropic properties in the finished film. Probably the most important automatic tool for the measurement and control of film thickness is an electronic device known as a *beta*-ray gauge (Chap. 8). Absorption of radiation is proportional to film thickness, and the sensing head transmits to a recorder a signal proportional to film thickness. The signal can be either a point reading or an integrated scanning average. In the recorder the signal is compared with a predetermined target value. Appropriate adjustments to process conditions may then be made manually or automatically to keep film thickness uniform.

A significant shortcoming of the tenter frame is its inability to accomplish simultaneous biaxial orientation. Its design is such that a fixed distance exists between adjacent tenter climps, permitting no longitudinal elongation of the film. A number of disclosures have been made in the patent literature for modified tenter frames and clamps which will permit simultaneous biaxial orientation (24–30). While one or more of these devices may have been successfully developed on a pilot-plant or even commercial scale, their usage is not widespread at the present time.

Although it was mentioned earlier that there are some disadvantages to orientation by the flat-film methods, it should be pointed out that there are general points in favor of this technique over the tubular process (17): (*1*) there is no division of the melt stream through the flat extruder die, whereas the spider legs of a tubular die divide the melt stream and can cause thickness differences in the extruded film; (*2*) two-stage stretching is readily obtainable when required (in this case, independent control of temperature and amount of transverse and longitudinal stretch is assured); (*3*) the quench process is easier since one can readily cool both sides of the film by chill rolls or water bath; (*4*) high stretching forces may be more easily accommodated than in many bubble processes; (*5*) higher running speeds for a given output are useful for rapid quench and high rates of stretch; (*6*) annealing is a considerably simpler operation and results in film which lies flat. (See also Chap. 8.)

IV. Variables Affecting Orientation

The general methods of effecting orientation were briefly reviewed in the previous discussion. It is also important to consider the variables of the process. The amount of orientation imparted to a film is dependent upon the stretching temperature, the amount of stretching, the rate of stretching, and the quench (31). While it is difficult to isolate the total effect of each of these variables, the general trend of the effects has been established.   It has been found that orientation will be increased by any one of the following: decreasing the stretching temperatures, increasing the amount of stretch, increasing the rate of stretch, and increasing the amount of quench.

The importance of uniform film temperature and its effect have already been discussed.

The extent of orientation may be increased by increasing the amount of stretch.   In order to provide certain minimum strength properties, the amount of stretch may be critical.   For example, in the case of polystyrene, ASTM standards require a minimum orientation ratio of 2:1 in each direction.   Lower ratios may result in a brittle sheet.   Orientation beyond 3:1 in each direction does not give appreciable additional improvement in properties; thus, 3:1 is considered optimum for general-purpose polystyrene sheet. Special applications where additional strength is required in one direction even at the expense of decreased strength in the other direction may require nonuniform stretching.

Generally speaking, increasing the rate of stretch will increase the amount of orientation.   Film being oriented at a low temperature must be stretched at a slow rate.   Increasing the rate of stretching would, if exaggerated, exceed the mechanical strength of the material and cause fracture.

At the other extreme, where a hot sheet is stretched at a low rate, molecular relaxation will occur and most of the orientation will be lost before the quench is applied.   In order to obtain a reasonable degree of orientation, the rate of stretching must vary as the temperature of the sheet.

A quench is vital in that it is used to freeze the molecules in their strained configurations.   The amount of quench needed is dependent upon the temperature of the sheet and the rate of stretch, but in any case, the amount of quench must be adequate to lower the temperature of the film sufficiently below its transition point to preclude molecular relaxation and disorientation, and to limit crystallite growth.

For crystallite-forming polymers the oriented crystalline form is what is desired in the finished film.   It is then transparent, relatively tough, and has other advantages such as controlled heat shrinkage. The unoriented crystalline film is generally brittle and lacks transparency. Certain techniques have been perfected in the past for achieving the proper oriented product when making fibers and film. These techniques might be termed "classical" procedures for orientation (17). The orientation procedure for crystallite-forming polymers such as polypropylene would have the following sequence.

(a) Heat the polymer to above the melting point to destroy crystallinity, e.g., melt extrude.

(b) Quench to preserve the amorphous condition, insofar as possible, and to facilitate subsequent orientation.

(c) Orient by stretching at a temperature somewhat above the second-order transition temperature but below the crystalline melting point.

(d) Anneal if desired to obtain minimum heat shrinkage. For this the oriented film is restrained from shrinkage (or shrinkage is controlled) during heat treatment at a temperature usually, but not always, above the stretching point.

For polymers of relatively low crystallinity, such as polyethylene terephthalate, the orientation procedure would be as follows.

(a) Heat the polymer to a temperature well above the second-order transition temperature so that the polymer is well into the viscoelastic region. Extrusion is thus carried out and maximum transparency is obtained.

(b) Cool to the optimum temperature (elastic region) for stretching.

(c) Orient by stretching at this temperature.

(d) Anneal if desired to reduce thermal shrinkage. Again the annealing conditions must be carefully controlled to prevent extensive disorientation.

References

1. C. P. Fortner, *Mater. Design Eng.*, **50**, No. 7, 94 (1959).
2. A. V. Tobolsky, *Properties and Structure of Polymers*, Wiley, New York, 1960, p. 61.
3. C. W. Bunn, "Polymer Texture," in *Fibres from Synthetic Polymers*, R. Hill, Ed., Elsevier, New York, 1953, Chap. 10, p. 257.
4. S. L. Aggarwal, G. P. Tilley, and O. J. Sweeting, *J. Appl. Polymer Sci.*, **1**, 91 (1959).
5. J. W. Ladbury, *Trans. Plastics Inst.*, **28**, 184 (Oct. 1960).

480   JEAN B. MAURO AND JOSEPH J. LEVITZKY

6. K. Richard, G. Diedrich, and E. Gaube, *Kunstoffe*, **49**, 671 (1959).
7. E. Studt and U. Meyer, U.S. Pat. 2,074,285 (Mar. 16, 1937).
8. H. Horn, Ger. Pat. 654,757 (Jan. 13, 1938).
9. M. O. Longstreth and D. W. Ryan, U.S. Pat. 2,695,420 (Nov. 30, 1954).
10. M. O. Longstreth and D. W. Ryan, U.S. Pat. 2,697,248 (Dec. 21, 1954).
11. M. O. Longstreth and D. W. Ryan, U.S. Pat. 2,770,007 (Nov. 13, 1956).
12. R. F. Pierce, U.S. Pat. 2,632,206 (March 24, 1953).
13. H. H. Tornberg, U.S. Pat. 2,433,937 (Jan. 6, 1958).
14. F. H. Reichel, U.S. Pat. 2,346,187 (Apr. 11, 1944).
15. M. R. Gerow, U.S. Pat. 2,720,680 (Oct. 18, 1955).
16. W. J. Johnson, U.S. Pat. 2,668,324 (Feb. 9, 1954).
17. J. Jack, *Brit. Plastics*, **34**, 312 (1961).
18. K. L. Knox, U.S. Pat. 2,718,666 (Sept. 27, 1955).
19. V. Veiel, A. Prietzschk, and H. Schnell, Union So. Africa Pat. 4642/58 (Dec. 8, 1958).
20. W. H. Ryan, U.S. Pat. 2,505,146 (Apr. 25, 1960).
21. P. S. Blatz, Ger. Pat. 1,084,021 (June 23, 1960).
22. F. E. Wiley, R. W. Canfield, R. S. Jesionowski, and J. Bailey, U.S. Pat. 2,412,187 (Dec. 3, 1946).
23. J. Bailey, U.S. Pat. 2,297,645 (Sept. 29, 1942).
24. W. R. Tooke and E. G. Lodge, Jr., U.S. Pat. 2,923,966 (Feb. 9, 1960).
25. S. M. Clark, U.S. Pat. 2,778,057 (Jan. 22, 1957).
26. F. P. Alles and K. A. Heilman, U.S. Pat. 2,728,941 (Jan. 3, 1956).
27. A. Fior and L. Malfetti, Union So. Africa Pat. 4229/58 (Nov. 5, 1958).
28. G. I. M. Bloom, W. A. Holmes-Walker and D. J. H. Sandiford, Brit. Pat. 849,436 (Sept. 28, 1960).
29. W. A. Holmes-Walker and D. J. H. Sandiford, Brit. Pat. 840,819 (July 13, 1960).
30. W. S. Thompson, U.S. Pat. 2,719,323 (Oct. 4, 1955).
31. M. L. Evans, *Package Eng.*, **4**, 42 (Aug. 1959).

CHAPTER 11

WEB-COATING EQUIPMENT AND APPLICATIONS

GEORGE L. BOOTH

Dilts Division, The Black-Clawson Company, Inc., Fulton, New York

I. Introduction

This chapter will deal with the coating of a web and the term *coating* in itself is generic in that there are many different ways in which an object or material can be coated. For the purpose of this chapter, however, web coating will be defined more explicitly as the process of uniformly applying a liquid or plastic material in a controlled manner to a moving web in order to develop desired characteristics in or on the web.

To avoid confusion as to what constitutes a web, we shall define a web as any material capable of being handled from an unwinding parent roll through a process and then of being rewound in roll form, cut into sheets, or converted into the consumer product. Normally, a web will fall into one of the following categories:

1. Paper
2. Paperboard
3. Cellulosic films
4. Plastic films
5. Textiles
6. Metals
7. Metal foils
8. A laminate of one or more of the above

A sheet of material cannot be considered as a web, although many of the processes that are to be discussed can be adapted to sheet-fed coating operations.

There are many reasons for the coating of web materials which include: printing quality improvement, brightness development, decorative properties, promotion of adhesion to another material, color change, adhesive for lamination or gluing to another web or object, gloss development, antiblocking characteristics, slip characterisitics, antiadhesion characteristics, film casting on a carrier web, intelligence recording (e.g., office copy paper, photographic film, magnetic tape), and barrier properties (e.g., grease holdout, reduction of water vapor transmission, and reduction of gas flow through a web).

This chapter will be divided into two parts: (*1*) the various types of coaters and the mechanics involved therein (1–6) and (*2*) the application of coaters to various types of coating classifications.

II. Methods and Equipment

A. Squeeze Roll Coater

The squeeze roll coater is one of the basic methods of coating in use today. It can be either a two-roll or a three-roll squeeze system, as shown in Figure 1.

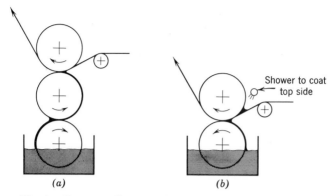

Fig. 1.   Squeeze roll coater: (a) three-roll; (b) two-roll.

The principle of operation is based on squeezing the web material between two rolls in order to control the amount of coating applied to the web.   With a three-roll squeeze coater, only one side is coated, since the bottom nip formed by the furnish and applicator rolls is used as a premetering system for the coating to be applied in the nip formed by the backup and applicator rolls.   The only means of mechanically controlling the coating weight is by the amount of pressure applied in this nip.   Other means of coating weight control include solids and viscosity adjustment of the coating, coating and web temperature, and hardness of the rolls in the squeeze roll system.   In some cases, there is a differential speed between the respective pairs of rolls which also will influence the amount of coating applied to the web.

The vertical roll arrangement is not the only method used.   In some cases, the furnish roll is offset either obliquely from the applicator roll or on the same horizontal plane with the applicator roll.   There are cases in which the furnish roll becomes, in effect, the metering roll and runs in the same rotational direction as the applicator roll for the metering of coating.

Uses of the three-roll squeeze coater include paraffin wax operations, priming of a web prior to an extrusion coating operation with polyethylene, adhesive applicator for wet lamination, and lacquer coatings such as cellulose acetate butyrate.

The roll construction is such that each nip has one resilient roll.

In the case of the three-roll squeeze coater, the center roll can be rubber-covered, which would satisfy this requirement, or the furnish and backup rolls could both be rubber-covered to satisfy the same requirement. In some cases, the nip is made up of two metal rolls, although this is not common.

Figure 1 also shows the two-roll squeeze coater, which follows in principle almost every aspect of the three-roll coater. The primary differences between the two is that there is no premetering of the coating applied to the bottom side of the two-roll coater and the very common use of a shower pipe to apply coating to the top side of the sheet so that there is a simultaneous two-side coating application to the web.

It is normal to have the two rolls mounted horizontally for the simultaneous two-side coating application.

The two-side squeeze coater is used for the application of starch size and lightly pigmented coatings on paper and paperboard machines. The coater can be used for adhesive applications, paraffin wax applications, and other relatively low-viscosity coatings. It is common practice to follow a two-roll squeeze coater with a smoothing device on each side of the web following the coating application. This is done only to remove the roll pattern generated by film splitting of the coating between the web and the roll upon emerging from the nip.

In some special cases, a four-roll squeeze coater is used when premetering is desired for coating application to both sides of the web.

B. Kiss Coater

There are basically four types of kiss coaters in use. In almost no case is a kiss coater used as a complete coating machine in itself, since the means of metering and smoothing the coating is insufficient.

The principle involved is simple. The web is run over the applicator roll without any backup behind it. The amount of wrap around the applicator roll and the viscosity of the coating are the two primary factors governing the amount of coating applied to the sheet.

In Figure 2, four common types of kiss applicators are shown. The single-roll applicator is the most basic of the systems and almost

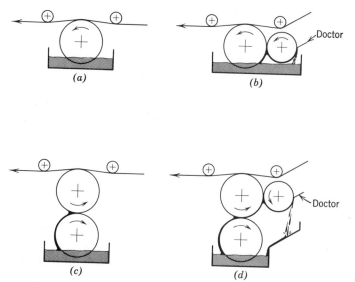

Fig. 2.   Kiss coater: (a) single roll; (b) two-roll system; (c) two-roll squeeze meter; (d) three-roll.

always has a metering or smoothing device, or both, following it. The applicator roll is usually run in the web direction at a speed less than the web speed.   There are cases in which the applicator roll is run faster and, of course, there are cases in which the applicator roll runs in the reverse direction to web travel.

A modification of the system is the two-roll kiss applicator, one roll being used as a metering roll in conjunction with the applicator roll.   A system of this type gives relatively accurate control of coating application to the web.

The same holds true with the squeeze meter system shown in Figure 2c.   In this case, squeeze metering of the coating formed in the two rolls will limit the amount of wet film thickness going up to and being applied on the web.

A refinement of the two-roll system is shown in Figure 2d—the three-roll kiss applicator, which gives a slightly more accurate metering of the wet film being carried on the applicator roll to the web.   The real advantage of such a system is that the overall machine line speed has relatively little effect on the splash or coating thrown from the applicator system, since it is possible to run the furnish roll at a reduced speed with respect to the applicator roll.

Roll construction in these various systems will be metal or rubber-covered or a combination of both. In the case of the two-roll reverse metering system, normally both rolls are metal with fairly close tolerances, in order to effect good, uniform metering. In the case of the two-roll squeeze metering system, one roll usually is rubber covered in order to form a squeeze-metering nip. In the case of the three-roll system, the roll materials will vary with the application, although the metering and applicator rolls should be of metal construction with fairly close tolerances.

As stated earlier, the kiss applicator system is almost always followed by a metering or smoothing device. These devices may be an air doctor, a flexible blade, a wire-wound or smoothing rotating rod, a rigid knife, or a small-diameter smoothing roll running in contact with the web. The system is quite sensitive to web tension and solids, as well as to the condition of the web itself. A wide variety of coatings can be applied by using the kiss applicator system. It is somewhat limited by the coating viscosity that can be used. The metering device which follows the kiss applicator system will influence the viscosity range of the coating.

C. Dip Roll Coater

The dip roll coater is usually employed for a saturation or impregnation process. It is a relatively simple coater to operate from a mechanical standpoint, in that the material runs into the coating

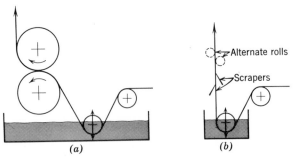

Fig. 3. Dip roll coater: (a) dip and squeeze; (b) dip and scrape (or smooth).

bath and comes out immediately into the metering system. Two typical dip coaters are shown in Figure 3. In one case, a web is dipped into the solution and immediately run through a squeeze nip where excess coating is removed and a certain amount of impregnation occurs, owing to the squeeze of the nip itself. A coater of this type is ordinarily used for phenolic impregnation and in the manufacture of waxed papers such as bread wrappers.

In the other part of the diagram the dip-and-scrape (or smooth) system is shown. The web is run through the coating bath, and upon emerging from it, is scraped on each side with a blade. It can be metered by use of rolls mounted one on each side of the web. In some cases, a combination of scrapers and rolls is used, the scrapers being employed primarily to limit the amount of coating being brought up to the rolls or nip.

The dip-and-scrape system is used for phenolic resin saturation as well as for latex saturation of papers such as gasket materials.

One requirement of the dip coater is to have a roll which will move upward out of the coating solution to allow the web to be threaded. It is also necessary to remove the web from the coating solution on machine stops, unless the web is material that will not soften and break in the solution.

In almost all cases of dip coating, both sides of the sheet are coated. Saturation of the web will occur if the web is absorbent. Under some conditions, it is possible to coat the web on only one side by using a dip method of application. This is accomplished by wrapping the dip roll more than 180° or positioning it so that over half of the roll is out of the coating solution. In this way, the coating will not tend to pocket between the web and the roll and coating application on one side is possible. Coating only one side of a web is cumbersome with this method and should be discouraged.

D. TRANSFER ROLL COATER

Transfer roll coaters have been used successfully since about 1935 for the coating of paper on the papermaking machine. During the period of development of the transfer roll coater, the large paper mills were very secretive in the methods employed for the purpose of coating paper on a machine. There are eight systems commonly employed, yet all methods use basically the same principle, that is, metering of the coating through a series of nips and

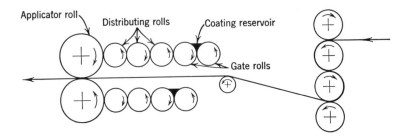

Roll	Diameter, in.	Rubber hardness, P&J ball ⅛ in. diameter
Applicator	32–48	55–60
Distributing	18–20	160–180
Gate	18–20	90–100

Fig. 4.   Consolidated transfer roll coater.

finally application of the coating to the web (14–19). Although eight systems commonly are used, only two such systems will be described.

The first system is the original transfer roll coater called the Consolidated roll coater or the Massey coater. The coater is shown in Figure 4 for simultaneous two-side application of a mineral-pigmented coating to a paper web. The machine is a transfer roll coater having several squeeze nips in which the coating is continuously split and worked from one roll to another. Many different versions of this coater are used today in which some rolls are chromium plated, some are rubber covered, and some oscillate and run at differential speeds. Normally, the gate rolls run at a slow speed, the two applicator rolls running very close to web speed. In all cases of simultaneous coating application to both sides, one roll is run at a speed slightly greater than web speed in order that the sheet will follow the faster roll and thus prevent flutter of the sheet emerging from the nip, which would cause bar marks on the freshly coated sheet.

The coater is used for water-based mineral-pigmented coatings to develop printing quality in the sheet. There are cases where functional coatings can be applied with the transfer roll coater, but these are definitely limited.

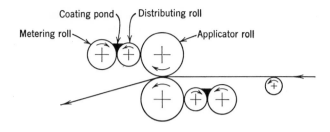

Roll	Diameter, in.	Rubber hardness, P&J ball ⅛ in. diameter
Applicator	32–42	50–75
Distributing	10	Metal chromium-plated
Metering	24	35–50

Fig. 5.   KCM Transfer roll coater.

The machine handles mineral-pigmented coatings of up to 65% solids and has been known to operate in excess of 1600 fpm.

Another version of the transfer roll coater is shown in Figure 5. The machine is commonly referred to as the KCM coater (Kimberly Clark-Mead) and is used again for identically the same purpose as the Consolidated coater.   The main differences between the two transfer roll coater systems are that the KCM coater uses variable-diameter rolls, as shown, and much higher nip pressures which can be up to 250 lb/linear in.   The ratio of roll diameters for metering efficiency and smoothing efficiency is employed in this coater. Other than these minor differences in metering systems, the end use, the products and coatings, and other characteristics are essentially the same.   The user of a coater system such as this must be very careful both in formulation and in operation to keep the roll pattern on the coated sheet to a minimum.   This is done through the metering system, coating formulation, coating tack, and other variables.

E. Gravure Roll Coater

Gravure coaters are among the most commonly used web coating machines today.   They are simple to operate and deposit a relatively uniform wet coating film.   The speed range of these units is variable, so they can be used in conjunction with other processes within one continuous web handling system.

The principles involved in gravure coating are simple. A knurled or etched roll having a uniform pattern is used as the metering system in conjunction with a doctor blade. Excess coating is applied to the roll either through a furnish roll or by dipping a gravure roll in the coating after which the excess is doctored off the surface of the roll, leaving the interstices of the gravure cells filled with the coating. The wet coating is then applied directly to the web in the case of direct gravure coating, or, in offset gravure, is transferred to a resilient roll and then to the web through a nip furnished by the top roll and a resilient roll.

Within a limited range, the coating weight can be changed through adjustment of coating solids. For a given gravure pattern, the wet coating thickness applied is always the same and the solids therefore must be adjusted to modify the dry coating thickness for a given gravure roll. The usual method of changing the coating weight is to change the gravure cylinder itself. Deckle control on a gravure coater is accomplished by undercutting the web backup roll or the transfer roll in the case of offset gravure operations.

Figure 6 shows a two-roll, direct gravure coater and a three-roll, offset gravure machine. It should be understood that in both these cases an additional roll can be used as a furnish roll to supply

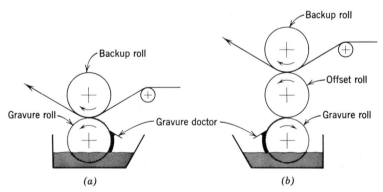

Fig. 6.   Gravure coater: (a) direct gravure; (b) offset gravure.

the gravure cylinder with sufficient coating. This is done in the case of a high-speed operation in order to cut down coating throw and splash in the pan. The furnish roll can be run at a greatly reduced speed with respect to the other two or three rolls in the system, which must run nearly at, or exactly at, the web speed.

Figure 7 shows patterns of engraving which are typical of those available for selection of the proper gravure pattern. The triangular cell is commonly recommended for offset gravure operation, the quadrangular cell is used for direct gravure operation, and the helicoid pattern is normally used for relatively viscous materials such as hot-melt coatings.

Another type of cell not shown on this chart is a hexagonal cell having a flat bottom which is used for mineral-pigmented coatings.

In the case of offset gravure coaters, it is not unusual to employ two such units for simultaneous coating application. This installation is found in the mineral-pigmented coating industry for the most part.

The gravure coater is used for priming or lacquer coating of

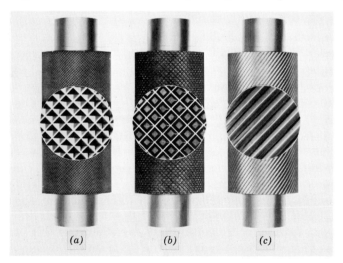

(a)   (b)   (c)

Fig. 7. Types of engraved applicator rolls: (a) pyramidal cell for offset application, usually for light- and medium-weight coatings; (b) quadrangular cell for direct application, using a doctor blade; (c) triangle-helix cell for heavy applications, using a doctor blade. (Courtesy Modern Engraving and Machine Co., Hillside, New Jersey.)

aluminum foil and other metal foils, application of adhesion-promoting salts in polyethylene extrusion coatings, hot-melt coatings such as polyethylene–wax blends, solvent system silicone emulsions, magnetic tape manufacture, nitrocellulose coatings, paraffin wax for slip promotion, shade cloth manufacture, and adhesive application in wet lamination processes.

F. REVERSE ROLL COATER

The reverse roll coater is the most versatile coater in use today. It is also one of the most difficult coaters to handle both from an operational and maintenance standpoint. The range of products that can be manufactured by a reverse roll coater is almost limitless.

A principle of reverse roll coating is to premeter an amount of coating by using a preset orifice between two accurately ground and precisely held rolls and then to apply the coating to the web immediately after metering. In the case of both metering and applicator nips, the rolls must be running in a counter direction to each other in order to achieve a true reverse roll operating condition.

Figure 8 shows four commonly employed reverse roll coaters. The pan-fed three-roll reverse roll coater is a relatively simple coater to operate. The coating is contained in a pan and has no means of edge deckle control. This can be a disadvantage as well because relatively thin web materials cannot be satisfactorily processed by the three-roll pan-fed reverse roll coater since coating will transfer from the applicator roll to the backup roll and cause problems. In the case of solvent vehicle coatings, the rubber backup roll is susceptible to attack by solvents, manifest by deterioration or swelling or both. This condition cannot be tolerated for long because the rubber roll will begin to chafe on the edges and eventually will be destroyed.

The pan-fed four-roll reverse roll coater eliminates this problem by permitting the use of edge doctors to maintain deckle within the width of the web being coated. It is also possible to reduce the speed of the furnish roll well below that of the applicator roll, thus preventing excessive agitation of the coating in the pan.

The four-roll coater has more exposure area for the coating and therefore can present problems when highly volatile solvents are

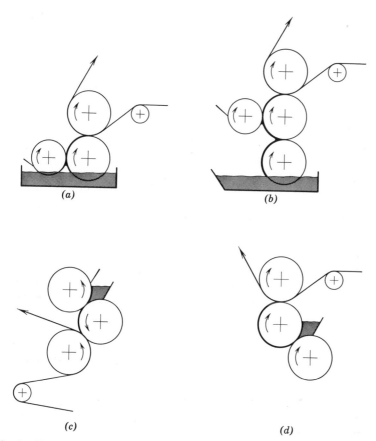

Fig. 8.   Reverse roll coater: (a) pan-fed three-roll; (b) pan-fed four-roll; (c) nip-fed three-roll; (d) nip-fed three-roll.

used.   It is possible to enclose the coater with shields to reduce solvent loss.

The nip-fed three-roll reverse roll coater is a very commonly employed unit.   Two versions of this coater are in use.   One has the coating puddle in the top nip formed by the metering and applicator rolls.   The rubber backup roll is in the bottom position and, as such, sometimes presents problems, in web lead takeoff from the rubber roll to the dryer.   Visual observation of the coating on the web is also restricted. The bottom-nip-fed reverse roll coater eliminates these problems and thus is more desirable. The coating exposure on the applicator roll is about twice as long as that for the top-nip-fed reverse roll coater, however, and can potentially cause problems with highly volatile solvent coatings.

Regardless of the type of reverse roll coater used, the rolls forming the metering nip, namely, the applicator and metering rolls, must be very precisely ground. Chilled cast-iron chromium-plated rolls are used almost exclusively in this position because of their inherent structural stability and close-grained metal. It is possible to obtain rolls that have virtually no dynamic runout or nip variation. Both antifriction and hand-scraped plain bearings are used in reverse roll coaters. Bearing accuracy must be the best obtainable in order to maintain precision in the metering nip.

The applicator roll usually is run at a speed in excess of web speed in order to deposit a smooth coating on the web. The wipe ratio, which is the ratio of applicator roll to web speed, will range from 1:1 up to 3:1 in many cases. The wipe ratio is almost never less than 1:1.

The metering roll speed normally will be from 1/3 to 1/50 of the applicator roll speed and, in some isolated cases, the metering roll may not be rotated except for the cleaning of streaks out of the metering nip.

When a furnish roll is used, it will run from about 15 to 60% of the applicator roll speed.

The drive system employed for reverse roll coaters is somewhat involved because both the furnish roll, when used, and the metering roll are equated to, and, in fact, under control of, the applicator roll speed. For instance, a coating machine may be running with a wipe ratio of 2:1 at a speed of 100 fpm and have a metering roll speed of 15 fpm, and a furnish roll speed of 30 fpm. If the machine line is increased to 200 fpm, the applicator roll then will be rotating at a linear speed of 400 fpm, the metering roll will run at 30 fpm, and the furnish roll at 60 fpm. Now, it may be desirable to reduce the wipe ratio to 1.5:1, in which case the applicator roll through its mechanical drive or electric drive control would be reduced to 300 fpm. It is imperative that, when this happens, the furnish roll be reduced to 45 fpm and the metering roll to 22.5 fpm. Such systems are commonly available and in general use with reverse roll coaters.

The products manufactured on reverse roll coaters are almost limitless. For example, artifical leather, shade cloth, office copy papers, magnetic tape, silicone coatings, vinyl dispersion coatings, cardiograph paper, electrophotographic coatings, and others are all manufactured on reverse roll coaters. The reason for the diversity of products lies in the extreme versatility of the reverse roll coating principle. It must be thoroughly understood, however, that this type of coater requires a better than average operator with a complete understanding of the machine in order to operate it satisfactorily. In comparison with other types of coaters, maintenance is high because of the cost of the chilled cast-iron rolls and their susceptibility to damage through carelessness on the part of the operator or mechanical repair crews. Perhaps a good principle to follow in the selection of a coater is to try to determine a way whereby a product can be made without employing a reverse roll coater and to use a reverse roll coater only as a last resort. Even when this is the case, it will be necessary to employ a reverse roll coater for a wide range of products.

G. CALENDER COATING

Coating of a web material by using a plastics calender is used to a limited degree and only in the case when the equipment is already available. The use of such an expensive unit for the coating of a web is not economically sound, although it is done because of market considerations and the fact that the user may have a plastics calender in his plant.

The plastics calenders generally fall into two groups: the inverted L calender and the Z calender. They are shown in Figure 9. The rolls in a calender will measure from 24 to 30 in. in diameter and are made of chilled cast iron. The rolls are precision ground and in most cases are hand ground at the startup of the machine in order to give the precise gauge control required for various plastic films. From the two nip-forming rolls containing the plastic material to the final roll in the calender stack, speeds and temperatures increase progressively. When a coating operation is to be performed, the web material—almost always a textile—is preheated and fed into the final nip where the calendered plastic

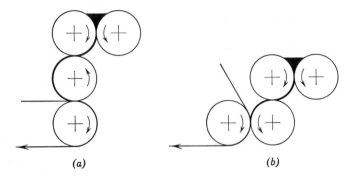

Fig. 9.   Plastics calender coating: (a) inverted L calender; (b) Z calender.

film is applied in the form of a coating. Frequently, the fabric has been precoated for adhesion promotion of the plastic film to the textile. An operation of this type is normally used for vinyl coating of textiles or in the manufacture of fabric-based friction tape. Since the coating material is always carried in its own plasticizer and at elevated temperatures, no solvent is required.

H. Extrusion Coating

Extrusion coating is one of the most commonly used means of processing pelletized thermoplastic resins for the coating of a web. The system which converts pellets into a coating on a web consists of three fundamental units: extruder, die, and laminator section.

The extruder is a heated zone tube of very hard and abrasive-resistant metal called Xalloy. A screw revolves in the extruder barrel, pressurizing slowly, heating, and feeding the resin in molten form to the die. The extruder receives heat from two basic sources—internal and external. The internal heat is developed through compression of the resin and work done on it, while that from external sources comes from heaters on the outside of the barrel. The external heaters can be of the resistance or induction type. As the pellets flow through the extruder, they become compressed and heated, and form a hot molten resin which is screened at the end of the extruder through a breaker plate and a screen pack. There are many parameters to be considered for an extruder, such as the length-to-diameter ratio which sometimes exceeds 28:1, the depth of the screw flight which can range from 0.175 to 0.090 in., and a lead on this screw which is quite variable. Means are incorporated in more modern extruders which chill the extruder from a maximum temperature of 600° down to less than 300°F, in case the process has to be stopped. Since most thermoplastic resins will oxidize or degrade at elevated temperatures, it is imperative that the temperature be reduced

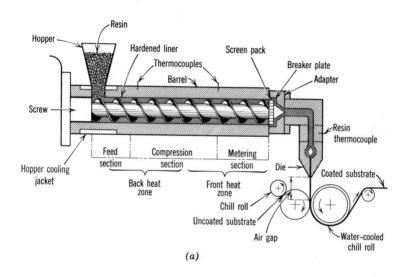

(a)

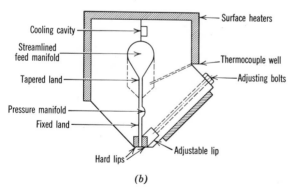

(b)

Fig. 10.   Extrusion coating: (a) extruder and die; (b) die cross section.

rapidly. Casting by melt extrusion is considered in detail in Chapter 8.

Many types of dies are in use, such as the blown-film die, wire-coating die, pipe-forming die, and sheeting die. In the extrusion coating process, the sheeting die is the only one used and therefore will be considered. It consists of a pair of heated plates through which the molten thermoplastic resin is fed and emerges in sheet form through a slot orifice for lamination to the web. Normally, the sheeting die has a slot orifice of 0.020 in. for most coating resins. It is heated by either resistance or induction heating sources and is adjustable in width to the slot orifice. It is also possible to profile the sheeting die in order to maintain a uniform film thickness in the process. Both extruder and sheeting die are shown in Figure 10. Deckle control in the die is of great importance and is a relatively difficult operation mechanically. Leakage is one of the basic problems of deckle control in a sheeting die. As long as this leakage does not affect or impair the process, it can be considered adequate.

The laminator shown in Figure 11 has the basic function of pressing molten thermoplastic resin, in extruded sheet form, to the web material. The pressure in the nip can range up to 250 lb/linear

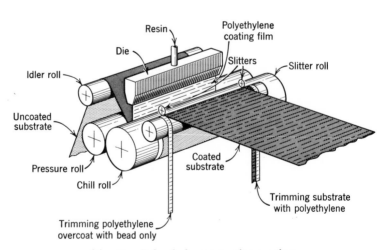

Fig. 11.   Polyethylene extrusion coating.

in. in order to obtain satisfactory adhesion of the thermoplastic resin to the web. In many cases, it is necessary to prime-coat the web material to promote adhesion of the thermoplastic resin to the web. Adjustments are built into the laminator section to allow the molten resin to touch first either the chromium-plated chill roll or the fed web material when entering the nip.

The laminator section consists of a rubber backup roll and a chromium-plated, chilled roll of large diameter for reducing the temperature of the molten resin from 600°F or more to a temperature at which it can be stripped from the chill roll, i.e., at 120–140°F.

The rubber pressure roll usually has a silicone rubber covering in order to prevent adhesion of the hot thermoplastic film. It is cooled internally for protection and can be undercut so that the plastic can be extruded beyond the web edge. The plastic extruding beyond the web is then trimmed and reclaimed.

The rubber roll is backed up by a chromium-plated chill roll to reduce the surface temperature of the rubber roll. Since rubber is heat sensitive, it is necessary to keep the surface temperature of the rubber-covered roll below the critical degradation point of rubber.

Today, milk cartons are almost exclusively coated with polyethylene and they represent a large tonnage of extrusion-coated product. In addition to this, one-serving sugar bags and many frozen-food wrappers are extrusion-coated, as well as a multitude of wrapping papers and plastic films. The science of extrusion coating continues to grow and, with the advent of new and different polymers, it can be expected that the extrusion coating process will increase in popularity and use.

There are isolated cases in which a hot solvent coating can be extruded. This is primarily for decorative purposes rather than for the functional purpose which has been discussed. The speed of extrusion coating operations depends on the capacity of the extruder, the thickness of the extruded film laminated to the web material, the material handled, and the ability of the thermoplastic resin to stretch or neck down without developing pinholes.

Although there are disadvantages to this method of coating, its growth is assured because of its ability to handle thermoplastic resins of high molecular weight.

I. Curtain Coater

The concept of passing a web material or sheet under a coating curtain is relatively old. During the late 1950s, the idea for coating of sheets or die-cut blanks was developed and perfected to a considerable degree.

Two basic systems of curtain coaters are in common use. One is the pressure type of feed system where the coating is held under a controlled pressure either above or below atmospheric pressure and the other is a weir type of curtain which is open to the atmosphere and maintains a curtain only by the flow of coating over the weir. The two systems of curtain coating are shown in Figure 12.

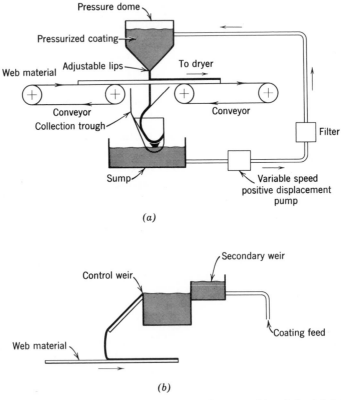

Fig. 12. Curtain coater: (a) pressure feed coater; (b) weir feed detail.

In the case of the pressure curtain, coating is pumped at a controlled feed into a pressurized chamber, generally V-shaped, at the bottom of which is a set of adjustable lips which form the curtain that drops by gravity downward toward the web. In most cases, the pressurized chamber will have an air cushion on top of it so that the chamber can have some resiliency and maintain nearly uniform pressures. Under a strictly hydraulic condition, it is much more difficult to maintain uniform pressure.

The open or weir feed curtain coater simply maintains an accurate weir over which the coating flows. Here, again, the coating is fed at a precise rate in order to maintain uniform flow over the weir.

In both types of curtain coaters, it is necessary to have a means of edge deckling which normally takes the form of a thin wire dropping vertically from the feed system into the gutter, along which the curtain will form very much as a soap bubble forms around a ring. In this way, the amount of neckdown is controlled for the curtain. Excess coating, in the case of sheet feed, drops into a gutter which is designed to reduce the amount of air entrainment and to allow the coating to flow back to the reservoir whence it is again pumped back to the curtain coater reservoir. Since air entrainment can be a very critical problem with this type of coater, extreme care is used in reducing to a minimum the amount of air entrainment. A self-cleaning filter is incorporated in the coating feed system to insure clean particle-free coating.

The curtain coater must have a coating which will form a film in its wet or liquid stage. The coating can be carried in water, in an organic solvent, or be a hot-melt system, since all three can be made to form a wet film.

The curtain coater has found its greatest use in the corrugated box industry and the insulation board industry. Since it is ideally suited to the coating of irregular surfaces and sheets, it is extensively used in these industries.

Attempts have been made to use this coater in continuous web-fed systems. Because of the problems in feeding and controlling the coater over a long period of time, it has not been used extensively for such processes.

Some of the operating adjustments include the liquid level or pressure in the coating reservoir, the opening of the lips forming the gap through which the coating flows, the speed of the web or sheets passing through the curtain, and the height of the curtain-forming device from the material to be coated.

It is interesting to note that some installations have dual coating heads so that one can be pulled out of coating position to be cleaned while the other is in operation.   Shutdown time for cleaning is held at an absolute minimum under these conditions.

J. KNIFE COATER

Knife coaters are among the oldest and most commonly used means of performing coating operations.   There are many different types of knife coaters and knife contours, depending on the ultimate product being manufactured.   Web material, coating material,

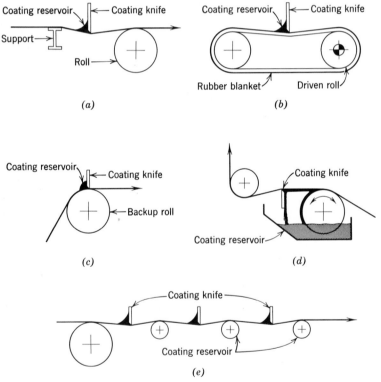

Fig. 13.   Knife coaters: (*a*) unsupported knife; (*b*) knife over blanket; (*c*) knife over roll; (*d*) inverted knife; (*e*) sequential knife.

specifications for the finished product, and speed of operation all influence the type of knife coater system and knife contour to be used. Generally, knife coaters are ideally suited to short runs because of the very minimum cleanup required. They do have the disadvantage of not operating at high speeds, and lack the ability to apply a relatively heavy thickness of wet film. On the other hand, a knife coater can be designed to coat a porous web material satisfactorily with a minimum of strike-through of coating to the opposite side.

The principle of knife coating is basically very simple in that the coating is simply "buttered" onto the web. Highly viscous materials can be handled satisfactorily with knife coaters.

Five fundamental types of knife coaters are shown in Figure 13. They are the unsupported knife coater, well suited for relatively strong webs; the knife over blanket, which is ideally suited for relatively weak webs; the knife over roll coater, which maintains a minimum amount of coating on the web; the inverted knife, which again is well suited for porous web materials; and the sequential knife coater, which is shown as having three unsupported knives. The primary advantage of the sequential knife coater is its ability to apply a fairly heavy thickness of wet film, mainly because there are several knives in a row.

Coating feed to the knives, as shown, can be either through a pump, or by manual hand dipping and feeding of the coating to the puddle directly behind the knife. Edge deckle control is maintained with a set of relatively simple edge dams that attach to the knife holder and can be moved in to suit the web width being run.

An extremely wide range of products is made with the knife coater, including magnetic tape, imitation leather products, cast films using a carrier web which is then stripped off the carrier web upon drying, pressure-sensitive tapes, industrial fabrics, fabric-backed wall coverings, shade cloth, inexpensive shoe materials, and nitrocellulose coatings for both barrier and decorative properties.

K. ROD COATER

There are many forms of rod coaters available (20,21). All have one or more features that are desirable, perhaps, but the coating

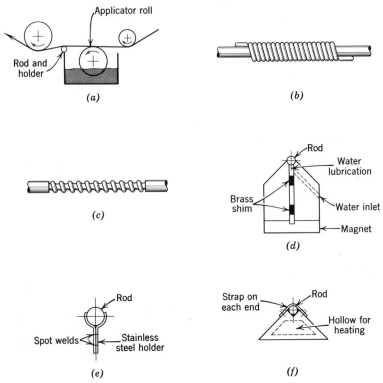

Fig. 14.   Rod coater and components: (a) basic rod coater; (b) wire-wound rod; (c) grooved rod; (d) magnetic holder; (e) Warner holder; (f) simple holder.

principle is not changed to any great degree.   Basically, the coater consists of a coating applicator, usually a single roll, as shown in Figure 14, followed immediately by a slowly rotating rod which both meters and smooths the coating prior to leaving the coater.

There are three metering rod designs used for this coating operation: namely, a smooth stainless steel or chromium-plated rod, a rod having wire wound around it (commonly called a wire-wound rod), and a grooved rod which actually has very fine grooves in it.   Rods will range from $\frac{1}{4}$ to $\frac{3}{4}$ in. in diameter.

It is possible to have more than one roll in the applicator system. In reference to kiss coaters, almost any one of the kiss coaters previously described would serve adequately as the coating appli- cator preceding the rod metering device.   In some cases, the entire applicator and rod system is mounted on a raise–lower carriage in such a manner that the complete coater can be removed from contact with the web.   With a system of this type, it is also possible to vary the degree of angle contact with both the applicator roll and the rod.   The applicator roll will have from 1 to 4 in. of laydown of the web on it with the attack angle being anywhere from 0 to 15°.

The rod will have from 10 to 30° contact with the web for the metering and smoothing operation.   The smaller the angle, the greater is the amount of wet film coating left on the sheet. In the case of wire-wound rods, the wrap angle is not the major influence in coating weight—rather, the size of the wire wrapped around the rod will influence the amount of coating left on the sheet.   The same holds true for a grooved rod; the larger the groove, the greater the amount of coating that is left on the web. The smooth rod is influenced mostly by the amount of web wrap and angle around it.

The kiss coater and rod metering device are very sensitive to tension changes and therefore require very good tension control in the web through the coater system in order to maintain uni- formity of coating weight.

Several holders are in common use.   The magnetic holder is ideally suited for mineral-pigmented coatings as it has a water purge system which both lubricates and cleans the rod during operation.   Changing of the rod is very simple.   It is simply dis- engaged from its drive and removed from the magnetic holder and a new one is inserted in its place.

The Warner holder, which has been used extensively, is also an excellent means of holding a rod.   It is a dry operation and it obtains its lubrication from the coating itself.   The major problem here is that the rod will sometimes pop out of the holder and cannot be reinserted without causing damage to either the rod or the holder.   The replacement of this rod involves a complete change of the rod holder and the rod itself.

The simple holder takes many different forms, but basically is held at each end outside of the web material being coated.   It relies on both a deep groove and the downward force vector from web tension to hold the rod in the groove itself.   This particular

506 GEORGE L. BOOTH

holder lends itself quite well to heating for use in hot-melt coatings. Again, changing the rod is a simple matter in that the two end clips are loosened and the rod is removed and replaced with a new rod.

The products made with the rod coater include hectographic reproduction paper, carbon papers, mineral-pigmented coatings, prime coatings for various multiple coating operations, and waxed paper products.

L. Spray Coater

Spray coaters are used to a limited degree in web coating operations. It is generally conceded that if there is another satisfactory means by which coating can be applied to a web, the spray coater will not be used because of the problems of misting, overspray, patterning, and maintaining coating uniformity. Spray coating is ideally suited to the coating of irregular surfaces, sheets, and irregular objects that are three-dimensional.

Two types of spray coaters are commonly used. The first is called the Bracewell spray coater and is used almost exclusively for mineral-pigmented coatings on printing papers. In this process, coating is literally thrown off an extremely high-speed rotating roll, so that the coating leaves in spray form and is then transferred to a second roll which repeats this process to reduce the particles in size. From the second high-speed rotating roll, the coating is then thrown to the web material being passed directly over it.

The other type of spray coater is the gun type, which is mounted on a carriage with several spray guns and traverses the web being passed underneath. Normally, more than one carriage on which several guns are mounted is used to maintain product uniformity and to reduce the problem of shutdown when one spray plugs.

The Bracewell spray coater is used only for pigmented coatings having a water vehicle, the product being high-quality printing papers. The gun-type sprayer is used for shingles, irregularly surfaced objects such as insulation board, and, in some cases, corrugated sheets and die-cut shapes.

M. Air Doctor Coater

The air doctor coater was developed and perfected in the paper industry during the 1930's at the S. D. Warren Company, Westbrook,

Maine (12). Its development was primarily for use in pigmented-mineral coatings used in high-quality printing papers. Since then, many different designs of air doctors have become available (7–11). The products manufactured by using the air doctor have increased substantially. During the 1950s, the air doctor coater was developed and applied for use with solvent coatings and it has been used successfully in this operation ever since.

The principle of operation of the air doctor coater is one wherein an excess amount of coating is applied to the web by using a kiss coater system; excessive coating is carried by the web around a roll and past the air doctor which issues a high-speed narrow jet of air for doctoring off the excess coating. Air doctor pressure will range from 3 oz/in.^{2} to 8 lb/in.^{2}. Air jet velocity at these pressures will sometimes approach the speed of sound.

The web material absorbs the liquid from the coating, and a filter cake deposit or semidried coating is left at the surface of the web. It is at this interface of the filter cake and the still-fluid coating that the shear and removal of excessive coating occurs.

There are several problems inherent in a system of this type. One of these is the need to employ a coating of low viscosity in order to allow the air doctor to operate satisfactorily. Misting and spray from the air doctor can cause problems, unless the mist-laden air is satisfactorily trapped and cleaned to remove the coating particles from it. The coating loss in a system of this type can become substantial. High-speed air doctors require efficient air flow, an excellent applicator system, and an excellent separation system for removal and reclamation of excess coating.

Speed of operation is now 2000 fpm because of better applicator systems and more efficient separation. There are several types of air doctors in use which include those having similarly contoured lips, those having dissimilarly contoured lips, and those which are basically a very simple nozzle for air brush work.

Figure 15 shows a typical cross section of two air doctors and also a conventional air doctor system using a two-roll kiss applicator followed by the air doctor metering system.

A great number of products is made with the air doctor, including high-quality printing papers, office copy papers, functional coatings, decorative coatings, barrier coatings, and gloss development coatings of the lacquer type. New and different uses for the air doctor are

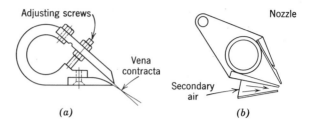

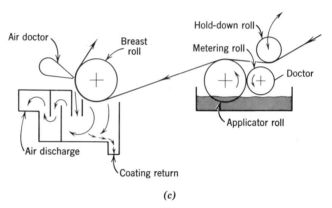

Fig. 15.   Air doctors: (a) Warren-Dilts air doctor; (b) Pomper air doctor;
(c) air doctor coater.

continually being found.   Because of the noncontacting feature
of the air jet, its utility is second only to that of the reverse roll
coater.

N. Flexible Blade Coater

A flexible blade coater is one of the newer coating systems now
in use.   Its original development was in the waxed paper field, primarily for bread wrapper (13).   Since then, it has been used for fine
paper production, publication grade paper, paperboard, and has
now begun to see development in the use of functional coatings.
The speed of operation of a flexible blade coater is in excess of
4000 fpm.   It appears to be limited only by the web handling and
drying systems that can be developed for its extremely high speed.

The principle of operation involves the application by some means of an excess of coating and the doctoring off of this excess with a flexible blade. The blade itself is from 0.010 to 0.030 in. thick and is loaded against the sheet which in turn is backed up by a rubber-covered roll. The normal pressure on the blade is from 1 to 5 lb/linear in. and is sufficient to bend the flexible blade so that it acts very much in the manner of a spring. The amount of coating deposited on the sheet is small, thus setting a limitation to this method of coating—especially with polymer and functional coatings where a relatively heavy coating weight usually is required. Further, in relation to functional coatings, the flexible blade acts very much like a road grader and will tend to fill the low spots of the web, leaving the high spots virtually free of coating. Because of this, there is not a uniformly deposited film of coating. This can be a real disadvantage in functional coatings.

There are six commonly used methods of blade coating today, as shown in Figure 16. They are the uniflex, the Champflex, the conventional trailing blade, the Flooded Nip, the Flexiblade, and the Controlled Fountain blade coater. The primary difference among these coaters is the means by which the coating is initially applied to the web. All of them use the flexible blade for final metering and smoothing of the coating.

In the case of the uniflex blade, the coating is roll-applied, using any one of several conventional kiss coater systems. The web is then passed around a rubber-covered backup roll for doctoring by the blade. In the case of the Champflex coater, the coating is also kiss-applied and metering is carried out by a slowly rotating rod rather than by a sharp blade; however, the rod is mounted on a flexible blade material in such a manner that there is spring in the blade itself.

The trailing blade coater uses a puddle or pond contained against the web, normally in the 3 o'clock position. Means are incorporated in this design to adjust the pond downward. Pond level is of importance in a coater of this type. Edge dams are required in order to contain the coating in the pond and inside the edge of the web. The trailing blade is operated, however, with the edge dams out beyond the paper width so that the complete paper width is coated.

In the Flooded Nip blade coater, the coating is applied by a roll

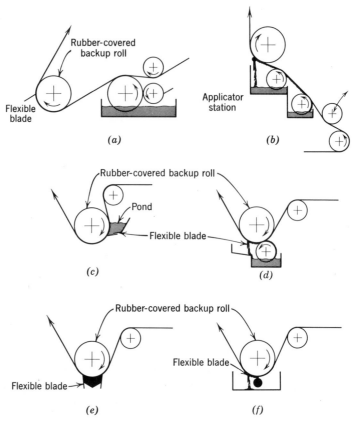

Fig. 16. Flexible blade coater: (a) uniflex blade; (b) Champlex; (c) trailing blade; (d) Flooded Nip blade; (e) Flexiblade; (f) Controlled Fountain blade.

held at a short distance from the large rubber backup roll. The nip through which the web passes is completely flooded so that an excess of coating is applied to the paper as it passes through. The runoff from the blade is collected and returned to a screening system prior to reuse.

In the Flexiblade coater, the coating is completely pressurized in a coating chamber so that the backup roll forms the top side of this pressure chamber. The paper runs around the backup roll and through the pond, exposing the web to the coating which is doctored off by the flexible blade which forms one side of the pond.

The last commonly used blade coater is called the Controlled Fountain blade coater.   In this case the coating is applied through the use of a fountain feed system which is in effect an extruder lightly loaded against the web.   Coating is picked up by the sheet moving past the orifice formed by the backup roll and the fountain. The web then passes by the blade, which doctors off the excess coating.

The conventional angle for these blades is 45° to the tangent line; however, most coaters have means of changing the angle from 30 to 70°, depending on the coating, the product being made, speed of operation, thickness of blade, rheology of the coating, and the diameter of the backup roll.

The flexible blade coater is now being used for functional coatings and the manufacture of products other than printing papers. Its full usefulness is being realized in other fields than those of mineral-pigmented coatings.

Products made at the present time include printing grade coatings, gloss development coatings, barrier coatings, reproduction coatings, and to some degree certain types of solvent vehicle coatings.

O. Brush Coater

The use of brush coaters has been on the decline for many years. The use of a flat bed or rotary brush actually was the beginning of the mineral-pigmented coating industry as it is known today. For the most part, brush coating has been limited to this particular product.

The principle of brush coating is basically simple.   The coating is applied either with a roll or with a brush, after which the freshly coated sheet is passed underneath either flat-bed brushes or rotating brushes in order to smooth and meter the coating applied to the web. The coating solids and speed of operation of the brush coaters are definitely limited.

Two designs of brushes are used in the brush-coating process: the flat-bed brush which oscillates in a transverse manner across the web over a stroke of approximately 1 in., and the rotating brush which normally will run at a speed differential to the web, and in most cases, in reverse linear direction to web travel.   Brushes are made from either natural hair or synthetic materials.

Products made by brush coating include mineral-pigmented coatings for printing purposes, and to a very limited degree, certain types of shade cloth.

P. Vacuum Coating

Vacuum coating of webs is of limited use.   The process consists of subjecting the web to a heated, vaporized coating, such as a metal element in a vacuum, the pressure approaching 10^{-3} mm Hg. The web material is passed over the sublimed metal and deposition of metal by condensation occurs on the web.   As yet, no truly good continuous method of vacuum metallizing has been developed.

There are three types of vacuum metallizing systems to be considered.   First, the batch coating method involves evaporation of metals or metallic components from a stranded tungsten filament. Since the life of the tungsten filament is less than 5 min, this system can be used only for batch coating of three-dimensional objects rather than for a web.

The second method is called the semicontinuous coating system. A roll of paper is placed within the vacuum chamber, which is evacuated, the paper is passed over a large bulk source of evaporating material, and is coated as it passes through.   The roll is then rewound.   At the completion of this roll, the chamber must be entered to remove the roll and to insert a new, uncoated roll for the process.   This is rather cumbersome, although it is today the most commonly used vacuum coating method.

The third system is air-to-air coating, which basically is similar to the preceding system except that the web passes through a complete seal from the outside of the vacuum chamber into the chamber and over the bulk source of material, and then back out of the chamber to be rewound.   In this system, continuous unwinding and winding equipment can be used; hence, the process does not have to be stopped.

At present, certain papers and plastic films are coated by the vacuum coating process.   It is still in the developmental stage, however, and is an expensive method of coating.   The obvious advantage is the ability to deposit a very thin coating by the vacuum process. Work is being done to increase the coating weight potential of this system. The coating which is produced can be considered only for decorative purposes, although some products are now being produced which do give barrier properties.

Q. Electrostatic Coating

Electrostatic coating is nothing more than the charging of coating particles with one polarity and directing them to a web passing through the chamber either on a drum or over a metal plate which carries the opposite charge. The charged coating particles are attracted to the opposite polarity, but with the sheet in between, the charged particles are deposited on this web. It is then necessary to fix these particles on the sheet (1) by fusion of a thermoplastic resin carried in the charged coating particles, (2) by an adhesive on the sheet, or (3) by precoating the sheet and leaving it slightly wet so that the particles will deposit on the sheet. A discontinuous coating is normally applied in this manner, which may or may not be an advantage.

There are three modifications of electrostatic coaters: the process developed at Battelle Memorial Institute several years ago, the Huber process developed for the carbon paper industry and which has seen only limited use, and the electrostatic coating of abrasive papers.

As the technology has developed, today only abrasive paper is consistently made with an electrostatic coater and in this process the electrostatic coater is only a small part of the entire process line.

As concerns limitations, especially with resinous coatings, the hazard of explosion is serious, owing to the very fine airborne particles. The cost can be prohibitive because of the use of expensive binders for the coating. It is difficult to deposit a heavy film on the web, and this constitutes another inherent limitation.

R. Bead Coater

Bead coaters can be considered similar in some respects to the kiss or, perhaps, dip-roll coater. Bead coaters are used almost exclusively for photographic emulsion coatings.

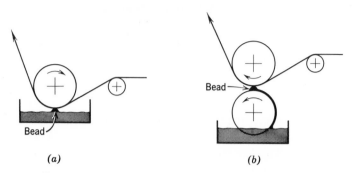

(a)   *(b)*

Fig. 17. Bead coater: (a) single roll; (b) two-roll.

In Figure 17, two types of bead coaters are shown. One such coater relies on a very uniform level of coating in the coating pan, whereas the other system relies on reasonably accurate rolls held at a predetermined gap and through which the paper runs to pick up the coating from the bead.

Normally the sheet will run about 0.005 in. away from the coating or the roll forming the bead and it is necessary to strike the bead in the initial coating operation. Once the bead is formed between the liquid and the sheet, the operation can proceed.

Surface tension, percentage of solids, and viscosity of the coating, as well as web speed, all directly determine the coating weight applied.

Coating deposits, although precise, are in a low range.

III. Application of Coaters and Coating Methods

To set the application of coaters into proper perspective, we have prepared a chart, Table 1. All of the commonly used coating methods are listed therein, together with some of the more important characteristics of these methods with respect to types of coatings, web materials, products, and mechanical and other limitations. In generalizing to the extent necessary to present such information in tabular form, there are several areas in which it is necessary to know the extent to which generalization has been carried out.

In the case of viscosity, consideration has been given to the use of a Brookfield viscometer, using the spindle and speed generally required for the viscosity range indicated.

In the case of operational speed, the methods listed generally operate at a speed less than that indicated. For the most part, a 75% factor would be reasonable to consider for a good, conventional operating speed. There are rare cases where the maximum operational speed may exceed that which has been listed for a given coating method.

In the case of percentage of solids, it must be realized that hot melts and certain organic coatings can be 100% solids. In most cases, the percentage of solids will be considerably less than 100% for any given process.

For standardization, the coating weight application has been given in grams per square meter as the units of measurement, rather than in the units of measurement normally used in a given industry. For example, textiles are almost always measured in ounces per square yard for the coating applied, but this would be cumbersome and confusing because in many cases it would be necessary to have more than one unit of measurement for a given method.

Perhaps of greatest importance for the reader is to understand that there are always exceptions to a general classification of coating method applications. However, the aim is to give an overall view of the many coating applications available.

TABLE
Operating Characteristics, Applications,

Factors that apply to coatings

Type	Maximum speed, fpm	Maximum solids, %	Viscosity,[a] centipoises	Maximum weight, g/m^2	Vehicles	Types
Squeeze roll	1500	100	15–500	12	Water and volatile solvents	Waxes, latexes, hot melts
Kiss roll	1000	100	15–500	12	Water and volatile solvents	Adhesives, waxes, hot melts, latexes
Dip roll	1000	100	15–300	40	Water and volatile solvents	Waxes, latexes, saturants
Transfer roll	1800	65	500–4000	12	Water	Mineral-pigmented starches
Gravure roll	2000	100	30–2000	15	Water, volatile solvent, plasticizer	Lacquers, latexes, vinyl dispersions, washcoats, hot melts, mineral-pigmented
Reverse roll	900	100	200–40,000	400	Water, volatile solvent, plasticizer	Almost unlimited
Calender	500	100	5000–100,000	400	Plasticizer	Vinyl dispersions, rubber compounds
Extrusion	2000	100	50,000–400,000	75	Self	Polyolefins, sarans, vinyls, polyamides, polycarbonates

1
and Limitations of Coaters and Coating Methods

Web material	Application	Limits
Paper, paperboard, textiles	Waxed paper	Coating precision, deckle control, viscosity, sensitive coating weight change by mechanical method is difficult
Paper, paperboard, textiles	Almost always used as applicator unit followed by metering device	Roll pattern, imprecise coating weight, tension sensitive
Paper, paperboard, cellophane	Almost always followed by metering device. Common use as saturator or for coating both sides simultaneously	Very little control over coating pickup, difficult to thread or stop
Paper, paperboard	Printing surface	Roll pattern
Paper, paperboard, filled textiles, plastic films, metal foils	Decorative shade cloth, magnetic tapes, heat-seal labels, primers	Cumbersome to change coating weight, deckle control pattern
Paper, paperboard, filled textiles, films, metal foils, casting paper	Almost unlimited	Requires intelligent operator, fairly high maintenance
Filled fabric	Textiles, friction tape	Expensive machine requires intelligent operator. Coating to be considered an adjunct, not prime use
Paper, paperboard, textiles, plastic films, metal foils	Packaging materials, special barrier structures	Must be hot coating, difficult to start and stop

(continued)

TABLE

Factors that apply to coatings

Type	Maximum speed, fpm	Maximum solids, %	Viscosity,[a] centipoises	Maximum weight, g/m^2	Vehicles	Types
Curtain	1200	100	500–20,000	100	Water, volatile solvent, plasticizer	Lacquers, hot melts, decorative
Knife	500	100	500–50,000	100	Water, volatile solvent, plasticizer	Lacquers, hot melts, vinyl dispersions, starches,
Rod	800	100	15–800	25	Water, volatile solvent	Hot melts mineral-pigmented latexes, sarans
Spray	500	55	15–100	50	Water, volatile solvent	Paints, mineral-pigmented latexes
Air doctor	2000	55	15–500	40	Water, volatile solvent	Mineral-pigmented, latexes, sarans, lacquers, functional
Flexible blade	4000	65	50–40,000	15	Water, volatile solvent	Mineral-pigmented latexes, functional
Brush	300	40	15–300	40	Water, volatile solvent	Mineral-pigmented paints, latexes
Vacuum	800	100	Solids	6	Self	Metals
Electrostatic	500	100	Solids	20	Electric charge	Vinyls, gums, hard particles
Bead	100	30	15–100	5	Water, volatile solvent	Photographic emulsions

[a] Viscosity relates to Brookfield viscometer using the spindle(s) and speed(s)

1 (*continued*)

Web material	Application	Limits
Boxboard blanks, paper, paperboard	Boxboard blanks, barrier coatings	Must have film-forming coating, coating weight precision can be erratic
Textiles, paper, paperboard	Industrial fabrics, tapes, shade cloth, rugs	Difficult to control coating precision, slow, used primarily in textile field
Paper paperboard, plastic films	Hectographic inks, waxes, printing papers, decorative, functional papers	Tension-sensitive nonuniform film
Metal foils, wood, paper, textiles	Wood shingles, metal siding, printing papers, functional	Large overspray, messy, slow, nonuniform, low viscosity
Paper, paperboard, plastic films, metal foils	Printing papers, functional and decorative	Air handling difficulty, airborne mist, low viscosity
Paper, paperboard	Printing papers, functional	Coating weight, coating scratches, requires superior base stock
Paper, paperboard, metal foils	Printing grades, metal coatings	Slow, messy, low viscosity
Paper, paperboard, plastic films, metal foils	Rigid plastic shapes, decorative	Roll-to-roll operation, vacuum maintenance
Paper, paperboard	Gummed papers, abrasive papers	Some explosion hazard, difficult to control, requires wet precoated paper
Plastic films, paper	Photographic films	Slow, difficult to change coating weight

necessary for viscosity range indicated.

520 GEORGE L. BOOTH

References

1. R. H. Mosher, *The Technology of Coated and Processed Papers*, Remsen Press Div., Chemical Publishing Co., New York, 1952.
2. *Plastics Engineering Handbook*, Reinhold, New York, 1960.
3. R. J. Jacobs, "Fundamentals to Consider in Selecting Coating Methods," *Paper, Film, and Foil Converter*, February–July (1963).
4. G. L. Booth, *Mod. Plastics*, **36**, No. 1, 91 (Sept. 1958); **36**, No. 2, 90 (Oct. 1958).
5. G. L. Booth, *Tappi*, **39**, No. 12, 846 (Dec. 1956).
6. *Tappi*, Monograph No. 28, New York, 1964.
7. R. W. Phelps, "Air doctor adjusting brackets," U.S. Pat. 2,981,224 (Apr. 25, 1961).
8. A. W. Pomper, "Blade support for reverse roll coater," U.S. Pat. 2,842,092 (July 8, 1958).
9. A. W. Pomper, "Air doctor," U.S. Pat. 2,679,231 (May 25, 1954).
10. E. P. Olszowka, "Dual air doctor," U.S. Pat. 2,981,223 (Apr. 25, 1961).
11. W. R. Penrod, "Air doctor," U.S. Pat. 2,940,418 (June 14, 1960).
12. K. E. Terry, "Air doctor," U.S. Pat. 2,139,628 (Dec. 6, 1938).
13. A. R. Trist, "Machine for the coating of webs of paper and like absorbent material," U.S. Pat. 2,368,176 (Jan. 30, 1945).
14. J. J. O'Connor, R. H. Savage, and H. C. Schwalbe, U.S. Pat. 2,565,260 (Aug. 21, 1951).
15. P. J. Massey, "Methods of coating paper," U.S. Pat. 1,921,369 (Aug. 8, 1933).
16. G. D. Muggleton, "Paper coating," U.S. Pat. 2,772,604 (Dec. 4, 1956).
17. G. Haywood and C. S. Dayton, "Methods of coating paper," U.S. Pat. 2,560,572 (July 17, 1951).
18. H. W. Faeber, "Coating machine," U.S. Pats. 2,456,495 (Dec. 14, 1948) and 2,555,536 (June 5, 1951).
19. W. A. Zonner, "Paper coating machine," U.S. Pat. 2,645,199 (July 14, 1953).
20. D. B. Bradner, "Methods of applying coating materials to paper," U.S. Pats. 2,229,620 and 2,229,621 (Jan. 21, 1941).
21. W. J. Montgomery and W. P. Taylor, "Apparatus for coating paper," U.S. Pat. 2,676,563 (Apr. 27, 1954).

CHAPTER 12

PROPERTIES AND METHODS
OF IDENTIFICATION OF
COMMERCIAL FILMS

PETER M. HAY

Sandoz, Inc., Hanover, New Jersey

This chapter is designed to present a survey of those properties of packaging films which serve to distinguish them from other classes of packaging materials, and to describe the range of these properties among commercially available films. A second purpose is to outline methods of identification of the principal films used in packaging. In many cases identification can be accomplished by rapid measurement or estimation of the physical properties of strength or clarity. More definitive identification is made by chemical and instrumental tests.

Some consideration will also be given to the combinations of properties which make certain films unique and which allow us to classify other films in groups having similar characteristics. Mention will be made here of the matching of property combinations to specific applications. This section is directed mainly at readers unfamiliar with films and plastics. Detailed information on film testing may be found in Chapter 13. Properties and applications of individual films are discussed in Volume II.

I. Properties

A. Strength

Many images are called to mind by the word *strength* but the main idea, common to all, is resistance to stress. There are many technical definitions of strength which have to be examined individually and then in combination to render a picture of the strength of a packaging film.

1. Tensile Strength

Tensile strength is a property taken as fundamental to all functional materials and the same is true for packaging films. To be

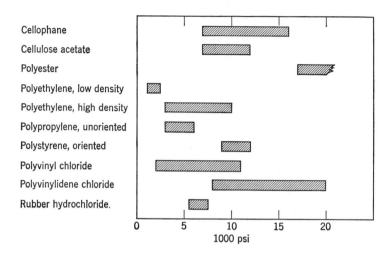

Fig. 1. Comparative tensile strengths of films.

of any use, a film must not break when subjected to a load. It will be found that packaging films vary widely in tensile strength from 1200 psi for polyethylene to over 17,000 psi for oriented polyester, but the tensile strength of all of them is adequate. A graphic comparison of tensile properties of commercial films is given in Figure 1. It can be seen that a wide range of tensile strengths exists among available films and even within one class.

To gain a better understanding of tensile behavior, we should look at the way in which tensile stress builds up in reaction to stretching (technically known as strain). The diagram in Figure 2 shows the two basic types of tensile behavior for packaging films.

In type A, stress builds up as a linear function of strain and reaches a certain maximum stress, S_1, at elongation E_1, where the specimen breaks. It should be mentioned that pure type A behavior is rarely seen in practice. Usually the line curves downward slightly before the break. Two packaging films showing this behavior are cellophane and polystyrene.

In type B, stress increases with strain but the line curves definitely downward. At elongation E_2 the stress reaches a peak S_2 and then decreases with increasing strain. This peak is conventionally

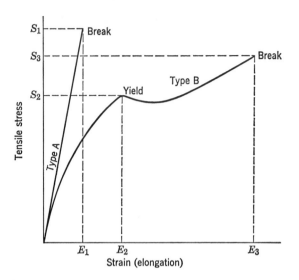

Fig. 2.   Two types of tensile behavior.

known in plastics and film usage as the "yield point" and differs in meaning from the same term as used in the metals and ceramics fields.   In our diagram, we show a continued extension with an unchanging tensile stress ending in an increase in stress and the breaking of the specimen at an elongation E_3.   The stress at the break point, S_3, is often higher than at the yield point, but not necessarily so.

With type A behavior, there is no difficulty in determining tensile strength; obviously it is S_1, the force at break divided by the unit area of the specimen.   With type B behavior, there are two tensile strengths: S_2, the yield strength, and S_3, the break strength. Another term used in relation to tensile strength is "ultimate strength."   This is the strength at the end of the test and thus is another term for tensile strength at break.

Tables of film properties and manufacturers' literature commonly list a single value for tensile strength without qualification.   One may usually assume that the reported value is the maximum value. For a material with type B behavior it may be the yield strength or, if the tensile curve has the shape shown in Figure 2, the break or

ultimate strength. Almost universally, strength values are calculated on the average or minimum cross section of the specimen at
the start of the test, although the thinning and narrowing which
occur during the test may greatly reduce the cross-sectional area.
Some theoretical studies have taken this into account but it is not
yet widely used in the practical literature.

2. Elongation

Closely associated with tensile strength is the property of elongation, the meaning of which has already been defined. As with tensile strength, elongation may be measured for some films at two
points, yield and break. A high elongation can be an asset in a
packaging film because elongation or stretching can relieve the stress
resulting from an applied load so that rupture of the film does not
occur. Films with low elongations must have fairly high tensile
strengths to sustain reasonable loads and be accepted in packaging.
This can be seen in Figure 3, which shows the ranges of elongation
for the same films whose tensile strengths were given in Figure 1.

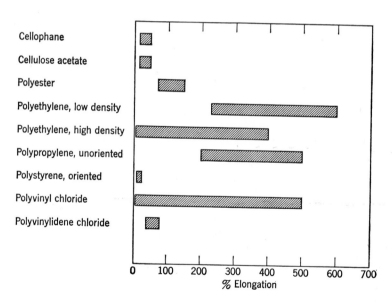

Fig. 3.   Comparative elongations of films.

Most of the films of low elongation are found to have higher tensile strengths.

Films having elongations of several hundred per cent are not ordinarily stretched more than a few per cent in actual use, although some of them, like polypropylene, may be stretched intentionally during manufacture to induce orientation. This prestretching reduces the amount of residual elongation and often converts a film of type B behavior into one showing type A behavior.

3. Energy to Break

Another distinguishing characteristic of films which may be derived from tensile stress–strain diagrams is energy to break. If we take the summation of (stress) $\times$ (Δ strain) increments up to the break point, expressing stress in units of force (e.g., lb) and Δ strain in units of extension (e.g., in.), the result is the area under the stress–strain curve, in units of in.–lb. For softer, more ductile materials, like polyethylene, where a larger volume of material is being stretched and broken, the energy may be fairly high, even though the stress is low. For more rigid and brittle materials showing type A behavior, a very small volume of material actually undergoes stretching and failure, and it is often true that the energy to break a high-tensile but brittle material is surprisingly low.

The methods used for determining the area under the curve all involve some approximations. A tensile tester equipped with an extensometer attached to the specimen will record a true stress–strain diagram, but most testing, done without an extensometer, gives a record of stress versus clamp separation. Any deformation of the specimen within the gripping area of the clamps is not included in the record of strain. Even when the extensometer is used on the specimen, irregularities in the specimen and in its deformation will spoil the validity of the assumption that the true strain is simply the elongation.

Nevertheless, a graphic integration of the area closely corresponds to the energy consumed. Some electronic strain gauge testers can feed the tensile force and extension data directly into an integrator attachment which constantly sums the (force) $\times$ (Δ extension) values up to the end of the test.

A more inexact but economical way of estimating energy is to take half the product of the breaking force times the extension at break. With reference to Figure 1, this would be equal to $\frac{1}{2}(S_1 \times E_1)$ for type A behavior, or exactly equal to the area under the stress–strain curve. For type B tensile behavior, the values of $\frac{1}{2}(S_2 \times E_2)$ for energy to yield and $\frac{1}{2}(S_3 \times E_3)$ for energy to break are both smaller than the true values. Still there are occasions when these approximations can be useful.

To repeat: a tough, ductile material has a high energy to break while for brittle materials it is low.

4. Effect of Testing Rate

The tensile testing we have referred to is generally carried out on a specimen 1 or 2 in. long stretched at the rate of 1–20 in./min. If the rate of extension is increased ten- or one hundredfold, either by speeding up the separation of the clamps holding the specimen or by shortening the length of the specimen, the shape of the stress–strain curve will be different, more like our "type A" film. A film which has "type B" behavior at 20 in./min may show "type A" behavior at 100 in./min testing speed. The change in behavior will be toward higher values of tensile strength and lower elongations at break. On the other hand, if the rate of extension is very slow, films will be likely to show lower tensile strengths and higher elongations.

The reasons for these changes are not hard to see. A polymeric film is made up of macromolecules and some of the forces holding the molecular segments in this state can sustain only very small loads. When these weak restraining forces are broken, the molecular segments slip past one another and the mass flows like a viscous liquid. The viscosity of this liquid is very high and it permits only a very slow extension, sometimes known as creep. If the specimen is extended at a great enough rate, viscous flow cannot relieve the stress and the load is taken up by the stronger forces holding the polymeric mass together. This results in the buildup of high elastic tensile stresses roughly proportional to the strain, with the eventual breaking of the specimen when the strength is exceeded.

In packaging film materials tested at intermediate rates of extension, both viscous and elastic responses take place simultaneously and the resultant curve is intermediate between "type A" and "type B." Such behavior is commonly called "viscoelastic."

5. Effect of Temperature

When films are tested at higher and lower temperatures, there are changes in stress–strain behavior very similar to changes dependent on rate. Lowering of the temperature has the same effect as an increase in the rate—and for a similar reason. As the temperature is lowered, the internal viscosity of the polymer increases and the rate of flow decreases. Thus, creep is slower and is a less important component in the stress–strain behavior, which becomes more like "type A."

There are limitations to the above relationships as a result of transitions within the polymeric material which occur at specific temperatures. The most important of these is the melting transition. As the temperature is raised, most polymers exhibit a sudden drop in viscosity with a small further increase in temperature. With some polymers, the melting range will be broader than that of others, and cellophane decomposes without showing an observable melting point.

Another transition, the second-order or glass transition, is observed in some polymers at a temperature below the melting point. As the temperature is lowered past the glass transition temperature, the polymer usually becomes stiffer and more brittle. Polypropylene, for instance, shows such a transition at about 0°C but some sensitive physical tests show several other minor transitions as well. It is to be expected that the stress–strain curve will change its character at these transition temperatures.

6. Directionality

Packaging films ordinarily do not show the same physical property behavior in all directions. During production a certain amount of orientation of the structure takes place, either through flow of the polymer in the direction of film movement or from tensions set up by shrinkage. Certain films, most notably, polyester, polystyrene, and polypropylene, are most useful when produced with biaxial orientation. When some films are stretched in both the machine and cross-machine directions while being processed, new internal structures develop, giving films with physical properties that differ from films of the same polymers in uniaxially oriented or unoriented conditions.

Whatever the method of processing, the effects of increased orientation on tensile properties are usually an increase in breaking strength and a decrease in ultimate elongation. With those films that are oriented in the machine direction, tensile strength will be higher in the machine direction and elongation will be higher in the cross-machine direction.

7. Impact Strength

The properties discussed above are relatively easy to measure by pulling on a strip of film. When in use, a film will enclose a package or contents and will be subjected to stresses in many directions, complicating any analysis of strength. The property of impact strength becomes important in packaging films when the dropping of a package rapidly imposes a complex pattern of stress on various parts of the film.

Some of the basic understanding of impact strength can be gained from high-speed, one-direction tensile testing, but for a test with practical application the shock must be imposed on an area of film in all directions. Several test methods of two kinds are useful for this.

In the dart drop, ball drop, or pendulum impact (1) tests a piece of film is clamped on all sides and suddenly struck in the center by an object. Sensitive timing devices can be used to determine the loss of energy suffered by the impacting object as it breaks the film. Alternatively, a series of tests can be run, using a range of impact energies and velocities to establish the impact value below which film will not break and above which it will.

A popular test with more direct applicability is the sandbag drop test (2). In this test, bags made from the film under test are filled with sand and dropped from various heights to establish impact strength.

The effects of orientation on impact strength are what might be expected from tensile behavior. Impact failure usually is in the direction having the lower tensile strength and results in a split in the film parallel to the direction of higher tensile strength.

In composite films, such as laminates and coated films, impact may cause failure of a film which has high elongation and tensile strength. The layer of ink on a printed film may act in the same manner as a coating. If one component of the structure, such as an expanse of printing ink, has low elongation at a high stretching rate, it may take all the impact load at first. Then, when that layer breaks, all the stress is suddenly transferred to a very small area of the rest of the film, causing it to break, even though that second component could have withstood the shock if it had been spread over a larger area.

8. Flexibility

Packaging films are known for their flexibility. There are several aspects to this property. There must be durability to repeated flexing, folding, or creasing. In general, softness, as in polyethylene, is advantageous but automatic packaging machinery demands stiffness, as in cellophane, for fast machine speeds.

The fundamental measure of stiffness is the modulus of elasticity, which relates deformation to stress. On the conventional stress–strain diagram the modulus is the slope of the line starting at the origin. If type B behavior occurs, the modulus is estimated from the tangent to the first portion of the stress–strain curve. The higher the modulus, the stiffer the film.

Flexibility, both in terms of durability and modulus, depends on temperature. Below the glass-transition temperature flexibility is less, in both senses.

9. Tear Strength

Another kind of failure is from tearing. We can differentiate two kinds of tear behavior—edge tear and continuation of tear. Everyone is familiar with the fact that cellophane shows a very high resistance to the start of a tear at an edge but a very low resistance to the continuation of a tear once it has been started. Other films show various combinations of these two properties. As with other properties, tear strength varies with orientation and direction in the film.

10. Combinations of Strength Properties

We have examined the important strength properties individually and have seen some relationships among them. For a balanced appraisal of packaging films all properties must be kept in mind and it is usual to find that the maximum in all properties cannot be found in any one film. Certain strength properties are needed for each application and generally the film which is selected delivers the best compromise or balance of properties at the lowest cost.

B. TRANSPARENCY

A prime characteristic of packaging films is transparency. It is this property which makes film superior to paper and foil because of the visual attractiveness of a package. Ideally, a packaging film should be invisible and should permit total transmission of light rays unabsorbed and undistorted. This is not realized because of four kinds of effects: glossy reflection, light scattering, distortion, and light absorption.

1. Glossy Reflection

A light ray passing from air into a film is always partly reflected when it enters a transparent medium with a different index of refraction. Although the interface may be perfectly smooth, about 4% of the light will be reflected as glossy or specular reflection. Another 4% is lost by reflection as the light ray passes through the second interface back into the air. This means that the maximum light transmission can be only 92% (3). Still, such a film is considered to have excellent clarity or transparency.

2. Light Scattering

In any transparent body there will be greater or smaller amounts of extremely small discontinuities which may be foreign particles, voids or bubbles, or, as in some polyolefin and nylon films, a "graininess" associated with spherulitic aggregates of a semicrystalline nature. The total effect of these minute discontinuities is the scattering of light in all directions to give an effect known as haze. This effect, unlike the glossy reflection, helps to obscure the view of the package contents and hence is always objectionable.

3. Distortion

Another way in which films depart from ideal clarity is through distortion by striations or gels. These irregularities in the surface have no sharp boundaries and do not scatter light, but deflect light rays in the same manner as prisms and lenses. The defect of distortion is not so noticeable in an overwrap situation where the object viewed is in direct contact with the film as it is in the packaging of loose contents. In the latter case, the deflection of light rays coming from an object deep within the package causes its image to be indistinct and distorted. Inspection of a film at arm's length emphasizes distortion as one views objects many feet away.

4. Light Absorption

A film with no haze or distortion may still lack good transparency because of color. Color results from the absorption of light from part of the visible spectrum by chromophoric groups in the film. Some correction for this absorption can be made by adding a dye or pigment to absorb light from the rest of the spectrum. For instance, blue is added to balance yellow. The limitation in this correction is that total light transmission is still lower, and rather than restoring total transparency, the correcting color results in a gray color. One way to "add white" to a colored film is to use an optical brightener. These dyes have the property of absorbing ultraviolet light and converting it by fluorescence to light in the visible range. The visible spectra of various brighteners differ; some give bluish-white while others may produce a pink-white fluorescence.

5. Optical Quality

The overall optical quality of a film will depend on the freedom from color in its raw materials and the stability of color during processing and use, the cleanliness and freedom from contamination during processing, and the smoothness and flatness of its surfaces (4). Except for the inherently frosty or grainy films, it is not uncommon to find plate-glass clarity with light transmission over 90%.

C. IMPERMEABILITY

Packaging films are used mostly to surround a package completely, and while a film may serve to contain loose contents and protect package contents from contamination by dirt or liquids, it also can provide a flexible, transparent barrier to vapors of various kinds. The three most important categories of vapors are water vapor, atmospheric gases, and the organic vapors which constitute odors and flavors.

Several structural and compositional factors must also be considered in any discussion of permeability. These would include the effects of coatings and plasticizers in the manufactured product and the effects of changes in the film composition through loss or gain by migration.

1. Water Vapor

Many products need to be protected against the gain or loss of moisture and some films do an excellent job of blocking the transmission of water vapor. These would include coated cellophane, polypropylene, polyethylene, polyvinylidene chloride (Cryovac and sarans), and polyester films. It is well to note, though, that even the most impermeable of these films has a measurable permeability and there there is as yet no flexible, transparent tin can.

Other products need to breathe so as to avoid condensation of water or the growth of mold, for instance, in fresh vegetables. Films of high water-vapor permeability like polystyrene, cellulose acetate, and certain grades of cellophane are suited to these conditions.

The rate of transmission of water vapor will depend on the driving force or vapor pressure gradient across the film. The fastest rate would be shown with completely dry, moisture-absorbing contents within a sealed package held at 100% relative humidity. For less extreme differentials, lower rates will be observed, and if the relative humidity inside and outside the package are equal, there will be no transmission through even the most permeable of films.

2. Atmospheric Gases

Oxygen, carbon dioxide, and nitrogen within a package often must be controlled. Oxygen hastens the breakdown of organic materials in foods and oxidation results in effects such as rancidity of the oil in baked or fried foods. Some packaging applications require a film which has a very low rate of oxygen transmission. In some of these a vacuum pump is used to remove all of the air from around the contents; then a seal is made. With time, small amounts of oxygen will permeate the film but with a good oxygen-barrier film such as Cryovac or saran, and many laminations, the shelf life of oxygen-sensitive foods is prolonged. For further protection some systems flush the package with nitrogen as well.

A different situation exists for the packaging of fresh meat. Here a high rate of oxygen transmission is required to maintain the bright red color of red meats. If the oxygen supply is too low, the meat will turn brown or gray and, although safe to eat, will be unattractive in the market. Special grades of cellophane and rubber hydrochloride have been developed for this purpose which have high oxygen permeability but still have the low water-vapor transmission needed to avoid drying of the meat. Very thin films of polyethylene have also been used for fresh meat packaging.

This phenomenon of high transmission for oxygen combined with low transmission of water vapor seems paradoxical, but it exists, not only for meat-wrap cellophane, but for polyethylene as well. The reverse characteristics apply to nylon films which have a relatively high permeability to water vapor but a low permeability to oxygen, nitrogen, and carbon dioxide. Other films have high transmission rates for all gases, including water vapor.

This has been explained by a consideration of the size, shape, and polarity of the permeating molecules and their interaction with the polymer chain segments of the film. A thorough treatment of the barrier properties of films will be given in Volume II.

3. Odors and Flavors

Other vapors which are controlled by films are the organic compounds that impart flavor and odor. A packaging film can protect package contents from either the absorption of unwanted odors or the loss of volatile flavoring ingredients. Among good barriers to organic vapors are cellophane, saran, and vinyl. Cellulose acetate and polyethylene are poor odor and flavor barriers unless coated with a good barrier material.

4. Structural Effects

Some good barrier films, such as polyvinylidene chloride, are composed of one homogeneous layer. Other good barrier films are produced by applying thin coatings of barrier material to base films of poor barrier characteristics, or by extrusion coating or laminating heavier layers together to form a composite.

Composites often show permeabilities lower than expected from the known permeabilities of the two separate components. This may be the result of interface effects which make it difficult for permeating molecules to diffuse from one layer across the boundary into the next layer, or it may be that the layer of adhesive provides an additional barrier to transmission (5).

One structural factor to be dealt with is the closing of the package. Heat, solvent, or adhesive seals may often not be complete. Under some conditions this does not impair the package protection, but in a cycling condition of alternate high and low humidity and temperature, water vapor and air may actually be forced through the gaps in the seal area. For critical applications a complete seal is a necessity.

5. Compositional Factors

The films we have been considering differ in their inherent barrier characteristics because they have chemically different compositions. Some, like cellophane, cellulose acetate, and vinyls, contain plasticizers as manufactured. Plasticizers generally increase the permeability of films and the permeability will vary with the plasticizer content.

If a film should absorb some compound from its environment, it will also become more permeable. This explains the high oxygen transmission of meat-wrap cellophane. In the normal state, below 35% relative humidity, cellophane is relatively impermeable to oxygen, but when it absorbs water, as from a piece of meat, it becomes swollen and its permeability to oxygen increases tremendously.

The same reasoning accounts for the fact that a cellophane-polyethylene laminate will show a greater permeability to water vapor when tested with the cellophane toward the high humidity than when the polyethylene is toward the moisture. When next to the moisture, the cellophane absorbs more water and contributes greater permeability to the laminate.

D. OTHER PROPERTIES

In addition to the main properties discussed in the three categories above there are other subordinate, but still essential, characteristics to consider. Most of these have to do with the practical handling and use of the film and might be referred to as functional rather than physical properties. Those to be discussed include the surface properties of slip and blocking, the properties of sealability, shrinkage, adhesion, chemical and aging resistance, and static electricity.

1. Slip and Blocking

Packaging films vary widely in their surface characteristics. Some are hard and smooth, some soft and smooth, and some soft and rough.   Each surface has its inherent advantages and disadvantages.   In unrolling and using a film there must be no problems in controlling the movement of the film.   If the film is not blocked on the roll, it will unwind without drag.   If it is too loosely wound or slips on itself too easily, of course, there will be other problems such as telescoping on the roll.

As a film passes through a packaging machine and in the handling of bags and overwrapped packages, there must still be low friction between film surfaces and no tendency to bind or stick under pressure.

Surface characteristics are controlled by introducing roughness and lubricants to the surface.   Materials providing microscopic roughness are often added to the surface of a film so as to minimize the close approach of large areas of soft surfaces which cause sticking when the air is forced out from between sheets.   Close examination reveals Liesegang rings where the air has been excluded, and it is virtually impossible to slide the films on each other under these conditions.   The most familiar material exhibiting this phenomenon is the household product Saran Wrap, which takes advantage of its clinging characteristics in use.   In most applications, however, this is not desirable and small amounts of fine powder are sometimes added to the surface or mixed with a film coating to keep film surfaces separated (cf. Chapter 9).

Lubricants are added to films which are inherently sticky and soft. A very thin layer of oily lubricant on the surface of a film helps lower the coefficient of friction.   Too much lubricant, however, can seal the layers of film together and give oil spots, so the amount must be closely controlled.

The major burden of controlling surface properties is carried by the film manufacturer but there are some measures to be applied by the film user.   One is the control of humidity to suit the film being used.   Another is the controlled application of dry powders, such as starch, to films while they are being rewound or printed.

Slip is most often tested by a hand test, using thumb and finger pressure.   More precise evaluation is done by measuring the coefficient of friction.

2. Static Electricity

Static propensity is another miscellaneous property which affects the efficiency of automatic film handling.   Here again the manufacturer adds certain agents to control static but the user can exercise further influence in the control of humidity or in the use of antistatic devices.

Antistatic devices all function in the same way, by ionizing the air so that the air ions can neutralize charges on the film. Some work by means of radioactive sources, whereas others use high voltage to accomplish the same thing.

Most films carry and build up charges and the ones which give the least trouble with static are those which discharge themselves most rapidly. Antistatic additives to films probably function as conductors to permit charges to bleed off.

3. Sealability

Three methods of sealing are heat, solvent, and adhesive. The first one is by far the most used. The best heat sealing is done with a coated film having a coating with a melting point well below the melting point of the base sheet. Heating two or more layers of film under pressure softens the coatings and fuses them, leaving the rest of the film undistorted. Heat sealability is often tested by making seals under controlled conditions and testing their strength.

Some films, like polyethylene, are heat sealed by fusing the entire seal area under pressure and cooling it again. This process requires more care than with coated films because until the entire seal area has cooled, it cannot be stressed without tearing or distortion. Special heat-sealing devices have been developed to handle these totally thermoplastic seals.

Solvent sealing can be done best on coated films whose coatings can be softened by solvent. The seal areas are pressed together and then in time the solvent evaporates, leaving a seal. Solvent sealing does not work well when the solvent attacks the entire film.

Adhesive sealing can be done on any film, provided the adhesive is selected to suit the film used. It operates much in the same way as solvent sealing, except that new material is added to bond one film surface to another.

4. Shrinkage

Many films will shrink when heated as a result of a relaxation of the orientation built into the film. This is an advantage in certain types of packaging wherein heat is applied after wrapping, causing the film to shrink and conform closely to the shape of the contents. For reliable shrink packaging the manufacturer must control the orientation. One can test for orientation by carrying out a shrinkage test under standardized heating conditions.

5. Adhesion

A film which is to be printed or sealed with adhesive may be tested by applying ink or adhesive and attempting to pull it off after drying (6). In the case of polyolefin films, surface treatments by flame or electric discharge are used to improve adhesion. It is commonly found, though, that overtreatment for adhesion may produce areas of film which do not heat-seal easily; therefore, the properties of ink adhesion and sealability are best considered together.

6. Chemical and Aging Resistance

Resistance to aging and chemical attack is often ignored in packaging films but occasionally becomes very important. Examples of chemical attack are weakening and delamination on exposure to water or food juices. Packages made with initially strong seals may leak or burst if attack by water or package contents weakens the seals.

Aging is a property frequently encountered with products of long shelf life. Plasticized films may become brittle through loss of plasticizer or the effects of sunlight, or fluorescent lighting, or air may cause polymer degradation with color change and loss of strength. The usual way to determine the aging properties of a film is to place the film—or preferably the entire package—under accelerated test conditions and observe the time to failure. Acceleration usually involves raising the temperature. If the accelerated effect of sunlight is sought, a carbon arc light can be used (7). Accelerated aging tests often provide good prediction of the effects of normal exposure but there are enough examples of poor correlation to suggest the value of caution in relying exclusively on accelerated tests.

E. STANDARD TEST METHODS

Most of the properties discussed in this section are tested for by standard methods as discussed in detail in Chapter 13. For convenience, the most important properties and the main or preferred test methods, as published by the American Society for Testing and Materials (8), are given below.

Property	ASTM method
Tensile strength	D 882-61T
Elongation	D 882-61T
Impact strength, dart	D 1709-59T
Flexibility	D 1043-61T
Tear strength, edge	D 1004-61
Tear strength, continuation	D 1922-61T
Reflectance	D 791-61T
Haze	D 1003-61
Distortion	D 881-48, D 637-50
Color	D 791-61T
Water vapor transmission	E 96-53T
Gas transmission	D 1434-58
Coefficient of friction	D 1894-61T
Blocking	D 1893-61T

II. Identification

Many packaging films and film types are available today, made from many different materials and by many processes. At first glance, a film may not be easy to identify because all films are transparent to some degree and flexible, but certain distinguishing characteristics can be used to establish quickly the identity of an unknown film sample. The characteristics may be related to physical, thermal, and chemical properties. In addition, there are instrumental methods of identification, the most useful being spectral absorption tests. Composites or laminates are more difficult to characterize until the layers have been separated and examined individually.

Physical, thermal, and chemical tests can be applied in various schemes or sequences to single out the identity of an unknown film. One such scheme is found in the *Modern Packaging Encyclopedia* (9). Another similar scheme appears in the British journal *Plastics Technology* (10).

A. Physical Property Tests

Properties which can be qualitatively estimated without instruments are stiffness, stretch, tear, and optical quality.

Three ways to judge stiffness are feel, drape, and sound. Polyethylene, saran, some vinyls, and some grades of Pliofilm are soft to the touch and limp, and they make little noise when rattled or crumpled. Polypropylene, cellulose acetate, cellophane, polyester, nylon, some vinyls, and some grades of Pliofilm are harder to the touch and stiffer; when crumpled they make more noise. Oriented polystyrene is very stiff and has a metallic ring when rattled.

The property of stretch separates films into two classes. The films which do not stretch in the hands or stretch only a minor amount are cellophane, oriented polystyrene, cellulose acetate, saran, nylon, and polyester. The films which stretch a great deal are polyethylene and polypropylene. Vinyl and Pliofilm may be more or less extensible, depending on the grade.

Tear strength, as judged on a film which has been nicked on the edge with a knife or the teeth, is low for cellophane and cellulose acetate, a little higher for polyester and nylon, and very high for polyethylene, polypropylene, vinyl, Pliofilm, and saran.

Optical quality ranges from sparkling clear and glossy for oriented polystyrene, cellophane, and cellulose acetate to clear for polypropylene, polyester, saran, and vinyl to dull but transparent for polyethylene and Pliofilm, to hazy for nylon, fluorocarbons, and Cryovac.

B. Thermal Property Tests

Tests involving heat or flame may be classified as melting, flammability, and burning odor.

Most films melt. Those which do not are cellophane and fluorocarbon films. Cellophane will burn in a flame with the odor of burning paper, while fluorocarbon will not.

The flammable films are cellophane, cellulose acetate, oriented polystyrene, polyethylene, and polypropylene. Intermediate in flammability are nylon, polyester, vinyl, saran, and Pliofilm. These films usually burn when held in the flame, but do not continue to burn when the flame is removed.

The burning odor may be classified as follows:

Odor	Film
Burning paper	Cellophane
Acetic acid	Cellulose acetate
Illuminating gas	Polystyrene (also gives sooty flame)
Burning wax	Polyethylene
Burning hair or feathers	Nylon
Sweet, esterlike	Polyester
Acrid, sharp	Vinyl, saran, Pliofilm

C. CHEMICAL TESTS

Among simple tests which can be called chemical are the Beilstein copper wire flame test for halogens, the pyridine–potassium hydroxide test for vinyl and saran, and solubility tests in various solvents.

To test for halogens a copper wire is heated in a flame until it produces no more color, and then a sample of the film material is applied to the hot wire. When the wire is again placed in the flame, a green color will appear if the film contains chlorine or fluorine. Saran, vinyl, rubber hydrochloride (Pliofilm), and fluorocarbon films give a green color. Some of the coatings used on nonhalogen films contain halogen, and care must be taken to distinguish between halogen in coatings and in the base sheet. Saran and vinyl may be identified by the Wechsler test (11) using pyridine and methanolic potassium hydroxide. In this test, vinyl gives a pale brown color and saran a dark brown to black color. Pliofilm does not react.

Some films, most often cellophanes, are coated with nitrocellulose lacquer. A spot test for nitrocellulose uses a solution of diphenylamine and concentrated sulfuric acid (12). When these are mixed on the surface of the film, a blue color forms. Saran coatings give a positive Wechsler test. Some cellophanes are uncoated; they are readily detected by their rapid water absorption. When touched to moistened lips, uncoated films will stick.

Solvent tests can be used to aid identification but distinctions are not so clear-cut. Cellophane and fluorocarbon films are unaffected by organic solvents. Polyethylene, polypropylene, and polyester will not dissolve in any cold solvents but may be swollen and softened by a variety of solvents. Films which will dissolve in specific cold solvents are as follows:

Film	Solvent
Vinyl	Tetrahydrofuran
Saran	Tetrahydrofuran
Cellulose acetate	Acetone
Pliofilm	Chloroform
Polystyrene	Toluene
Nylon	Formic acid

Some films are soluble in more than one solvent and the rate of solution will vary, depending on the solvent.

D. Other Identification Tests

Another identification test is odor. Vinyls, cellulose acetate, and saran contain plasticizers with distinctive odors. Pliofilm has a rubber odor like a child's balloon. No film is completely odorless, but the other films have less odor.

Density is a property which can be of assistance in identifying polyolefins. Polypropylene has a density of 0.900–0.905 g/cm^3. The lowest-density polyethylene has a density of 0.917 g/cm^3; other polyethylenes used as films range upward in density to about 0.94 g/cm^3. Density in this range can be estimated by mixing ethanol and water in such proportions that a small piece of film will remain suspended. The density of the film is the same as the density of the solution; the latter can be determined with a hydrometer or pycnometer. When many determinations of density are required, a density gradient column is useful (13).

Other specific tests exist for certain films but we shall not try to treat them extensively here. As appropriate, they will be included in Volume II.

E. Instrumental Tests

Well-equipped analytical laboratories have spectrophotometers which can be used to identify films. The ultraviolet and infrared absorption spectra of the unknown film are compared with spectra previously obtained on known films. A spectroscopist who interprets the spectra will be able not only to establish the identity of the film but also can often detect the presence of impurities or additives.

Distinctive physical properties such as tensile strength and the stress–strain curve can also be useful for identifying films.   For this purpose, a recording tensile-strength testing machine is necessary.   Study of the tensile curve will often disclose additional facts about the physical structure of the film, such as its orientation during manufacture.

References

 1. K. W. Ninnemann, *Mod. Packaging*, **30**, No. 3, 163 (Nov. 1956).
 2. *Commercial Standard CS 227-59*, U.S. Dept. Comm., Washington, D.C., 1959.
 3. D. B. Judd, *Color in Business, Science, and Industry*, Wiley, New York, 1952, p. 332.
 4. P. M. Hay, C. P. Evans, and K. W. Ninnemann, *Mod. Packaging*, **33**, No. 3, 179 (Nov. 1959).
 5. L. E. Simerl, *Food Technol.*, **7**, 256 (1953).
 6. W. S. Kaghan and F. J. Lindsey, *Soc. Plastics Engrs. J.*, **20**, 1305 (1964).
 7. *ASTM Recommended Practices D 1499-59T and E 188-61T*.
 8. American Society for Testing and Materials, Philadelphia, Pa., *ASTM Standards*, issued annually in 32 parts.
 9. *Modern Packaging Encyclopedia*, McGraw-Hill, New York, issued annually in November.
10. V. Bird, *Plastics Technol.*, **9**, 40 (1963).
11. H. Wechsler, *J. Polymer Sci.*, **11**, 233 (1953).
12. A. G. Roberts, *Anal. Chem.*, **21**, 813 (1945).
13. *ASTM Method D 1505-60T;* R. F. Boyer, R. S. Spencer, and R. M. Wiley, *J. Polymer Sci.*, **1**, 249 (1946).

CHAPTER 13

MEASUREMENT OF PHYSICAL PROPERTIES OF FLEXIBLE FILMS

KARL W. NINNEMANN

Fabricated Products Div sion
Allied Chemical Corp., Morristown, N.J.

I. Introduction

As man began to use more accurate tools to measure supposedly identical objects, he probably discovered that exact correspondence of the tested characteristics was a rarity, that the objects were indeed not "as like as two peas." Moreover, when using the same instrument to measure the same characteristic of the same object, he probably did not obtain the same answer every time. He must have learned that variability is inherent in nature, that it is certainly present in scientific measurement, and that he had to repeat his measurements many times because they were affected by factors which he could not control and of which he might not even be aware. Therefore, in order to establish the average attributes or performance characteristics of an object or material, he had to measure these characteristics over and over again. The procedure he followed each time is the test.

A. Objectives of Testing

Testing is the set of experiments or methods used to disclose the nature of a material with respect to one or many of its physical or mechanical properties (1,2). There are probably three objectives to testing which warrant spending the amounts of money annually spent on this activity. The first is research, and in the course of this work theories are developed which must be proved or disproved by tests in the laboratory, the results of which are then used to formulate plans and programs or to determine courses of action. On the basis of the test results the direction which the research activity is to take may thus be set, greater effort may be indicated, or a project may be discontinued.

Process development is another major objective of testing. Results of tests on a product may suggest possible changes in a chemical process, for example. One raw material may be shown to result in one product having better or more suitable characteristics than might be obtained with another. The data may further show or verify the result of changes in the machines or mechanical process. Also, the variability that is found may be used to develop the specifications for the product: the limits within which the product can be manufactured.

Finally, tests may be used for product evaluation. Once specifications have been set, the product must be measured to determine its level of quality. Moreover, its performance in actual service must be compared against the specifications. In the case of a polymer film intended for a packaging application, whether for food or soft goods or hardware, its use depends on whether it will fulfill certain general requirements specific to its intended application.

B. What Is Tested

If flexible films are intended for use in packaging applications, there are a number of physical properties they must have in order to do a satisfactory job. These characteristics fall into four basic groups.

The first are those physical-mechanical properties which answer the question, *How strong is it?* They describe the ability of the material to do its job of containing and protecting, and consist of such characteristics as tensile strength and the various moduli of elasticity, stiffness, impact strength, and tear strength. These properties are related to the chemical structure and, in part, to the mechanical process used to fabricate the film.

There is a second class of mechanical properties which frequently form the basis for the selection of the material. They are in general combinations of the basic physical properties and are determined by means of end-use tests because the films have been selected to contain and protect a product. These characteristics are determined in tumble tests, drop tests, vibration tests, and other practical or empirical methods, and generally differentiate between films only on a qualitative basis.

A third category of essential properties is the transmission properties. These, too, are related to the chemical structure of the film and are important determiners of the packaging application. Water vapor, gas, and odor transmission fall into this class, and the selection of a film for a given packaging application depends on whether the packaged product can tolerate a high or low rate of any of these. Clarity of the material may also fall into this class of properties.

Finally, there is a group of properties which may be defined as production properties. They are those properties which characterize how well the material handles when being finished, that is, when it is put into final form at the end of the manufacturing process, or when actually being used. Such properties as friction, abrasion resistance, heat sealability, printability, crease resistance, blocking, and a number of others fall into this category.

Various methods used to measure all of these different properties of flexible films are described in the following sections of this chapter. This is not intended to be an exhaustive study of all possible methods which may be used to determine each characteristic. Rather, it represents many of the methods which the author has successfully used, and is intended to serve as a starting point for a research man who in the course of investigating flexible films is embarking on a new physical-test program.

Many of the methods used to measure the physical properties of flexible films were developed in the paper industry and were subsequently adapted for these new materials because of their packaging applications. Reference should be made to methods such as the *ASTM Standards* (3) and the *Tappi Standards* (4). A handy reference for the source of non-standard methods and new methods is the *Bibliography of Papermaking and U.S. Patents* (5) compiled by J. Weiner and published by TAPPI. Development of standard methods is carried on constantly by these two organizations as well as by those technical societies whose basic interest lies in the plastics field.

II. Testing that Measures Strength of Films

A. Tensile Strength

The use of various plastic films and fibers, or of paper and cellophane, is usually determined by the mechanical properties of these materials rather than by their chemical behavior (6). Mechanical properties are those properties which are related to the deformation of the material by applied forces. They are usually measured by means of a tensile test or tests for stiffness, impact strength, flexibility, and others. The tensile test probably has been used more than the others because in the minds of most people it answers the question, "How strong is it?" The number and different kinds of tensile testers available indicate the preference for this test.

Tensile values have been used for a long time to indicate the general quality of a material for several reasons (7). First, there is a large quantity of data available and, above all, persons in research are familiar with them since they have been used for defining the quality of many materials other than polymer films. Tensile data can be obtained with relative ease. Finally, although the two main attributes of a material measured in the tensile test, namely, ultimate strength and elongation, are very important, there are several other fundamental relationships which describe how the mechanical properties are related to the chemical structure and which can therefore be used to determine whether changes or improvements have been made in a material during the course of development work.

Since the tensile test measures fundamental physical characteristics, it is one of the most reliable ways to prove the properties of a material, providing that whoever performs the test concentrates on details to produce accurate and well-defined results. While the data may not always be suitable for design purposes without modification, they are useful for setting specification and help to insure that the material has a certain minimum quality when measured in a certain way (see Chap. 12, Sect. I-A).

1. Tensile Testers (1,7,8)

The general procedure in the tensile test is to grab the specimen at both ends and apply a force over its entire width until it is pulled apart. The stresses should be uniform over the width of the test specimen and parallel to one another and to the direction of loading in order to obtain a pure tensile strength value. There are several different kinds of test equipment, the major ones of which employ constant rate of loading, constant rate of elongation, constant rate of powered grip separation, and constant rate of stress. This variety in tensile testing machines probably results from the number of different materials which have been and must now be tested, and one must decide which is the right one for his purpose. The first tensile testing instruments were constructed to handle very large loads. They were not suitable for testing the new plastic materials because they lacked accuracy at the lower breaking loads obtained with these substances.

Constant Rate of Loading. Here loading causes the elongation. The effect of applying the force to the test specimen is to extend it until it eventually breaks. The inclined-plane instrument employs a constant rate of loading. It consists of a tilting table, the angle of which can be changed during the test, and a carriage which moves along this table (Fig. 1). As the carriage moves it applies a load to the test specimen, and the angle of the table is changed to produce the constant rate of loading. The rate at which the sample grips separate is controlled by the strength of the material. This rate will probably vary from sample to sample because of differences in the stress–strain characteristics of the test specimens.

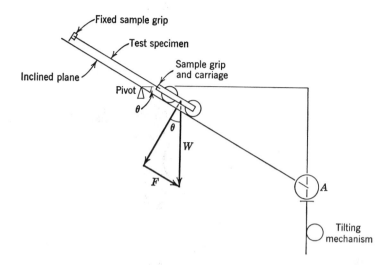

Fig. 1. Constant rate of loading tensile tester. The plane is tilted by dropping the end at A. If F is the force parallel to the plane, the load on the specimen is directly proportional to the sine of the angle of inclination. $F = W \sin \theta$.

Although the rate of sample grip separation is variable, inertial effects noted in pendulum-type testers are absent.

There is not much interest in constant rate of loading tests, probably because the instrument is limited in terms of loading capacity and elongation.

Constant Rate of Powered Grip. One of the most commonly used types of tensile testers is the type which uses a pendulum to measure the applied load (Fig. 2). In this instrument both sample grips move an appreciable amount during the test, the lower or powered sample grip moving at a constant rate and the movement of the upper sample grip depending on the stress–strain characteristics of the sample being tested. The rate of grip separation is therefore variable as is the rate of straining. Only when the material being tested has high elongation is the rate of grip separation approximately equal to the rate of motion of the lower, powered grip.

When using a tensile tester with a pendulum type of load meas-

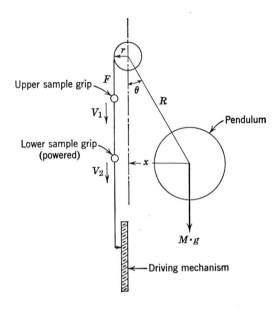

Fig. 2. Constant rate of powered grip tensile tester. The small pulley is pulled around, swinging the pendulum from vertical. The force F acting on the pulley is equal to the tension in the test specimen.

uring system, the following characteristics of the instrument should be kept in mind. Although the rate of motion of the lower, powered sample grip is constant, the rate of loading and the rate of strain vary, depending on the properties of the test specimen. The inertial effects of the pendulum may also be a source of error because the pendulum has an angular velocity which depends on the rate of movement of the lower jaw and the elongation of the test specimen. It therefore possesses kinetic energy and will continue to move until this energy is dissipated. The recorded load will be large by an amount which depends on the angular velocity of the pendulum at the moment of break. The error is greatest for small loads. This may be reduced by using a slower rate on the driven sample grip. If the test specimen has a low amount of elongation, the error may be large because the velocity of the upper sample grip will be close to that of the lower, driven sample grip, and the pendulum will have a high angular velocity. If the breaking load is low, it is always preferable to perform the test on an instrument of lower capacity or to reduce the load range of the instrument.

There also is an error due to acceleration of the pendulum when the pendulum tensile tester is used. As a load is applied to the test specimen at the beginning of the test, the specimen will stretch, but no load will be recorded. More and more tension will be developed in the test specimen as the lower sample grip moves downward until the tension developed overcomes the inertia of the pendulum, at which point the pendulum will begin to move. However, the force at this time is greater than that required to keep the pendulum moving at the angular velocity which corresponds to the linear velocity of the upper sample grip, and the pendulum will overshoot the point to which it should move. An increased load will be recorded, when actually the test specimen shows signs of slackness.

Constant Rate of Extension (9). In this type of test (Fig. 3) the tension on the test specimen is zero at the beginning of the test. As the bottom sample grip moves downward, the specimen is

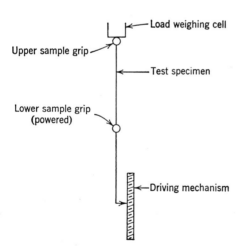

Fig. 3. Constant rate of extension tensile tester. As the lower, powered grip is driven downward, the tension in the test specimen increases from zero to the breaking point.

stretched and the tension on it increases until it breaks. The type of tester differs from that described in the previous section in that the movement of the top sample grip, which is attached to the load weighing system, is not perceptible. Since the bottom sample grip is driven at a constant rate and the top sample grip barely moves, the two grips separate at an almost constant rate.

The load or force weighing system in an instrument of this type is electronic, consisting of strain gauges in one of several different configurations, depending on the manufacturer. These are mounted in load cells which are interchangeable, depending on the load range. The strain gauges produce signals corresponding to variations in the load. The signal is amplified and operates the pen of a high-speed graphic recorder.

Instruments of this type have a number of advantages. First, they are free from inertial errors and friction because the load is measured directly by a low-deflection electronic sensing device without any mechanical levers, weights, or pendulums. Second, they test under conditions of virtually constant rate of extension, the deflection of the beam to which the strain gauges are fastened being extremely small. Finally, they are very versatile, allowing for the combination of a wide range of scale sensitivities, cross-head drive speeds, and recorder-chart magnification ratios.

There are, however, some disadvantages to the strain gauge type of instrument. They usually have a high initial cost and they require the services of an expert for maintenance and repair.

Constant Rate of Stress. Some tensile test instruments have manual controls which permit adjustment of the load on the test specimen to produce a constant rate of stress. Instruments of this kind are not very common.

In all of these instruments it is very difficult to align the test specimen in the sample grips, and it is therefore necessary to use self-aligning grips. It should not be assumed that self-aligning grips will prevent bending moments from developing during the application of the tensile load, and the technician should always be alert to this possibility.

2. Definitions

There is frequently a great deal of confusion when defining terms used in the tensile test. The following definitions are the more common ones used when measuring the tensile properties of films. It is advisable, however, to refer to methods such as *ASTM D 638* (10) and *D 882* (11) or any others, for definitions which are specific to the given method and for the many definitions not included in this list.

Stress is the force exerted on a body that tends to deform its shape. It is defined algebraically as the ratio of the applied force to the cross-sectional area.

$$\text{stress} = \frac{\text{applied force}}{\text{cross-sectional area}}$$

In the tensile test, the force exerted on the body is a tensile force, and the stress is therefore referred to as *tensile stress*. When film is tested, the minimum original cross-sectional area within the gauge boundaries is used to calculate stress, and not the area at the time a given stress exists. True stress is calculated from the area at the time the given stress exists.

Tensile strength is the maximum tensile stress which a material can sustain. It is calculated from the maximum load during a tension test—regardless of whether or not this load occurs at rupture—and the original cross-sectional area of the test specimen, if film.

Strain is the change resulting from an applied force in the size or shape of a body, with reference to its original size or shape. Thus, when a load is applied to a tensile-test specimen, some stretching takes place. The amount will vary with the initial length of the specimen. Strain is the term used to relate this elongation to the initial length of the test specimen.

$$\text{strain} = \frac{\text{total elongation}}{\text{gauge length}}$$

Gauge length is the original length of that portion of the test specimen over which strain or change of length is determined. When films are tested, this is usually taken to be the initial distance between the sample grips. For thicker sections, when the familiar dumbbell-shaped test specimen is used, gauge marks are made on the narrow portion of the specimen. These marks are a distance apart which is less than the distance between the sample grips at the start of the test. Some type of extension indicator or extensometer is frequently used in this case, a technique which is not feasible when testing thin films.

Elongation is the increase in the gauge length of the test specimen. It is expressed as units of length in calculating strain. For films, elongation is usually expressed as the percentage of change of the original gauge length and refers to the elongation at the breaking point of the test specimen.

Stress–strain curve is a diagram in which the corresponding values of stress and strain are plotted against each other, with the values of stress as ordinate and values of strain as abscissa values. The stress–strain curve enables a more direct comparison to be made between different types of materials and structures than does the load–elongation curve, described below, since its shape is governed by the molecular structure of the material. (See p. 523 ff.)

Load is the weight (in grams or pounds), used like gravitational units of force, applied to a specimen in its axial direction which causes a tension to be developed in the specimen. It is common practice to omit the word "weight" and quote the load in grams or pounds.

The *load–elongation curve* describes the behavior of the test specimen from zero load and elongation up to the breaking point. It is produced when the load on the test specimen is plotted against the elongation and may actually be represented by the recorder chart of the instrument.

The *breaking load* is the load in grams or pounds at which the specimen breaks. *Breaking extension* is the extension of the specimen in percent at the breaking load.

Yield point represents one of the major characteristics of the test material. It is that point in the stress–strain curve in which the curve begins to bend and beyond which the material no longer behaves like a spring. Large extensions are produced by relatively small increases in stress and most of the stretch in the test specimen beyond this point is not recoverable. Thus, the yield point is the first stress in a material, less than the maximum attainable stress, at which an increase in strain occurs without an increase in stress. Certain materials have a stress–strain curve in which the yield range is of a gradual curvature, and the yield point cannot be easily selected. Various methods exist for selecting the yield point in these instances. Reinhart, however, suggests that the problem of developing a method for determining and reporting yield strength values for plastics is complicated by the wide variation in the shapes of stress–strain curves obtained with various plastics and their time dependence, particularly at higher strain levels (7).

Young's modulus is an important characteristic which can be obtained from the initial portion of the stress–strain curve, starting at zero stress and strain. Perfectly elastic materials are ones in which the strain is proportional to the stress. They are said to obey Hooke's law; that is, they behave as perfect springs. A given load stretches them a definite amount; when the load is removed, they shrink back to their original length. The first portion of the stress–strain curve is usually fairly straight and indicates the linear relationship between stress and strain; that is, the material behaves like a spring and obeys Hooke's law. In this region, when the load is removed, the material recovers its original length or very nearly so.

Young's modulus for a material which obeys Hooke's law is defined as the ratio of tensile stress to tensile strain.

$$\text{Young's modulus} = \frac{\text{tensile stress}}{\text{tensile strain}} = \frac{\text{force per unit area}}{\text{stretch per unit length}}$$

It is the tangent of the angle between the initial part of the stress–strain curve and the horizontal axis.

3. Interpretation of Tensile Test Data

As indicated from the number of definitions above—only a part of those included in the various standard test methods—a wealth of information is available in the results of the tensile test. The test material can be characterized in general from the shape of the load–elongation and stress–strain curves and from the yield point, the breaking load and extension (12). For example, soft, weak materials have a low modulus, low tensile strength or breaking load, and only a moderate elongation to break. Hard, brittle materials have a high modulus and quite high tensile strengths or breaking loads. They break, however, at small elongations and do not have a distinct yield point. Hard, strong materials also have a high modulus and a high tensile strength. They elongate somewhat more than do the hard, brittle materials. The curves for this kind of material often look as though the test specimen broke about where a yield point might be expected. Soft, tough materials have a low modulus, a marked yield point, and very high elongations. They have moderately high breaking strengths. Finally, hard, tough materials have a high modulus, a definite yield point, high tensile strength, and large elongation, which proceeds in a characteristic manner during the test.

4. Tensile Test Procedure

To obtain reliable results in the tensile test a number of technical details must not be overlooked, since reliability of the test results depends on the sample characteristics as well as on the test machine.

The tensile properties may vary, depending on the thickness of the specimen and the method of preparation. For example, the specimens must be cut with utmost care, as very slight imperfections or nicks in the edges are sources of failure in the form of tears rather than sharp breaks, resulting in low tensile strengths and elongations. As indicated previously, the speed of testing, type of sample grips used, and the manner of measuring the extension also must be considered. In regard to the last item mentioned, soft, highly extensible films will frequently cold-draw back some distance into the clamp.

For testing films, the specimens consist of strips of uniform width carefully cut as indicated above. The nominal width depends in part on the strength of the material being tested, but should not be less than about 0.2 in. or more than 1.0 in. Normally samples 1 in. wide are practical. The specimens should be gauged for thickness, and any should be discarded which vary by more than about 10% over their length. Since most polymer films are anisotropic, a set of test specimens should be cut in each of the principal directions. Five to 10 specimens are usually sufficient to obtain reliable results.

The initial grip separation depends on the material being tested. *ASTM Method D 882* (11) recommends at least 2 in. for materials having a total elongation at break of 100% or more, and at least 4 in. for materials having a total elongation at break of less than 100%. The gauge length selected for highly extensible films is of course limited by the maximum travel distance of the movable sample grip. The speed of testing is the rate of separation of the sample grips. *Method D 882* recommends that this be calculated from the required initial strain rate and depends in part on the type of instrument being used: static weighing or pendulum weighing. For normal day-to-day laboratory procedure, a standard rate can be selected and used repeatedly on the same family of materials.

Performance of the test itself is quite simple. The specimen is placed in the sample grips of the instrument. Usually the upper grip is suspended from a pivot of some kind, and it is therefore desirable to fasten this end of the specimen in place first. The specimen is aligned vertically in the grips and they are tightened evenly and firmly so that no slipping occurs. The tester is started and the load versus the extension is recorded by the instrument. The desired values of load and elongation are selected from the chart and necessary calculations made.

5. Other Methods for Young's Modulus

In the previous discussion it was indicated that the tensile test measures two main attributes of a material: ultimate strength and elongation. However, even though several other fundamental properties which relate the mechanical characteristics and chemical structure can be calculated from the stress–strain or load–elongation curves, there are other instruments which can be used to measure these properties directly. Young's modulus is probably the most important of these relationships, and two specific methods for determining it are described. Whereas the tensile test is a destructive test which is usually affected by local sample variations, these tests are nondestructive and have the advantage of being insensitive to sample variation and speed of testing.

Dynamic Modulus. Hansen (13) describes an apparatus which was developed especially for determining the dynamic modulus in tension of thin cellulosic and other polymeric films. He selected this apparatus in order to study the effect of orientation on the elastic properties of the films, as these effects are very important in packaging film performance. They are difficult to study by other methods such as the use of a torsion pendulum which can detect directional effects only if the width and thickness of the specimen are similar in magnitude.

With the dynamic modulus only very small pieces of experimental film are required, and these are easily adaptable to performing a series of experiments under a variety of conditions.   In this method the film specimen, in the form of a narrow strip, is treated as if it were a spring with spring constant, K. A mass, m, is suspended from the lower end of the spring and the resonant frequency of this system in longitudinal vibration, f_e, is given by eq. (1).

$$f_l = (1/2\pi)(K/m)^{\frac{1}{2}} \qquad (1)$$

The spring constant K of this strip of film is equal to EA/l where E is Young's modulus of the material parallel to the long dimension of the specimen, A, is the cross-sectional area of the specimen and l is the length of the specimen.   Substitution in eq. (1) gives

$$f_l = (1/2\pi)(EA/lm)^{\frac{1}{2}} \qquad (2)$$

The film is held at the end opposite the weight by a clamp which is mounted in the needle holder of a phonograph recording head. This arrangement permits a longitudinal vibration of variable frequency to be impressed on the specimen. The frequency is varied until resonance occurs, and the modulus may then be calculated from this resonance frequency, the cross-sectional area, the length of the specimen, and the mass which it supports.   One of the difficulties in this test is the determination of the cross-sectional area of the test specimen.   In order to overcome this problem, the phonograph recording head is used to excite the transverse mode of vibration in the test specimen by vibrating its upper end in the direction perpendicular to the plane of the film.   The following equation relates the transverse resonance frequency of a string, f_{tr}, to the length l, the tension of the string, T, and the linear density, μ.

$$f_{tr} = (1/2l)(T/\mu)^{\frac{1}{2}} \qquad (3)$$

In this case tension is equal to the gravitational force on the specimen (the product of mass, m, and the gravitational constant, g), and the linear density is equal to the usual volumetric density ρ times the cross-sectional area of the test specimen, A.   Again, substitution gives

$$f_{tr} = (1/2l)(mg/\rho A)^{\frac{1}{2}} \qquad (4)$$

From eqs. (2) and (4) the modulus can be calculated directly from the density, the length of the specimen, and the two resonant frequencies without calculating the cross-sectional area explicitly, as follows:

$$E = [(4\pi)^2/g]\rho l^3 (f_l f_{tr})^2$$

Apparatus. Vibrator. This is a commercially available phonograph recording head with an internal resistance of 500 Ω. An oscillator with sufficient output is required to drive the recording head without additional amplification.

Upper Clamp. The upper clamp consists of a steel rod, $\frac{1}{8}$ in. in diameter, with one end turned down to fit in the needle holder of the recording head, and the larger end slotted to hold the test specimen. The sample is held in place by pulling the two faces of the slotted end together by means of a screw through the center of the rod in the plane of the slot.

Lower Clamp Assembly. Two small aluminum blocks held together by small screws comprise the lower clamp. They contain two eyelets, one on either side of the clamp, to receive the hooks which support the weight.

Weight. A 2-in. 1/4-20 bolt with a stack of washers held between two nuts makes a suitable weight. One nut is turned up against the head of the screw and serves to hold the suspension hooks in place. The weight can be varied by changing the number of washers on the screw.

Mount for Vibrator. This is an L-shaped block with a shaft extending backward from one leg to permit its rotation and a guide rod for the jig to mount the test specimen. The vibrator is suspended from the other leg by two U-bolts. Rotating the block allows the clamp to be pointed downward for the determination of the transverse resonance frequency or horizontally for the longitudinal resonance frequency.

Frame. The frame should be constructed of heavy steel plate and rest on vibration mounts so that the sample is not influenced by external vibration.

Jig. To mount the specimen in the apparatus so that it is properly aligned, a jig is necessary.

Procedure. Carefully cut the test specimen 2 mm wide by 60 mm long. A special template may be used to insure proper preparation of the test specimen which should be handled carefully to prevent moisture and oil from being absorbed from the fingers. Place the specimen in the jig to attach the bottom clamp. Then slide the jig with its sample onto the guide rod, fasten the other end of the test specimen in the upper clamp, and remove the jig.

Permit the lower clamp to swing freely and time the period of its oscillation. Then, by vibrating the upper end of the specimen horizontally, measure the resonant frequency of the transverse vibration. Finally, vibrate the upper end of the specimen vertically and find the resonant frequency of the longitudinal vibration.

From the first result calculate the length, l, of the specimen. Use the second to find the cross-sectional area. Finally, use the third measurement along with the other two to find the modulus of the specimen.

Sonic Modulus. It may be assumed after studying the dynamic modulus test that it is a tedious test to perform. For this reason, it is not a completely suitable test, especially if it should be desirable to determine the effects of low temperature on the modulus. The sonic test has the advantage of simplicity (14).

In order to establish a method for determining Young's modulus of an elastic (or viscoelastic) material, we have to answer the following question: When a strain is produced at some point in an elastic material, what happens throughout the rest of the material? For a rod it can be shown that if the rod is stretched at some place in the direction of its length and then released, the disturbance will move along the bar with a velocity

$$c = (E/\rho)^{\frac{1}{2}}$$

constituting what is called an elastic wave. The velocity, c, is the velocity with which sound would travel along the bar, for sound is propagated in this manner; E is Young's modulus; and ρ is the density.

The sonic tester has two transducers, one fixed and one movable. An electronic network produces a spike voltage of 10^{-5} sec duration which is transmitted to a ceramic piezoelectric transducer resonant at 10,000 cps.

Simultaneously, an internal timing circuit is initiated. Attached to the sending transducer is a round tipped metal probe which touches the test specimen and vibrates parallel to the surface.

A longitudinal sonic wave is thus propagated in the test material in the direction of vibration of the probe and a sonic shear wave is propagated normal to the direction of vibration in the contact plane. The sending transducer is shocked into oscillation once every 5 msec, which is a low enough pulse repetition rate to allow each mechanical pulse to dampen out completely before the succeeding pulse is propagated. A second transducer, a measured distance away and properly oriented, receives the sonic pulse (either compressional or shear mode) and converts it into an electrical signal which closes the timing circuit. The elapsed time is registered on a recorder and can be plotted as a function of distance between the probes to give a linear relation. The inverse slope of the line is the sonic velocity.

In the case of plastic films certain boundary conditions are dictated by both theory and practical aspects. Commercial films are in all probability not isotropic in regard to their physical properties, and anisotropy and symmetry may have to be taken into account. The geometry of the film may also have to be considered since it consists of a plane with thickness very small in comparison to its planar dimensions. Considering these conditions and assuming that the axes of symmetry coincide with the axes of manufacture, the relation between sonic velocity and Young's modulus for an anisotropic, thin film can be shown to be

$$V_L = [E/\rho(1-\sigma^2)]^{\frac{1}{2}}$$

where V_L is the velocity of the longitudinal wave, E is Young's modulus, ρ is the density of the film, and σ is Poisson's ratio for the film.

Poisson's ratio for stretched cylinders or prisms is the ratio of the lateral contraction to the longitudinal extension when there is a longitudinal stress but no lateral stress acting. The accuracy of the determination is usually limited by the accuracy of the transverse strain measurements because the percentage of errors in these measurements are usually greater than in the axial strain measurements. The standard methods for the determination of Poisson's ratio at room temperature generally use two pairs of extensometers: one pair to measure the longitudinal strain and the other to measure the transverse strain. In a determination of this ratio for thin films, strain gauges and extensometers impose local stresses which are unacceptable.

Poisson's ratio can also be determined using the sonic tester, although the author has not developed the exact method. Poisson's ratio is apparently dependent on the geometry of the test specimen which must be given special consideration in the sonic test. Unless specimens of the same shape are tested, it is necessary to determine this ratio in order to arrive at the absolute value of Young's modulus. The sonic velocity, and therefore also the modulus, are responsive to the degree of orientation of the test specimen. The device can thus be used to determine the anisotropy of the film.

Most of the work to date on the determination of the physical characteristics of polymeric materials by means of the sonic modulus test has been confined to fibers, and work is only now beginning on the test of film material. It is therefore not possible to describe a test procedure. Sample size and configuration will depend on the equipment used—whether the film is tested as a strip which can be accomplished on a commercially available instrument or as a sheet on an instrument designed for that purpose.

B. Stiffness

Stiffness as a property of polymer films is of considerable importance to the flexible packaging industry because this characteristic greatly influences the performance of the film when used on various kinds of packaging machines. In general, stiffness may be considered to be the resistance of an object to almost any type of mechanical distortion. However, in packaging applications it is usually considered to be the resistance of the film to bending, which is not a simple property but rather one which depends on two other properties, the thickness of the film and the inherent stiffness of the substance of which the film is made.

In the development of new flexible films the quantity desired from a stiffness test is not only the stiffness of the finished film, which may be dependent on the processing variables as well as the thickness, but also the inherent stiffness of the material comprising the film; that is, it is necessary to determine the relationship between the stiffness of the film and the combination of inherent stiffness and thickness. In principle, this may be done by testing films of different thicknesses made of the same material. However, it is not safe to assume that the inherent stiffness of the material will be the same in film varying widely in thickness because of the different conditions which may exist during its fabrication.

Stiffness may be measured in several different ways. Clark (15) describes seven types of tests: cantilever types, beam, column, folding; modified cantilever; pendulum, either simple or torsion; and a method he developed for paper which is also a modified cantilever but is sufficiently different to be classed separately. Clark further defines two properties which he believes are measured: (1) stiffness, which is the ability of the material to support its own weight and (2) rigidity, the property of the material which resists an applied force, or its flexural resistance. If these two properties are synonymous, then several additional methods are available, for, as flexural rigidity, stiffness can be calculated from the dimensions of the test specimen and Young's modulus of the material. Thus, stiffness may be determined from data obtained in the tensile test or in any of the various tests used to determine the modulus. This may not always be the most precise method, however, since most polymeric films are usually inhomogeneous and anisotropic.

Young's modulus is probably the most widely used measure of inherent stiffness of a single material because it is nearly independent of dimensions and stress for small strains. The determination of Young's modulus in the tensile test has been described above. In the case of composite materials, i.e., laminations consisting of two or more different materials in which the properties are not well defined, it may be more realistic to measure the flexural rigidity.

There are several easy methods for measuring stiffness which are quite direct. When using methods of this kind, however, one should determine whether there is a thickness effect by comparing the desired test with another stiffness test whose thickness dependence is already known or may be derived from first principles. In this type of comparison one assumes that the measures of inherent stiffness determined by both methods are the same, and the extent to which this assumption is justified will be indicated by the degree of correlation ultimately found between the two methods.

1. Stiffness Test Methods

Handle-O-Meter (16–18). The Handle-O-Meter is one of the simplest tests available, although its use is generally limited to thin films (Fig. 4). It has been shown to be a good measure of flexural

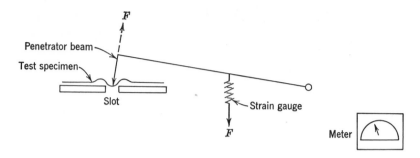

Fig. 4. Schematic drawing of the Handle-O-Meter Stiffness Tester.

rigidity, since each small element of the sample actually undergoes an extension or compression in an amount determined by the modulus of the material.

The principle of the test is simple; the method consists of measuring the force required to push a film specimen into a slot of a given width over which it has been laid. The instrument thus consists of a metal plate with a slot that extends completely across the plate. The width of this slot is normally 5 mm but may be wider, depending on the thickness of the test specimens. For the sake of consistency, the width of the slot should be constant for films of the same thickness or range of thicknesses which the instrument can accommodate. When the instrument is in operation, a horizontal bar parallel to the slot moves downward through the slot.

In performing the test, a square sheet of film, 8 in. × 8 in., is placed on the metal plate so that the center of the sheet lies across the slot. To determine the machine-direction stiffness, the sheet is oriented with the machine direction perpendicular to the slot; in the transverse-direction test, the transverse direction of the sheet is perpendicular to the slot. When the instrument is turned on, the horizontal bar is lowered automatically through the slot against the resistance of the film. The force exerted on the bar as it pushes the test specimen through the slot is sensed by a strain gauge mounted on the beam which holds the bar, and is indicated by a meter. The meter dial is calibrated directly in grams of force. As the bar is lowered, the force is observed to increase, pass through a maximum, and then decrease. The maximum value of the force is taken as the Handle-O-Meter stiffness of the material. The limit of the scale of the meter is 50 g. If the maximum reading is higher than this, the specimen is cut in half in the direction perpendicular to the slot and the measurement is repeated. This result is then multiplied by 2 so that it will correspond to an 8 in. × 8 in. sheet. If the stiffness is still too high, the specimen is cut in half again and the result multiplied by 4, and so on.

Morton and Marks (19) have made an extensive study of the Handle-O-Meter as a method for measuring the stiffness of films. Using the 8-in. × 8-in. test specimen they found that the Handle-O-Meter values depended on four factors: the flexural rigidity of the film; frictional resistance between the film and the surface of the instrument (which was particularly important for films having a high charge of static electricity); friction of the film against the edges of the slot; and finally, the leverage exerted by the weight of the overlapping film. The latter, they believe, was the most important because when allowance was made for this gravitational pull the expected dependence of film stiffness on the cube of the thickness was found.

Reproducibility of the results varies from film to film, depending on the uniformity of thickness of the film being measured. For most of the films tested, the range of six measurements was within ±15% of the mean.

Cantilever. Clark (15) suggests that the oldest and most obvious test for stiffness is that of using the material as a cantilever. In a test such as this, one end of a strip of film having parallel sides is clamped in a horizontal plane and the droop of the strip under its own weight is measured at the free end as a measure of stiffness— either by measuring the distance through which the free end has dropped, by measuring the angle which the tip makes with the edge of the clamps, or by measuring the length of the horizontal projection of the bent strip. The horizontal projection of the test strip may be lengthened until the free end of the strip drops a preselected amount, or to a fixed angle, and then measuring the length of overhang.

Beam. Tests of this type are frequently used for heavier sections. A strip of the test sample is placed across two supports which have been suitably spaced and the beam thus formed is then loaded at its center. The deflection produced in the beam is measured. This test may be performed in several ways: the beam may be subjected to a constant load and the deflection determined at that load; the beam may be deflected a constant amount, the load required to obtain the deflection being measured; and the maximum lead which the beam is able to support can be determined. The Handle-O-Meter is a method which determines the maximum load which the beam, in the form of a thin film, can support.

Modified Cantilever. The modified cantilever tests are represented by the Taber and Gurley instruments and the Clark Softness-Stiffness tester

The Taber test measures flexural rigidity or stiffness by determining the bending moment required to deflect the free end of a vertically clamped test specimen 15° from its center line when the load is applied 5 cm away from the clamp. This test is applicable for testing a variety of materials such as plastics, paper, cardboard, light metallic sheet, foil, wire, and other flexible materials up to $\frac{1}{8}$ in. in thickness. The test is described in detail in *Tappi T 489* (20).

The Gurley Stiffness Tester measures the bending moment in both directions which a short strip of paper will withstand before bending to a certain degree by means of a small weighted pendulum. The balanced pendulum is pivoted at its center with weights which may be clamped 1 in., 2 in., or 4 in. below the center, in holes provided for that purpose. The pointer moves parallel to a scale, which is graduated in both directions. In the test, the pointer is deflected by a sample of the test material pressed against its top end. The length and width of the specimen strips are variable and a wide range of materials can be tested.

The Clark Softness-Stiffness Tester (15) can be used with considerable success to measure the stiffness of very soft films such as low-density polyethylene. The test consists of finding the length of a piece of test material which, when clamped at one end, will just fall from one side to the other when the clamped end is rotated through an angle of 90°. The instrument consists of a rotating clamp formed from two rollers which are held together by spring pressure, one of which may be slowly turned by a worm drive. The line of contact of the two rolls forming the clamp is the axis of rotation of the clamp which can be rotated in either direction by means of a small electric motor.   A pointer attached to the frame of the clamp indicates the relative angular position of the framework against a rotatable circular scale.

Test specimens are cut in each of the principal directions about two inches wide and of a convenient length.   The edges of the strip should be parallel.   One end of the test specimen is placed in the clamp, and the overhanging length should be such that when the clamp is rotated from one side to the other through 90° at a speed of 1 rpm, the strip does not yet fall over to the other side.   The length of the overhanging strip is shortened slightly by means of the worm drive, and the clamp assembly is rotated in the opposite direction again through an angle of 90°.   The operation is repeated after the length of the overhanging specimen has been shortened, and the clamp is rotated through 90° until the length of the test specimen is reduced to the point where it will just fall from one side to the other in the 90° rotation.   The effective length is then measured from the line of contact of the rolls to the end of the test strip.

The instrument and test method are described in *Tappi T 451* (21) and equations are given for calculating rigidity, stiffness, and softness.   The test is very sensitive to small differences in the stiffness of soft and thin materials, but care must be taken to insure that the test specimen is completely free of static.

C. Flexibility

In the stiffness tests described previously, stiffness was also defined as "flexural rigidity" which can be determined from the dimensions of the test specimen and Young's modulus.   Although the stiffness tests may be bending tests whereas the modulus is determined in simple tension, the relationship exists because each small element of the film in a bending test is actually undergoing an extension or a compression, the amount of which is determined by the modulus as measured in a tensile test.   Similarly, the ability of the film to withstand repeated flexings or foldings in a flexibility test must in part be related to the modulus in the same way, for each small element of the test specimen in the region of the fold is alternately stretched and compressed.   Thus, one can surmise that a film having a low modulus, low tensile strength, and yield point, with low elongation, will exhibit poor flexibility.

Another film with low modulus but which may have a slightly higher tensile strength and very high elongation will have high flexibility. On the other end of the modulus spectrum, a film with a high modulus, no yield point, but quite high tensile strength and low elongation, will have low flexibility; however, a film with high modulus, high yield point and tensile strength, and high elongation will have very high flexibility.

1. Fold Endurance and Flex Tests

Schopper Folding Endurance Test. This test is used more or less universally as the standard test for establishing the folding endurance of paper. The test requires test specimens 15 mm wide by 100 mm long. The test specimen is held at each end by sample grips which apply a constant tension to the material. A slotted blade, driven back and forth by a motor, catches the specimen at its midpoint and folds it back and forth at a rate of 120 double folds a minute until it breaks. The number of double folds is recorded on a counter. The sample grips will accommodate a test specimen up to 10 mils in thickness. The method is described in detail in *Tappi T 423* (22).

M.I.T. Folding Endurance Test. The M.I.T. folding endurance tester consists of a fixed sample grip and an oscillating sample grip mounted on a disk so that the top of the grip is at the center of the disk and the bottom is at the edge of the disk, i.e., the face of the sample clamp lies on a radius of the disk. When the disk oscillates, the face of the clamp moves through an angle of approximately 135° from the vertical. The lower sample grip changes its direction of motion at a rate to provide 175 double folds per minute. The upper clamp is spring loaded, and loads of up to 1.5 kg can be applied to the test specimen. A counter records the number of double folds required to fracture the specimen. The counter is disconnected immediately upon failure of the material and the tester automatically stops. Sample grips of different opening widths are available in 10-mil increments.

While this test, as well as the Schopper test, may not always be applicable to polymeric films because of the number of folds required to break the test specimen, one should check the effect of the thickness of the test specimen. For example, 1-mil nylon film may withstand over 200,000 double folds in the M.I.T. test before failing, but 10-mil nylon may fail after fewer than 10,000 double folds.

The M.I.T. fold test is frequently used as a specifications test for paper as well as for other sheet materials. It is described in *ASTM D 2176* (23).

Stress–Flex Tester (24). The stress–flex test seems to be a good empirical test of flexibility and durability of films which have a moderately high modulus with not very great elongation. Horst and Martin, who developed the stress–flex tester, point out that there has long been a need in the packaging industry for a single test which would predict in the laboratory the commercial package durability performance of flexible packaging films. They had observed that certain films which were apparently satisfactory based on extensive and exhaustive testing by standard physical tests sometimes exhibited poor durability in use. They further noted that these films would perform poorly in a hand-flex test, which is commonly used by a pressman or the operator of a packaging machine. He grips a piece of the film in both hands with the thumbs parallel and several inches apart. Then moving his hands together until his thumbs almost touch, he moves his hands back and forth, keeping his thumbs parallel, alternately stressing and flexing the film specimen.

This is in essence also a description of the stress–flex test apparatus which consists of clamps which grip a film specimen, with dimensions of about 4 in. by 7 in., in the same manner as do the hands in the preceding description. One clamp is fixed to a framework and the other is free to move on a second parallel member of the framework. As the framework rotates, the movable clamp falls until it is restrained by the film specimen. The film is subjected to a flexing action and an increasing stress at the end of each stroke because of the elongation of the material. The framework revolves at 38 rpm, subjecting the specimen to 76 stresses and flexes per minute. Initially, the movable clamp falls a distance of approximately 4 cm, the impact energy depending on the weight of the clamp.

In this test the specimen is subjected to complex forces which defy precise definition. Failure of the film is related to repeated impact which stretches the sample to its breaking point and the initiation and development of pinholes, either or both. Factors which can affect the test values but which can be readily checked and controlled are the distance between the clamps, the weight of the movable clamp, and the velocity of the clamp. The latter is a function of the falling distance, friction, and the speed of rotation.

Sample dimensions are an important consideration also, since the distance the movable clamp can fall depends on the width of the test specimen. The number of stress–flex strokes increases as the sample width decreases because the distance of fall of the movable clamp is reduced. It is therefore necessary to prepare the test specimens carefully.

The major drawback to the stress–flex test is the wide range of test values. A sample of standard-weight cellophane tested repeatedly over a period of weeks had average values ranging from 19 to 33 strokes and standard deviations ranging between 7 and 17. It is obviously advisable to check whether the number of specimens tested is sufficient to give the desired degree of reliability.

Horst and Martin believe that the stress–flex tester represents a major step toward providing a laboratory tool for predicting the commercial durability performance of a film. However, the interpretation and intelligent application of stress–flex data require a thorough knowledge of all factors affecting both the test itself and the durability performance of the film.

The Gelbo Instrument (25). Since the most important property of a water-vapor-proof barrier material is its low water vapor transmission rate, the Gelbo tester was designed to offer a means for predicting the performance of materials under application and use conditions in this regard. Thus, this instrument is not intended to measure flexibility directly but rather the effects of flexing on the continuity of the film, namely, the formation of pinholes and breaks in the region of the creases which develop.

In studying the various uses of flexible barrier materials, Gelber and Bowen noted that the barrier material in most applications was subjected to fairly sharp creasing followed by a force along the crease in a direction roughly perpendicular to it. They attempted to simulate this creasing and flexing action which they have accomplished in the Gelbo tester. The test specimen is made up as a cylindrical sleeve which is fitted over the heads of the machine and clamped in place. One head is stationary and the other is rotated while it is moved back and forth. The rotating motion of the movable head is controlled by a groove in the shaft to which it is attached. The groove follows a pin on a stationary bearing housing. Initially the movable head had a stroke of 6 in. with a continuous rotation of 180°. The length of the stroke can be decreased by $1\frac{1}{2}$ in. The test specimen is 8 in. $\times$ 12 in. and is sealed on the 8-in. side to form a cylinder. The 6-in. stroke of the tester provides $3\frac{1}{2}$ in. of twisting action and then $2\frac{1}{2}$ in. of pushing action. Since the tester and test method were originally designed for moderately heavy film combinations, the author found that the test is not generally applicable to softer film types because pinholes do not form except after a great deal of abuse. In the case of those film types which are abused by the flexing action of the tester, the number of pinholes in the test sheet increases markedly as the number of cycles increases. The speed of the flex cycle in the range tested appears to have only a negligible effect on the number of pinholes generated.

Other Flexibility Tests. One can conceive of other flexing actions which might be incorporated in a test device. For example, the principle of the Gelbo tester could be used to measure flexibility directly rather than indirectly by means of a permeability test. Instead of the large tube used in the Gelbo test, a much smaller tube can be used, this tube being alternately compressed and extended in the direction of its major axis until failure. The number of cycles is a measure of the flexibility of the film. At the beginning of the test the tube is just slightly inflated, and when a pinhole forms as a result of the flexing action, the excess air or gas escapes reducing the pressure to normal. This slight change in pressure can be used to actuate a pressure switch to stop the test instrument. Again, a test such as this is not always satisfactory for very soft films because they either do not fail or they require too long a time to complete the test, an indication in itself of very great flexibility.

Conti (26) describes a test similar to the stress–flex test. This device was also designed to reproduce to some extent the hand flexing test. For test samples, sheets are cut $7\frac{1}{4}$ in. by 10 in., the 10-in. length cut in the machine direction. The jaws of the tester grip opposite edges of the sample and move in a parallel plane, but in opposing directions. An operator familiar with the test can immediately detect the slightest rupture by sound, but in practice it is best to stop the tester periodically to inspect the film. A counter records the number of stress and flex cycles.

Several other flex testers are in common use for testing rubber and fabric materials, leather cloth, belts, tires, and other articles composed of fabric bonded together by rubber compounds. Some of these methods require molded test specimens. They are not generally applicable to testing films.

In summary, the major difficulty of any flexibility test of polymeric films results from the very flexible character of the material. The film does not fail, even after long periods of repeated folding or flexing. Other tests, such as the determination of the modulus or of stiffness, may therefore be better measures of flexibility than the tests described. If any of the foregoing tests of flexibility are used, care should be taken in the preparation and conditioning of the test specimens, especially if the material has a special affinity for water since any variation in softener content, including water, can significantly change the test results.

D. Impact Strength

Not long ago the selection of polyethylene film for use in many packaging applications became more frequent because of its ability to withstand severe handling conditions. Unfortunately at that time it had a weakness which was probably caused by process variables and which was manifest in its lack of impact strength. As the number of applications for this novel polymer film increased, there was a corresponding increase in the number of instances in which this particular weakness was apparent.

Almost every user of polyethylene developed some sort of test by which to estimate impact strength because he was vitally concerned with the ability of the packaging material to withstand shocks of various degrees. These tests generally took the form of tests of increased severity in which polyethylene bags were filled with a quantity of sand, for example, and dropped repeatedly from a given height, or through a series of heights, until failure. This was a time-consuming job and not very practical, for the manufacturer of the film needed an accurate test, as well as one in which he could sample the film at the beginning of a new mill roll, test it quickly, and make the necessary adjustments in the extrusion conditions for maintaining a high impact-strength film. Various tests were tried, and these took the form of falling-ball or dart tests and pendulum-impact tests, probably because similar tests were used in other industries.

Thus, the purpose of the impact test, as it developed for polyethylene film, is to determine the resistance of the film to failure under conditions of high-velocity impact. Nielsen (6) points out that the characteristics measured by most impact tests are quantities hard to define in scientific terms, but they have great practical importance. He describes various tests which have been used primarily for the evaluation of heavier sections. Here we describe the most common impact tests used on film.

1. Falling-Ball or Dart Tests

The falling-ball or dart test has taken several forms in regard to the manner of measuring the impact strength of the test specimen (Fig. 5). In some instances the combination of weight of the dart and height of drop is selected to insure that the test specimens will always be ruptured. The dart is decelerated by the test specimen, and its kinetic energy after it has passed through the specimen is calculated from the velocity by measuring the time of flight between two disjunctors, each consisting of a light source and photoelectric cell placed immediately below the test specimen, which start and stop an electronic counter chronograph. The difference between the potential energy of the dart at the point from which it is dropped and the kinetic energy after it has passed through the film is the amount of energy absorbed by the film and hence its impact strength. In a more elaborate setup, a second pair of disjunctors may be located just above the test specimen to determine the velocity of the dart immediately before it strikes the film specimen. The impact strength is determined as just described but with the initial energy determined from the measured velocity of the dart just before it strikes the film.

The method adopted by ASTM Committee D-20 on Plastics, *ASTM Method D 1709* (27,28), does not require an expensive chronograph arrangement to determine the dart velocity. In this test the dart is dropped from a constant height, but the weight of the dart is increased in equal increments from the minimum just light enough not to rupture any test specimens to a maximum just

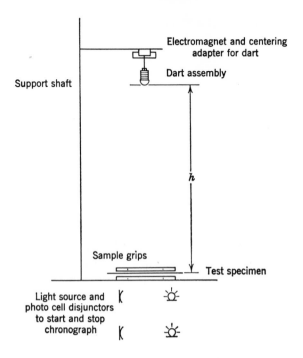

Fig. 5.   General configuration of drop impact testers.   For Method A of *ASTM D 1709* the height of drop, *h*, is 66.0 cm.

heavy enough to rupture all of the test specimens.   Impact failure weight is the dart weight at which half of the test specimens fail.

Two variations in the method are described: one which is intended for films with impact strengths of 300 g or less, and the other for films with impact strengths of greater than 300 g and up to a maximum of about 1300 g.   The difference between these two tests is in the initial weight of the dart.   In the test for lighter-weight films, the dart consists of a hemispherical head 3.81 cm in diameter fitted with a shaft 11.43 cm long and 0.64 cm in diameter to accommodate removable weights and a locking collar to hold the weights in place.   The total weight of this assembly without any additional weights is 32.0 g, including the locking collar.   The shaft is attached to the center of the flat upper surface of the hemisphere so that its long axis is perpendicular to the surface.   It is made of aluminum but has a long steel tip at the end so that it can be held for dropping by a magnet.   Additional weight may be placed on the shaft by removing the locking collar.   The dart is suspended by the magnet, and when suspended, the distance between it and the film surface is 66.0 cm.   For testing heavier films the dart consists of a hemispherical head 5.08 cm in diameter fitted with an identical shaft and locking collar, the total weight of which is 320 g.

Apparatus.   The free-falling dart impact tester consists of a table with a hole in the center and to the back of which is fastened a shaft or pole for supporting the magnet used to drop the dart. The sample grips which are located over the hole are circular with an inside diameter of 12.7 cm.   The lower half of the clamp is rigidly mounted to the table so that the plane of the test specimen is horizontal and at right angles with the pole which supports the magnet.   The upper half of the clamp moves up and down by means of small air cylinders or any other suitable mechanical method, but this part of the clamp must be able to maintain sufficient contact with the lower half of the clamp to hold the film sample in place during the test without any slipping.   To avoid slipping, rubber gaskets may be cemented to the faces of the sample grips.   This, of course, minimizes any variations in the thickness of the test specimen.

The magnet should be capable of supporting a weight of 2 kg. It is mounted in a suitable bracket which can be moved up and down the pole to the desired height.   It should extend outward from the pole over the sample clamp and should contain a centering hole to insure that the dart is located over the center of the test specimen when suspended from the magnet.

In those instances when the dart ruptures the film, some arrangement should be made for catching it beneath the table to prevent the head from becoming nicked or scratched or to avoid any other surface irregularities.

Procedure.   Cut the test specimens approximately 7 in. wide by at least that long so that they can be easily placed in the sample grips.   Care should be taken to insure that the test specimen is flat in the clamp, free of any folds, and that it covers the entire clamp surface in order to avoid slippage.   Any slippage of the test specimen in the clamp will result in high values.   Select a dart weight which should break the film sample approximately half the time and adjust the dart to this weight. Energize the magnet and insert the steel tip of the shaft into the centering hole. Then release the dart by deenergizing the magnet. The operator should attempt to catch the dart immediately after it strikes the test specimen to avoid multiple impacts if the dart bounds off the film. Examine the film samples to determine whether the dart has ruptured it. Repeat this procedure on a new specimen for a total of 10 test specimens. Then change the dart weight by uniform increments of 5 g or more and test a minimum of 10 specimens at each dart weight.

The weight increment should be selected to give a minimum of three points between the no-failure weight and the 100% failure weight. The impact failure weight can be calculated by plotting the data on probability graph paper with the weight of the dart on the linear scale and the per cent failure on the probability scale. Draw the best fitting straight line through the points and select the missile weight for the test sample from the intersection of this line with the 50% probability line.

The impact failure weight can also be calculated from the formula:

$$W_F = W_L + \Delta W (S/100 - \tfrac{1}{2})$$

where W_F is the impact failure weight or, by definition, the weight at which half of the specimens fail; W_L is the lowest missile weight, in grams, according to the particular increment being used, at which failure of all specimens occurs (10 specimens); ΔW is the weight increment being used; and S is the sum of the percentage of breaks at each missile weight from the weight at which no failures occur up to and including W_L.

It should be noted that the graphic method does not require that the 0 and 100% failure weights be determined. Moreover, the weight increments need not be uniform. When using the above formula to calculate the impact failure weight, equal weight increments must be used. The two methods give approximately equivalent values.

2. Pendulum Impact Tests

Most references on mechanics treat the subject of the physical pendulum at some length. Its history as a device for measuring energy is old, dating back at least to the early 1700's when Benjamin

Robins used it for measuring the energy and velocity of a bullet. Modern industry has used the pendulum, too, in the Izod and Charpy testers for the measurement of certain strength character-istics of metals and, more recently, of plastics. Many tensile testing devices measure the load being applied to the test specimen from the displacement of the pendulum bob as indicated previously, and the Izod tester has even been adapted for use as a high-speed tensile tester.

In considering a test for the impact strength of films, it was natural because of this history to study ways of using a pendulum. The physics of the instrument for this use is straightforward and corresponds to that of the falling-ball test. A value of impact strength is obtained by measuring the residual energy of a falling mass, the pendulum in this case, after it has ruptured the test specimen. In the case of the pendulum tester, the motion of the mass is along the arc of a circle because of the invariable length of the pivot arm. This residual energy can be measured in a number of ways. The easiest device consists of a pointer which is moved by the pendulum along a calibrated scale. Other devices for indicating the height to which the pendulum moves after being decelerated by the test specimen are possible, such as an electric timer actuated either by a sliding contact or a pair of photoelec-tric cells.

A pendulum impact tester should be able to rupture the wide variety of packaging films or sheet materials available at present. If the shape of the pendulum is properly designed, this can be accomplished by a simple addition of weights in order to obtain the desired range of impact strengths without sacrificing the sen-sitivity of the instrument in the low-strength range. The impact-ing head should be easily removable for replacement with heads of various shapes or sizes and its position should be adjustable for ease of alignment and calibration of the instrument.

The device for indicating the impact strength should be one which adds little friction to the system to be overcome by the pendulum, as well as one that is not affected by its own inertia. For this reason, the pointer is not the most satisfactory indicator. Moreover, the pendulum should be supported by the best possible bearings to have minimum friction and in a manner to eliminate any sidewise motion. These bearings should be oilless to enable low-temperature testing. It is highly desirable that the tester should be usable in any atmosphere.

The location of the sample clamp is also important. It should be so located that the vertical line through the point of suspension and the center of gravity of the pendulum hanging at rest lies in the plane of the film specimen when clamped in place. With the pendulum in this vertical position, the impacting head should just touch the film specimen. The specimen clamp should be positive in its action, holding the film securely; for if the film slips in the clamp upon impact, high values will result. If the clamp is operated by a foot switch, both hands of the operator are free for inserting the test specimen.

Although of minor consideration, the pendulum release mechanism should not be overlooked, as the manner in which the pendulum is released to start its swing may be a source of variability in test results. Finally, the instrument should contain all necessary safety devices.

A pendulum impact tester which was designed to include these specifications has been built by the author (29). The impacting head is suspended below the major mass of the pendulum in order to obtain a value of impact strength which would be most nearly representative of the resistance of the film to the impacting head regardless of its shape. The catch for supporting the pendulum in its starting position is located a sufficient distance above the impacting head so that the only objects passing through the test specimen are the impacting head itself and a short length of pendulum above it which has been sharpened to form a knife edge. The latter tends to slice its way through the softer materials and only a small percentage of the measured impact strength may be considered to be due to this cutting action. In the case of polyethylene film up to 6 mils in thickness, this amounts to about 5% of the measured strength. In the case of materials more brittle than polyethylene, such as paper or cellophane, the fraction of measured impact strength owing to the cutting action is negligible, since these materials shatter on impact.

In the design of the pendulum, if the mass is distributed symmetrically around a line through the axis of rotation and the center of gravity, a change in the mass changes only the equivalent length. This symmetry was obtained by designing the pendulum in the shape of a sector of a circle with a portion of the rear half of this sector removed to compensate for the weight of the impacting head and knife from which it is suspended. Weights were designed having the same size as the sector. These may be attached to the pendulum for providing several test ranges. Any number of different weights may be made and the instrument calibrated accordingly. Regardless of its mass, the pendulum may always be considered a simple pendulum for the purpose of calibration.

The pendulum is fastened to a heavy hub on the steel supporting shaft. This shaft rotates on stainless steel bearings and any sidewise motion has been eliminated. Switches for actuating the timer on the basis of which the instrument is calibrated and the automatic return mechanism are also mounted on this shaft.

The release mechanism of the pendulum is actuated by means of a push-button microswitch. As a safety measure for the protection of the operator, the pendulum cannot be released without first closing the sample clamp. This is done by pressing a foot switch, leaving both hands of the operator free for handling the test specimens. The clamp opens automatically after the pendulum has been returned to its starting position. The clamp itself makes use of five small air motors mounted in a circular pattern, and it has a very positive action which holds the film sample firmly in place.

Procedure. The pendulum impact tester should be standardized at the beginning of the test. This can usually be accomplished by checking the cross test level mounted on the instrument base and adjusting the level of the instrument, if necessary, by means of leveling screws. The position of the striking head with respect to the film specimen, depending on the type of tester being used, should also be checked when the pendulum is hanging at rest, and any necessary adjustments should be made.

Cut the test specimens to the proper size. Since many films have a directional character, it is imperative to cut the test specimens always in the same direction. Experience has shown that it is desirable to cut the specimen so that the machine direction of the material is perpendicular to the plane of the pendulum when inserted in the clamp. With the instrument standardized as recommended, place a test specimen between the faces of the sample grip with the top edge of the film lined up with the top edge of the clamp. Hold the film taut, but do not stretch or distort it; then close the jaws. Set the pointer or whatever means is used to determine the impact strength to the proper setting for the beginning of the test, release the pendulum, and allow it to rupture the specimen. Record the value, remove the ruptured specimen, and repeat the steps of the test until the desired number of specimens has been tested.

It is apparent that this is a simple and rapid test which makes it ideal for both research and quality control purposes.

The Elmendorf Tear Tester has also been adapted for measuring the impact strength of films. The results obtained with this tester can be reported as pounds per square inch or grams per square centimeter using the proper factor.

3. Impact Fatigue

The pendulum impact test and, in general, the dart drop or falling-ball test are methods in which the film is subjected to an impact in excess of that which it can withstand. In practice it is probably more likely that the film will be subjected to a series of impacts which, initially, are not of sufficient force to result in failure. Rather, failure is due to fatigue from repeated impacts. Although a repetitive test could be performed using the equipment described above, it is obvious that such a test would be very time consuming. However, there are at least two test devices for performing a repetitive test.

Impact Fatigue Tester. This device controls the rate at which stainless steel balls drop on a rigidly held piece of film or sheet material. It consists of an inclined plane which holds a number of balls and which can be raised to allow the balls to be dropped from any height between 8 in. and 17 in. The test specimen is held by means of a clamp located on the base of the instrument so that the balls strike the sample as they are released, one after the other, at the bottom of the incline. By use of varying heights of drop and balls of different sizes, a great variety of conditions can be obtained. When the sample fails, a gate prevents the remaining balls from falling.

Frag Tester. Although this device was originally intended for testing paper, it has more recently been used with equal success on various packaging films. In essence, the test consists of a container open at both ends, the bottom of which can be closed off by the test specimen. The film is not stretched tightly across the bottom but is fastened in place with a controlled amount of drape. A number of steel balls are then placed in the container. The entire assembly consisting of container, film specimen, and balls is then raised by a motor to a predetermined height and allowed to drop in free fall. This action is repeated until fracture of the test specimen occurs and the balls fall through, actuating a cutoff switch. The tester is so designed that the ruptured sample may be removed, another inserted in its place and then the dropping assembly inverted to start the next test. Arnold has stated (30) that for paper the number of drops increases with the cube of the impact strength.

E. TEAR STRENGTH

It is rather difficult to start a tear in most polymer films; however, if the film has been notched at some point along an edge, a tear can usually be propagated with relative ease from this point. Thus, the most common tear tests consist of measuring the force required to continue a tear rather than the force required to initiate the tear because the film is more likely to fail when the former force is low.

Many of the tests for determining the physical properties of plastic films have their origin in the paper industry, and the Elmendorf tear test—originally a strength test for paper—is one of these, dating back to the 1920's. This test measures the average force required to propagate a tear through a specified length of paper, plastic film, or nonrigid sheeting. The term *specified length*, is of importance when testing films having varying degrees of orientation and extensibility because many films do not tear in a straight line over the specified length of the test specimen. However, provisions have been made in the method to compensate for the tendency of the tear to turn obliquely across the test specimen toward the transverse direction. The value of the test thus has been retained for ranking the relative tearing resistance of thin polymer films, and several variations have been developed which use the same configuration of the test specimen but measure the propagation tear resistance on one of the more sensitive tensile testing instruments.

1. Elmendorf Test

The Elmendorf tear tester is a pendulum impulse testing instrument. It consists of a stationary clamp which is mounted on an upright on the base of the instrument and a movable clamp mounted on a pendulum preferably formed by a sector of a wheel or circle. The pendulum contains a nearly frictionless bearing and swings on a shaft mounted perpendicular to the upright member of the base. There is a stop catch located on the bottom of the base for holding the pendulum in a raised position and then releasing it to perform the tear. When the pendulum is in the raised position, the faces of the clamp on the pendulum and those of the stationary clamp on the upright are in the same plane.

The pendulum also contains a scale for registering the maximum arc through which it swings when it is released. This arc is indicated by a pointer which the pendulum carries with it on the return portion of the swing. The scale is graduated from 0 to 100% of the machine capacity.

The instrument actually measures the work done in tearing through a fixed length of test specimen equal to 4.3 cm. The tearing force, however, has to be applied through twice that distance, or 8.6 cm, because the two halves of the test specimen are being pulled in opposite directions. Since the 100% capacity of the tester is 137.6 g-cm and the total length of tear is 8.6 cm, each unit of the scale is equivalent to 16 g. Thus, the tear strength of a test specimen can be shown in grams by multiplying the scale reading by 16.

The test specimen is cut to provide a 43-mm-long radius of tear beginning at a notch 20 mm deep (Fig. 6). This configuration is used for polymer films rather than the original rectangular specimen 63 mm in length by at least 76 mm wide in order always to obtain the 86 mm of tear length regardless of the direction taken by the tear. Specimens should be cut and tested in both principal directions of the material. As in all of these methods it is advisable to measure and record the thickness of each of the test specimens to insure that deviations in the measured characteristic are not the result of gauge variation. Raise the pendulum to its starting position. The two sample grips, one stationary and the other on the pendulum, are now side by side. Fasten the straight, bottom edge of the test specimen into the clamp so

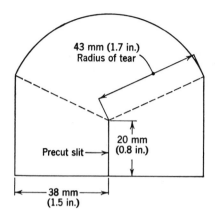

Fig. 6. Constant-radius test specimen for use with the Elmendorf Tear Tester.

that the midpoint of this dimension is centrally located in the space between the two sample grips. The 43-mm-long curved portion of the test specimen will now extend above the clamps. This portion should be laid over backward in the direction of the pendulum pivot. Using the knife mounted on the upright below the sample grips, make the 20-mm-deep slit in the specimen. Then release the pendulum and tear the specimen. Lightly catch the pendulum after it has completed its return swing so as not to disturb the position of the pointer, and record the tear value from the scale. Return the pendulum and the pointer to their starting positions and insert the next test specimen.

The ASTM method (31) suggests that inspection be made of the tear. If the tear is through the constant-radius section with an approximate angle of 60° on either side of the vertical line of intended tear, record the pointer reading to the nearest 0.5 unit. If the line of tear was more than approximately 60° from the vertical, reject the reading and test an extra specimen in its place. If rectangular specimens are tested, reject all specimens that tear obliquely more than 9.5 mm from the vertical line of intended tear. Test extra specimens to replace those rejected.

The tester has its maximum accuracy when the tear values lie between 20 and 60 on the pendulum scale. If thin materials are being tested which have very low tear strength, it may be advisable to test several specimens together and divide the result by the number of specimens. However, if the material tears obliquely and in opposite directions, the test will have to be performed on single specimens even if the readings are low.

2. Tear Tests Using Tensile Testing Machines

A test incorporating the same configurations of specimen used in the Elmendorf test can also be performed on one of the more sensitive tensile testing machines. For films which have exceedingly low tear strengths, such as cellophane, this modification of the method is not only desirable, but necessary, as tear strengths may amount to only three or four grams, well below the sensitivity of the Elmendorf tester. The specimen must be notched before placing it in the sample grips which are brought almost together at the initiation of the test. One "ear" of the specimen is fastened into the upper sample grip and the other in the lower sample grip. The test then proceeds in the same manner as a tensile test. The tear strength is recorded on the strip chart of the instrument recorder. Data obtained in tests using rates of jaw separation of 20 to 50 in. a minute have correlated well with results obtained on the Elmendorf tester.

Another test, often referred to as the Graves test (32), uses a somewhat different sample geometry. In this test the shape of the

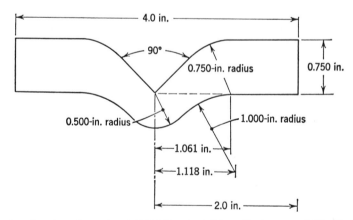

Fig. 7.   Test specimen for *ASTM D 1004*, "Tear Resistance of Plastic Film and Sheeting" (32).

specimen is designed to concentrate the stress in a small area (Fig. 7). Tear resistance is the maximum stress recorded, usually found at the beginning of tearing. A low rate of loading of 2 in. per minute has been selected for this procedure with an initial jaw separation of 1 in.

The same rules apply here as in the previous tests. Single or bundled specimens may be used, depending on the tear strength, although if a very sensitive load cell is available with the tensile tester, single specimens having very low tear strength can be tested.

3. Puncture Propagation of Tear

Each of the tear tests described has been concerned only with the propagation of the tear. A puncture-propagation test has been developed in an attempt to assess the ability of film or sheet materials as well as those of paper and various fabrics to resist failure resulting from snagging hazards, such as a polyethylene bag engaging a nail or a garment catching on a hook. The instrument, developed by Patterson and Winn (33), is unique in that it measures the combined effects of puncture and tear.

The tester is in a sense a modification of the dart drop test, except that a carriage having a sharply pointed puncture- and tear-propagating probe replaces the dart. The carriage, which falls a distance of 20 in. before the initial point of puncture, is guided by a pair of rails or guide channels during its fall so that the probe always engages the test specimen in the same way. The sample holder is vertical, and the top is curved away from the rails. Thus, when the test specimen is draped over the sample holder and clamped in place, the probe snags the film or sheet at a point always the same distance down from the edge of the specimen.

The probe is a piece of rod $\frac{1}{8}$ in. in diameter which extends 1.56 in. out from the face of the carriage. It has a conical point with a 30° included angle. A probe with these dimensions easily penetrates the test material in a relatively short distance so that the maximum portion of the tear will be propagated by the rod.

An arbitrary minimum tear length of 1.57 in. was selected, and if a tear of length less than this is obtained the next heavier carriage is used.

Procedure. The test requires a specimen 8 in. × 8 in. which is clamped in place on the curved face of the sample holder. The carriage, with a weight great enough to continue the tear through a distance of at least 1.56 in., is placed in the release mechanism at a height to give the required 20 in. of drop before the initial point of puncture by the probe. The carriage is released, falls, and the probe engages and tears the test specimen. The length of this tear is reported.

The length of the tear is actually a measure of the energy absorbed by the test specimen in stopping the probe as it moves through the material. For example, the potential energy of the carriage before release is equal to $w(h + l)$, where w is the weight of the carriage, h is the height of drop of the carriage, and l is the length of tear. The work done on the test specimen is equal to the force exerted by the probe propagating the tear, $f \times l$, where f is the force exerted by the probe. Since the carriage is brought to a stop by the film, all of the energy is absorbed in the propagation of the tear. Therefore

$$w(h + l) = fl$$

and

$$f = wh/l + w$$

In the normal operating range of the tester, Patterson and Winn found that the values obtained were almost entirely ascribable to the tear resistance of the material, and that there was no significant influence of puncture. The tester was capable of characterizing a wide range of materials including films and papers in terms of their ability to resist tear.

III. End-Use Tests to Measure Mechanical Properties of Films

In a packaging application, the first function of the wrapping material—in this discussion, a polymer film—is to contain and protect a product of some kind, a bag of potato chips, candy, apples, or an overwrapped carton of cookies or tea. The film must resist breakage resulting from any of a number of different causes. Although the standard laboratory tests of physical properties, such as tensile and impact, may give some insight into the ultimate behavior of the film in the use application, it is difficult to obtain from them a really precise measure of the complex forces to which the package is often subjected. For this reason, a number of empirical tests have been developed to measure the film properties in simulated but exaggerated use conditions.

Although package testing has developed considerably over the past 25 years as a means of estimating the efficiency of the package in the laboratory, this work has been concerned with cartons and shipping containers rather than with films.   The drum test (34) was one of the first used, probably evolving from a test in which a carton was rolled down a flight of stairs.   And so a tumble test also seemed to be a good place to start with the package testing of film.   It should be kept in mind that the tests described here are not universally accepted as official technical society methods; however, with slight variations from laboratory to laboratory they are as universal as any.

A. Tumble Test

This test has a distinct advantage over the drop test briefly mentioned earlier in that the product in the bag is contained in some sort of revolving drum after the test bag breaks and is thereby prevented from being scattered all over the floor.

The author does not know of a commercially available tumble tester for the packages envisioned here, but one can be constructed to meet the requirements of the testing laboratory.   For example, the drum tester might be designed to be a small replica of those used for testing shipping cartons which come in sizes ranging from 7 to 14 ft in diameter.   These drums are usually hexagonal and contain a series of baffles and guides so that when they revolve the package is caught up and dropped about six times during one revolution.

The size of the drum might be reduced substantially by making it square.   In this way the package makes only 4 drops per revolution instead of 6.   Thus, if a 2-ft drop is desired, the diameter of the drum is reduced from 4 to 2 ft.   Moreover, the smaller sized drum can consist of a number of compartments providing space for as many as 8–10 test bags to be tested simultaneously.

The speed of revolution of the drum must not be fast.   The test bag must be given plenty of time to fall from one baffle to the next.   If the drum revolves too fast, the bag does not fall at all but is held against the side of the drum by centrifugal force.

Bags used in a tumble test can be made by hand or by machine. If they are handmade, it is advisable to form them on a rectangular tube or forming mandrel so that they will be uniform in size.

Selection of the contents to be used in the bag is also important. It must be of a density sufficient to insure breakage of the film after a moderate number of drops—moderate, that is, as defined by the need. Beans, polyethylene pellets, and rice make ideal contents for the bag.

If the height of drop in the tumble tester is not changed, that is, if the dimensions of the drum are not changed, it has been found that larger numbers of drops will be obtained with smaller bags containing a smaller amount of product than with large bags containing a larger amount of product. This is not unexpected. An ideal size of bag is approximately $1\frac{3}{4} \times 2\frac{1}{2} \times 3\frac{1}{4}$ in. high which holds about $\frac{1}{2}$ lb of rice. The size of the bag should be adjusted to the quantity of product used or vice versa so that the film is tight but not stretched around the product.

Procedure. Fabricate and fill the bags with the product, either by hand or on a packaging machine if one is available. If the bags are fabricated by hand, this can be done on a rectangular plastic mandrel of the proper size. For example, if a bag of the above size is desired, the rectangular tube should have a dimension on one side of $1\frac{3}{4}$ in. and $2\frac{1}{2}$ in. on the other side. The length of the tube should be sufficient to accommodate the predetermined quantity of product. Cut the test specimens from the film sample about 2 in. longer than the desired length for this size of bag, in the machine direction and about $\frac{1}{2}$ in. longer than the perimeter of the tube in the transverse direction.

Place one of the shorter ends of the film specimen along the center line of one of the wide faces of the rectangular tube, and wrap the other end all the way around until it overlaps the first end of the film. Seal this end down to the film with pressure-sensitive tape, allowing the tape to extend about an inch beyond either end of the film. Readjust the film around the form, centering the taped seam on the wide face of the form.

Slide the film tube up or down so that when the two narrow sides are folded in toward the center of the tube they just touch, thereby forming two triangular tabs, one of which has a piece of the tape extending beyond its vertex. Fold the other in to the center of the tube first, and then bring down the second with the tape and press the tape in place. It may be necessary to place a second piece of tape across the bottom of the bag just formed but at right angles to the first piece. This will insure that the bottom of the bag will not fail.

Stand the tube upright on the bottom of the bag, fill the tube with the product, and holding the bag, remove the tube, leaving the product behind in the bag. Form the top of the bag and tape the tabs down in the same way that the bottom was closed. In order to make identical bags it is convenient to have a second tube which has the exact outer dimensions of the bag and into which the bag and rectangular forming mandrel can be placed before filling the bag with the product. This second tube aids materially in forming the top closure of the bag and insures that all have the same height.

Regardless of how the bags are made, whether by machine or by hand, they should be allowed to stand for a period of time to permit the contents to come to equilibrium with the room conditions. This is surely necessary if the film itself is sensitive to the laboratory conditions.

At the end of the conditioning period, place the bags in the tumble tester. If each bag is inserted by letting it slide down to the bottom of that side of the tumble drum, this should be counted as the first drop. With all bags inserted and the drum closed, start the tester. Record the number of drops to break for each bag.

There is, of course, no reason for not testing a commercial package in the same way. This could consist of a small carton, such as a package of cigarettes or a deck of playing cards overwrapped with the test film. This package would necessarily be of a size which could be accommodated by the tumble tester. The four-sided drum as described could handle a small bag of potato chips or pretzels as an example but not the larger sizes found in the supermarket.

As in all of the tests of this kind, the forces involved are complex and any mathematical analysis is difficult, if not impossible, to perform. The major drawback of the test is the wide variation in the numbers obtained, but this may actually represent the wide variation in the behavior of the material undergoing test. This conclusion is substantiated to some extent by the good agreement obtained among several operators in the same laboratory when testing the same material, even though one can conceive of substantial differences occurring as a result of fabricating the test bags by hand.

B. Package Drop Test

Like the falling dart impact test, the package drop test is a time-consuming one to perform, especially if it is done package by package through a range of drop heights. A somewhat faster method is to place a number of packages in shipping containers, drop the individual container once or several times, depending on the material, and then inspect the bags or overwrap of cartons or product in each of the containers for failure.

It might be advisable at this point to define the word *package* as we are using it in this discussion. In its normal usage *package* probably denotes the complete assembly consisting of container plus dividers, the devices which separate the space within the container into two or more spaces (cells, compartments, or layers), plus contents. Containers are the receptacles such as bags, barrels, drums, boxes, or crates used in commerce for packing, storing, and shipping commodities. A carton is a closed receptacle used in interior packaging, made of various kinds of boxboard, such as pasteboard, fiberboard, strawboard, etc. The term *carton* should be applied to interior boxes and not to a shipping container. The *packages* we are discussing are indeed containers, but they usually form the contents of a larger shipping container and thus are not containers by the above definitions. They may be small bags, an overwrapped product, or overwrapped cartons. The definitions for container and carton are those given in *ASTM D 996* (35).

There are various sources of variability in this test which probably cannot be overcome. The manner in which the bags are placed side by side in the container, the container itself, and the way in which the container is dropped can be controlled to some extent. Of these the drop is the most likely candidate for control, as an automatic drop table is available commercially.

The package drop tester is a simple instrument consisting of a heavy steel base about 4 ft square on which a steel pole about 6 ft in length is mounted upright. The drop table assembly rides up or down on this pole so that the height of drop can be adjusted from a minimum of 12 in. to the height of the pole. The drop table extends out from the pole over the steel base insuring that the package will always drop on a hard, solid surface. It is formed from two machined aluminum leaves, each of which pivots downward in the form of a trap door on a shaft extending forward, parallel with the floor, from the rear bracket which supports the table on the pole. These drop leaves are supported initially in a horizontal plane by quick-release support blocks, and they are spring loaded so that when they are released they fall away quickly, allowing the package to fall without obstruction. Electric solenoids actuate the support blocks and these, in turn, are operated from a foot treadle.

In practice, the test packages are placed in a container which is standard for the product being used in the test. For example, if the test material is intended as a new wrap for potato chips, it would be fabricated into bags of a size normally used for potato chips, and these in turn would be packed in the standard potato chip container. This container is then placed in the desired position on the drop table for a flat, edge, or corner drop. The flat drop may be made on any of the four sides or the two ends of the container, and similarly for the corners and edges. The exact manner of drop is usually selected after several test drops are made to determine which drop or combination of drops results in a desirable spread of damage to the packages. After the carton is dropped, each of the bags in the carton must be examined for damage.

The estimate of the damage is qualitative, based on some arbitrary scale of damage. Bags which develop only a few pinholes, while still acceptable, are not as good as bags which show no sign of damage. And bags which contain several tears, say about $\frac{1}{2}$ in. long, are not as acceptable as bags which have only one or two very short tears. One can thus set up a grading table of about ten grades between acceptable and unacceptable—no pinholes, tears, or other damage, to complete failure—and after grading the materials used in the test, ranking them according to their ability to resist the kind of abuse that might occur in handling containers of the product.

Reference should be made to the standard test for shipping containers, *ASTM D 775* (36), for guidance in selecting the drop pattern which might be used.

C. Vibration Test

The drop test is often combined with a vibration test with the idea of better simulating use conditions. For if the container with its product, whether packaged in bags or cartons or merely overwrapped, is shipped only a short distance by truck or rail, the contents would undergo considerable vibration and impact—similar to that of the drop test—during the transport. It is probably impossible to duplicate the exact kind of vibration that occurs in a truck or rail journey, and consequently a standard type of vibration with respect to amplitude and frequency has been accepted.

The vibration tables used in the packaging industry for these tests are different from the vibration fatigue testing equipment used in a large measure for testing electronic components and the like, which are used in the space programs. They are of much lower frequency and usually consist of a bed which is driven up and down by two eccentrics at each end connected in phase with each other. The table thus describes a circular harmonic vibration with an amplitude of 1 in. for containers. This amplitude is usually fixed and the frequency varied between 100 and 300 cycles/ min, depending on the equipment. For assessing the packages in the container, an amplitude of 1 in. may be too great to obtain a meaningful evaluation of many packaging films. Consequently, a vibration table having variable amplitude as well as variable frequency is desirable, if one is available.

For the testing of flexible packaging materials the container should be fastened to the vibration table so that no point of it can be lifted off the table. Also, the package strapping should be strong enough so as not to yield, regardless of the frequency used. As in the drop test, the test packages are placed in a standard container for the product being used in the test. The container is fastened to the vibration table with the straps, and the table is operated for a certain period of time at the desired amplitude and frequency. Finally, at the conclusion of the vibration period, the packages are examined for damage and grade, using, for example, the same grading system used in the drop test.

Sometimes the container with packages wrapped in the various test materials is subjected to both drop and vibration before examining. Under these conditions both drop and vibration tests must be somewhat less severe than those used if the grading were done at the end of each test. Grading of the test bags is a tedious operation, and therefore the single grading seems somewhat more practical.

Grading of the films can be expedited by examining the test bags between polarized plates, as the stress points show up clearly. An examination box can be fabricated without much difficulty in the laboratory.

A vibration test for containers is published in *ASTM D 999* (37) and refers briefly to the procedure just described.

D. Summary

Package testing, Paine points out, is a relatively new science, as indeed is the entire area of packaging film testing (38). There is still considerable debate as to the efficacy of performing physical tests rather than packaging tests of the type outlined which will probably continue for some time, at least until such time as sufficient data are available to determine the correlation between physical test data and actual use data.

The package tests probably derived from shipping tests in which containers were sent on a journey to a customer or a sales office in another city and back to the plant or laboratory. At the end of this journey each container and its packages were examined for failure and the results correlated with laboratory tests. Unfortunately this was not a satisfactory method, since conditions encountered during a trip would vary from one time to the next, and there was no guarantee that the same route between the two locations would be used. As a result, data obtained in the shipping test often were incorrect or at least resulted in wrong judgments. In addition this kind of testing is expensive in both time and money.

Analysis of what happened in these shipping tests through the use of impact and vibration measuring devices seemed to indicate that the tumble, drop, and vibration tests performed in the laboratory are a good substitute for the shipping test. But even these laboratory tests cannot be performed under entirely reproducible conditions. The sequence of tumbles and falls in the tumble test cannot be programmed; cartons used in a drop or vibration test vary, as does the product used in the individual bags; and as in any test, if a reasonable degree of reliability in the test data is desired, then a large number of containers must be tested to obtain this reliability, since variation in the amount of damage from one carton to the next will usually be very large. Unfortunately, this amount of testing is also expensive, and as a result decisions are usually based on too few data.

Finally, of the tests described, only the tumble test provides quantitative data. These are probably easier to analyze than the qualitative data of the drop and vibration tests. Certainly the tumble test does not require the long hours in examining the test materials for pinholes and tears. The tumble test would thus seem to have a slight advantage.

IV. Gas Transmission Tests of Films

A discussion of testing without reference to transmission tests would be incomplete; even though this subject is discussed at some length in Volume II, it is included here because the tests used in the average testing or control laboratory have a certain elegance in their simplicity even though they do not possess the scientific refinement of the equipment used by Barrer and other investigators in the development of our current knowledge (39).

The process of permeation of a gas or water vapor through a film consists of three steps: adsorption of the water vapor by the surface of the film adjacent to the humid atmosphere, diffusion through the film, and desorption or evaporation from the surface of the film adjacent to the dry atmosphere. This point of view on the diffusion of gases through solids was proposed as early as 1866 by Graham.

The most generally accepted treatments of this subject are based on three asumptions:

1. The diffusion is in a steady state.
2. The concentration gradient through the film is linear.
3. The diffusion takes place in one direction only.

Under these conditions Fick's first law of diffusion holds, which states that the flux of the diffusing gas or water vapor will be proportional to the concentration gradient.

$$q = -D \ dc/dx \tag{1}$$

Here q is the amount of the diffusing gas in milliliters, D is the diffusion constant, and dc/dx is the rate of change of concentration with thickness of the film.

If this equation is rewritten and integrated, we obtain

$$q = D(C_1 - C_2)/l \tag{2}$$

where C_1 and C_2 are concentrations of the gas at the two surfaces of the film and l is the thickness of the film.

It is more practical to express the driving force in terms of the partial pressure of the gas rather than concentration. The relationship given by Henry's law is applicable if it is assumed that there is equilibrium between the gas and the first layer of film:

$$C = Sp \tag{3}$$

where S is the solubility coefficient and p is partial pressure of the gas.

Substituting eq. (3) in eq. (2) gives:

$$Q = DS(p_1 - p_2)/l \tag{4}$$

If we then make the following substitutions in eq. (4), $P = DS$, $q = Q/At$, and $p_1 - p_2 = \Delta p$, where P is the permeability coefficient, Q is the amount of gas or water vapor transmitted through the film (in milliliters), A is the area in square centimeters, t is time in seconds, and Δp is the pressure gradient of the gas across the film (in millimeters of mercury), we obtain the equation

$$Q = PAt\,\Delta p/l \tag{5}$$

Equation (5) states that the amount of water vapor diffusing through a film is proportional to the vapor pressure and inversely proportional to the thickness of the film. It assumes that both D and S are independent of concentration, that is, that there in no interaction between the film and the diffusing gas or water vapor. In the case of certain films, this interaction does take place. Because of the finite diffusion rate of the solute in the film, there is an interval before the steady state is established. The situation during this period is described by Fick's second law. If x and $x + dx$ represent two planes in the film at distances x and $x + dx$ from the high-pressure surface, and if the rate of gas permeation at x is q ml/sec, and at $x + dx$ is $q = (dq/dx)\,dx$, then the amount of gas retained per unit volume of film is dq/dx. This is equal to the rate of change of concentration:

$$dq/dx = -\,dc/dt \tag{6}$$

Differentiating eq. (1) with respect to x and substituting eq. (6) gives

$$dc/dt = -D \, d^2c/dx^2 \tag{7}$$

The general solution to eq. (7) has not yet been found, but under certain boundary conditions it may be shown that

$$Q = Dct/l - cl/6 \tag{8}$$

Thus, the amount of gas or water vapor permeating the film increases linearly with time once the steady state has been reached as indicated in Figure 8.

If the linear portion is extrapolated back to $Q = 0$ where the intercept $t = \gamma$, we have

$$Dc\gamma/l = cl/6 \tag{9}$$

$$D = l^2/6\gamma \tag{10}$$

The value γ is called the time lag and is the time required to reach the steady state. The determination of γ thus provides an experimental basis for calculating the diffusion constant D.

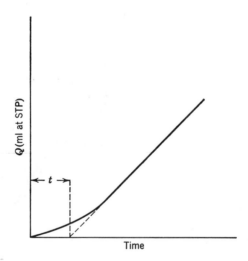

Fig. 8.   Change in rate of permeation with time.

Much of the experimental work on the permeability of sheet materials has been performed using the vacuum technique of Barrer. In this method the high-pressure side of the test cell is connected to the gas source, a trap, and manometer, and the low-pressure side with a McLeod gauge via a solvent trap. This method permits measuring both P and D in the same experiment, the latter taken from time lag.

A. WATER VAPOR TRANSMISSION

The method described briefly above does not lend itself to routine or control testing, especially if the diffusing gas is water vapor. In this case one of a variety of tests is performed in which the permeability rate is determined from the increase or decrease in weight of the test specimen. These tests have been standardized and are in wide use, such as *ASTM E 96* (40) and *Tappi T464* (41).

1. Weight Loss Test (42)

A test specimen of standard size seals off the top of a small metal cup which contains a quantity of water, normally about 55 ml. The test cup is weighed initially, placed in an oven controlled at prescribed conditions of temperature and relative humidity, conditioned for a period of time, or consecutive periods of time, and then reweighed. The weight loss is a measure of the transmission rate.

Test cups may be assembled in several different ways. The film specimen may be sealed to the top of the cup with wax, for example, by dipping the rim of the cup in molten wax, allowing the excess to drain off and then pressing the cup against the film, or by holding the film in place with a template and pouring the molten wax into the space between the template and the rim of the cup to form a ring.

In the first method the cup is fabricated upside down and the water must be put into it by some means other than pouring. This can be accomplished as follows. Place a test specimen down on the table so that the surface which in use will be exposed to the high relative humidity is up. On top and in the center of the test film place a small square of cardboard, such as backing board or blotter. Then obtain a small metal cup with a capacity of 20–25 ml, fill this little cup with distilled water, and place it on the card-

board square.   Aluminum cups or thimbles are ideal for this purpose.   Hold the larger test cup upside down and dip the rim into
molten wax, drain off the excess wax, and press the rim down onto
the test specimen over the thimble containing the water.   Allow
the wax to harden; during this time it is advisable to place a weight
on top of the cup (actually the bottom side of the test cup which
is now facing up) to insure a perfect seal between the rim of the cup
and the test specimen.   When the wax has hardened, lift the assembly consisting of film, cardboard square, thimble containing water,
and test cup rapidly but smoothly upward and over so that the
cup is held film-up at the end of the maneuver; the hand actually
turns 180° in a small semicircle.   The water is thus spilled into
the bottom of the cup, and the test specimen is ready for its original weighing.   With very little practice the cup can be turned
upright without splashing any water on the test specimen.   The
rim of this kind of test cup is flared out slightly to allow a small
wax bead to form between the film and the lip of the cup.   An
aluminum cup $2\frac{1}{2}$–3 in. deep is a cup of appropriate size.   The
diameter should be such that it is easy to grasp.   These cups can
be fabricated easily by spinning.

A commercially available aluminum test cup has a heavy, flat
ring around the rim with about eight small threaded holes spaced
around it.   A rubber gasket ring is placed on this aluminum ring,
the test specimen on top of the gasket, and a second aluminum

TABLE 1
Change in Pressure with Temperature
(Oven relative humidity, 0–5%)

Oven temperature		$p,$
°F	°C	mm Hg
110	43.3	65.9
100	37.8	49.2
90	32.2	36.1
80	26.7	26.3
70	21.1	18.8
60	15.6	13.3
50	10.0	9.21
40	4.44	6.29

ring over this.   The gasket, test specimen, and second aluminum ring all have holes which correspond to the holes in the lower, heavy ring.   The assembly is held together with thumbscrews inserted through the holes and screwed into the lower ring.   This cup and the one in which the sample is sealed in with a wax ring are fabricated upright, and this may be an advantage for the technician.   However, few test specimens are lost as a result of leaks when using any of these test cups.

Since the weight change is directly proportional to the vapor pressure gradient across the film, the test is usually run at conditions providing the greatest driving force and thus the shortest test period.   The relative humidity in the cup is 100%.   If the oven is maintained at 100°F and 0–5% relative humidity, the driving force across the film will be about 49.2 mm Hg.   The

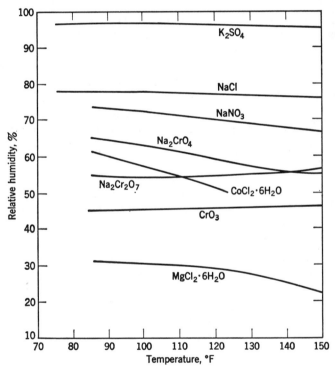

Fig. 9.   Change in equilibrium humidity over various saturated salt solutions with change in temperature.

driving force can be reduced by lowering the temperature as illustrated in Table 1, or by replacing the distilled water in the cup with one of a number of different salt solutions as shown in Figure 9. If the latter course is used, a salt solution must be selected which does not corrode the metal cup.

2. Weight Gain Test

This type of test is the one most generally used in the packaging industry when evaluating flat, undamaged film samples. It reverses the procedure of the previous tests in that the weight of calcium chloride or some other desiccant packaged inside the test film increases in proportion to the amount of water vapor passing through the film. In general, the same types of test cup can be used as those described in the previous section; but it is preferable to use one such that the space between the film and the calcium chloride is minimal. The standard TAPPI test cup (41) of approximately 5-in. diameter provides for a space of about $\frac{1}{4}$ in. between the film and the top surface of the calcium chloride. This has been found to be very satisfactory.

It is also common practice in many laboratories to invert the test plate so that the desiccant is in direct contact with the film; if this practice is followed, however, it is advisable to place a sheet of tissue immediately under the sheet of film to avoid scratching or other damage which might result in an erroneously high permeability value.

When the 5-in. TAPPI test dish is used, treat each specimen as follows. Fill a 50-ml beaker (level with the top edge) with calcium chloride, pour the salt into a numbered permeability dish, and smooth it to obtain an even depth of salt over the bottom of the dish. Center a test specimen cut to the proper size on the top of the dish in contact with the flange of the dish but not touching the calcium chloride in the bottom. If the dish is to be inverted during the test so that the calcium chloride is in direct contact with the test film, place a sheet of tissue, such as facial tissue, cut to the same dimensions as the test film into the dish first, and then place the test film on top of the tissue. For films which have a sidedness with respect to their permeability characteristics, determine how the material will eventually be used and expose the correct side to the contents of the dish (42). Such films may also be tested twice so that the effect of side is determined.

In order to hold the test specimen while sealing it in place with molten wax, use a circular template having an edge tapering at about a 45° angle to a smooth bottomed surface which has a just slightly larger diameter than the inside testing diameter of the permeability dish.   Lubricate the beveled edge of the template with a suitable silicone, but take care to keep it away from the bottom edge of the template where it could deface the film specimen. Center the template over the test specimen and lower it so that it just rests on the innermost edge of the flange of the dish.   An annular space is thus formed between the edge of the template and the outside edge of the dish.   Pour molten wax into this space, filling it so that the level of the wax just approaches the top of the template.   Allow the wax to cool until the template can be readily removed.   Inspect the surface of the test specimen to make sure that no dirt or fingerprints have defaced the test surface.   If the surface is marred in any manner, discard the test specimen and prepare another.

B. Gas Permeability (39)

The many new packaging techniques which require drawing of a vacuum in the package, gas flushing, or sterilization have made it necessary to learn more about the gas permeability characteristics of the many available flexible barrier materials when deciding on the varied uses to which these materials can be put.   Since so many of these uses require low permeability rates, it probably is a natural development to determine the gas transmission characteristics under conditions approximating those encountered in the specific packaging applications, since the most useful information would be obtained under those conditions.   Consequently, the best known methods seem to duplicate conditions encountered in vacuum packaging—the packaging technique in which all air is drawn from the package before it is finally sealed.

Simply described, these methods involve the permeation or diffusion of a test gas through a film specimen into an evacuated manometer.   The driving force across the film may be 1 atm total pressure, or any desired level of pressure, although 1 atm seems to be a sufficient pressure gradient.   The equipment is generally of the test cell–manometer type, the film specimen being placed in the cell and covering the entrance hole to the manometer, which is subsequently evacuated and then filled with mercury to indicate the pressure of the transmitted gas. There are several cells available which are commonly used, such as the Dow cell, the method for which is described in *ASTM D 1434* (43), the Linde cell (Volume II) and the Henderson-Wallace cell (44). With minor variations, all three function in approximately the same way.

As the gas passes through the test film into the capillary section of the manometer, the mercury level therein is forced downward. After an initial time period during which volatile substances in the film may be pulled into the capillary leg of the manometer, a constant rate of transmission is reached. When this constant rate of transmission is achieved, a plot of the height of the mercury column in the capillary leg as a function of time results in a straight line. The gas transmission rate of the film specimen can be calculated from the slope of this line (Fig. 8).

1. Test Procedure

Since each of these test cells differs slightly from the other, only a very general description of how the test might be performed with one of the cells is given. First, it is important to have clean manometers and mercury since either dirty mercury or a dirty capillary or both will cause the mercury to cling to the walls of the manometer and produce erratic results. Depending on the frequency with which the gas permeability test is performed, cleaning of the mercury and manometer system about once every three months or so is sufficient.

To begin, all of the mercury is transferred into the reservoir of the manometer system. The test specimen is then placed in the cell. Generally these cells have a very shallow depression which just accommodates a piece of filter paper. This provides a passage for the gas which is permeating the film to reach the manometer. A perfect seal must be made between the film and the two halves of the test cell to avoid any leakage in the vacuum side of the system. If stopcock grease is used anywhere in the system, extra care must be taken to avoid getting any on the test area of the film, as the gas permeability of some types of film may be changed as a result.

After the cover of the cell is mounted and tightened in place, the manometer is evacuated. This usually takes about 5 min. While the vacuum pump is still running, it is good practice to tighten the wing nuts which hold the top of the cell in place further at this time to insure an airtight seal between the film and the bottom part of the cell. Close the stopcock to the vacuum pump, shut off the pump, and remove the rubber tubing from the manometer. Then fill the manometer with mercury. Depending on the construction of the manometer, this can be done by tipping the test assembly and spilling the mercury from the reservoir into the manometer. Record the initial level of the mercury in the capillary at this time. If the mercury does not remain stationary, a leak probably exists. If so, discontinue the test, and repeat the entire procedure.

If no leak exists, slowly admit the test gas to the system. If a pressure differential of 1 atm is desired, the flow of the gas over the test specimen should be such that the gas can barely be felt as it leaves the escape port in the cover plate. Record the height of the mercury column in the capillary leg of the manometer at regular time intervals, depending upon the test specimen. For example, for 1.5-mil low-density polyethylene, film readings should be taken at intervals of a few minutes because of the high permeability rate; however, for 2-mil saran, the time interval between readings may be several hours.

2. Calculations

The gas transmission rate can be determined graphically if certain instrument constants are known. These are usually determined in the initial calibration of the instrument and must be rechecked each time the cell is dismantled when the manometer is cleaned. At such times, if a check indicates that the critical volumes have not changed, no further calibration is necessary.

To determine the permeability rate, plot the readings of the height of the mercury column in the capillary leg of the manometer as a function of time on millimeter graph paper. In that portion of the curve where a constant transmission rate is indicated, draw a straight line through as many points as possible. From the slope of this line determine the drop in height of the mercury column in millimeters for a convenient time period. Substitute this height in millimeters and the time in hours and fractions of hours in the following equation.

$$\text{ml gas/m}^2\text{-24 hr} = K \times \Delta p/\text{time (hr)}$$

Here Δp is the change in height of the mercury column in milli-meters, and K is a cell factor predetermined for each test cell.

The cell factor is calculated as follows. The volume of gas at standard temperature and pressure per square meter per day is given by the formula:

$$Q = (V/A)(\Delta p/760)(273/T_k)(24/t)(10,000)$$

where Q = milliliters of gas, STP/m²-day
  V = total free volume (in ml) above the mercury level in the capillary plus the free volume of the filter paper
  A = area of the test sample, cm²
  Δp = change in pressure in the manometer as measured by the change in height (in mm) of the mercury column in the capillary
  T_k = ambient temperature,°K
  t = time, hr

Determine the value of each term in the above expression. Accurately measure the inside diameter of the circular ridge of the metal test cell in which the filter paper fits and calculate the area (this area is constant for each test cell and need be determined only once). Measure the free volume of the filter paper disc by water absorption. Calculate the free volume in the capillary from the bore diameter and the distance from the top of the mercury level to the top of the cell base. A slight difference in the initial height of the mercury level (in millimeters) has a negligible effect on the total calculations, and an average value can therefore be determined for this distance. The total free volume, V, is equal to the sum of the free volume of the filter paper and the free volume of the cap-illary. Make the necessary substitutions and calculate Q.

There are two assumptions which must be made when calculating Q: (1) the free volume is constant; (2) the change in pressure in the manometer, Δp, can be measured directly from the drop of the mercury level in the capillary, disregarding any rise of the mercury level in the other two legs of the manometer. Calculations indi-cate that the error introduced by these two assumptions is small and less than the reproducibility of the method.

3. Other Methods

With the ever increasing use of the gas chromatography, the recent advances in the determination of gas permeability have incorporated this instrument in the method. Its use is obvious. It permits a precise measure of the permeation rate of one gas or several gases simultaneously, or of a gas in the presence of water vapor. The test cell is not markedly different from the ones just described except that provisions must be made for extracting the gas sample for analysis.

Several methods have been developed which are described as isostatic. A true isostatic test probably does not exist, as there must always be a pressure gradient of the test gas across the film sample.

C. Odor Tests

This is a highly specialized field and it is mentioned here only because it is another area of film permeability that should not be ignored. There are many packaging uses which require excellent odor barriers, and therefore tests must be performed at some time to select the best qualified materials for those uses. One can almost say without fear of contradiction that this is done on a trial and error basis.

The author has performed odor penetration tests in which various odoriferous substances were packaged in pouches made of the test material. The pouches were placed in clean bottles which were sealed with aluminum foil for a predetermined period of time. At the end of that time the bottles were opened and sniffed for any sign of the odor. A number of test specimens for each sample of film and odor were made so that the test could be continued to the first sign of odor penetration. These are subjective tests, but reliable information can be obtained from them if they are conducted carefully.

The odor tests can be set up in various ways to improve the rating of the test substances. Samples may be paired, or the triangular system may be used. If the number of samples is large, a ranking method is probably the best, although ranking does not give an indication of relative quality. Another method involves determination of the minimum length of time required for the odor to penetrate, a threshold test. This is a variation of the threshold test used in taste testing.

Gas chromatographic methods have also been tried, and new methods incorporating this instrumentation no doubt will be developed and improved with time.

V. Production Properties of Films

A. HEAT SEAL

With very few exceptions, packaging operations in which flexible packaging materials are used involve heat sealing. It is the characteristic of heat sealability which permits these packaging operations to be performed at rates up to several hundred packages per minute. Good, poor, or no heat sealability of a film therefore determines whether or not it can be used for a specific application.

Over the years many polymer films have been introduced and many more will be developed. They may have excellent characteristics, the correct physical properties for packaging applications, but, in general, the ones which will fail will primarily have one serious deficiency: they will resist bonding by such conventional means as the application of heat, adhesives, or chemicals.

1. Factors Influencing Heat Seal (45)

At first glance it would appear that there should be no serious difficulties in obtaining satisfactory heat seals when using heat-sealable packaging materials. However, a closer examination of many packaging operations will indicate that a number of factors other than the heat sealing characteristics of the material are involved. For example, the pressure required to hold the layers of film together for sealing may be limited by the type of product and, in most packages, there are areas where several layers of the packaging material must be sealed together, in some instances with no backing force or, at most, with only the weight of the package as the backing force. The application of additional heat in these cases is of little value if the layers of the film are not contiguous.

Futhermore, the amount of heat available may be limited by the capacity of the heating elements, by the rate of heat transfer of the sealing bar and its coating, or possibly even by the type of product being packaged. More heat can usually be obtained, of course, by increasing the dwell time, that is, the time during which heat is applied to the film; however, this in turn may result in an economic problem, since the number of units handled per minute must be decreased. The heat transfer of the film, variations in the film itself, the large variety of packaging machines, their variability and individual idiosyncrasies, and a host of other problems are all a part of this larger problem of heat sealing.

2. Heat Seal Test Requirements

These considerations make it obvious that if any changes or improvements are to be made in the heat-sealing qualities of various polymer films used in packaging applications, or if the results of control tests in the manufacturing process are to give an accurate estimate of these qualities, heat-sealing tests are required which must be able to simulate the normal range of sealing conditions encountered on packaging machines and maintain them under close control. It must be possible to obtain the variety of conditions demanded for packaging small objects or light ones, soft or fragile products, or any of the myriad items packaged daily in these heat-sealable materials. In short, it should be possible to cover the entire range of sealing conditions beginning with the minimum ones—the combination of sealing temperature, pressure, and dwell time at which seals can barely be formed.

Testing procedures, then, are required which quantitatively measure the characteristic *heat sealability*. Multiple layers may be involved or just two sheets, both types of heat seal being equally common. In a test incorporating the former, sufficient heat is required to seal through the layers, pointing up differences in heat transfer, sealing temperature, or even the manner in which the layers of film are held together. If only two layers of film are used, the sealing temperature can be selected to duplicate the condition to which the innermost layers of a multiple-layer seal are subjected, or any other temperature.

Thus, in development work on the heat-sealing characteristics of a polymer film, the estimate of sealability will be based on the measurement of the force required to peel apart the layers of film forming the seal. These have been bonded together by the application of heat while they were held together under pressure which may range from little or nothing to many pounds per square inch for a time sufficient to soften or melt the material so that it may be pressed or fused together.

3. Heat-Sealable and Non-Heat-Sealable Films

Heat-sealable films are considered to be those films which can be bonded together by the normal application of heat, such as by conductance from a heavy heat-resistant metal bar containing a heating element. Nonheat-sealable films cannot be sealed this way; however, non-heat-sealable films can be made heat sealable by coating them with heat-sealable coatings. In this case the two facing coated surfaces become bonded to each other by the application of heat and pressure for the required dwell time.

Certain non-heat-sealable films can sometimes be bonded together by other methods, such as by electronic or ultrasonic welding devices. In both methods, internal heat is localized at the interface of the two materials by the energy transducer. Therefore, the film is not weakened in the vicinity of the heat seal—in this instance, seam or weld—and shrinkage of oriented films does not occur.

Heat-sealable films are represented by polyethylene and polypropylene. Cellophane is a coated heat-sealable film. The vinyls, saran, Mylar, and acetate can be welded together more satisfactorily by the other methods and therefore should be designated as non-heat-sealable in terms of the definition used here.

4. Heat Sealers

It is difficult to evaluate results of heat-seal experiments if test seals are fabricated at optimum conditions, the high sealing temperature and pressure and the long dwell time being carefully selected to give maximum seal strength. These conditions are unrealistic in terms of modern packaging techniques because machines normally operate at speeds over 100 units per minute, and newer models will be capable of operating at 300–400 units per minute or faster. It becomes evident that the success of any work concerned with improving the heat-sealing characteristics of a film will depend to a great extent on a sensitive heat-seal test. For this purpose a heat sealer is required by means of which sealing temperature, dwell time, and pressure can be accurately regulated. Some investigators believe that one should also be able to change the method by which the heat is generated, that is, hot bar, band, thermal impulse, or in some instances even wire, knife, air, or flame. There are obviously many methods, and each has its advantages or disadvantages. Because of the number of methods, one method of heat sealing probably will not completely define the heat-sealing characteristics under all conditions. Since so many packaging operations depend on heat conductance from a constant-temperature surface, this would seem to be the best type of heat sealer to use for a bench test.   Moreover, it permits the best control of sealing conditions.

5. Heat Conductance Sealers (46)

Very few manufacturers make heat sealers which are intended for laboratory use.   A commercial sealer of this type usually has only one electrically heated jaw, the temperature of which is controlled by means of a good thermostat or indicator–controller. In this case, the second jaw will eventually reach an equilibrium temperature which is usually substantially less than the sealing temperature and depends on the number of seals made and the rate of fabrication.   The second or backing jaw often has to be covered with or consists of rubber or similar material to distribute the pressure evenly and to aid in smoothing out the film in the sealing area.   To achieve better control, the unheated backing jaw may be water cooled, or it may be heated to the same temperature as, or some lower temperature than, the first jaw so that sealing through the sheets of film is accomplished in exactly the same way every time.   The temperature range of the sealing bar ought to be at least 200–500°F.   The temperature setting should be quickly adjustable by the operator on a calibrated scale.

Dwell time should be controllable to fractions of a second by a good electrical timer.   The timer should be easy to adjust so that the same settings can be made repeatedly.   Action of the heat sealer should be initiated by a foot pedal so that the operator may have both hands available for holding the film between the sealing jaws.   Once the operator sets the sealer in motion, the timer controls the length of the sealing cycle.

Pressure holding the layers of film together may be achieved by several means.   The commercial laboratory sealers usually depend on air pressure both to operate the jaw and to control the maximum pressure applied to the sheets of film.   Using the same air motor for these two purposes is unsatisfactory, however, when the sealing pressure is very low, i.e., only a few ounces per square inch. In this case, the line pressure has to be reduced to a level which is usually insufficient to operate the air motor satisfactorily.

A laboratory sealer (Fig. 10) has been designed to overcome this difficulty, although it is not available commercially.   It reproduces

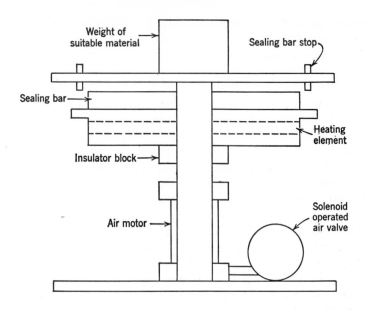

Fig. 10. Schematic diagram of a laboratory heat sealer which provides control of sealing temperature, pressure, and dwell time. Control units are not shown.

as simply as possible the sealing action used in many packaging machines to form the bottom seals of a package and is built in the form of a small table containing an opening through which the sealing bar can move freely upward. In its rest position, the sealing bar is approximately $\frac{1}{2}$ in. below the top of the table, and it is moved up into its sealing position and down again by means of a small air motor. This distance below the table is sufficiently great so that even after prolonged periods of operation at a high sealing-bar temperature the increase in temperature of the metal plate forming the top presents no problem, barely becoming warm to the touch.

The sealing element is constructed in two parts. The lower portion, consisting of a steel bar about 6 in. long with a square cross section of approximately 1 in., contains a hole $\frac{1}{2}$ in. in diameter at its center extending almost through its entire length. A heater of appropriate size is inserted in this hole. The second part forms the surface which comes into contact with the heat-sealable packaging material. This bar can be made of aluminum or any other suitable metal and may be Teflon-coated or left bare. It is fastened to the larger bar containing the heater and thus is readily interchangeable with bars of other sizes or shapes, some of which may actually reproduce those of various packaging machines if desired. The entire assembly, consisting of the steel bar with its heating element and the sealing bar, is in turn fastened to the air motor through a heat-insulator block. The sealing bar may be made to move upward by other means than an air motor, such as a solenoid actuating a cam-and-lever arrangement.

The height to which the sealing bar may move is controlled by several screws in the metal plate which act as stops. They should be set to allow the sealing-bar assembly to rise approximately $\frac{1}{16}$ in. above the surface of the table. The sealing pressure is controlled by dead weight, the sealing bar lifting a weight of a predetermined size as it moves above the surface of the table. Consequently, it is possible to make heat seals for testing purposes at a variety of sealing pressures that range from a minimum equal to the weight of the upper piece of film to a maximum of several pounds per square inch. As the sealing bar moves through the opening in the table top upward against the specimen, it lifts the packaging material and the known weight placed on top of it.

In practice, the technician will normally place a piece of chipboard between the packaging film and the weight, insulating the latter from the sealing bar and insuring constant conditions during the sealing operation. The chipboard may represent, for example, the paper tray used under a cake.

Temperature and dwell time may be regulated in the same way as for commercial heat sealers, that is, by means of a temperature controller and an electric timer. For an accurate estimate of the temperature of the surface of the sealing bar, either in this instrument or in one of the commercial types, a thermocouple should be located in the portion of the bar where seals are generally made and as close to the sealing surface as possible.

The electric timer is started by a microswitch adjusted so that the sealing cycle is timed from the instant the sealing bar first touches the test specimen. The timer, in turn, operates whatever mechanism is required to move the sealing bar into position and back again at the end of the sealing cycle.

In heat-sealing instruments of the types just described, each of the three sealing variables—pressure, sealing temperature, and dwell time—is controlled independently of the others and it is thus possible to make heat seals at an almost infinite number of sealing conditions.

6. Electrical Impulse Sealers

In the electrical-impulse heat sealer, a high current is sent through a resistance wire for a short period of time. The wire, usually flat and about $\frac{1}{8}$ in. wide, is mounted on one of the jaws. The current heats the wire to the desired temperature in a predetermined time. The temperature of the wire is controlled by the amount of current which is fed to it. Control is usually achieved by means of an accurate, substantial transformer and is indicated on a voltmeter. It should be possible to duplicate the exact settings at any time to make similar seals.

Dwell time of the heating impulse is controlled by the same type of electrical timer used in the heat-conductance sealers described above. However, in this type of instrument there must also be means for controlling the length of the cooling period. A second timer is therefore used to control the total time for both heating and cooling of the resistance wire to allow the film to harden under pressure to prevent deformation of the film in the sealed area.

In commercial instruments, one of the jaws may be water cooled to prevent excess heating and promote rapid cooling. When sealing very heavy films, both of the jaws may contain resistance wires, just as both bars may be heated in the conductance type of sealers.

Pressure to hold the two or more layers of film together is usually achieved by means of air cylinders, as described previously. Since the seal in this instance is made by melting the film, the main purpose of pressure applied through the jaws is to hold the layers of film together during the heating and cooling cycles so that they are under constant tension until the film has had an opportunity to set. This type of sealer may be used on films which have been oriented, for example, or films which have been fabricated in such a manner that they might shrink or expand when heat is applied during the heat-sealing operation.

7. Electronic Sealing (Dielectric Heating)

Certain polymer films cannot be sealed or bonded together satisfactorily by means of the heat-conductance sealers, that is, using the conventional hot jaw type of heat sealer. In these cases, dielectric heating is used in which a high-frequency current is passed through two or more layers of the film. The sealed area takes the shape of top and bottom dies or bottom bed which form the electrodes in the system. These dies are usually made of brass and are shaped to the same contour as the pattern to be sealed. The top die is clamped to a holding bar which in turn is moved up and down against a stationary lower die by means of an air cylinder. The bottom die or electrode is the ground of the circuit. With the layers of film in place between the two electrodes, a high-frequency current is passed between the two electrodes through the polymer film, heating and liquefying the plastic in the shape of the die. At the same time, the pressure which the electrodes exert on the material helps to bring about thorough fusion and bonding. The seal is complete when the current is shut off.

8. Ultrasonic Sealing (47)

In this method a piezoelectric or magnetostrictive transducer converts a high-frequency electrical signal into a mechanical vibration of the same frequency. Certain nonmetallic minerals, such as quartz, Rochelle salt, and tourmaline, conduct electricity; they have the ability to undergo mechanical deformation, or change in dimension, when an electrical potential is placed across the crystal between two of its faces. That is, the crystals vibrate when they are placed in an alternating electrical field. This phenomenon is called piezoelectricity, and crystals that have these properties are called piezoelectric crystals. Piezoelectric transducers make use of this phenomenon.

In a similar way, magnetostrictive transducers use the property of certain metals to undergo minute expansions and contractions when placed in an oscillating magnetic field. These metals are nickel, iron, cobalt, and their alloys.

Ultrasonic sealing equipment consists basically of an ultrasonic generator, a transducer, either piezoelectric or magnetostrictive, a conical welding head, and an anvil. The conical welding head is a focusing tool which amplifies the mechanical compressions and rare-factions of the transducer. The layers of material to be welded or sealed together are placed between the welding head and the anvil. The film is guided under the welding head either by means of an automatic feed or manually in the same manner that a piece of fabric is guided under the needle of a sewing machine.

The ultrasonic frequency should be the natural frequency or a harmonic of the natural frequency of the focusing tool in order to have constant intensity. Good bonds can be obtained by carefully selecting the ultrasonic intensity and adjusting the pressure of the welding head on the film and the feed rate.

The weld probably results from high-frequency working of the plastic film by the vibrating welding head which produces enough internal heat at the film surfaces to cause the layers of material to bond. As in electronic sealing, this internal heat is so localized that oriented films do not undergo any change in dimensions. The character of the seal is again a function of the temperature, pressure, and dwell time, the latter expressed in terms of the rate at which the film is fed between the welding head and anvil.

Ultrasonic seals will generally be as strong as the film from which they are made, breaking in the unsealed portion of the film near the weld line when tested in a tensile testing machine. Occasion-ally a sample will break along the weld line, probably because of a decrease in thickness of the material caused by the vibrating weld-ing tip.

9. Heat-Seal Testing

The strength of heat seals, or of seams made in polymer films by impulse sealers, electronic welding, or ultrasonic welding, may be determined by measuring the force required to pull apart the pieces of film which have been sealed together, either in a dynamic load test or a static load test, in peel or in shear. The static load test tends to be qualitative whereas the dynamic test performed on some type of tensile testing machine is quantitative, assigning a number to each seal tested.

Seals made under exactly the same conditions of sealing tem-perature, pressure, and dwell time and tested in both shear and peel tend to be weaker when tested in the latter manner. There-fore, seals are normally tested in peel because this is the way in which they are more apt to fail in a packaging application. The magnitude of the strength of a seal fabricated under marginal conditions of sealing temperature, pressure, and dwell time is an indication of the ease or difficulty which may be encountered when using the material on a high-speed packaging machine.

Apparatus. The heat sealer may be of any type, but the ones described in the section on heat conductance sealers are most practical for laboratory work. The heat sealer should have means for accurately controlling the sealing temperature, pressure, and dwell time. A tensile testing machine sufficiently accurate for measuring small loads is required for the dynamic test. For the static test, a framework of wood or metal from which the weighted heat seal specimens can be suspended without touching each other, clamps for attaching the test specimens to the framework and weights to the test specimens, and a set of weights are required. Since seal strength is normally determined for heat seals 1 in. wide, the clamps should be 1 in. in width with a sufficient bearing surface between the clamp faces so that they will not slip off the specimen when under static load.

For either type of test, ovens or especially conditioned rooms may be required for conditioning the heat seal specimens prior to testing to determine the effect of atmospheric conditions and temperature on the strength of the seal. The ovens should be large enough, of course, to accommodate the framework from which the static test specimens are suspended. Specimens which are being conditioned prior to measuring the seal strength in a tensile tester should be hung separately from a suitable stand so that the air can circulate freely about them (see Sect. VI).

Procedure. Cut a sufficient number of sheets from the film sample to be used as test specimens. Since these should be of a size convenient to handle in whatever type of heat sealer is used, the size may vary. Place the sheets in a stack so that the same side of each sheet is up; thus, the bottom of the top or first sheet will be adjacent to the top of the second sheet. The bottom of the second sheet will be adjacent to the top of the third sheet, and so on through the stack. If the film material has a machine direction, be certain to align the test specimens in the same direction. It is desirable to cut the specimens in one direction to a different length from that for the other so that the sheets do not become turned at right angles to each other. Take the first pair of sheets from the stack and place these on a piece of chipboard cut to the same dimensions. For multiple-layer seals, place 5–8 sheets in a stack, depending on the thickness of the film, being careful to maintain the same orientation of each piece with respect to the other as indicated above.

Place the specimen stack and the chipboard in the sealer with the film facing the heated sealing bar and the machine direction of the film perpendicular to the length of the sealing bar, and make the seal. After the sealing bar returns to its initial position, remove the film and chipboard. Place the chipboard aside to cool, and proceed with the next seal until enough specimens have been made to give the required degree of reliability. This has to be determined on the basis of experience and will probably vary with the type of material being tested.

From the center portion of the transverse width, cut a seal 1 in. wide (or any other desired width, but all seals must be cut to the same width). Test the seal strength on the tensile tester and record the values obtained for each of the seals. If seals are cut wider than 1 in., the results should be converted to seal strength per 1-in. width, dividing the measured value by the width of the seal. In the case of multiple-layer seals, obtain a value for each layer. The top layer, i.e., the one closest to the sealing bar during the sealing operation, is numbered one; the successive layers are numbered accordingly.

For the static test, make the seals from two pieces of film in the same manner as described above. Center a clamp on one leg of the test specimen and suspend it from the framework so that the other leg hangs freely. Hang the test specimens so that adjacent specimens do not touch. Then gently attach a weight by means of another clamp to the free leg of each specimen in such a manner as not to load the seal initially with an impact. Each weight should be centered on the bottom leg so that it acts in a vertical direction at the midpoint of the specimen width. Let the weights hang for a stipulated time. At the end of this time period detach them and examine each seal for evidence of failure, whether simple separation or failure of the bond between the coating and base sheet, in the case of a coated material.

The above procedures may be repeated with change in the sealing temperature, for example, in 25°F increments above and below the initial temperature, until a maximum temperature is obtained at which heat-seal strength decreases, and a minimum temperature at which no seal is made. Sealing temperatures for most packaging films range from 225 to 450°F.

The report should include the average seal strength, preferably in grams per inch of width, for the number of seals tested. For multiple-layer seals, average the seal strength for each layer. In addition, the report should include the type of heat sealing machine used, and the heat sealing conditions of temperature, pressure, and dwell time used in preparation of the test specimens. Finally, any pertinent information regarding the type of seal failure should be noted. In the case of static tests, the weight used and the time the load was permitted to act on the seal should be recorded.

Seal strengths may deteriorate with age, especially if the seals are exposed to conditions of high relative humidity. It is therefore sometimes desirable to condition heat-seal test specimens at various conditions of temperature and relative humidity to determine whether or not degradation occurs with age under these conditions. If the film is sensitive to water, its heat sealability may also be affected. In general, dryer films have greater heat-seal strengths when sealed at the same conditions, probably because less water vapor is entrapped between the layers resulting in bonding of a greater area of the film.

Precision. The difference in magnitude of seal strength between seals made on packaging machines and those made in the laboratory should not be unexpected. Tests of a number of heat sealing devices have shown considerable variability to occur when using different sealing bars, even though these were made as nearly identical as possible. This variability is probably caused by differences in heat transfer of the bars and of their coating, if they are coated. All causes of variability need to be determined to eliminate them, if possible, since interlaboratory agreement is highly desirable. Moreover, this factor must also be considered when comparing film characteristics in actual machine operation. Interlaboratory agreement may always be difficult to obtain as long as coated sealing bars are used.

B. Friction

In nature all surfaces are actually rough to a greater or less degree, and because surfaces differ, they exert different forces on particles or objects resting on them which act parallel to the surface and resist sliding motion. This force is called friction. It differs from adhesion which bodies have for each other, since adhesion is independent of the forces pushing two surfaces together, whereas friction depends on the normal thrust.

There are two experimental laws of friction. First, the ratio of the frictional force to the total normal thrust or force for two surfaces just about to slip over each other is constant for a considerable variation in the normal force. By definition, the coefficient of friction is as follows:

$$\mu = \text{frictional force/normal force}$$

The second law states that for a given normal force, the force of friction is approximately independent of the area of surface contact. Obviously, both laws have their limitations, for the coefficient of friction depends on the nature of the surfaces in contact and varies greatly as these vary (see Chap. 9).

The character of the motion, rolling versus sliding, for example, affects the coefficient; after the motion has been established, the coefficient may change further. Friction in sliding motion is less than static friction. We may be concerned with both sliding and static friction, for in the use of polymer films in the form of moving webs, especially in packaging applications, they are subjected to start-and-stop motion.

1. Importance of Friction

We normally talk about friction, but the property that is of most interest is the opposite one, slip. The characteristic of good slip is indicated by a relatively low coefficient of friction, about 0.2 to 0.5, without any indication of periodic sticking.

When a soft, flexible film is used in a packaging operation, or is subjected to some type of conversion process, it moves through a machine where it is subjected to a number of different forces (48). Some of these are normal forces which press the film tightly against stationary plates to form it into bags or around cartons. Along the edges of these plates, for example, the film can be subjected to very great localized pressures which can stop the forward motion of the entire web if the film should stick. As the film moves through the machine, it must slide over other machine parts until it is finally ejected from the other end as a finished bag or along with a carton or product of some kind as an overwrap. If it is ejected as a bag, it must slide over the other bags which have already been stacked at the end of the machine. And if it is the overwrap of a package it must slide along a surface of one kind or another to the next step in the packaging or packing process. Sticking at any point along the way stops the entire operation.

Friction is also important in the manufacturing process itself. Good roll formation depends on the right level of friction. If the coefficient of friction is too low, the roll may telescope during handling or shipping. If friction is too high, it may cause the film to buckle on the roll. Thus, changes in the manufacturing process must also be related to their effect on the surface characteristics of the material.

2. Friction Tests

Inclined Plane. The most common measure of the coefficient of friction is the familiar classroom experiment of the inclined plane (Fig. 11). The simple case is that of a rough surface in the form of a plane inclined at an angle θ on which a body of mass m is placed. The forces present are the normal reaction $R_1 = mg \cos \theta$, and the frictional force, F, acting along the plane in a direction which is opposite to the possible motion of the body.

F can vary from 0 to μR_1, the latter value being the limiting case just as slipping is about to occur. The angle at which this does occur is called the angle of repose and is given by

$$x = \arctan \mu$$

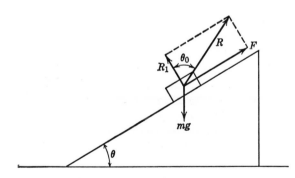

Fig. 11.   The coefficient of friction is equal to the tangent of the angle of repose.

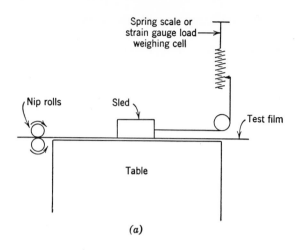

(a)

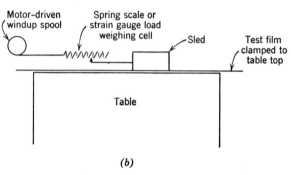

(b)

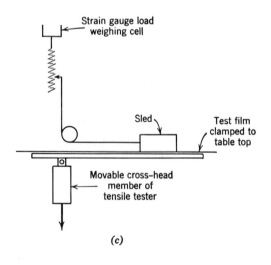

(c)

When F attains its maximum value,

$$F = \mu R_1 = \mu mg \cos \theta_0 = mg \sin \theta_0$$

and

$$\mu = \tan \theta_0$$

which is the angle of the plane at which slipping just begins. Since the angle θ_0 can be determined fairly accurately, this provides a method for determining μ.

Inclined-plane friction testers are available in which a metal plate is equipped with a spring clamp to hold the test specimen. The angle of the plate with the horizontal is increased until a weight, placed on the test specimen, just begins to slide. The coefficient of friction for the material against various other materials may be determined, or the weight may be wrapped in the film itself so that the coefficient of friction for the material against itself is measured. The latter method appears to be most satisfactory.

Moving Web or Moving Sled. The use of the inclined plane sometimes requires careful judgment on the part of the technician to decide when the weight or sled first begins to move. To overcome this problem the film specimen can be pulled under a sled or rider, restrained by a thread, the opposite end of which is fastened to a spring scale or to a strain-gauge type of load-weighing cell. The continuous tangential force thus is recorded. This average restraining force must be estimated, either by observing the average position of the pointer on the spring scale or from the load–time chart if a load weighing cell is used, and the coefficient of friction can be calculated in the normal fashion:

$$\mu = \text{average tangential force/normal force}$$

Also, the sled can be made to move over the film which is fastened to a stationary table. This can be accomplished in various ways, and instruments are available for both moving-web and moving-sled tests (Fig. 12 a,b). A clever device for performing the latter type of test makes use of the more elaborate tensile testing machines (Fig. 12c). A table with a light, low-friction pulley mounted at one end fastens, at the pulley end, to the lower, movable cross-member

Fig. 12. (a) Coefficient of friction test in which the web moves. (b) Coefficient of friction test in which the sled moves. (c) Coefficient of friction test in which a tensile tester with a movable lower cross-head member is used to move a sled and measure the tangential force. As the cross head moves down, the string is shortened, pulling the sled toward the pulley.

of the tensile testing machine. The sled is fastened to the strain-gauge weighing cell by a thread which passes under the pulley. As the cross head moves down, the length of the thread from its point of tangency with the pulley to the sled is decreased, causing the sled to be pulled over the film fastened to the table. The tangential or restraining force is measured by the load weighing cell and recorded on the recorder chart. The calculation of the coefficient of friction is the same as that for the moving web test.

Test Procedure. Cut one test specimen about 4 in. wide by 25 in. or so long in the machine direction, the length being determined by the length of the stationary table or the length of specimen over which the friction is to be measured in the moving-web test. Cut a second specimen to the proper size required by the dimensions of the sled and wrap this piece of film around the sled so that when it is placed on the long test specimen, the machine directions of each coincide. Be certain that the film specimens, both on the sled and on the table, remain flat during the test.

If the moving-web test is used, thread one end of the long specimen through the nip roll and set the drive to give the desired speed. This is normally about 6 in. per minute, but depends on the type of information required. In both the moving-web and moving-sled tests place the sled on the film so that no load is applied to the load measuring device. Start the drive—either film or sled. Note the original deflection of the load measuring device, which is used to calculate the coefficient of static friction, and the average deflection of the load weighing cell, which is used to calculate the coefficient of sliding friction.

ASTM Method D 1894 (49) is one of the few published methods for determination of this coefficient.

3. Seizure

As already indicated, operation of high-speed packaging machinery requires that the packaging material slide freely over itself or the component parts of the machine. Failure to slide, especially at points of high localized pressure, is seizure. It is the tendency of the moving web to bind, either with itself or the surfaces of the packaging machine with which it comes in contact. Unfortunately, this tendency to bind and stop moving is not always apparent in the data obtained by the standard coefficient of friction tests because too often one looks only at the coefficient and fails to study the variation in the restraining force from point to point in the film. Moreover, the coefficient is usually determined at relatively low normal forces, and although the coefficient does not vary markedly with an increase in the normal force, the coefficient of friction test is performed at a normal force well under those representative of pressure points on packaging machines.

The only difference, then, between a test for seizure and the determination of the coefficient of friction is that in the former test the normal force is increased almost to the point where the tangential force becomes equal to the normal force and the film surfaces no longer slide freely over one another but periodically adhere to each other and then release.

Seizure Test Procedure. The seizure test may be performed in the same manner as the coefficient of friction test except that the sled must be modified in such a way as to accommodate large loads. For this reason it appears advisable to perform the test using a moving web. Moreover, as the normal force increases, the rate at which the test specimen is pulled under the sled must be markedly reduced from that used in the friction test. A film speed of 0.1 in. per minute has been found to be satisfactory.

For some films seizure does not occur until the normal force reaches 50–60 lb/in.^{2}, making the standard friction test equipment unsuitable for use in the seizure test. Several different types of test equipment have been used. A very accurate research tool consists of a lathe bed (50) in which the screw is used to move the sled over the test film at speeds as low as 0.001 in. per minute.

A simpler laboratory test can be developed using one of the more elaborate tensile testing devices. The sample clamp consists of a stationary face and a smaller movable face actuated in a direction perpendicular to the larger face by means of a small air motor so that the pressure between the two faces can be increased by a controlled amount. Both faces should be made in a manner to allow covering them with the test material. This clamp is attached to the lower movable cross head of the tensile testing machine (Fig. 13).

Fasten one end of a wider piece of the test film in the upper clamp, which in turn is attached to the strain-gauge weighing cell. Thread the specimen between the faces of the clamp, and apply minimum pressure to the air cylinder so that the test specimen is

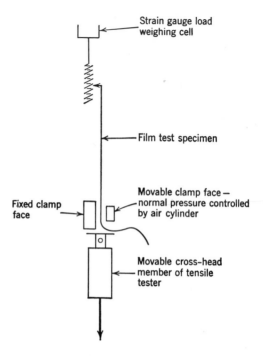

Fig. 13. Schematic diagram of a method for measuring the seizure of films using a tensile tester with a movable lower cross-head member.

held between the clamp faces. Start the cross head to descend until the pen trace on the recorder chart indicates that the tangential force is constant. Increase the air pressure stepwise on the cylinder, and thus on the jaw faces, until either the test specimen no longer slides between the clamp faces, binds completely, and fails in tension, or until a desired maximum level is obtained.

Increases in pressure should be clearly marked on the chart, as these values may be used to calculate the coefficient of friction, if desired. Since the normal force is applied to both sides of the test specimen, the actual tangential force is equal to one-half the recorded tangential force. Calculation of the coefficient of friction proceeds in the normal fashion.

The more important observation is the character of the pen trace. A material which exhibits little tendency to bind is characterized by a pen trace which is smooth and parallel to the base line of the recorder chart paper. A markedly saw-toothed trace indicates a surface which has a tendency first to stick and then to slip and suggests the possibility of difficult performance on a packaging machine. Little or no tendency to stick at normal forces up to approximately 55 psi is very satisfactory performance.

C. Coating Weight

Many polymeric materials are applied to substrates by one of several different methods to provide the substrate, which is in all probability also a polymer film with characteristics which increase its utility. The cellophane base sheet is coated with a polymer lacquer to make it heat sealable and water-vapor proof. Other polymeric films such as polypropylene may be oriented to improve their strength properties, and although these materials may be heat sealable, the application of heat to form the seal or weld causes the film to shrink back to its original dimension, making the tensilized film unsatisfactory for many packaging applications. Materials of this type require coatings to heat seal them without deformation. Still others, such as nylon films, are coated to provide them with lower or improved gas permeability rates, making them into better gas barriers ideally suited for vacuum packaging applications.

If it is intended to use a polymeric material in a coating application, the material must eventually be applied to the base sheet in order to assess its value as a coating, and of course the optimum coating thickness or weight must be determined. Although one can calculate the approximate amount of coating that has been laid down on the substrate, eventually a more exact determination usually becomes necessary, if for no other reason than establishment of the relationship between coating thickness and the desired characteristic, gas transmission rate, heat sealability, etc.

Coatings can usually be stripped from the base sheet by some method. In the case of films which have water-sensitive substrates, heating in water at boiling temperature for 10–15 min is usually sufficient to loosen the coating. Sometimes it may be necessary to heat the test specimen in a pressure cooker to raise the temperature just a few more degrees and thereby loosen the coating. With a little practice, the coating can then be stripped from the substrate, the excess water blotted off, and the swatch of coating then dried in an oven to constant weight. If the coating is applied to both sides of the web, it may be advisable to keep the coating removed from each side separate as a measure of variation in the coating process. Coating weights may be shown as pounds per ream, grams per square meter, or grams per 100 in.[2]

If the coating formulation contains water-soluble ingredients such as plasticizers, these may be leached from the coating as a result of soaking in the hot water. When new formulations are used, they should be cast as unsupported films and tested for extractables. Once it has been determined that certain ingredients are lost and the extent to which they are lost by boiling, or cooking in water, appropriate corrections can be made or correction factors determined.

In some cases the coating can be dissolved from the base sheet by use of a suitable solvent. This technique is possible only when the base sheet is not also soluble in the same solvent. The method is simple. Carefully measure out about 0.1 m² of the test sample. Wash off the coating from this test specimen in the proper solvent (some agitation is usually required). After a sufficient period of agitation, based on experience, remove the sample, place it in fresh solvent, and agitate it again to make certain that all of the coating has been removed. Combine this solvent with the original quantity from the first washing. Using additional fresh solvent, rinse the test specimen and beaker once more and combine this with the solvent from the other two washings. Evaporate the solvent to dryness and weigh the residue. Multiply the weight of the residue by the proper factor to obtain the average coating weight in the desired units. The coating should be removed from at least three test specimens to obtain a reliable average.

There are various laboratory instruments such as ultraviolet or infrared spectrometers which can be used in some instances in the measurement of coating thickness. These methods require that the coating material absorb light at the proper wavelength and that the base sheet not interfere at this wavelength. If these requirements are established, calibration curves can be developed and the coating weights or thicknesses then determined without removing the coating from the base sheet. This is distinctly advantageous, especially if the coatings are anchored to the base sheet by anchoring resins and are therefore very difficult or impossible to remove. However, just as interference from the base sheet must be determined, in the case of anchored coatings the interference of the anchoring resin or primer must also be established. Instrumental methods are unsatisfactory if any interference exists. It should also be borne in mind that these instrumental methods are not primary methods, in that absorbance must first be related to coating thickness.

Depending on the instrument used and whether sufficient space is available in the area provided for inserting the test sample into the instrument, it may be possible to make a sample holder which will accommodate a strip of the coated film. In this case the strip can be fed through the instrument, automatically, if desired, and the variation in thickness of the coating across the web or along a certain lane in the machine direction can be measured. This gives an immediate picture of the uniformity of application of the coating and allows a determination as to whether anomalies in the physical characteristics being measured are a result of variations in coating thickness or are characteristic of the material being evaluated.

D. CLARITY AND GLOSS

Most polymer films now in use are almost crystal clear, and any that have clarity which does not approach this kind of glasslike clarity immediately become the victims of competition from this direction. Thus, one can say almost categorically that unless a polymer film is perfectly clear when examined visually, it is not satisfactory for most packaging applications.

The purpose of this section, therefore, is simply to state that there are instruments and standard tests available for measuring the clarity and gloss of packaging films. One of these methods is *ASTM D 1003* (51). The stated scope of this method is to measure the light-transmitting properties and from these, the light-scattering properties of planar sections of transparent plastics. It is thus intended for the evaluation of thicker specimens and has not necessarily given good correlation with visual tests of film.

ASTM Method D 1003 uses the hazemeter or the pivotable-sphere hazemeter which are commercially available or, for the second method described, a recording spectrophotometer. An integrating sphere is used in connection with these methods which should be of a diameter such that the total area of the ports does not exceed 4.0% of the internal reflecting area of the sphere.

An old standby method is to use a Snellen eye chart and to view this from a given distance through the film, which is usually held at arm's length. From the standpoint of packaging films, this is not a satisfactory method for several reasons. The distance from which the chart is viewed, and the way the film is held, probably variable from one test to the next, do not necessarily duplicate the conditions met in most packaging applications in which the film is in direct contact with the product. Moreover, when the film is held at this distance any graininess or other imperfections such as die lines or striations and fish eyes have a tendency to disappear. Finally, the light which illuminates the chart originates behind the film. In the packaging situation, the light passes through the film, illuminates the contents and then is reflected back through the film to the eyes of the viewer.

If a new film appears to have the necessary qualifications for use as a packaging film, it is advisable to test it for clarity by some method other than the two just mentioned. Light from the incident beam will be lost for various reasons. Some will be reflected from the surfaces of the material as from a mirror. It will be scattered in all directions by the roughness of the surface or from the interior by the film's composition. Imperfections such as streaks or fish eyes will cause some of the light to be refracted. Some light will be absorbed by color bodies. A device is needed which could measure this loss and at the same time reproduce what is observed when viewing a wrapped package, that is, in order to see what is inside the package light rays must first enter the package, be reflected from the product inside the package, and pass back through the packaging material again to the viewer. Such an instrument has been described (52).

A collimated light beam is directed at an angle of 45° to the test film. It passes through the film and is reflected from a first-surface mirror to a Weston "Photronic" photocell. The beam is cut originally by an aperture to a round beam $\frac{3}{4}$ in. in diameter, and on the photocell side it is reduced even further to $\frac{1}{2}$ in. in diameter.

A film strip $3\frac{3}{4}$ in. wide is mounted on spools so that readings can be taken consecutively on many parts of the specimen. The sample could be fed automatically through the instrument, providing a continuous record of film clarity if desired. The photocell potential is measured on a strip-chart potentiometer recorder.

Bending the light beam accomplishes two purposes. By having the light pass through the test film twice, losses are intensified, thus making it easier to obtain a more accurate measurement and a more compact instrument. In order to measure the clarity of a film sample, the instrument is first adjusted to a zero point with no film in the sample holder. The zero point will be a point about 90% of the full scale deflection. Then the sample is inserted in the sample holder, and the amount of light reaching the photocell is measured again. This value will be less than the one recorded with no film interrupting the light beam because of the losses mentioned previously. The ratio of the second reading to the first times 100 is the percentage of light passing through the film and reaching the photocell. The clarity of the film is thus compared with the clarity of air as measured by an undistorted beam of light.

This method has been found superior for evaluating films used for overwrap purposes, as compared with conventional instruments in which the light passes through the film only once. However, it is not suitable for evaluating films used in window glazing or where distant objects are viewed through the film.

E. Dimensional Stability

This does not appear to be an important test until one considers that shrinkage of a film being used as an overwrap in the amount of only a few per cent may be enough to crush a carton or cause the film to rupture. Especially when characterizing a new film, one ought to establish how changes in climate effect changes in the dimension of the film, that is, changes in both temperature and relative humidity. With the development of shrink packaging techniques, the ability of a film to shrink tightly around a package as a result of the application of heat must also be measured.

Tests for measuring dimensional stability tend to be based on an arbitrary set of conditions. If a film is sensitive to changes in relative humidity, for example, it might be subjected to periods of conditioning at both extremes, wet and dry, with measurements being made at the beginning and end of the test as well as at some intermediate point or points. While this may not duplicate actual use conditions, it does provide some measure of how the film will behave.

The normal procedure is to cut strips of film in both the machine and transverse directions. Although the length of these strips may be limited by the size of the film sample, the longer the strip, within reason, the better to reduce errors in measurement. Strips 40 or 50 cm long are satisfactory. Gauge marks may be scribed on the sample and the distances between them measured as indicated above, or the actual length of the strips themselves can be used.

For films which are sensitive to water, exposure to varying conditions of humidity at constant temperature may be sufficient. For example, the specimen may be conditioned initially at a standard condition; 73°F and 50% relative humidity might be selected. The conditioning period should be long enough for the material to reach equilibrium. At the end of this period, the distance between the gauge marks would be measured as the initial length. Subsequently, the sample would be placed in a high relative humidity condition, say 90%, for a specified period of time but not necessarily of a duration to achieve equilibrium and measured again. Then it might be moved to a very dry condition such as 0–5% and the measurements made again, and finally replaced in the initial condition to equilibrium, the final measurements of length being made at the end of this conditioning period.

Since the specimen in all probability will not return to its original dimension, the difference between the initial and final lengths might be defined as the permanent set. The difference between the lengths at the high and low relative humidity conditions might be described as the deformation.

Several pieces of laboratory equipment are available for conditioning and measuring the expansion or shrinkage of paper, and they can also be used for testing hygroscopic polymer films as well. They are convenient because they usually measure several specimens at one time and without removing them to change the relative humidity condition (53). The strip specimen is cut to the specified length and is fastened between two clamps in the instrument. The specimen is usually preloaded and the measuring device is set to read zero. The test material is then subjected to various relative humidity conditions and the changes measured by means of a micrometer which is an integral part of the apparatus. These devices markedly speed up the test procedure.

Other films, such as the heat-shrink types are subject to change when exposed to high temperatures. *ASTM Method D 1204* (54) is a satisfactory method in this instance. Two specimens are cut 10 × 10 in. with a template, one specimen from either of the two transverse edges and one from the center. The midpoints of each side of the specimens are marked for use as reference marks. The specimens are properly conditioned, if required. Each specimen is placed on a sheet of heavy paper 15 × 15 in. which is lightly dusted with talc and covered with a second piece of paper. Then the sandwich is fastened together with paper clips without restraining the test specimen.

This sandwich is placed horizontally in an oven at the temperature and for the length of time applicable to the material being tested. The sandwiches should not be stacked because this may restrict movement of the film between the papers. At the end of the conditioning period, the samples are removed from the oven, allowed to cool to room temperature, and removed from the papers. The distance between the opposite edges at the reference marks is then measured to the nearest 0.01 in. and the linear change, in per cent, calculated. A negative value denotes shrinkage and a positive one expansion. This test, or one performed in a similar manner, is of importance for determining the shrink properties in connection with this type of packaging application.

F. SUMMARY

There are obviously many more tests which can be used to characterize and classify films, for packaging applications or for any other use; but the ones described are considered to be those which give the greatest amount of practical information for the work involved. Some of them are really quite refined physical tests, and if carefully and properly conducted, tend to show relationships of film strength and certain other physical properties to the chemistry of the material. The use type of test, while still measuring the strength of the material, measures it as a combination of the forces to which the film may be subjected in use. These tests frequently are empirical, and the forces are often very difficult, or impossible, to analyze.

There is great value in all of this testing, however, for the numbers generated are the guide posts in a research plan or a quality control program. Obviously, any standardization of test procedures especially through the assistance of the various technical societies will help provide a better understanding of the use properties of films and make it possible to achieve meaningful comparisons among laboratories.

VI. Conditioned Environments

By definition, conditioning as it pertains to a test specimen or a group of test specimens is the exposure of the specimens for a stipulated period of time to the effects of a prescribed atmosphere. A prescribed atmosphere is one which is maintained at a specified temperature, relative humidity, and in some instances, atmospheric pressure. The length of the conditioning period may depend on whether equilibrium is reached between the material and the atmosphere. The prescribed atmosphere is the standard atmosphere, and the standard condition is that condition which the specimen reaches when its temperature and moisture content are in equilibrium with the temperature and relative humidity of the standard atmosphere.

A. Reasons for Using Conditioned Atmospheres

In the preceding discussions of test methods the statement was occasionally made that tests should be performed on conditioned samples or at standard conditions. Certain materials are sensitive to variations in temperature or relative humidity; and in order to insure that test results on such temperature- or water-sensitive samples are comparable, it is desirable, if not necessary, to bring them to the same level of conditioning, generally in equilibrium with the atmosphere. This does not necessarily mean that the characteristics of the specimens are identical at this condition, but rather that similar materials when exposed to the same set of atmospheric conditions may behave in the same or different manner, even in regard to the taking up of water from the atmosphere.

Most polymer films, regardless of their final use, must also maintain their special properties over a wide range of conditions. Thus, films in packaging applications will be used at both low and high temperatures and exposed to dry or humid conditions at both temperature extremes. They are used in winter and summer, simultaneously in the north and in the south. Moreover, the manufacturing process may be upset by the seasonal changes in the weather. Some films may be more extensible in summer than in winter because of the greater quantity of water vapor in the air in summer, and thus they tend to break or even shatter in the normally dryer indoor air of winter.   Changes in the manufacturing process become necessary to compensate for this film difference.   If the film reacts rapidly to fluctuations in atmospheric conditions, even diurnal variations may be enough to produce changes in film properties.   Because of this, it seems vitally important to test not only at a standard condition, but even at various reproducible conditions which may be selected to simulate use conditions so that the material may be completely characterized.

B. Atmospheric Variables Which Should Be Controlled

As already indicated, the primary conditions which should be controlled are temperature and relative humidity.   Depending on the final use of the film, it may even be necessary to control pressure, but for most packaging applications this does not yet appear to be essential.

1. Temperature

The effects of changes in temperature are usually apparent as changes in strength and toughness.   Films generally lose their extensibility at lower temperatures and eventually become brittle. Conversely, an increase in temperature may change the character of the film so that its stress–strain curve changes from the hard-brittle type to hard-tough to soft-tough.   Other properties for which we have not discussed test methods may also change with temperature, such as creep and stress relaxation.   Nor is the rela-

TABLE 2
ISO Recommended Standard Atmospheres

Temperature °C	Temperature tolerance	Relative humidity, %	Relative humidity tolerance
20	Normal ±2°C (if closer	65	Normal ±5% (if closer
27	tolerance is required,	65	tolerance is required,
23	±1°C may be specified)	50	±2% may be specified)

tionship between permeability and temperature linear for a number of materials. One can thus see the need for the selection of standard test temperatures, and most of the technical societies involved with testing have selected conditions which have become standard for their industries. The Coordination Committee on Atmospheric Conditioning for Testing, of the International Organisation for Standardisation (ISO) recommends standard atmospheres for testing materials that are used in conditions approximating ambient atmospheres, Table 2 (55).

2. Relative Humidity

The effect of humidity on many polymer films is first apparent in the hand, or feel, of the material. Films which are susceptible to changes in humidity take up or give off water in direct proportion to the increase or decrease in relative humidity, defined as the ratio of the water vapor pressure to the saturation vapor pressure at that ambient temperature $\times$ 100. Films conditioned at the same relative humidity but at different temperatures will eventually reach the same moisture content, but the one conditioning at the lower temperature will require the longer conditioning period since the driving force, the vapor pressure, approximately, at the ambient temperature, is proportionately lower.

Water vapor in the atmosphere may be measured in a number of ways (56–59). The amount of water in a measured volume of air may be determined by means of the Karl Fischer method, or by passing the air through a drying tower and measuring the increase in weight of the desiccant. The sling psychrometer, consisting of a wet bulb and dry bulb thermometer mounted together on a metal strip and attached to a handle so that they can be twirled in the air to ventilate the wet bulb, is the most commonly used instrument. In order to determine the relative humidity, the psychrometer is twirled in the air until the equilibrium wet bulb and dry bulb temperature readings are obtained, usually for about 2 min. Relative humidity is read from a temperature–humidity chart or from the psychrometric tables which take atmospheric pressure into account. Direct reading hygrometers or hygrographs are also available as well as several electronic "dew cells." The first practical device of this latter type was developed in the early 1940's by F. W. Dunmore of the National Bureau of Standards. These devices are often used with recorder controllers to monitor and regulate the amount of moisture in the air.

The dew cell, a dew-point sensing device, consists of bifilar electrodes wound on a spool coated with a conductive salt solution such as lithium chloride (bifilar refers to a winding which has two separate conductors wound side by side). The two wires are connected to the secondary coil of an integral transformer. The bifilar electrodes are not interconnected but depend on the conductivity of the atmospherically moistened lithium chloride for current flow since the salt solution will tend to change the electrical conductivity of the entire assembly (60). The electrodes are basically a resistance thermometer and thus are also sensitive to temperature changes. Being sensitive to both temperature and humidity simultaneously, the dew cell may even be calibrated in per cent relative humidity. The major shortcoming of this type of device is its limited range.

Standard humidity conditions also have been included. Because the water content of some films changes rapidly with changes in relative humidity, such as nylon 6 and uncoated cellophane, it is necessary to try to control the relative humidity as closely as is possible, although this kind of control is sometimes difficult to achieve.

3. Atmospheric Pressure

Because of its limited effect on packaging film test results, the control of atmospheric pressure in the laboratory will not be discussed. As packaging becomes a growing part of space technology, the effects of changes in pressure will no doubt require investigation and the eventual need for pressure-controlled test chambers.

C. Conditioned Rooms

When designing a laboratory having a controlled atmosphere, it is essential to maintain the standard conditions uniformly throughout the room, that is, at all points in plan and elevation, without dead spots, either accidental or persistent. Maximum allowable deviation of the conditions from standard must be set. The normal and maximum work loads in the room must be established—such as the loads resulting from the extremes of climatic conditions of the area in which the laboratory is located; the number of persons who will be working in the room and the degree of their activity; the amount of electrical equipment, such as lighting, motors, and other heat-producing equipment; and transients, such as opening the door into the room from the outside, or heat gain from opening an oven door in the laboratory, product load, and heat and moisture transfer through the walls, ceiling, and floor which would depend on the type of insulation used. The standard conditions may be those adopted by industry or they may be based on individual requirements. Typical conditions may be similar to the following: 73°F and 50% relative humidity, to be controlled as closely as can be obtained but not less precise than ±1°F and ±2% relative humidity as a maximum deviation from standard after a period of one minute from a disturbance, and to average out within two minutes to not more than ±1% deviation in relative humidity. Sufficient fresh air makeup must be provided for the personnel involved.

The use of the extremes of climatic conditions for calculating work loads does not seem to be the usual practice. Rather, the average seasonal temperatures and humidities apparently are used. And consequently when an abnormally hot and humid period occurs, the air conditioning unit does not have the capacity to cool and to remove the excess water from the fresh air makeup and simultaneously maintain the desired room conditions. When making these calculations it cannot be stressed too much that it is the unusual variations that cause the trouble in trying to maintain control and not the seasonal averages.

In order to maintain control limits throughout the room and to reestablish them once they have been upset, a good rule is to withdraw air from the room and recycle it at a rapid rate of about one air change per minute. A substantial portion of this air would be bypassed through the humidifier and dehumidifier, the amount to be controlled by proportioning dampers. There should be a sufficient number of diffusion supply ports distributed evenly over the ceiling area to provide thorough distribution of the air without discomfort for the personnel.

Good design further suggests that the entrance to the return duct be located near the door and close to the floor. This location will help to minimize disturbances in the conditions caused by frequent opening and closing of the door, as air entering the room through the open door is immediately drawn into the return air stream and conditioned as part of the fresh air makeup. The use of air locks through which the personnel must pass to enter the chamber are not a necessity in every case. If such locks are used, the conditions maintained in them should be carefully studied in relation to the conditions in the test chamber. For example, the dew point temperature of an air lock for a high-humidity room should be higher than that of the conditioned room in order to avoid condensation of water vapor on the walls of the air lock every time the door into the conditioned room is opened.

1. Construction

One of the most difficult problems in the construction of laboratory facilities having a humidity- and temperature-controlled atmosphere is concerned with keeping moisture out of the insulation. Differences between the relative humidity of the room and the surroundings produce a vapor pressure gradient between the inside and outside which forces vapor into the insulated space. For rooms having a normal temperature and low relative humidity, the resulting problem is one of insufficient air drying capacity. In the case of a high-humidity atmosphere in the test chamber the problem manifests itself as water, generally on the floor at the base of the walls; if there is considerable leakage, the floor becomes completely wet, both inside and out of the chamber. In rooms having temperatures below freezing, as warm air leaks into the insulated space from the outside, its temperature is finally reduced to that of the dew point. The water vapor condenses, and eventually the insulation becomes saturated with water.

Moisture tends to increase the thermal conductivity of the insulation. A commonly accepted rule is that a 1% gain in weight because of water will increase the conductivity of the insulating material by 5%. To give some idea of the magnitude of this problem, insulated trailers in the trucking industry gain hundreds of pounds in weight from the condensing of water vapor in the insulated space between the walls. These trailers must then be brought into the garage and their interiors heated to drive out the water and dry the insulation and walls, requiring several days' time.

2. Wall Surfaces

There is no known material from which the walls of a conditioned room can be constructed which is absolutely impervious to water vapor, according to the United States Quartermaster Corps (61), including foil or even thin sheet steel. Welded construction using either of these materials appears to be the best, but it is necessary to test all seams to insure perfect welds with no pinholes. This test can be made by placing a rubber gasketed glass cup over the seam, producing a vacuum in the cup by means of a small pump and looking for soap bubbles to appear over the weld which has been previously coated with a soap solution. Steel must of course be painted or coated to prevent rust.

Specially shaped rubber gasket material is now available which can be used with sheet metal instead of welding to reduce the cost of construction. The edges of the sheet metal are slid into the molded rubber gasket, and the sheets butted tightly together. In this type of construction great care must be taken to get good contact between the metal sheets and the gasket material, or paths will be provided for leakage through the wall section.

Plastic sheeting of various kinds, especially the heavier gauges of polyethylene, and asphalt-impregnated papers, although effective, are by far, poorer vapor barrier materials than sheet metal, aluminum, or steel. When installing these materials no holes or tears can be permitted, as a hole, however small, is an open window to water vapor.

The type of wall covering must be selected with great care in relation to the conditions in the chamber. If the relative humidity in the room is high in comparison with that of the atmosphere surrounding the room on the outside, a substantial vapor pressure gradient will exist across the wall, driving moisture through the wall—unless it is absolutely water-vapor proof—to the outside atmosphere. A layer of vinyl 10 mils thick sprayed onto a concrete block wall in this instance would not be satisfactory because the entire structure would be too porous, and maintenance of a painted outer wall surface would be almost impossible after a short period of operation.

3. Insulating Material

Selection of the insulating material will be based on its cost and its effectiveness, which is demonstrated by the K factor of the material. K, the coefficient of thermal conductivity, is the amount of heat transmitted per hour through a square foot of the material 1

in. thick, when the surface temperatures on the opposite sides of
the material differ by 1°F. The K factor of still air, for example,
is approximately 0.18; of reflective aluminum, 0.23; and of corkboard, 0.27. In addition to insulating value and cost of the insulation, one must also consider cost of construction. Insulating
materials such as corkboard and the foamed plastics ought to be
secured to the studs by tying. Successive layers should be embedded in mastic and additionally secured with hardwood skewers.
The insulating boards should be of a size to fit snugly between the
framing timbers, and the framing timbers should not come in contact with the inner and outer wall material, whether this be a metal
skin or Portland cement plaster commonly used in the construction
of coolers.

The author found the combination of still air and aluminum foil
to provide very effective insulation in the construction of a low-
temperature room, although certain difficulties developed as will
be pointed out. The aluminum used had a nominal thickness of
0.006 in. It was secured to wood battens, $\frac{3}{4}$ in. thick by 2 in. wide,
placed on 16-in. centers. The joints between the sheets of foil were
covered with aluminum tape. This appears to have been the weakest point of this construction, for considerable leakage of air through
the walls eventually occurred, and when the room was maintained
at a below-freezing temperature for an extended period of time,
water would drain from the walls for several days whenever the
temperature was raised to some point above freezing. This type
of construction obviously requires air-tight joints in the inner and
outer wall coverings. In that event, some provision must also be
made for breathing of the walls because of expansion and contraction of both the inner and outer walls which occurs during changes
in the temperature both inside and outside the room. These
breather ports should contain cartridge type dryers with a valve
assembly so that moisture can be expelled from between the walls
but trapped in the dryers when air is being drawn in.

This experience further suggests the possibility of building the
room in a sort of tray which contain a drain connected to the sanitary system of the building. In this case, provisions would have
to be made along the bottom of the walls and in the floor of the
environmental chamber to permit any water trapped in the space
between the layers of wall and floor to drain into the tray when-

ever the temperature of the room was raised to a point above freezing.

4. Air Conditioning Equipment

Air conditioning and refrigeration engineers have the background and experience to recommend excellent equipment which can be used to obtain the desired conditions in the environmental test laboratory. The purpose of these paragraphs is to underline certain considerations which might otherwise be overlooked if there has been no experience in building rooms of this type, considerations which may require additional time and money to correct and the possible loss of many tests and testing time as well.

As already indicated, one should be certain that the capacity of the conditioning and air handling units is sufficient for the extremes of climate that occur, such as extended hot and humid weather. This is especially true if cooling and drying of the air is accomplished by means of a cooling coil. It therefore appears advisable when several rooms are being controlled, at as many different conditions, that each room have its own conditioning equipment and be completely independent of the others.

The equipment required for a high-humidity room in which the temperature is controlled above the normal ambient temperature of the building in which it is located is the least difficult. It consists of an air handling unit and heater with the necessary controllers. A room with a volume of about 1200 ft^3 can be heated to 100°F with a 1200–1500-watt strip heater placed in the air stream. Steam vented into the air stream from a pipe containing about a dozen ports is a simple source of water vapor and additional heat. A water spray will not add heat to the room air. Air ducts should be short and preferably noncorrosive, and the air handling unit should be fabricated from the same noncorrosive metal, such as aluminum or copper. The ducts will always have some condensate in them even if they are insulated, as the temperature of the return air will probably be lower than the dew-point temperature. It is a good idea to include the installation of a drain pipe connecting all the sections of the duct and running this to the sanitary drain of the building (half-inch pipe is sufficient for this drain).

A room conditioned at a very low relative humidity, say less than 15%, will require some kind of air drying equipment even if a cooling coil is used for this purpose as well as for sensible cooling. Here personal preference is involved, but one should again be concerned with capacity as well as with ease of operation and maintenance. Dryers using the solid desiccants such as silica gel seem to be less prone to operation and maintenance problems.

For coolers and cold rooms, a good defrost system is important and should be included in the system. If necessary, two sets of coils should be installed so that one can be defrosted while the other is in use. Since the cooling coils will probably be located in the room itself, the defrost system can be either automatic or manual, depending on how much the room is used and the amount of traffic in and out. However, the requirement of being able to defrost should not be dismissed. Capacity of the refrigeration unit should be based on the summer climatic extremes to insure that the same low temperatures can be obtained in the hot summer months as well as during the other seasons of the year.

The degree of control is dependent on the design of the air handling equipment as well as on the capacity of the system. According to the author's experience, rooms which were conditioned with the equipment running continuously 24 hr a day incorporating such devices as balance loaders to insure a constant load on the machinery at all times had the smallest variation in temperature or humidity, or both. The heating coil was always hot and the cooling coil always cold; when conditions needed to be corrected, a series of proportioning dampers opened and closed, changing the amount of air flowing over one coil or the other to heat or to cool as required.

D. Conditioned Cabinets

Over the period of the past 10 or 15 years, a number of manufacturers have entered the field of environmental cabinet construction. Cabinets of all sizes and providing almost any condition or set of conditions are now available. Some are sectional so that a number of sections can be joined together to form a room of any size. They usually are complete with all conditioning equipment and instrumentation and can be set up where space is available.

Since these cabinets or walk-in environmental chambers are carefully engineered, many of the problems one might encounter when trying to design and build a room can be avoided. It is certainly advisable to make a careful study of rooms versus environmental cabinets and chambers before embarking on a difficult construction program.

E. OTHER METHODS FOR CONDITIONING
AND TESTING SMALL SAMPLES

Draft Recommendation Number 597 of the International Organisation for Standardisation describes methods for maintaining constant relative humidity in small enclosures by means of aqueous solutions (62). The procedures are intended for conditioning small quantities of plastic materials prior to test and for such tests as may be carried out entirely within a small enclosure, e.g., electrical tests.

The method discusses the relationship between the surface area of the solution and the size of the chamber. It points out that equal distribution of humidity in a small enclosure can be promoted by relating the smallest side of the surface area of the solution with the height of the enclosure. If the enclosure is without air circulation, its height should not exceed the length of the smaller side of the surface of the solution. If the enclosure has air circulation, its height may be up to 1.5 times the smaller side of the surface area of the solution.

Two methods are given for obtaining the desired relative humidity in the enclosure, viz., saturated salt solutions and glycerol solutions. When saturated salt solutions are used, the atmosphere over the solution at constant temperature will be maintained at a definite relative humidity for every salt solution. It is essential to have an excess of the solid salt to insure a saturated solution. Similarly, various relative humidities can be achieved over solutions of glycerol. The concentration of glycerol solutions in terms of the refractive index at 25°C for the desired relative humidity is given by the formula:

$$(R_1 + A)^2 = (100 + A)^2 + A^2 - (H + A)^2$$
$$R_1 = 715.3(R - 1.3333)$$

where A is $25.60 - 0.1950T + 0.0008T^2$, H is the relative humidity in per cent, R is the refractive index of the glycerol solution, and T is the temperature of the solution between 0 and 70°C.

The refractive index of the glycerol solution must be checked regularly since loss of water from the solution to the material being conditioned will reduce the relative humidity over the solution.

A suggested conditioned enclosure has a volume greater than 14,000 cm^3. The box consists of a rectangular enclosure of hard polyvinyl chloride with a jacket of thermal insulation. A fan moves a current of air over the saturated salt solution which is at the bottom of the enclosure after the air has been heated to the required temperature by an electric heating element. A baffle directs the air current upward and around over a thermostat which controls the temperature, wet and dry bulb thermometers for measuring the relative humidity, and finally along the holders for the test pieces. Near the fan there are water-cooled tubes which cool the air, but its temperature is brought back to the desired working temperature when it passes the fan and the heating element.

For very small samples the standard laboratory desiccators can also be used with the same salt or glycerol solutions that one would use in the chamber just described.

VII. Comparative Evaluation of Competitive Products

Up to this point our concern has been the testing of polymer films under standard conditions to find out how strong they are, how they might be improved, and how they might be used. As a concluding note, we should briefly discuss a large area of testing, the purpose of which is to assess the performance of untried films as alternates or replacements for those already in use. This testing is usually carried out as a field test, the actual application of the film in a packaging operation. Large portions of these tests are performed under conditions which cannot be controlled, and therefore results are sometimes very difficult to interpret. Nevertheless, the additional information secured may have sufficient weight to influence the selection of the new film to replace one which may have been the standard for many years.

There are many reasons why a new material might be selected, such as better appearance, improved strength, better machine performance, additional protection, and lower cost. That last mentioned carries great weight, especially if all other considerations are almost as good as the film currently in use. Thus, one has to set a minimum performance level which must be met when the cost of producing the new material is attractive. These tests will normally take some time to run, and if data are to be available on time, it may be necessary somehow to adjust their severity.

If the reason for the tests is to establish differences among several materials, the same tests apply; for unknown and new materials require complete analysis both in the testing laboratory and in the field. They must be checked on packaging machines, amount of sample permitting, and in every way that they might eventually be used. Consequently, a series of use tests must finally be carried out to determine the total effect of the packaging operation and distribution system on the film and the effect of the film on the packaged product, or whatever the use may be.

The final decision regarding the material will depend on the analysis of all data, and since cost plays such an important role, some compromise between risk of loss or damage to the goods and the price of the film will have to be made finally. Unfortunately, reliable correlation between laboratory tests and field performance is difficult to obtain, and it will probably take many more years of accumulating data before the desired degree of confidence is assured.

VIII. Conclusion

The important question in this entire discussion is how to establish the best material for the job. Testing, as we have seen, can be performed generally in two different ways. Finished packages can be subjected to conditions of use—vibrated, dropped, tumbled—at a variety of climatic conditions, hot or cold, high humidity or low. These tests are usually only qualitative, frequently subjective, and may test the durability of the contents rather than that of the packaging material. Often combinations of materials must be used, and control cannot be maintained on all of them. Variations therefore are due in many instances to unassigned causes. These tests may also be expensive as large quantities of an expensive product may have to be vibrated or bounced or dropped to destruction along with the relatively inexpensive test material. Also, all too frequently, because of this expense, the number of samples involved in the test is too small to provide statistically meaningful and reliable results.

In laboratory bench tests, the primary type of test, usually a single attribute is measured on a single test specimen. Since true physical characteristics are being determined, the basic structure of the substance will be defined. Measurements are quantitative and objective and lend themselves to mathematical analysis so that statistically reliable interpretations can be drawn. Data of this kind may also be suspect, since a sufficient number of tests is not always performed nor does a tensile test—for example, run at a rate of a few inches per minute—represent the total stresses to which a bag might be subjected in a drop. In general, however, the pitfalls in the way of correct packaging materials decisions appear to be fewer because test conditions are easier to control and large numbers of specimens can be tested.

There are as many different laboratory bench tests which can be used to evaluate a material as there are physical properties that need to be measured, and the essential ones have been described in the preceding sections. Since the stresses which cause failure in a package can be analyzed, they ought to be measurable in the physical test. The major question concerns what tests to perform, and the individual packaging problems and requirements dictate this.

The laboratory tests will usually show up specific shortcomings, since no material is so completely versatile as to satisfy every need. This kind of knowledge is necessary, and its proper use should help to avoid some of the unrealistic thinking that sometimes occurs when trying to select one material for a number of diversified applications. In short, knowledge gained from testing of the kind outlined previously in detail is well suited to establish the best material for the job.

There is no doubt that testing is expensive. The alternative to a well planned testing program is quite clear. Select something that seems suitable for the job, use it, wait to see what happens, and hope that the amount of losses sustained because the wrong material has been used does not exceed the cost of a well thought out test program.

References

1. J. E. Booth, *Principles of Textile Testing*, Chemical Publishing Company, Inc., New York, 1961.
2. J. R. Tucker, *Ind. Res.*, **5**, 45 (1963).
3. *ASTM Standards*, American Society for Testing and Materials, Philadelphia, Pa., issued annually in 32 parts.

648   KARL W. NINNEMANN

4. *Tappi Standards and Suggested Methods*, Technical Association of the Pulp and Paper Industry, New York, correct to end of 1966.
5. *Pulp and Paper Manufacture: Bibliography and U. S. Patents*, J. Weiner, Ed., Technical Association of the Pulp and Paper Industry, New York, 1942–.
6. L. E. Nielsen, *Mechanical Properties of Polymers*, Reinhold, New York, 1962.
7. F. Reinhart, *Modern Converter*, p. 36 (March 1965).
8. J. Fachet and T. T. P. Plimpton, Jr., *Modern Converter*, p. 13 (Jan. 1966).
9. H. Hindman and G. S. Burr, *Trans. ASME*, **71**, 789 (1949).
10. "Test for Tensile Properties of Plastics," *ASTM Method D 638-61T, 1964 Standards*, Part 27, p. 207.
11. "Tests for Tensile Properties of Thin Plastic Sheeting," *ASTM Method D 882-61T, 1964 Standards*, Part 27, p. 376.
12. T. S. Carswell and H. K. Nason, *Mod. Plastics*, **21**, No. 10, 121 (June 1944).
13. O. C. Hansen, Jr., T. L. Fabry, L. Marker, and O. J. Sweeting, *J. Polymer Sci., A*, **1**, 1585, 3666 (1963).
14. J. K. Craver and D. L. Taylor, *Tappi*, **48**, 142 (1965).
15. J. d'A. Clark, *Paper Trade J.*, **100**, 41 (1935).
16. R. N. Brownlee, *Pulp Paper*, **29**, 130 (1955).
17. O. C. Hansen, Jr., L. Marker, K. W. Ninnemann, and O. J. Sweeting, *J. Appl. Polymer Sci.*, **7**, 817 (1963).
18. T. C. Hendrickson, *SPE J.*, p. 95 (June, 1966).
19. D. H. Morton and A. Marks, *J. Sci. Instr.*, **42**, 591 (1965).
20. "Stiffness of Paperboard," *Tappi T 489m-60, Tappi Standards and Suggested Methods*.
21. "Rigidity, Stiffness, and Softness of Paper," *Tappi T 451m-60, Tappi Standards and Suggested Methods*.
22. "Folding Endurance of Paper," *Tappi T 423m-50, Tappi Standards and Suggested Methods*.
23. "Folding Endurance of Paper by the M.I.T. Tester," *ASTM Method D 2176-63T, 1965 Standards*, Part 15, p. 741.
24. H. C. Horst and R. E. Martin, *Mod. Packaging*, **34**, No. 7, 123 (March 1961).
25. P. A. Gelber and J. H. Bowen, Jr., *Mod. Packaging*, **25**, No. 5, 125 (Jan. 1952).
26. J. Conti, *Mod. Packaging*, **37**, No. 8, 216 (April 1964).
27. "Impact Resistance of Polyethylene Film by the Free-Falling Dart Method, *ASTM Method D 1709-62T, 1964 Standards*, Part 26, p. 236.
28. R. H. Supnik and C. H. Adams, *Plastics Technol.*, **2**, 151 (1956).
29. K. W. Ninnemann, *Mod. Packaging*, **30**, No. 3, 163 (Nov. 1956).
30. K. A. Arnold, 6th Tappi Testing Conf., Springfield, Mass., 1955.
31. "Test for Propagation Tear Resistance in Plastic Film and Thin Sheeting," *ASTM Method D 1922-61T, 1964 Standards*, Part 27, p. 642.
32. "Test for Tear Resistance of Plastic Film and Sheeting," *ASTM Method D 1004-61, 1964 Standards*, Part 27, p. 428.
33. D. Patterson, Jr., and E. B. Winn, *Mater. Res. Std.*, **2**, 396 (1962).

34. C. H. Bulmer, *Brit. Paper Board Makers' Assoc., Proc. Tech. Sect.*, **40**, 311 (1959).
35. Definitions of Terms Relating to Shipping Containers," *ASTM Method D 996-59, 1965 Standards*, Part 15, p. 351.
36. "Drop Test for Shipping Containers," *ASTM Method D 775-61, 1965 Standards*, Part 15, p. 191.
37. "Vibration Test for Shipping Containers," *ASTM Method D 999-63T, 1965 Standards*, Part 15, p. 366.
38. F. A. Paine, "Evaluation of Package Performance," in *Fundamentals of Packaging*, F. A. Paine, Ed., Blackie & Son, London-Glasgow, 1962, p. 257.
39. A. W. Myers, J. A. Meyer, C. E. Rogers, V. Stannett, and M. Szwarc, *Permeability of Plastic Films and Coated Paper to Gases and Vapors*, Monograph 23, Technical Association of the Pulp and Paper Industry, New York (1962).
40. "Tests for Water Vapor Transmission," *ASTM Method E 96-63T, 1964 Standards*, Part 30, p. 307.
41. "Water Vapor Permeability of Sheet Materials at High Temperature and Humidity," *Tappi T464m-45, Tappi Standards and Suggested Methods*.
42. K. W. Ninnemann and L. E. Simerl, *Pulp Paper Mag. Can., Tech. Sect.*, **63**, T-68 (1962).
43. "Test for Gas Transmission Rate of Plastic Sheeting," *ASTM Method D 1434-63, 1964 Standards*, Part 27, p. 497.
44. R. Henderson and G. A. Wallace, *Food Technol.*, **10**, 636 (1956).
45. K. W. Ninnemann, *Mod. Packaging*, **31**, No. 3, 171 (Nov. 1957).
46. J. H. J. Stocker, *Brit. Plastics*, **39**, 149 (1966).
47. J. A. Radford, *Package Eng.*, **8**, 54 (1963).
48. H.-E. Roder, R. Heiss and G. Schricker, *Package Eng.*, **5**, 42 (1960).
49. "Test for Coefficients of Friction of Plastic Films," *ASTM Method D 1894-63, 1964 Standards*, Part 27, p. 603.
50. H. P. Sponseller and F. M. Gavan, *Mater. Res. Std.*, **3**, 992 (1963).
51. "Test for Haze and Luminous Transmittance of Transparent Plastics," *ASTM Method D 1003-61, 1964 Standards*, Part 27, p. 420.
52. P. M. Hay, C. P. Evans, and K. W. Ninnemann, *Mod. Packaging*, **33**, No. 3, 179 (Nov. 1959).
53. "Moisture Expansivity of Paper," *Tappi T 447x-59, Tappi Standards and Suggested Methods*.
54. "Measurement of Changes in Linear Dimensions of Nonrigid Thermoplastic Sheet or Film," *ASTM Method D 1204-54, 1964 Standards*, Part 27, p. 462.
55. "Standard Atmospheres for Conditioning and Testing Materials," *ASTM Method E 171-63, 1964 Standards*, Part 30, p. 496.
56. A. Wexler and W. G. Brombacher, "Methods of Measuring Humidity and Testing Hygrometers," *Natl. Bur. Std. Bull. 512*, U.S. Government Printing Office, Washington (1951).
57. R. W. Worrall, *J. Environ. Sci.*, **8**, 27 (1965).
58. R. E. Fishburn, *Automation* (Jan. 1963).

59. E. J. Amdur, *Instruments and Control Systems* (June, 1963).
60. Bull. 90-1716, Minneapolis-Honeywell Regulator Company, Minneapolis, Minn. (1962).
61. L. F. Zsuffa, in *Low-Temperature Test Methods and Standards for Containers —A Symposium*, E. C. Myers and N. J. Leinen, Eds., Natl. Acad. Sci., National Research Council, Washington, 1954.
62. Draft ISO Recommendation No. 597, International Organisation for Standardisation, General Secretariat, Geneva, 1963.

CHAPTER 14

THE ANALYTICAL CHEMISTRY
OF CELLULOSIC FILMS

F. J. REIDINGER

Olin Film Division, Olin Mathieson Chemical Corp.,
Pisgah Forest, North Carolina

651

In this chapter, only cellulosic films of commercial value and importance are covered. For most cellulosic films, methods of analysis have been developed over a long period of time and are described in the literature. Standard methods for the determination of basic characteristics of cellulosic films are widely published and it is not intended to repeat all of these methods. It is rather the author's intention to describe some of the less well-known methods for the analysis of the finished film product. For cellophane, for instance, only very few methods of analysis have been published in the past. There are also instrumental methods which in the past few years have been more widely used and applied to cellulosic films. Special applications include the characterization of pyrolysis products of cellulosic films by gas chromatography and determination of trace quantities of substances in cellulosic films which could not be performed previously because of the insensitivity of existing methods. There are methods which can be applied to all cellulosic films as they determine a characteristic which is inherent in all cellulosic films.

I. Sampling

In order to obtain accurate and reproducible results in the analyses of cellulosic films, good sampling procedures are of the utmost importance. Since most cellulosic films are more or less hydrophilic, the importance of proper sampling procedures can hardly be emphasized enough. Methods are available for the sampling of paper which can be applied to cellulosic films (1,2). These methods do not give specific procedures, but they emphasize that the specimens to be tested shall be placed immediately after sampling in an airtight container. In the case of rolls, care shall be taken to select sheets that are not damaged. It is good practice to discard the first three layers of the roll to be sure that a unit sample in good representative condition is obtained. The specimens shall be cut from sheets taken across the full width of several unharmed layers. A sufficient number of specimens from each unit sampled should then be arranged consecutively in rotation to form a representative sample.

It is good practice to randomize sheets of film of equal size into several stacks each containing 10–15 sheets, depending upon the analysis to be performed. The stacks are then sealed in glass jars and saved for analysis. When the specimen is analyzed, a sufficient quantity for each analysis shall be cut from one end of the entire stack. The test specimen should thus contain a small portion of each of the 10–15 sheets.

II. Determination of Moisture in Cellulosic Films

Most characteristics of cellulosic films are greatly affected by the amount of moisture contained in the film. Unprotected films can lose or gain moisture very rapidly and the protection of coated films depends upon the permeability characteristics of their coating. The determination of moisture is therefore one of the most important analyses for almost all cellulosic films. There are three basic methods for the determination of moisture in cellulosic films: the oven drying method, the Karl Fischer method, and the toluene distillation method.

The oven drying method is applicable only when no other volatile substances besides water are present in the test material, which is seldom the case with cellulosic films. If the equipment is available, the Karl Fischer method is the most convenient and fastest and can be performed with very small amounts of sample. No interferences are expected from cellulosic films unless the samples contain unusual plasticizers, such as amines. The toluene distillation method can be used only when at least 50 g of sample is available.

A. Oven Drying Method (3)

In this method, a sample taken with extreme care to avoid any change in moisture content is transferred quickly to a weighed container, preferably to an airtight glass weighing bottle. The specimen is then weighed in the closed container to obtain its wet weight. It is then dried together with the container, but preferably outside of it, in an oven for about 2 hr at 105 ± 3°C. The specimen is then, if possible in the oven, replaced in the original container. This is then closed and cooled to room temperature and weighed. The specimen is then reheated for another hour at 105 ± 3°C and the bottle is closed in the oven, cooled to room temperature in a desiccator, and weighed. This drying and weighing process is repeated until the difference in weight between two successive weighings is not more than 0.1% of the weight of the specimen.

B. Karl Fischer Method

The Karl Fischer method is described in detail in *ASTM D 1348-61* (4). This method is most helpful for samples containing nonaqueous materials, volatile at 110°C, since such substances interfere in the oven drying method. One has to be aware that anhydrides, alkalies, large amounts of aldehydes and ketones, amines, and some metal oxides interfere with the Karl Fischer method. Only small amounts of sample are needed. The determination can be performed in two ways, either by adding an excess of Karl Fischer solution to the sample and after a 15–20-min extraction, a back-titration with a standardized water–methanol solution, or by a direct titration with Karl Fischer solution. The end point is best detected electrometrically but it may be satisfactorily determined visually.

An alternative Karl Fischer method was developed by Scopp and Evans (5) using a visual end point. With some modifications, this method was found to be very useful for the determination of moisture in cellophane and other cellulosics. The big advantage of this method is the possibility of vigorous stirring which aids in the extraction of moisture from the sample and therefore shortens the time of analysis. It also eliminates a common interference in the electrometric method, which involves pieces of film winding around the platinum electrodes, giving erratic end point readings.

C. Toluene Distillation Method

The third basic method is described in detail in *Tappi, T 484m-58* (6). Although this is an accurate method, it requires at least 50 g of sample and more than 1 hr for the complete extraction of water from the sample. Such large sample weights are not always available and the time requirement is about 3–4 times that of the Karl Fischer method.

Besides these three basic methods, other interesting methods appear in the literature which seem worth mentioning. Ermolenko and Gusev (7) studied the infrared spectra of celluloses, containing various amounts of adsorbed water and found a linear dependence of the intensity of the band at 1640 cm^{-1} and the water content. Kajanne (8) extracted a cellulosic sample in a closed container with a measured amount of pyridine and calculated the moisture content of the resulting mixture from the refractive index of the solution.

III. Cellulose Nitrate

Mainly because of its inflammability, cellulose nitrate is not used as widely as a film-forming polymer as it had been some years ago, but it still plays an important part in the coatings field.   The degree of nitration is a characteristic property of cellulose nitrate and therefore the determination of the nitrogen content is the most important analysis.   The complete nitration to a nitrogen content of 14.16% is not done commercially.   Usually the nitrogen content ranges from 11 to 13%.

The standard method for the determination of the nitrogen content of cellulose nitrate is *ASTM Method, D 301-56* (9), the nitrometer method.   In this method, cellulose nitrate is decomposed with concentrated sulfuric acid and the evolved nitrogen oxide gas is measured in a gas-measuring tube.   From the amount of gas evolved, the nitrogen content of the sample can be determined.   The apparatus has to be calibrated frequently with potassium nitrate.

Many authors (10–14) claim that the nitrometer method is hazardous because of the large amount of mercury involved.   Although all agree that it gives excellent results, it requires an experienced operator and is rather time consuming.   Many other methods have been developed for the determination of cellulose nitrate.   Volumetric methods, based on the transnitration of salicylic acid (13,15) or ferrous chloride reduction (12,14), are much used.   One of these (12), is of interest here because it is claimed that by this method cellulose nitrate can be determined in the presence of alkyd resins and plasticizers.   Direct titrations with ferrous sulfate (16,20) or ferrous ammonium sulfate (17–19) have the advantage of speed and simplicity.   Reduction of cellulose nitrate followed by the Kjeldahl method (21) was found to give excellent results, but the method is rather slow.

Several colorimetric methods have been developed.   Three will be described in detail because of their specific applicability to cellulose nitrate lacquers or plasticized cellulose nitrate.

A. Direct Colorimetric Determination of Cellulose Nitrate in Lacquers (22)

Principle.   The yellow color formed by reaction of cellulose nitrate with alkali in the presence of acetone is a direct measure of the amount of cellulose nitrate in the sample.

Procedure. A sample containing 40 mg or less of cellulose nitrate is transferred to a ground joint 125-ml erlenmeyer flask and diluted to 10 ml with acetone. Ten milliliters of 10% aqueous KOH are added along with a berl saddle and the sample is refluxed for 1 hr in a water bath under a water-cooled condenser. The sample is then promptly cooled to room temperature, transferred to a 50-ml volumetric flask, rinsed, and finally diluted to volume with a mixture of acetone and water (2 parts acetone to 1 part water by volume). The absorbance of the yellow solution is measured at 425 mμ and the nitrocellulose content of the sample is determined from a calibration curve prepared in like manner from pure nitrocellulose. Certain resin products, phenol–aldehyde condensates, and free aldehydes interfere slightly.

B. FERROUS SULFATE METHOD (23)

Principle. The dried sample is treated with a ferrous sulfate reagent and the resulting red-to-purple color is determined spectrophotometrically at 525 mμ. From a graph prepared with known nitrates, the nitrate content of the cellulose nitrate may be determined.

Procedure. The reagent is prepared by dissolving 0.5 g of $FeSO_4 \cdot 7H_2O$ in 25 ml of water and adding, with cooling, 75 ml of concentrated H_2SO_4. Potassium nitrate is used for standardization. A weighed quantity is dissolved in water and diluted to volume so that aliquots containing 0.5–2.0 mg of NO_3^- can be withdrawn into 25-ml erlenmeyer flasks having ground glass stoppers. These samples are dried in an oven at 105°C. Nitrocellulose samples are dissolved in acetone and dried at 60°C. To the dried sample, or standard, exactly 20.0 ml of the ferrous sulfate reagent is added. The flask is stoppered and allowed to stand for 30 min (90 min for nitrocellulose) with frequent agitation. The absorbance of the red-to-purple color developed is determined on a spectrophotometer at 525 mμ, using the ferrous sulfate reagent as a blank. The color is stable for several hours.

C. PHENOL METHOD (24)

Principle. The sample is treated with sulfuric acid and phenol. The resulting yellow color is determined spectrophotometrically at 360 mμ and is a measure of the nitrate content of the sample.

Reagents. *Phenol solution* (5%). Add 95 ml of water to 5 g of pure phenol.

Sulfuric acid. Add 300 ml of concentrated sulfuric acid to 100 ml of water and cool to room temperature.

Procedure. The sample and a known nitrate salt are prepared for analysis as in the previous ferrous sulfate method, except that smaller samples are used. Sample aliquots should contain 0.03–0.1 mg of NO_3^-. Exactly 10.0 ml of the 3:1 sulfuric acid is added to the sample in the standard flask. The flask is stoppered and allowed to stand for 30 min, with occasional agitation. Three drops of the 5% phenol solution are added slowly, with swirling of the flask after addition of each drop. The flask is stoppered and allowed to stand for 30 min, with occasional agitation. Three more drops of the 5% phenol solution are added slowly, with swirling of the flask after addition of each drop. The flask is stoppered and allowed to stand for an additional 30 min. The absorbance of the yellow solution is then determined at 360 mμ, using 10 ml of the 3:1 sulfuric acid to which three drops of the phenol reagent have been added as a blank. For highest accuracy, absorbance readings should fall between 0.2 and 0.6. From a graph prepared with known nitrates, the nitrate content of the nitrocellulose may be determined.

D. Other Methods

The determination of cellulose nitrate by infrared spectrometry has been investigated by several authors (10,11,24). Rosenberger and Shoemaker (10) have determined cellulose nitrate by infrared spectrometry in the presence of other cellulosics, modified resins, phenol–formaldehyde resins, phthalic alkyd resins, aldehydes, alcohols, and plasticizers by using the absorption band at 11.92 μ. The sample was dissolved in acetone and a portion of that solution was placed in a liquid cell. Acetone was used as a blank. A straight-line calibration curve was obtained by plotting absorbance readings at 11.92 μ against the cellulose nitrate content of standards, over a range of 0–5%.

Levitsky and Norwitz (11) have determined cellulose nitrate by infrared spectrometry, using tetrahydrofuran as a solvent and measuring the absorbance at 6.0 μ.

Clarkson and Robertson (25) have published a refined calculation for the determination of nitrocellulose by infrared spectrometry.

Ettre and Varadi (26,27) have investigated the pyrolysis products of cellulose nitrate by gas chromatography. With the use of a specific pyrolysis apparatus and different chromatographic columns, they resolved and identified almost 100% of the decomposition products. They also found no appreciable changes in the composition of the breakdown products over a wide range of temperatures. This pyrolysis–gas chromatographic technique offers not only a way for the qualitative identification of cellulose nitrate and possibly other cellulosics but also a method for the quantitative determination of cellulose nitrate via its decomposition products.

IV. Cellulose Acetate

Cellulose acetate is the most widely known ester of cellulose and may be found in photographic films, transparent sheeting, surface coatings, thermoplastic transparent films, and others. It is odorless, tasteless, and nontoxic. Depending on its degree of substitution (DS), it can either be extruded or cast from solvents. Its properties depend upon the acetyl content, plasticizer content, and other additives. It is characterized mainly by its DS, i.e., the average number of acetyl groups per anhydro-D-glucose unit; its degree of polymerization (DP), i.e., the average number of anhydro-D-glucose units per molcule; and by the distribution of unesterified OH groups among the primary 6-position, and the secondary 2- and 3-positions.

A. Determination of Combined Acetyl or Acetic Acid Content

The determination of combined acetyl or acetic acid content has been exhaustively investigated and reported (28–32). Basically, there are three wet chemical methods: saponification in suspension (Eberstadt method), saponification in solution, and the acid distillation method of Ort.

1. Saponification in Suspension (Eberstadt Method)

Reagents. *Ethanol*, 75% by volume. Denatured alcohol (formulas 2B, 3A, or 30) may be used.
Standard aqueous hydrochloric acid, 0.5N.
Standard aqueous sodium hydroxide, 0.5N.

Procedure. The sample is ground in a Wiley mill or other suitable grinder so that 100% will pass a No. 20 sieve. About 1 g of the sample is dried in a weighing bottle at $105 \pm 3°C$ for 2 hr. The bottle is then stoppered and cooled in a desiccator. After the bottle with the sample has been weighed to the nearest milligram, its content is transferred to a 250-ml erlenmeyer flask, and the bottle is reweighed to determine the exact sample weight. To each sample 40 ml of 75% ethanol is added. Blank determinations are carried through the complete procedure.

The flasks, loosely stoppered, are heated for 30 min at 50–60°C. Then 40 ml of $0.5N$ sodium hydroxide solution is added to each flask, and all are again heated for 15 min at 50–60°C. The flasks are stoppered tightly and allowed to stand at room temperature for 48 hr. In case the acetyl content of the sample is over 43%, or the sample is hard and horny, the flasks are allowed to stand for 72 hr. At the end of this reaction time, the excess sodium hydroxide is titrated with $0.5N$ hydrochloric acid, using phenolphthalein as indicator. An excess of about 1 ml of $0.5N$ hydrochloric acid is added and the sodium hydroxide is allowed to diffuse from the regenerated cellulose for several hours, or, preferably, overnight. Disappearance of the pink color is an indication of the complete neutralization of sodium hydroxide. The small excess of hydrochloric acid is then very carefully titrated with $0.5N$ sodium hydroxide to the pink phenolphthalein end point. During this titration, the sample has to be shaken vigorously to extract the remaining traces of hydrochloric acid from the regenerated cellulose. The addition of $0.5N$ sodium hydroxide has to be continued until the faint pink color persists.

Calculation. The acetyl and acetic acid content are calculated as follows:

$$\% \text{ acetyl} = \frac{[(B_1 - A_1)N_1 + (A_2 - B_2)N_2] \times 43.05 \times 100}{\text{Sample weight in g} \times 1000}$$

$$\% \text{ acetic acid} = \frac{[(B_1 - A_1)N_1 + (A_2 - B_2)N_2] \times 60.05 \times 100}{\text{Sample weight in g} \times 1000}$$

where

$B_1 =$ ml of HCl required to titrate blank
$A_1 =$ ml of HCl required to titrate sample

N_1 = normality of HCl solution
A_2 = ml of NaOH required to titrate sample
B_2 = ml of NaOH required to titrate blank
N_2 = normality of NaOH solution
43.05 = equivalent weight of acetyl
60.05 = equivalent weight of acetic acid

The degree of substitution (DS), i.e., the average number of acetyl groups per anhydro-D-glucose unit of cellulose, can be calculated from the acetyl content (33):

$$N_a = \frac{3.86a}{102.4 - a}$$

where N_a is the average number of acetyl groups per anhydro-D-glucose unit of cellulose; and a is acetyl content, per cent.

2. Saponification in Solution (29,32)

Saponification in solution is claimed to be more precise but less widely applicable than the Eberstadt method. Samples with an acetyl content from 32 to 41% are dissolved in a mixture of acetone and water. Samples with a higher acetyl content can be dissolved in a mixture of acetone and dimethyl sulfoxide, or with heating, in a mixture of acetone and pyridine. Sodium hydroxide, approximately $1N$, is used for saponification. The reaction time for saponification is considerably shorter in the solution method than in the Eberstadt method. Samples with the lower acetyl content are allowed to saponify for 30 min and samples with high acetyl content are allowed 2 hr when dimethyl sulfoxide is used as the solvent or 3 hr when pyridine is used. After saponification, the excess sodium hydroxide is back-titrated with standard $1N$ sulfuric acid.

3. Acid Distillation Method According to Ort (28)

This method is useful for acetyl analysis when other nonvolatile acids or alkali-consuming groups are present. The sample is dissolved in 1:1 sulfuric acid and after addition of water, the volatile acids are steam distilled at a rate of about 600 ml/hr of distillate. The distillation is continued until a total of 1500–1600 ml has been collected. The entire distillate is then titrated with standard sodium hydroxide, using phenolphthalein as indicator. A modification of this method is also used to determine mixed cellulose esters.

B. INSTRUMENTAL METHODS

1. Determination of Acetyl Content by
Near-Infrared Spectroscopy (34)

The acetyl content of cellulose acetate in the range from 35–44.8% can be determined by near-infrared spectroscopy due to the absorbance of residual hydroxyl groups at about 1445 mμ. This absorbance of the first overtone of the hydroxyl group is inversely proportional to the acetyl content of cellulose acetate. The solvent for cellulose acetate in this analysis is pyrrole.

Procedure. The samples are dried for 2 hr at 105°C. Exactly 1.25 g is weighed into a glass stoppered bottle of about 60 ml capacity and redried for 30 min at 105°C. Then, 25 ml of pyrrole is added and the bottle is shaken until the sample is dissolved. It is then placed in a constant temperature bath at 25°C and allowed to come to equilibrium. To obtain reproducible results, it is imperative that the sample be at exactly 25°C. The solution of sample is then placed in a 5-cm stoppered quartz cell, and pyrrole containing exactly 5% carbon tetrachloride is placed in a matched quartz cell. With the sample in the sample beam and the pyrrole–carbon tetrachloride solution in the reference beam, the spectrum is scanned from 1300 to 1700 mμ on a Beckman Model DK-2, or an equivalent spectrophotometer. The absorbance at 1445 mμ is measured and the acetyl content is obtained from a predetermined calibration curve. It is claimed that average acetyl values determined by this method are in good agreement with the Eberstadt and the carbanilation methods.

2. Determination of Acetyl Content
by Infrared Spectroscopy (35)

Quantitative analytical data have been reported for the determination of the acetyl content of cellulose acetate by infrared spectroscopy. The sample is dissolved in a 9:1 mixture of methylene chloride–methanol and the absorbance, at a wavelength of 5.8 μ, is measured. In this method, the sample would have to be free of any other esters such as plasticizers.

C. Determination of Total Hydroxyl Content by the Ultraviolet Absorption Method (32,36,37)

The total hydroxyl content of pyridine-soluble cellulose acetate can be determined by the carbanilation method. The acetyl content can then be calculated by difference from the hydroxyl content.

Apparatus. *Beckman Model DU Spectrophotometer* (or equivalent instrument).

Special reflux tubes. These may be constructed as follows. A test tube approximately 20 × 150 mm is made from the outer part of a 24/40 standard-taper ground glass joint by closing the open end in a blast lamp. The tubing of the inner joint is drawn to a constriction just above the joint. The glass is cut at that point and an 80-mm length of tubing is sealed on to provide a bearing for a glass stirrer. The stirrer is made from a 4-mm glass rod with a semicircle at right angles to the shaft at the bottom and small enough to fit into the test tube. This unit acts as an air condenser, thus preventing loss of solvent by evaporation.

Reagents. *Acetone.*

Ethanol (denatured formulas 2B, 3A, or 30).

Pyridine (redistilled, of water content less than 0.05%).

Phenyl isocyanate.

Methylene chloride and methanol mixture (90:10 by weight). This mixture should have an absorbance of less than 0.2 at 280 mμ in a 1.00-cm quartz cell when measured against air. The solvents have to be redistilled if they do not meet this specification.

Procedure. The phenyl isocyanate must be used under anhydrous conditions. The sample, reflux tubes, pipets, and all other equipment must be dried thoroughly. In the special reflux tube, a sample of about 0.5 g is dried in an oven at 105°C for 2 hr. The tube is removed from the oven and 5 ml of pyridine is added. The completely assembled reflux apparatus is then placed in an oil bath at 115–120°C. The mixture is stirred occasionally until the sample has completely dissolved. With a pipet, 0.5 ml of phenyl isocyanate is added, the sample is stirred thoroughly, and refluxed for 30 min in the oil bath to complete the reaction.

At the end of the reaction time, the tube with the sample is removed from the oil bath, cooled, and the solution is diluted with acetone to the proper viscosity for precipitation. The amount of acetone used to dilute the sample is critical in acquiring a good precipitate. Samples having a low viscosity require little or no dilution. The average sample requires the addition of about an equal volume of acetone. The carbanilate is precipitated by pouring the solution into about 200 ml of denatured ethanol with vigorous stirring. The precipitate should be fluffy and white. Sticky precipitates indicate too little dilution. The precipitate is filtered through paper on a büchner funnel, with suction applied only as long as is necessary to remove the bulk of the solvent. Prolonged suction may cause the precipitate to redissolve in the pyridine.

The precipitate is transferred to a 4-oz screw–cap bottle containing 75 ml of denatured ethanol. The bottle is covered securely, and shaken for 30 min on a shaker at medium speed. The precipitate is then filtered again on a büchner funnel, pressing out as much of the liquid as possible with a glass stopper. The washing and filtering operation is repeated twice more or until the absorbance of the filtrate at 280 mμ is about the same as that of an ethanol blank. The precipitate is then allowed to dry in air for 1–2 hr at room temperature to assure the evaporation of the ethanol. Samples wet with ethanol may sinter or stick to paper or glass when dried at 105°C. The sample is then dried in an oven at 105°C for 1 hr and cooled in a desiccator.

A 0.1231-g sample of the dry precipitate is weighed into a 100-ml volumetric flask, 60–80 ml of the mixture of methylene chloride and methanol is added, and the flask is shaken occasionally until complete solution occurs. (For triacetate, the sample size is doubled and the observed absorbance is divided by two.) The solution is diluted to 100 ml with methylene chloride–methanol solution and mixed thoroughly. With a Beckman Model DU spectrophotometer, or equivalent instrument, the absorbance of the sample solution is measured in a 1.00-cm quartz cell at 280 mμ against the solvent mixture as a reference.

Calculation. For a sample weight of 0.1231 g, the carbanilate, c, is calculated as follows:

$$\% \text{ carbanilate} = c = \text{absorbance at 280 m}\mu \times 17.1$$

$$\% \text{ hydroxyl} = \frac{14.3c}{100 - c}$$

$$\% \text{ acetyl} = \frac{4480 - 65.1c}{100 - c}$$

assuming exactly three hydroxyl units to one anhydro-D-glucose unit; the number of hydroxyls per anhydro-D-glucose unit is equal to $(6.86 \times \% \text{ hydroxyl})/(40.5 + \% \text{ hydroxyl})$.

V. Cellulose Acetate Propionate and Cellulose Acetate Butyrate and Their Analysis by the Partition Method (38,39)

These cellulose mixed esters are well known for their toughness, transparency, and weather resistance. Their properties depend upon such factors as DS and plasticizer content. They are characterized by their total apparent acetyl content, their propionyl or butyryl content, and their hydroxyl content.

The analysis of mixed esters involves the following steps.

1. Determination of the apparent acetyl content:

For samples containing less than about 35% propionyl or butyryl.

For samples containing more than about 35% propionyl or butyryl.

2. Saponification and isolation of the acids:

For samples containing less than about 35% of propionyl or butyryl.

For samples containing more than about 35% of propionyl or butyryl.

3. Determination of the molar ratios of the acids.

4. Calculation of acetyl, propionyl, and butyryl contents.

A. DETERMINATION OF APPARENT ACETYL CONTENT

The determination of the apparent acetyl content or the total acyl content calculated with an equivalent weight of 43 is necessary since this value is required in the calculation of the actual acetyl, propionyl, or butyryl content. Two methods are described, one for low propionyl or butyryl content and one for high propionyl or butyryl content.

1. Method for Samples with Less than
about 35% Propionyl or Butyryl Content

Reagents. *Acetone* (reagent grade).
Dimethyl sulfoxide.
Pyridine (reagent grade).

Sodium hydroxide (aqueous solution, about 1*N*).
Sodium hydroxide (standard aqueous solution, 0.1*N*).
Sulfuric acid (standard solution, 0.1*N*).

Procedure. The sample is finely ground, well mixed, and dried in a weighing bottle for 2 hr at 105 ± 3°C. By difference, 1.9 g of the sample is weighed to the nearest milligram into a 500-ml erlenmeyer flask. A blank is carried through the entire procedure. If the sample is acetone-soluble, 150 ml of acetone and 5–10 ml of water are added. The flask is gently swirled to mix its contents, stoppered, and then allowed to stand with occasional swirling until solution is complete. Solution may be hastened by gently shaking on a suitable mechanical shaker.

If a sample of low propionyl or butyryl content is acetone-insoluble, it is dissolved by the following method. The sample is evenly distributed on the bottom of the flask, and 70 ml of acetone is added. The flask is swirled gently until the sample is wetted completely. The flask is stoppered and allowed to stand for 10 min. Then 30 ml of dimethyl sulfoxide is carefully added by pouring it down the wall of the flask to wash down any particles clinging to the side. The flask is stoppered and allowed to stand with occasional swirling until solution is complete. Solution may be hastened by gentle mechanical shaking. When solution is complete, 50 ml of acetone is added and the flask is swirled for 5 min.

To the dissolved sample, 30 ml of about 1*N* sodium hydroxide is added with constant swirling or stirring. The use of a magnetic stirrer is recommended. It is absolutely necessary that a finely divided precipitate of regenerated cellulose be obtained. The flask is stoppered and under continuous stirring 30 min are allowed for saponification when the solvent used is acetone or 2 hr when the solvent is dimethyl sulfoxide. At the end of the saponification period, 100 ml of hot water is added, washing down the sides of the flask, and the contents are stirred for about 2 min. The excess of sodium hydroxide is then titrated with 1.0*N* sulfuric acid, using phenolphthalein as indicator. The sample is titrated rapidly with constant stirring until the end point is reached and then an excess of 0.3 ml of 1.0*N* sulfuric acid is added. The mixture is stirred for at least 10 min with a magnetic stirrer. More phenolphthalein indicator is added and the small excess of acid is titrated with 0.1*N* sodium hydroxide to a persistent pink end point.

Calculation. The apparent acetyl content is calculated as follows:

Apparent % acetyl

$$= \frac{[(B_1 - A_1)N_1 - (B_2 - A_2)N_2] \times 43.05 \times 100}{W \times 1000}$$

where

A_1 = ml of sulfuric acid required to titrate sample
B_1 = ml of sulfuric acid required to titrate blank
N_1 = normality of sulfuric acid
A_2 = ml of sodium hyroxide required to titrate sample
B_2 = ml of sodium hydroxide required to titrate blank
N_2 = normality of sodium hydroxide
W = sample weight in grams

2. Method for Samples with More than about 35% Propionyl or Butyryl Content

Reagents. *Acetone–methanol mixture* (1:1 by volume).
Pyridine–methanol mixture (1:1 by volume).
Hydrochloric acid (standard aqueous solution, 0.5N).
Sodium hydroxide (aqueous solution, about 0.5N).
Sodium hydroxide (methanolic solution, about 0.5N).
Procedure. The sample is finely ground, well mixed, dried at 105 ± 3°C for 2 hr, and cooled in a desiccator. A sample portion of 0.5 g is weighed to the nearest 0.5 mg and transferred to a 250-ml, glass-stoppered erlenmeyer flask. The sample is dissolved in the appropriate solvent which depends upon the propionyl or butyryl content. At least two blanks are carried through the entire procedure.

Samples containing from about 35 to 45% propionyl or butyryl are dissolved in 100 ml of the acetone–methanol mixture. Aqueous sodium hydroxide and water are added in exact amounts in the following order from a buret or pipet and the contents of the flask are swirled vigorously during all additions: 10 ml of NaOH and 10 ml of water; 10 ml of NaOH and 5 ml of water; 20 ml of NaOH and 5 ml of water. The flask is stoppered and allowed to stand at room temperature for 16–24 hr.

Samples containing 45% or more propionyl or butyryl are dissolved in 100 ml of the pyridine–methanol solution. Thirty milliliters of methanolic sodium hydroxide is added slowly with agitation. Then 20 ml of water is added in 2 ml increments with agitation. The flask is continually swirled until the solution becomes turbid. It is then stoppered and allowed to stand overnight at room temperature.

After this saponification period, the excess of sodium hydroxide is back-titrated with 0.5N hydrochloric acid, using phenolphthalein indicator.

Calculation. The apparent acetyl content is calculated as follows:

$$\text{Apparent \% acetyl} = \frac{(B - A)N_1 \times 43.05 \times 100}{W \times 100}$$

where A is the number of millimeters of hydrochloric acid required to titrate sample, B is the number of millimeters of hydrochloric acid required to titrate blank, N_1 is the normality of hydrochloric acid, and W is sample weight in grams.

B. Saponification and Isolation of Acids

The apparatus and reagents listed below are used for both of the procedures that follow.

Apparatus. *Constant-temperature bath or oven at 40°C.*

Vacuum distillation apparatus. This consists of a 500-ml round-bottomed flask which is fitted with a very small capillary inlet tube, and a kjeldahl distilling head. This in turn is connected to a vertical condenser of which the lower end reaches within 7.5 cm of the bottom of a 500-ml side-arm distillation flask, used as the receiver. A condenser of the spiral type is recommended. The kjeldahl distilling head is fitted with a funnel or stoppered opening for the addition of extra water during the distillation. A water bath is provided for the heating of the distilling flask and an ice bath for cooling the receiver.

Reagents. *Ethanol, denatured* (formulas 2B, 3A or 30).

Phosphoric acid (1:14). Sixty-eight milliliters of 85% H_3PO_4 is diluted to 1 liter with water.

Sodium hydroxide (aqueous solution, about 0.5N).

1. Procedure for Samples of Propionyl or Butyryl Content Less than 35%

About 3.0 g of sample, not especially dried nor accurately weighed, is heated with 100 ml of 0.5N sodium hydroxide in a stoppered

500-ml round-bottomed flask in a water bath or oven at 40°C for 48–72 hr. About 50 ml of the 1:14 phosphoric acid is added to form monosodium phosphate, which liberates the organic acids from their sodium salts. The flask is connected with the distillation apparatus, and the solution is vacuum distilled to dryness, allowing a small stream of air to enter through the capillary inlet tube to avoid bumping. The receiver is cooled to 0°C. A portion of 25 ml of water is added to the residue and the vacuum distillation is again taken to dryness. This is repeated with a second 25-ml portion of water. Care has to be taken that during the distillation no phosphate ion is carried over into the distillate. It is not necessary to work with quantitative accuracy in this operation but it is important to obtain an aqueous solution of the acids in the same ratio as they occur in the esters. The volume of distillate and rinsings is diluted to 250 ml which usually results in a distillate of an acidity of 0.06 to 0.12N, the range desired for subsequent extractions.

2. Procedure for Samples of Propionyl or Butyryl Content More than 35%

About 3.0 g of sample, not especially dried nor accurately weighed, is placed in a 500-ml round-bottomed flask and 100 ml of denatured ethanol and 100 ml of 0.5N sodium hydroxide is added. The sample is allowed to stand stoppered at room temperature for 48–72 hr. At the end of this saponification period, the regenerated cellulose is filtered off and the filtrate collected in a 500-ml round-bottomed flask. The flask is connected to the distillation apparatus and all of the alcohol is removed by vacuum distillation. After distillation to dryness and release of the vacuum, the apparatus is rinsed with water and the distillate and rinsings are discarded.

To the residue in the distillation flask about 50 ml of 1:14 phosphoric acid and 100 ml of water are added. The distillation apparatus is reassembled and the distillation of the volatile acids is carried out as described in the preceding section.

C. Determination of the Molar Ratios of the Acids

Reagents. *Acetic, propionic, and butyric acids of highest purity.* *n-Butyl acetate* (this extraction solvent should be free of acidity and water and should contain not more than 2% of butyl alcohol).

Sodium hydroxide (standard aqueous solution, 0.1N).

Procedure. A 25-ml portion of the distillate as obtained in Section V-B-1 or -2 above is titrated with 0.1N sodium hydroxide solution, using phenolphthalein as indicator. The volume of NaOH required in the titration is designated as M. Then, 30 ml of distillate and 15 ml of n-butyl acetate are measured accurately into a separatory funnel. The mixture is shaken for about 2 min and then the two layers are allowed to separate for several minutes. The lower aqueous layer is drawn off and 25 ml of this solution is titrated with 0.1N NaOH solution. The volume of NaOH required in the titration of this portion is designated as M_1. The percentage partition ratio, K, of the acids in the distillate is calculated as follows:

$$K = (M_1/M) \times 100$$

Approximately 0.1N solutions of each, acetic, propionic, and butyric acids are prepared and their distribution ratios are determined in the same manner. The distribution ratios of the pure acids are calculated as follows:

$$K = M_1/M$$

where k_a can be used for the distribution ratio of acetic acid, k_p for propionic acid, and k_b for butyric acid.

For mixed acids the molar ratios are calculated in the following manner:

For acetic–propionic acid mixtures:

$$P = \frac{K - 100k_a}{k_p - k_a}$$

$$A = 100 - P$$

For acetic–butyric acid mixtures:

$$B = \frac{k - 100k_a}{k_b - k_a}$$

$$A = 100 - B$$

where P is mole % propionic acid, B is mole % butyric acid, A is mole % acetic acid, and K is % partition ratio of the acids in the distillate.

These constants have to be checked from time to time and must be determined by each operator for each supply of n-butyl acetate.

D. Calculation of Acetyl, Propionyl, and Butyryl Content

The percentages of the acetyl and propionyl contents in cellulose acetate propionate and the percentages of acetyl and butyryl content in cellulose acetate butyrate are calculated as follows:

$$\% \text{ acetyl} = A \times C$$

$$\% \text{ propionyl} = P \times C \times \tfrac{57}{43}$$

$$\% \text{ butyryl} = B \times C \times \tfrac{71}{43}$$

where C is the percentage of apparent acetyl content as determined in Section V-A-1 or V-A-2.

The total hydroxyl content of cellulose acetate propionate and acetate butyrate may be determined in the same manner as described under Cellulose Acetate, Section IV-C.

VI. Ethyl Cellulose

Of the cellulose esters, only ethyl cellulose and hydroxyethyl cellulose are of importance as film-forming polymers. Ethyl cellulose may be divided into two groups.

Ethyl cellulose with an ethoxyl content of 46–48% is a tough plastic with good low-temperature properties. This grade is mainly used for injection molding extrusion.

Ethyl cellulose with an ethoxyl content from 48–49.5% has a lower softening point, a better compatibility, a wider solubility, and a good water resistance. This grade is mainly used for coating applications.

For this reason, ethyl cellulose is mainly characterized by its ethoxyl content. All methods for the determination of ethoxyl content are based on the original procedure by Zeisel (40). The Zeisel procedure has been modified by Vieböck and Brecker (41) and later by Samsel and McHard (42). The Samsel and McHard procedure has been adopted with some modifications as the Standard ASTM Method (43) for determination of the ethoxyl content of ethyl cellulose.

A. Determination of Ethoxyl Content

Apparatus. *Distillation apparatus.* The all-glass distillation apparatus consists of a boiling flask of about 30-ml volume with a capillary side arm for admission of CO_2, an air condenser with trap, and a receiver. Two receivers may be used to assure complete absorption of the ethyl iodide.

Oil bath. This can be maintained at 145–150°C.

Reagents. *Absorption solution.* Five milliliters of bromine are dissolved in 145 ml of a solution containing 100 g of potassium acetate, 900 ml of glacial acetic acid, and 100 ml of acetic anhydride.

Phosphorus suspension. In 100 ml of water, 0.06 g of red phosphorus is suspended. If sulfur compounds are present in the sample, the phosphorus is suspended in a 5% cadmiun sulfate solution.

Gelatin capsules (size 0, available from Parke, Davis & Co.).

Hydriodic acid. Constant boiling mixture with water, 57%, sp gr 1.7.

Carbon dioxide. Commercial cylinder of CO_2 equipped with a suitable reducing valve.

Sodium acetate solution. Dissolve 220 g of anhydrous sodium acetate in 1 liter of distilled water.

Potassium iodide (reagent grade).

Formic acid (90%, sp gr 1.20).

Starch indicator solution (1%).

Sodium thiosulfate (standard solution, 0.1N).

Sulfuric acid solution (1:9 in distilled water).

Procedure. The finely ground sample is dried in an oven at 105°C for at least one half hour. The trap is filled about half full with the aqueous phosphorus suspension, or, if sulfur compounds are present in the sample, with the suspension of phosphorus in a 5% cadium sulfate solution. The receiver is filled with 19–20 ml of the absorption solution. If two receivers are used, the first is filled with 6–7 ml of the absorption solution and the second with about 13 ml. Into a gelatin capsule 50–60 mg of sample is accurately and rapidly weighed and the capsule is dropped into the boiling flask. The rapid weighing is important because oven-dry ethyl cellulose picks up moisture rapidly. A few small glass beads or boiling stones and 6 ml of 57% hydriodic acid are added to the flask. The flask is attached immediately to the condenser and the glass joint is sealed with a few drops of hydriodic acid. The side arm of the boiling flask is connected to the CO_2 source and a stream of gas at the rate of 2 bubbles per second is passed through the apparatus. The boiling flask is then immersed in the oil bath at 150°C and heated for 40 min.

After this reaction period, the contents of the receiver or receivers are washed into a 500-ml erlenmeyer flask, containing 10 ml of sodium acetate solution. The contents of the flask are diluted with distilled water to 125 ml and formic acid is carefully added, dropwise, with swirling to destroy excess bromine. When the bromine color has disappeared, 6 more drops are added. After 3 min of standing, 3 g of potassium iodide and 15 ml of 1:9 sulfuric acid are added. The liberated iodine is then titrated with 0.1N sodium thiosulfate, using starch indicator solution. A blank determination is made, using the same amounts of reagents and the same procedure as for the sample.

Calculation. The ethoxyl content is calculated as follows:

$$\% \text{ ethoxyl} = \frac{(A - B)N \times 7.51 \times 100}{\text{Sample weight in g} \times 1000}$$

where A is the number of milliliters of $Na_2S_2O_3$ solution required to titrate the sample, B is the number of milliliters of $Na_2S_2O_3$ solution required to titrate the blank, and N is the normality of the $Na_2S_2O_3$ solution.

B. OTHER METHODS FOR DETERMINATION OF ALKOXYL GROUPS

Cundiff and Markunas (44) have reported a simplified procedure for the determination of alkoxyl groups. The sample is decomposed with hydriodic acid, 55–58%, and the respective alkyl iodide is absorbed in pyridine. The formed alkylpyridinium halide, which acts as a very weak acid in this nonaqueous medium, is directly titrated with tetrabutylammonium hydroxide. The titration can be performed either potentiometrically or visually, using azo violet as an indicator. It is claimed that free iodine and sulfides do not interfere in this method so that the trap which has to be used in the other procedure can be eliminated.

Cobler, Samsel, and Beaver (45) have developed a gas chromatographic technique for the determination of mixed alkoxyl groups in cellulosics. The sample is decomposed with hydriodic acid by a modified Zeisel procedure and the formed alkyl iodides are collected in a cold trap at −80°C containing 2,2,5-trimethylhexane. An aliquot part of the solvent containing the alkyl iodides is then analyzed by gas chromatography.

VII. Hydroxyethyl Cellulose and Hydroxypropyl Cellulose

Hydroxyethyl and hydroxypropyl celluloses of a molar substitution of about 0.15 are becoming increasingly important as film-forming materials. These low-substituted celluloses are insoluble in water but readily soluble in aqueous alkali. Their molar substitution (MS) is defined as the average number of moles of ethylene or propylene oxide combined with the cellulose per anhydro-D-glucose unit (46). This number may be greater than three. The molar substitution can be determined by a modified Zeisel method, originally described by Morgan (47) and further modified for low-substituted materials by Lortz (48).

A. DETERMINATION OF ETHYLENE OR PROPYLENE OXIDE IN LOW-SUBSTITUTED CELLULOSE ESTERS

Principle. Hot constant-boiling hydriodic acid decomposes the hydroxyethyl unit quantitatively to ethyl iodide and ethylene. The hydroxypropyl unit is decomposed to propyl iodide and propylene, respectively. In distillation, the gases are swept by carbon dioxide into two separate traps, where the ethyl iodide (propyl iodide) is collected in a silver nitrate solution and the ethylene (propylene) is collected in a bromine solution. After completion of the reaction, the contents of the silver nitrate and bromine traps are titrated separately, the total alkoxyl content being the sum of ethyl iodide (propyl iodide) and ethylene (propylene) expressed as ethylene oxide (propylene oxide).

Apparatus and Reagents. *Modified alkoxyl apparatus.*

Hydriodic acid (sp gr 1.70, constant boiling).

Silver nitrate solution. Fifteen grams of silver nitrate is dissolved in 50 ml of water and then added to 400 ml of absolute ethanol. Several drops of concentrated nitric acid are added. This solution is standardized against $0.05N$ ammonium thiocyanate by the Volhard method.

Bromine solution. One milliliter of bromine is added to 300 ml of CP glacial acetic acid, saturated with 5 g of potassium bromide. This solution is standardized against $0.05N$ sodium thiosulfate. It is stored in a dark bottle and kept in the dark. It is standardized at least once a day during use.

Potassium iodide (10% aqueous solution).

Sulfuric acid (10% aqueous solution).

Sodium thiosulfate ($0.05N$ standard solution).

Ammonium thiocyanate ($0.05N$ standard solution).

Starch indicator (1% aqueous solution).

Ferric ammonium sulfate (saturated aqueous solution).

Cadmium sulfate (5% aqueous solution).

Procedure. The apparatus (Fig. 1) is cleaned and dried. Trap B is filled with a suspension of a small amount of red phosphorus in enough water to cover the inlet tube. If the sample to be analyzed contains sulfur, the phosphorus is suspended in 5% aqueous cadmium sulfate solution instead of water. Ten milliliters of silver nitrate solution is pipetted into the first absorption tube, C, 15 ml of bromine solution is pipetted into the spiral absorption tube, D, and 10 ml of 10% potassium iodide solution is placed in the final tube, E. A weighed sample of 200–300 mg of hydroxyethyl cellulose (hydroxypropyl cellulose) is placed in reaction flask A, together with a Hangar boiling granule and 20 ml of hydriodic acid. The flask is connected to the apparatus, a slow stream of carbon dioxide, (one bubble a second) is passed through, and the flask is heated slowly in an oil bath to 140–145°C.

The flask is kept at this temperature for a minimum of 40 min, and frequently longer. Two indications of the completion of the decomposition are the absence of any cloudy reflux in the condenser above the reaction flask and the nearly complete clarification of the supernatant liquid in the silver nitrate trap. Before completion of the reaction, the silver nitrate trap is heated to 70°C but not above, with a hot water bath to drive out any dissolved olefin.

At completion of the decomposition, tubes D and C are disconnected cautiously in that order. The carbon dioxide source is disconnected and the heat is removed from flask A. The spiral absorption tube, D, is then connected by its lower adapter to a 500-ml iodine titration flask containing 10 ml of 10% potassium iodide solution and 15 ml of water. The potassium iodide tube, E, is removed

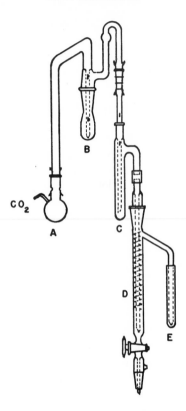

Fig. 1.   Apparatus for the determination of ethylene or propylene oxide in low-substituted cellulose esters.

and the side arm rinsed into it.   The bromine solution is allowed to run into the titration flask through the stopcock and the tube and spiral are rinsed with water.   The contents of the potassium iodide tube are added to the titration flask, which is then stoppered and allowed to stand for 5 min.   Five milliliters of 10% sulfuric acid are added and the solution is titrated at once with $0.05N$ sodium thiosulfate, using 2 ml of starch indicator solution.   The contents of the silver nitrate trap are rinsed into a flask, diluted to 150 ml with water, heated to boiling, cooled to room temperature, and titrated with $0.05N$ ammonium thiocyanate, using 3 ml of ferric ammonium sulfate solution as an indicator.

Calculations.   After the titrations of the contents of the bromine and silver nitrate traps are subtracted from the corresponding blank titrations, the following calculations are made:

$$\frac{\text{Difference in ml of } Na_2S_2O_3 \times N \times 2.203}{\text{Weight of Sample}} = \% \ C_2H_4O \text{ as } C_2H_4$$

$$\frac{\text{Difference in ml of } NH_4SCN \times N \times 4.405}{\text{Weight of Sample}} = \% \ C_2H_4O \text{ as } C_2H_5I$$

The corresponding factors for the calculation of propylene oxide are 2.904 and 5.808.

Although the modified Zeisel method is the best method for the accurate determination of the total ethylene oxide content of hydroxyethyl cellulose, it is rather time consuming.

B. Determination of the Ethylene Oxide Content of Hydroxyethyl Cellulose by a Precipitation Method

Muller and Alexander (49) have presented a precipitation method where the precipitating liquid was methanol.   It is claimed that the method is most useful in the ranges of 2–8% ethylene oxide content.   Relatively large variations in DP can be tolerated.

Procedure.   The neutral sample is finely ground and oven-dried. Exactly 0.5 ± 0.005 g of hydroxyethyl cellulose is weighed into a 125-ml erlenmeyer flask, equipped with a glass stopper.   With a pipet, 50.0 ml of water is added and the flask is shaken vigorously for 30 sec.   Then 50 ml of 14% sodium hydroxide are added with a pipet and the contents of the flask are stirred or shaken for 20 min, or until the sample has dissolved completely.

A 10-ml aliquot portion of this solution is transferred into a Lumetron sample tube and 3.00 ml of dry methanol is added.   The tube is stoppered, shaken 25 times, and placed in a water bath at $20 \pm 1°C$.   Exactly 2 min after the addition of methanol, the tube is removed from the water bath momentarily and inverted four times, removed again after a total of 5 min and inverted twice, and again removed after a total time of 8 min and inverted twice. The inverting has to be done gently to prevent a settling of the precipitate without the introduction of air bubbles.   Exactly 10 min after the addition of methanol the tube is removed from the water bath, gently inverted once more and the optical density is measured in a Lumetron Model 401 colorimeter equipped with a 420 filter. Other turbidimeters are equally suitable.

A blank solution is used which contains 5.0 ml of distilled water, 5.0 ml of 14% sodium hydroxide, and 3.0 ml of methanol. The optical density of the sample solution is related to the ethylene oxide content with the help of a predetermined calibration curve. The calibration is obtained with the aid of the conventional Zeisel–Morgan method. It is claimed that the method is reproducible to approximately 0.1% ethylene oxide substitution.

In contrast to the MS (molar substitution) of hydroxethyl cellulose, the DS (degree of substitution) is defined as the average number of cellulose hydroxyls substituted per anhydro-D-glucose unit (50). The degree of substitution is determined by phthalation of the hydroxyl group of the ethylene glycol unit. It was found by Senju (51) that the hydroxyl of the ethylene glycol unit is easily phthalated at room temperature within an hour, whereas the unreacted hydroxyl of cellulose does not react at room temperature even after 24 hr.

Determination of the hydroxyethyl group by gas chromatography also has been reported (52). Although this method was applied to hydroxyethyl starch, it might possibly be useful, with some modifications, for the determination of the hydroxyethyl group in hydroxyethyl cellulose. In this method, the sample is pyrolyzed at 400°C and the volatile products are determined by gas chromatographic analysis. The acetaldehyde peak is used as the analytical peak and its peak height correlates quantitatively with the hydroxyethyl content of the sample.

C. Determination of the Hydroxyl Content in Hydroxyethyl Cellulose by Phthalation

Principle. The sample is treated at room temperature with a measured amount of phthalic anhydride in water-free pyridine. After the reaction period, water is added to hydrolyze the excess phthalic anhydride and the sample is titrated with standard sodium hydroxide using phenolpthalein as an indicator.

Reagents. *Phthalic anhydride* (A.R.).
Pyridine (A.R.).
Sodium hydroxide (0.35N standard solution).
Phenolphthalein (1% alcoholic solution).

Procedure. From 1.0 to 2.0 g of sample is weighed into a clean, dry erlenmeyer flask and 25 ml of phthalic anhydride solution is added, prepared by dissolving 20 g of phthalic anhydride in 200 ml of pyridine (this solution has to be made up fresh daily). The flask is stoppered immediately and left standing for 1 hr. At the end of this time, 50 ml of water is added. The solution is mixed and cooled under a cold-water tap and titrated immediately with standard 0.35N sodium hydroxide solution, using phenolphthalein as an indicator. A blank determination is made in the same manner on the reagents used.

Calculation. The calculation of per cent hydroxyl content is as follows:

$$\% \text{ hydroxyl} = \frac{(B - A) \times N \times 17 \times 100}{W \times 1000}$$

where A is the number of milliliters of standard sodium hydroxide used to titrate the sample, B is the number of milliliters of standard sodium hydroxide used to titrate the blank, N is the normality of standard sodium hydroxide, and W is the sample weight in grams.

One of the potential by-products in the manufacture of hydroxyethyl cellulose is ethylene glycol. A method for determining small quantities of ethylene glycol (in the parts per million range) in hydroxyethyl cellulose solutions (53) is a modification of the Malaprade reaction.

D. Determination of Small Amounts of Ethylene Glycol in Hydroxyethyl Cellulose Solutions

Principle. The basic reaction adapted for the determination of ethylene glycol is the well-known Malaprade reaction (54). When ethylene glycol reacts with periodic acid, two moles of formaldehyde are formed for each mole of ethylene glycol. After elimination of excess periodic acid by precipitation with lead acetate and sodium bicarbonate (55), the formaldehyde is treated with chromotropic acid (4,5-dihydroxy-2,7-naphthalenedisulfonic acid) to form a purple color (56), the intensity of which is determined spectrophotometrically.

Apparatus and Reagents. *Centrifuge.*

Spectrophotometer, Beckman DU or equivalent.

Sulfuric acid (1:4 aqueous solution).

Phenolphthalein (0.1% ethanolic solution).
Periodic acid (1% aqueous solution).
Lead acetate (10% aqueous solution).
Sodium bicarbonate (10% aqueous solution).
Chromotropic acid (10% aqueous solution, freshly prepared).
Sulfuric acid (concentrated A.R.).

Procedure. About 1 g of hydroxyethyl cellulose solution is weighed to the nearest 0.0001 g into a small beaker. Several milliliters of water are added and the solution is neutralized with 1:4 sulfuric acid, using phenolphthalein as indicator. The neutralized solution is filtered through a fast filter paper into a 250-ml volumetric flask. The residue on the filter paper is washed thoroughly with water to a volume of about 200 ml. To the flask, 5 ml of 1% periodic acid solution is added, and the contents are mixed well and are allowed to stand for 1 hr. Then 5 ml of 10% lead acetate solution and 10 ml of 10% sodium bicarbonate solution are added. The contents of the flask are mixed well and allowed to stand for 30 min. After this time, the flask is filled to the mark with water and the contents are mixed thoroughly. Several milliliters of this solution are then transferred to a centrifuge tube and are centrifuged for 5 min. One milliliter of the clear solution is withdrawn and transferred to a 50-ml volumetric flask. For a blank, 1 ml of water is placed in a 50-ml volumetric flask and processed simultaneously with the sample. One-half milliliter of 10% chromotropic acid solution and then, slowly, 5 ml of concentrated sulfuric acid are added to the flask. The flask is heated on a steam bath for 15 min, cooled to room temperature, and brought to volume with water. The absorbance of the sample solution is read against a blank on a spectrophotometer at 570 mμ. From a predetermined calibration curve, the number of micrograms of formaldehyde is obtained.

Calculation. The ethylene glycol content is calculated as follows:

$$\mu\text{g ethylene glycol} = \mu\text{g formaldehyde} \times 1.0335$$

$$\text{ppm ethylene glycol} = \frac{\mu\text{g ethylene glycol} \times (250/\text{aliquot})}{\text{Sample weight}}$$

To determine ethylene glycol in hydroxyethyl cellulose crumbs or films, it is necessary to extract a 10-g sample with ethanol for

24 hr.   A sample of wood pulp of equal weight should be processed simultaneously as a control.   The extract is then concentrated to a volume of about 10–20 ml and analyzed in the same manner as described above.

Recently, a new gas chromatographic method for the determination of trace amounts of ethylene glycol in aqueous solution has been reported (57).   This method uses a flame ionization detector, a mixture of helium and steam as the carrier gas, and Teflon as the column support.   It is claimed that as little as 0.02 ppm of ethylene glycol can be determined.   This method might lend itself even to a better degree to the determination of ethylene glycol in hydroxyethyl cellulose because it is more specific and also has a higher sensitivity.   It was found that cellulose (and also hydroxyethyl cellulose) when extracted with ethanol, breaks down to a certain degree and one of the degradation products is formaldehyde. The formaldehyde interferes with the method described previously.   The gas chromatographic method is specific for ethylene glycol and there would be no interference from decomposition products of hydroxyethyl cellulose or any other hydroxy compounds such as glycerol or propylene glycol which might be present in the film as softeners.

VIII. Cellophane

Cellophane is still the most widely applicable film-forming polymer among the cellulosic polymers.   The cast, uncoated cellophane has only a limited application because of its hydrophilic nature, but coated cellophane is still one of the most versatile packaging materials.   There are nearly 100 different varieties on the market today.

The cellophane amendment to the Food and Drug Administration regulations (58) lists about 130 substances, which, with certain limitations, may be used in cellophane.   In addition, there are substances which are generally recognized as safe (59) and substances for which prior sanctions have been granted (60).   All these substances, an extremely vast number, could possibly be found in different types of cellophane.   As many as 10–15 substances may be found in one single type.   (A full discussion of these matters may be found in Chap. 16.)   This makes obvious the complexities of cellophane analyses and only substances which may be considered most important in the characterization of cellophane will be described in detail. The most important substances in the cellophane base sheet are water and softening agents. Determination of water, because of its importance for all cellulosic films, has been described separately in Section II.

A. Softening Agents in Cellophane

The most common softening agents in cellophane are glycerol, propylene glycol, triethylene glycol, polyethylene glycol of a molecular weight higher than 300, and urea. In some cellophane of Canadian and European orgin, ethylene glycol may be found. This softening agent does not have Food and Drug Administration acceptance in the United States.

The two most important reactions employed in the determination of softening agents are the Malaprade reaction (54) for substances with vicinal OH groups, as is the case in glycerol, ethylene glycol, and propylene glycol, and the oxidation with acid dichromate for triethylene glycol and polyethylene glycol. Although glycerol, ethylene glycol, and propylene glycol may be analyzed by dichromate oxidation in acidic solution, the reaction with periodic acid or sodium metaperiodate is more specific and allows futher characterization of the individual softening agents by analysis of the oxidation products. When oxidized with periodic acid, these glycols give the following reaction products:

Glycerol: 2 moles of formaldehyde and 1 mole of formic acid
Ethylene glycol: 2 moles of formaldehyde
Propylene glycol: 1 mole of formaldehyde and 1 mole of acetaldehyde

The difference in reaction products makes it possible to determine these glycols in the presence of each other (61).

1. Identification of Softening Agent Types

Before the softening agents can be determined quantitatively, the types of softening agents present in cellophane have to be established qualitatively. Softening agents with vicinal OH groups can be identified on the basis of their reaction products with periodic acid.

Reagents.   *Periodic acid* (1% aqueous solution).
Methyl red indicator (0.1% methanolic solution).
Morpholine (20% aqueous solution).
Sodium nitroprusside (5% aqueous solution).
Sulfuric acid (concentrated, analytical reagent).
Lead acetate (10% aqueous solution).
Sodium bicarbonate (10% aqueous solution).
Chromotropic acid (4,5-dihydroxy-2,7-naphthalenedisulfonic acid,
    10% aqueous solution).
Hydrochloric acid (1:4).
Barium chloride (10% aqueous solution).
Phosphomolybdic acid (10% aqueous solution).
Acidic dichromate reagent.   Dissolve 9 g of potassium dichromate
    in 1 liter of 55% sulfuric acid by volume.
Urease powder or tablets.
Phenolphthalein indicator (0.1% methanolic solution).

Procedure.   An aqueous extract of the cellophane sample to be tested is prepared as follows.   About 1–2 g of cellophane is cut into small pieces, placed in a 16-oz jar and 100–200 ml of water is added.   The mixture is vigorously blended in a mechanical blender for several minutes and then filtered.

For the detection of glycerol, about 10 ml of the filtrate is transferred to a 25-ml erlenmeyer flask.   A few drops of methyl red indicator are added and the solution is neutralized with either $0.1N$ hydrochloric acid or $0.1N$ sodium hydroxide so that the indicator just turns yellow.   A 1% periodic acid solution is neutralized with sodium hydroxide, using methyl red as indicator, to a slight alkaline end point.   About 10 ml of this solution is added to the neutralized sample solution.   If it turns red within a few seconds, the presence of glycerol is indicated.

For the detection of propylene glycol, about 10 ml of the filtrate and 10 ml of 1% aqueous periodic acid are mixed and left standing for a few minutes.   Then 5 drops of this solution are placed on a porcelain spot plate, and 5 drops of 20% aqueous morpholine and 5 drops of 5% sodium nitroprusside are added.   The solution is mixed with a small glass rod and if it turns blue within 30 sec, the presence of propylene glycol is indicated.

If either of the two tests described above is negative, but ethylene glycol may be present, it may be detected as follows.   About

10 ml of the filtrate and 1 ml of 1% periodic acid solution are mixed and left standing for a few minutes. Then 5 ml of 10% lead acetate and 10 ml of 10% sodium bicarbonate are added, mixed well, and centrifuged for several minutes. The clear liquid is decanted and about 1 ml is placed in a test tube. Several drops of 10% chromotropic acid solution are added and about 5 ml of concentrated sulfuric acid is very slowly and carefully poured down the wall of the test tube. If a deep purple color develops, the presence of ethylene glycol is indicated. A pale purple color may develop if a melamine– or urea–formaldehyde resin is present in the sample and should not be mistaken for an indication of ethylene glycol. If a mixture of glycerol and ethylene glycol is suspected to be present in the cellophane, only a quantitative analysis of the cellophane extract can reveal this combination.

For the detection of polyethylene glycol, 5 ml of the filtrate is transferred to a test tube and successively 1 ml of 1:4 hydrochloric acid, 1 ml of 10% barium chloride, and 1 ml of 10% phosphomolybdic acid are added, mixing thoroughly after each addition. If a flocculent greenish-yellow precipitate is formed immediately, the presence of a polyethylene glycol is indicated.

For the detection of triethylene glycol, 5 ml of the filtrate is transferred to a test tube and 5 ml of the acidic dichromate reagent is added. The solution is mixed well and heated in boiling water for several minutes. If the originally orange solution turned green, the presence of a softening agent which is oxidizable by acidic-dichromate is indicated. Only if the tests for glycerol, ethylene glycol, propylene glycol, and polyethylene glycol are negative, may the presence of triethylene glycol be assumed.

If one is in doubt, a large amount of cellophane may be extracted, filtered, and the filtrate brought to dryness on a steam bath. The syrupy residue which consists mainly of the softening agent can then be identified by infrared spectroscopy.

For the detection of urea, 1–2 g of cellophane is cut into small pieces, and placed into a 16-oz jar. About 200 ml of water, 0.1 g of urease powder or one urease tablet, and 1 ml of phenolphthalein indicator are added and the mixture is blended on a mechanical blender for 10 min. If the solution turns from colorless to pink, urea is present in an amount sufficient to be considered as a softening agent.

2. Determination of Glycerol, Ethylene Glycol, or Propylene Glycol by Periodic Acid Oxidation

Reagents. *Periodic acid reagent* (0.015*M*). In 200 ml of distilled water 2.75 g of periodic acid is dissolved, then diluted to 1 liter with glacial acetic acid.

Potassium iodide (20% aqueous solution).

Standard sodium thiosulfate (0.1*N*).

Starch indicator solution (0.5%).

Procedure. Aqueous extracts of cellophane are prepared as follows. About 1–1.5 g of cellophane is weighed accurately, cut into small pieces, and transferred to a 16-oz jar. Then, 250 ml of distilled water is added and the sample is vigorously blended for 15 min in a mechanical blender such as an Oster or Waring blender. After this extraction, the jar is cooled and the water-cellophane mixture is filtered through a fast filter paper. Of the filtrate, 25 ml is transferred by pipet into a 500-ml glass stoppered erlenmeyer flask, 50 ml of the periodic acid reagent is added, and the stoppered flask is left standing for exactly 15 min. Then the stopper and the sides of the flask are washed down with about 25 ml of distilled water, 20 ml of the 20% potassium iodide solution is added, and the sample is titrated immediately with 0.1*N* Na$_2$S$_2$O$_3$, using starch solution as the indicator. A blank is carried through the entire procedure, substituting 25 ml of distilled water for the sample solution. The volume of 0.1*N* Na$_2$S$_2$O$_3$ consumed in the sample titration should be within 75–80% of that consumed for the blank titration.

Calculation. The softening agent present is calculated as follows:

$$\% \text{ softening agent} = \frac{(B - A)(N)(F)(D) \times 100}{W \times 1000}$$

where

A = ml of Na$_2$S$_2$O$_3$ consumed in sample titration.

B = ml of Na$_2$S$_2$O$_3$ consumed in sample titration

N = normality of Na$_2$S$_2$O$_3$

F = equivalent weight of softening agent (for glycerol, F = 23.023; for ethylene glycol, F = 31.025; and for propylene glycol, F = 38.045)

D = dilution factor

W = sample weight in grams

*3. Determination of Glycerol, Ethylene Glycol, and
Propylene Glycol in the Presence of One Another*

Reagents. *Periodic acid reagent* (as in Section VIII-A-2).
Periodic acid (1% aqueous solution).
Potassium iodide (20% aqueous solution).
Standard sodium thiosulfate (0.1N).
Standard sodium hydroxide (0.1N).
Starch indicator solution (0.5%).
Methyl purple indicator solution.

Procedure. The aqueous cellophane extract is prepared in the same manner as described in Section VIII-A-2. The total amount of softener is determined iodometrically in the same manner as described in Section VIII-A-2.

DETERMINATION OF GLYCEROL

Of the filtrate, 100 ml of sample solution is transferred by pipet into a 500-ml stoppered erlenmeyer flask, 50 ml of 1% periodic acid solution is added, and the flask is left standing for 30–40 min. After this reaction period, the sample is titrated with 0.1N sodium hydroxide, using methyl purple as indicator. A blank is carried through the entire procedure, substituting 100 ml of distilled water for the sample solution.

Calculation. The amount of glycerol present is calculated as follows:

$$\% \text{ glycerol} = \frac{(A - B)(N)(F)(D)(100)}{W \times 1000}$$

where

A = ml of NaOH consumed in sample titration
B = ml of NaOH consumed in blank titration
N = normality of sodium hydroxide
F = equivalent weight of glycerol = 92.06
D = dilution factor
W = sample weight in grams

If a mixture of glycerol and ethylene glycol is present in the sample, the calculation for ethylene glycol is as follows:

$$\% \text{ ethylene glycol} = (T - G) \times 1.348$$

where T is the total content of softening agent, determined iodometrically, as described in Section VIII-A-2 and calculated as glycerol, and G is the glycerol content, determined acidimetrically, as described above.

If a mixture of glycerol and propylene glycol is present in the sample, the calculation for propylene glycol is as follows:

$$\% \text{ propylene glycol} = (T - G) \times 1.652$$

4. Determination of Glycerol, Ethylene Glycol, Propylene Glycol, Triethylene Glycol, and Polyethylene Glycol by Dichromate Oxidation in the Presence of Acid

All these softening agents in aqueous solutions can be determined by acidic dichromate oxidation when other oxidizable substances are absent (61–64). The rate of this reaction is directly dependent upon the concentration of sulfuric acid. All softening agents, with the exception of propylene glycol, are oxidized to carbon dioxide and water when refluxed with potassium dichromate and 55% sulfuric acid by volume. Propylene glycol is stoichiometrically oxidized to carbon dioxide, water, and acetic acid when sulfuric acid of 35% by volume is used. If the sulfuric acid concentration is higher, some of the acetic acid is further oxidized and too high results are obtained.

Reagents. *Sulfuric acid–potassium dichromate reagent.* This contains 9 g of potassium dichromate in 1 liter of 55% (by volume) sulfuric acid (reagent grade).

Potassium iodide (20% aqueous solution).

Sodium thiosulfate (standard solution, $0.1N$).

Starch indicator solution (0.5%).

Procedure. The aqueous cellophane extract is prepared in the same manner as described in Section VIII-A-2. Of the filtrate, an aliquot of exactly 10.00 ml is transferred by pipet into a 250-ml erlenmeyer flask fitted with a ground glass joint. Exactly 25.00 ml of the 55% sulfuric acid–potassium dichromate reagent is added and the solution is refluxed for exactly 10 min on a hot plate of at least 230°C temperature. After this reaction period, the flask is cooled, the content is diluted with 150 ml of distilled water, 20 ml of 20% potassium iodide is added, and the solution is titrated with $0.1N$ sodium thiosulfate, using starch solution as indicator. A blank is carried throughout the entire procedure, substituting 10 ml of distilled water for the sample solution.

Calculation. The percentage of softening agent is calculated as follows:

$$\% \text{ softening agent} = \frac{(B - A)(N)(F)(D)(100)}{W \times 1000}$$

where

A = ml of $Na_2S_2O_3$ consumed in the sample titration

B = ml of $Na_2S_2O_3$ consumed in the blank titration

N = normality of $Na_2S_2O_3$

F = equivalent weight of softening agent [for glycerol, $F = 6.576$; for ethylene glycol, $F = 6.205$; for triethylene glycol, $F = 5.01$; and for propylene glycol (when 35% sulfuric acid is used), $F = 9.512$. Commercial polyethylene glycol is usually a mixture of glycols of different molecular weight so that no equivalent weight can be calculated. F has to be determined experimentally by analysis of known solutions of the respective batch of polyethylene glycol.]

D = dilution factor

W = sample weight in grams

5. Determination of Glycerol and Triethylene Glycol in the Presence of Each Other

When both glycerol and triethylene glycol are present, the total softening agent content is determined by acidic dichromate oxidation as described in Section VIII-A-4, and the glycerol content is determined by periodic acid oxidation as described in Section VIII-A-2. The triethylene glycol content is then calculated as follows:

$$\% \text{ triethylene glycol} = (T - G) \times 0.762$$

where T is the percentage of total softening agent, determined by acidic dichromate oxidation and calculated as glycerol, and G is the percentage of glycerol, determined by periodic acid oxidation.

6. Determination of Urea

Aqueous solutions of urea are best analyzed according to a colorimetric method reported by Watt and Chrisp (65). Urea, when mixed with p-dimethylaminobenzaldehyde produces a yellow-green color, the intensity of which is determined spectrophotometrically.

Reagents.   *Urea* (Baker, CP).

Hydrochloric acid (concentrated, reagent grade).

Ethanol (95%).

p-Dimethylaminobenzaldehyde (Eastman No. 95).   This is the color reagent and is prepared as follows: 2.000 g of p-dimethylaminobenzaldehyde is dissolved in 100 ml of 95% ethanol, and 10 ml of concentrated hydrochloric acid is added.   This solution is of a yellow color; if any other color is produced, the reagent should be discarded.   The reagent can be stored up to a week.

Procedure.   *Preparation of Calibration Curve.*   Exactly 0.250 g of urea is dissolved in distilled water and diluted to 250 ml in a volumetric flask.   Aliquot portions of 1–6 ml are pipetted into 25 ml volumetric flasks, 10.0 ml of color reagent is added and diluted to volume with distilled water.   The samples are thoroughly mixed and allowed to stand for at least 10 min, to allow full color development.   The absorbance of these solutions is measured with a spectrophotometer at 420 mμ against a reagent blank which consists of 10.0 ml of color reagent diluted to 25 ml volume with distilled water. Pyrex cells of 1-cm path length are used.   A calibration curve is prepared where milligrams of urea are plotted as a function of absorbance.

Analysis of Samples.   An aqueous extract of the cellophane sample is prepared as described in Section VIII-A-2.   A 10.0-ml aliquot portion of the filtrate is pipetted into a 25-ml volumetric flask. Depending upon the urea content of the sample, a smaller aliquot portion may be used.   The procedure described for the preparation of the calibration curve is followed.   The number of milligrams of urea is read from the calibration curve and the urea content of the sample is calculated as follows:

$$\% \text{ urea} = \frac{U \times D \times 100}{W \times 1000}$$

where, U is the number of milligrams of urea read from calibration curve, D is the dilution factor, and W is the sample weight in grams.

Substances normally found in cellophane do not interfere with this procedure.

B. IDENTIFICATION AND DETERMINATION
OF ANCHORING RESINS

The most common resins used in cellophane manufacture are urea–formaldehyde types, melamine–formaldehyde types, and polyethyleneimine. The purpose of these resins is to aid in the adhesion of coatings to the cellophane base sheet. Since these resins are usually present in rather small amounts and such a wide variety of resins of those types is used, analyses are mostly limited to the identification of the basic resin type or to the determination of certain components of those resins. Both melamine and urea resins contain formaldehyde and the presence of formaldehyde in cellophane is the best indication that one or both of these resins are present. Melamine can be identified by the formation of the derivative melamine cyanurate, or it can be determined quantitatively in uncoated or polymer-coated cellophane by an ultraviolet absorption method. Urea can be identified by the formation of the derivative dixanthyl urea. Polyethyleneimine can be identified by a dye staining test or by the nitrous acid test. All of these resins can be determined quantitatively by the classical Kjeldahl method if no other nitrogen-containing substances are present.

1. Determination of Free or Extractable Formaldehyde (56)

Reagents. *Formaldehyde* (37% reagent grade).

Chromotropic acid (4,5-dihydroxy-2,7-naphthalenedisulfonic acid) purest grade, 10% aqueous solution.

Sulfuric acid (concentrated, reagent grade)

Procedure. *Preparation of Calibration Curve.* The actual formaldehyde content of reagent grade formaldehyde is determined by the sulfite procedure (66). A stock solution of 0.100 g of formaldehyde per liter is then prepared. From this stock solution, formaldehyde solutions of the following concentrations are prepared: 1.0, 2.0, 3.0, and 4.0 mg per 100 ml. One milliliter of each of these solutions is transferred to 50-ml volumetric flasks; 0.5 ml of the chromotropic acid solution and 5.0 ml of concentrated sulfuric acid are added. The contents of the flasks are gently mixed, heated in a steam bath for 15 min, cooled in ice for about 5 min, and diluted with distilled water to about 1 cm below the mark. The solutions are then allowed to cool to room temperature, diluted exactly to the mark, and mixed well. A blank is prepared in the same manner, substituting 1 ml of distilled water for the formaldehyde solution.

The per cent transmission of the purple solutions versus the blank is determined with a suitable spectrophotometer at 570 mμ using 1.0-cm Pyrex cells. A calibration curve is prepared on semi-logarithmic paper where the amount of formaldehyde contained in the 50-ml volumetric flask is plotted as a function of per cent transmission.

Analysis of Samples. About 1 g of exactly weighed cellophane is extracted in a suitable blender with 100 ml of distilled water for 15 min. The solution is cooled and filtered through a fast filter paper. Of the filtrate, 1 ml is used for analysis as described in the preparation of the calibration curve. From the calibration curve, the number of micrograms of formaldehyde is obtained.

The percentage of formaldehyde is calculated as follows:

$$\% \text{ formaldehyde} = \frac{A \times D \times 100}{W \times 1000 \times 1000}$$

where A is the number of micrograms of formaldehyde obtained from the calibration curve, D is the dilution factor, and W is the sample weight in grams.

2. Qualitative Determination of Melamine

Reagents. *Glacial acetic acid* (analytical reagent).
Ethyl ether solvent.
Sodium hydroxide (40% aqueous solution).
Cyanuric acid (0.1% aqueous solution).

Procedure. About 15 g of cellophane, cut into small pieces, is placed in a 32-oz jar, 300 ml of water is added, and the mixture is blended for 15 min. The cellophane–water mixture is then transferred to a 1500-ml erlenmeyer flask which is fitted with a 24/40 joint, 300 ml of glacial acetic acid is added, the flask is attached to a condenser, and the contents are refluxed for 1 hr on a hot plate. After this hydrolysis period, the mixture is filtered, and the filtrate transferred to a 1000-ml beaker. A few glass beads are added and the solution is concentrated by boiling to about 40–50 ml. This part is then transferred to a separatory funnel and extracted with ethyl ether. After separation, the lower, aqueous layer is drained into a small beaker, placed on a steam bath and concentrated to about 15–20 ml.   The solution is then neutralized with 40% sodium hydroxide, using pH paper, and 20 ml of 0.1% cyanuric acid solution is added.   If melamine is present, a fine, white precipitate of melamine cyanurate is formed.   At times it may take several hours for the precipitate to form.

3. Quantitative Determination of Melamine

In uncoated or polymer-coated cellophane, melamine can be determined by an ultraviolet spectophotometric method (67).   In nitrocellulose-coated cellophanes, interfering substances are present and therefore this method is not applicable.

Reagents.   *Hydrochloric acid* (approximately 0.1N aqueous solution.

Sodium lauryl sulfate. For the analysis of coated cellophane, about 0.5 g is added per liter of 0.1N HCl.

Procedure.   About 0.5 g of cellophane is exactly weighed, cut into small pieces, and placed in a 500-ml erlenmeyer flask which is fitted with a 24/40 ground joint.   From a 100-ml volumetric flask, exactly 100 ml of 0.1N hydrochloric acid is added, a few glass beads are added to the flask, and it is then attached to a condenser.   On a hot plate, the contents of the flask are refluxed for 30 min.   The flask is cooled and within 30 min, the absorbance of the solution is determined spectrophotometrically at 237 and 260 mμ, using silica cells of 10-mm light path length.

Calculation.   The percentage of melamine is calculated as follows:

$$\% \text{ melamine} = \frac{(A_{237} - A_{260})(100)}{(a_p)(W)(10)}$$

where A_{237} is the absorbance of the solution at 237 mμ, A_{260} is the absorbance of the solution at 260 mμ, a_p is the absorptivity of melamine, which is reported to be 79, and W is the sample weight in grams.

4. Qualitative Determination of Urea

Urea reacts with xanthydrol to form crystalline xanthyl or dixanthyl derivatives (68,69).   This reaction may be conveniently used for the qualitative determination of urea.

Reagents.   *Glacial acetic acid* (analytical reagent).

Xanthydrol (1% methanolic solution).

Procedure. About 15 g of cellophane, cut into small pieces, is placed into a 32-oz jar, 300 ml of water is added, and the mixture is blended for 15 min. The cellophane–water mixture is then transferred to a 1500-ml erlenmeyer flask, which is fitted with a 24/40 joint. Enough glacial acetic acid is added to make it a 10% solution. The flask is attached to a condenser and the contents are heated on a steam bath for 2 hr. After this hydrolysis period, the mixture is filtered into a beaker and 10 ml of 1% methanolic solution of xanthydrol and 50 ml of methanol are added. If the solution is milky, more methanol is added until it clears. The beaker is placed on a steam bath and if urea is present, a white, flocculent precipitate of dixanthyl urea will form within a short time.

5. Qualitative Determination of Polyethyleneimine (70)

On uncoated cellophane or coated cellophane, after removal of the coating, polyethyleneimine may be detected by a dyeing test or more specifically by a sensitive test which is based on the reaction with nitrous acid.

Reagents. *Eosin dye* (0.1% aqueous solution).
Sodium hydrosulfite (20% aqueous solution).
Sodium nitrite (10% aqueous solution).
Glacial acetic acid (analytical reagent).
Urea (10% aqueous solution).
Hydrochloric acid (concentrated).
Diphenylamine–sulfuric acid reagent. It is prepared as follows: 0.5 g of diphenylamine is dissolved in 30 ml of distilled water and 70 ml of concentrated sulfuric acid is added.

Procedure. *Dye Test.* A small sample of uncoated cellophane is boiled for 5 min in a 0.1% solution of eosin dye and subsequently washed in flowing water for 5 min. If polyethyleneimine is present, the cellophane is stained a bright scarlet. The intensity of the color is dependent upon the amount of polyethyleneimine present.

Test Based on Reaction with Nitrous Acid. A small sample of uncoated cellophane is placed in a beaker and 50 ml of freshly prepared 20% sodium hydrosulfite is added. After standing for 10 min at room temperature, the solution is poured off and the sample is rinsed twice with small portions of distilled water. The sample is then covered in a beaker with 50 ml of 10% solution of sodium nitrite. One milliliter of glacial acetic acid is added, and the mixture is stirred

and left standing for 10 min at room temperature.   The solution is
then poured off and again washed with two portions of distilled water.
   Any remaining nitrous acid is eliminated by covering the sample
with 50 ml. of a 10% solution of urea.   Two milliliters of concentrated hydrochloric acid is added and the mixture is heated to 50°C
for 10 min.   The solution is then poured off,. and the sample is
rinsed several times with distilled water and pressed between discs
of filter paper.
   The still-damp sample is placed on a glass plate lying on a white
background and several drops of the diphenylamine–sulfuric acid
reagent are placed on the sample.   If, after 30 sec to 3 min, a blue
color develops, the presence of polyethyleneimine is indicated.   If
the blue color develops immediately and with great intensity, it is a
sign that the nitrous acid has not been completely removed and the
test has to be repeated.

C. Identification of Coatings and Coating Ingredients

   Substances such as moisture, softening agents, and anchoring
agents are usually found in the cellophane base sheet.   But most
cellophane types found on the market are coated and therefore the
coating becomes an integral part of the cellophane and should be considered here.   A wide variety of substances are used to coat cellophane, and the composition of the coating influences greatly certain
characteristics of cellophane such as permeability to moisture and
gases, heat sealability, strength, odor, slippage, and many others.
Some of the most important substances which may be found in
coatings are listed below.
   Basic coatings polymers.   Nitrocellulose, polyethylene, polypropylene, polystyrene, polyvinyl acetate, copolymers of ethylene
and vinyl acetate, copolymers of vinyl acetate and vinyl chloride,
and a variety of copolymers of vinylidene chloride.
   Residual coating solvents.   Tetrahydrofuran, methyl ethyl ketone,
toluene, ethyl acetate, *n*-propyl acetate, isopropyl acetate, and
n-butyl acetate.
   Surface-active agents.   Ammonium, magnesium, sodium, and
potassium salts of lauryl sulfate, sodium dodecylbenzenesulfonate,
sodium dioctyl sulfosuccinate, sorbitan esters, and others.
   Plasticizers.   Hydrogenated castor oil phthalate, dibutyl phthalate, dicyclohexyl phthalate, diethylene glycol ester of the adduct of

terpene and maleic anhydride, bis(2-ethylhexyl) adipate, bis-(2-ethylhexyl) phthalate, dimethylcyclohexyl phthalate, triethyl acetylcitrate, tributyl acetylcitrate, ethylphthalyl ethyl glycolate, and others.

Waxes and release agents. Beeswax, carnauba wax, paraffin wax, spermaceti wax, pentaerythritol tetrastearate, stearamide, N,N'-ethylenebis oleamide, N,N'-ethylenebis stearamide, behenamide, erucamide, stearic acid, and others.

1. Identification of Basic Coating Polymers

The indentification of coating polymers is accomplished best and fastest with the help of infrared spectroscopy. Two excellent books by Hummel (71) and by Haslam and Willis (72) have been devoted to the identification of polymers by infrared spectroscopy and can be used as references. The voluminous collection of infrared spectra by Sadtler (73) is also a useful reference. With the help of attenuated total reflectance attachments to an infrared spectrograph, the identification of coating polymers of cellophane is possible without previous removal of the coating by mechanical means. The chemical analysis of coating polymers is exhaustively treated in the excellent book by Kappelmeier (74).

2. Determination of Residual Coating Solvents

Coatings can be applied to cellophane from aqueous dispersions, by extrusion, or from organic solvents. In certain types of coatings, it is difficult to remove completely the organic solvents, and methods have been developed for the determination of residual solvents in cellophane (75–77). Furthermore, limits have been set by the Food and Drug Administration (58) for the amount of residual solvents in cellophane in the order of 0.1% of each solvent.

Difficulties in the way of development of good and accurate methods for determining solvents are enhanced by the fact that no standards are available. No films are available which contain definite amounts of residual solvents and furthermore, these solvents are continuously released to the atmosphere. For this reason, duplicate analysis can be made at best with a precision of $\pm 10\%$. The complexity of this type of analysis may be realized by the entirely different approaches taken in the published methods. Nadeau and Neumann use a liquid extraction and subsequent distillation for the low-boiling solvents and pyrolysis for the high-boiling solvent (75). Phifer uses a liquid extraction in a sealed tube (76). These two methods aim at a complete recovery of residual solvents in cellophane. Others heat the sample in a closed bottle and analyze the gas phase above the sample (77). This method is not aimed at the complete recovery of solvents in the sample but determines only the easily releasable portion of solvents in cellophane.

3. Determination of Surface-Active Agents

Surface-active agents which may be found in cellophane are either anionic or nonionic. The most common is sodium lauryl sulfate which is an anionic surfactant. It can be determined quantitatively by a colorimetric method according to Moore and Kolbeson (78). Other anionic surfactants such as sodium dioctyl sulfosuccinate and dodecylbenzenesulfonate can be determined by the same method.

DETERMINATION OF SODIUM LAURYL SULFATE (78)

Reagents. *Methyl green solution* (0.5% aqueous solution).
Benzene (analytical reagent).
Chloroform (analytical reagent).
Glycine–hydrochloric acid buffer (pH 2.5). Glycine (7.5 g) and 5.8 g of sodium chloride are dissolved in distilled water and diluted to 1 liter. This solution is adjusted with hydrochloric acid, about 0.1N, to a pH of 2.5.
Buffer–water mixture. Equal volumes of buffer solution and water are mixed thoroughly.
Sodium lauryl sulfate (highest purity).
Procedure. *Preparation of Calibration Curve.* A standard solution is prepared which contains 5 µg of sodium lauryl sulfate per milliliter. From this solution, 5 aliquot portions containing from 10 to 50 µg of sodium lauryl sulfate are pipetted into 125-ml separatory funnels. Each of these aliquots is diluted to 20 ml with distilled water and 10 ml of the buffer solution, and 2 ml of the methyl green solution is added. The contents of the funnels are mixed thoroughly by swirling and then exactly 40 ml of benzene is added from a buret. The funnels are then shaken for 1 min, their contents allowed to settle, and swirled again to dissolve completely the dye–surfactant complex in the benzene. The water layer is then removed carefully so as not to lose any of the benzene layer. Then 15 ml of the water–

buffer mixture is added to the benzene layer, the funnels are again shaken for about 1 min, and the contents allowed to settle. The funnels are swirled gently and the water is allowed to settle out of the benzene layer which takes from 20 to 30 min. A blank, containing all reagents except sodium lauryl sulfate, is carried along throughout the whole procedure. With a clean, dry pipet a portion of the benzene layer is transferred to a 1-cm cell and with a suitable spectrophotometer the absorption at 615 mμ is measured against the blank. A calibration curve is prepared by plotting absorption of the benzene layer against concentration of sodium lauryl sulfate. Beer's Law is obeyed in the range from 10 to 50 μg of sodium lauryl sulfate.

Analysis of Sample. Depending upon the amount of sodium lauryl sulfate expected in the cellophane sample, from 0.3000 to 1.0000 g are weighed, or, if the results are reported in milligrams per square meter, from 100 to 300 cm^2 of cellophane are taken. The sample is cut into small pieces, placed in a 16-oz jar, 250 ml of distilled water is added, and the contents are vigorously blended for 15 min. The jar is then cooled to room temperature and the water–cellophane mixture is filtered. An aliquot of from 5 to 20 ml is placed in a 125-ml separatory funnel. If the aliquot is less than 20 ml, it is diluted with distilled water to 20 ml. One then proceeds as described above for the preparation of the calibration curve. Because of the sensitivity of the method, clean equipment is of utmost importance. It is advisable to wash all glassware thoroughly with chloroform.

Calculation. The quantity of sodium lauryl sulfate is calculated as follows:

$$\% \text{ sodium lauryl sulfate} = \frac{A \times F \times 100}{W \times 1000 \times 1000}$$

$$\text{Sodium lauryl sulfate, mg/m}^2 = \frac{A \times F_1 \times F}{1000}$$

where

A = μg of sodium lauryl sulfate obtained from the calibration curve

F = dilution factor = (250/aliquot)

F_1 = area factor = 10,000/cm^2 of sample taken

W = sample weight in grams

4. Identification of Plasticizers

Plasticizers used in cellophane coatings as well as in other cellulosic polymers are best indentified by extraction and subsequent analysis of the extract by infrared spectroscopy. Since cellophane coatings may contain more than one plasticizer, the plasticizers in the extract can be separated by programmed gas chromatography and subsequently indentified by infrared spectroscopy or by retention time. These instrumental methods are by no means substitutes for the chemical analysis of plasticizers which are described in detail in books by Hummel (71), by Haslam and Willis (72), and Kappelmeier (74). In recent years, gas chromatographs have been perfected and column packings have been made available which can be operated at temperatures up to 400°C. This makes feasible the separation of high boiling plasticizers, and because of the high sensitivity of the instruments very small samples can be used.

Several authors have described the separation of plasticizers by gas chromatography. Dibutyl phthalate, benzyl butyl phthalate, and dibenzyl phthalate have been separated on a 9 × ¼-in. stainless steel column packed with Embacel, 80–125 mesh, coated with 25% silicone grease (79). The column was operated at 235°C. Esposito (80) separated and determined quantitatively dimethyl phthalate, diethyl phthalate, dibutyl phthalate, dibutyl sebacate, butyl benzyl phthalate, di(2-ethylhexyl) phthalate, tricresyl phosphate, and di(2-ethylhexyl) sebacate on a 6-ft. × ¼-in. copper column, packed with acid- and alkali-washed Chromosorb W coated with 20% silicone grease. The column was operated at a starting temperature of 210°C and programmed at 4°C/min up to 290°C. Zulaica and Guiochon (81) separated the same plasticizers on a 2 m × 4 mm i.d. copper column packed with glass beads which had been previously etched by concentrated hydrofluoric acid, and coated with 0.5% silicone gum SE-30. These authors analyzed plasticizers in plastics, volatilizing the plasticizers from the plastic material in a pyrolysis unit which was fitted to the injection port of the gas chromatograph.

Very small sample weights in the range from 3 to 7 mg were used. The successful demonstration of this technique makes it appear feasible that small samples of cellophane can be injected directly into a gas chromatograph with a solid sample injector and plasticizers and other volatile substances can thus be determined either qualitatively or quantitatively.

We have separated dibutyl phthalate, tributyl acetylcitrate, dicyclohexyl phthalate, and tricyclohexyl citrate on a 2-ft × ¼-in. stainless steel column, packed with Diatoport W and coated with 5% silicone gum SE-30. The column was operated at a temperature of 140°C and programmed at 20°C/min to 340°C.

A generally applicable procedure for the identification of plasticizers would be to extract several grams of sample with carbon tetrachloride, evaporate the solvent, and obtain an infrared spectrum of the residue.   If only one plasticizer is present in the sample, it usually can be identified by this procedure.   If, however, it is obvious from the infrared spectrum that a mixture of plasticizers is present, the residue can be diluted with a few milliliters of carbon tetrachloride, and the plasticizers can be separated by gas chromatography, using the proper column and operating conditions.   If the separated plasticizers cannot be identified by their retention times when compared with those of known plasticizers, it is possible to trap the effluents of the gas chromatograph and identify them by infrared spectroscopy.

5. Identification of Waxes and Release Agents

Waxes are present in cellophane coatings to improve permeability characteristics and release agents are present to improve the machinability of the film on packaging machines.   A generally applicable procedure for the identification of these substances is as follows. Several grams of cellophane are extracted with hot carbon tetrachloride.   The extract is brought to dryness and about 10 ml of methanol is added.   The mixture is cooled in an ice bath, filtered, and washed with several milliliters of cold methanol.   The residue is dried in air, and an infrared spectrum is obtained by melting a small amount of the substance on a salt plate.   Most waxes and release agents are precipitated by cold methanol.   If, however, methanol-soluble substances are suspected, the filtrate is concentrated to about 10 ml and 50 ml of water is added.   If the solution turns cloudy, several grams of aluminum sulfate is added to aid in precipitation.   The mixture is filtered, washed thoroughly with cold water and dried in air.   The residue is then identified by infrared spectroscopy.

Ludwig (82,83) reported gas–liquid chromatograms of carnauba wax, montan wax, ouricury wax, a microcrystalline wax, and a

Fischer-Tropsch wax. These chromatograms show clearly that a gas chromatographic technique can be extremely helpful in the identification of waxes in cellophane.

References

1. *ASTM 1966 Standards*, Part 15, Standard Method of Sampling Paper and Paperboard, *ASTM Method D 585-62*, American Society for Testing and Materials, Philadelphia, p. 78.
2. *Tappi T 400 ts-64*, Sampling Paper and Paperboard.
3. *ASTM 1966 Standards*, Part 15, Standard Method for Moisture Content of Paper and Paperboard by Oven Drying, *ASTM Method D 644-55*, American Society for Testing and Materials, Philadelphia, p. 119.
4. *ASTM 1966 Standards*, Part 15, Standard Method for Moisture in Cellulose, Sections 17–25, *ASTM Method D 1348-61*, American Society for Testing and Materials, Philadelphia, p. 488.
5. H. A. Scopp and C. P. Evans, *Anal. Chem.*, **28**, 143 (1956).
6. *Tappi, T 484m-58*, Moisture in Paper and Paperboard by Toluene Distillation.
7. I. N. Ermolenko and S. S. Gusev, *Vysokomolekul. Soedin.*, **1**, 1462 (1959).
8. P. Kajanne, *Paperi Puu*, **39**, 391 (1957).
9. *ASTM 1966 Standards*, Part 15, Standard Specifications and Methods of Test for Soluble Cellulose Nitrate, Sections 7 and 8, *ASTM Method D 301-56*, American Society for Testing and Materials, Philadelphia, p. 48.
10. H. M. Rosenberger and C. J. Shoemaker, *Anal. Chem.*, **31**, 1315 (1959).
11. H. Levitsky and G. Norwitz, *Anal. Chem.*, **34**, 1167 (1962).
12. W. E. Shaefer and W. W. Becker, *Anal. Chem.*, **25**, 1226 (1953).
13. J. D. Mullen, *Anal. Chim. Acta*, **20**, 16 (1959).
14. R. H. Pierson and E. C. Julian, *Anal. Chem.*, **31**, 589 (1959).
15. H. Stalcup and R. W. Williams, *Anal. Chem.*, **27**, 543 (1955).
16. T. Murakami, *Bunseki Kagaku*, **9**, 100 (1960).
17. A. F. Williams and J. Brooks, *Proc. 1958 Intern. Symp. Microchem., Birmingham Univ., 1958*, 430 (1959).
18. R. Leclercq and J. Mathe, *Bull. Soc. Chim. Belges*, **60**, 296 (1951).
19. W. C. Easterbrook, *J. Appl. Chem.*, **9**, 410 (1959).
20. F. C. Bowman and W. W. Scott, *Ind. Eng. Chem.*, **7**, 766 (1915).
21. T. E. Timell and C. B. Purves, *Svensk Papperstid.*, **54**, 328 (1951).
22. M. H. Swann, *Anal. Chem.*, **25**, 1504 (1957).
23. M. H. Swann, in *Chemical Analysis of Resin-Based Coating Materials*, C. P. A. Kappelmeier, Ed., Interscience, New York, 1959, p. 325.
24. L. P. Kuhn, *Anal. Chem.*, **22**, 276 (1950).
25. A. Clarkson and C. M. Robertson, *Anal. Chem.*, **38**, 522 (1966).
26. K. Ettre and P. F. Varadi, *Anal. Chem.*, **34**, 752 (1962).
27. K. Ettre and P. F. Varadi, *Anal. Chem.*, **35**, 69 (1963).
28. L. B. Genung and R. C. Mallatt, *Ind. Eng. Chem. Anal. Ed.*, **13**, 369 (1941).
29. C. J. Malm, L. B. Genung, R. F. Williams, Jr., and M. A. Pile, *Ind. Eng. Chem. Anal. Ed.*, **16**, 501 (1944).

30. L. B. Genung, *Anal. Chem.*, **22**, 401 (1950).
31. Progress Report by Subcommittee on Acyl Analysis, Division of Cellulose Chemistry Committee on Standards and Methods of Testing, *Anal. Chem.*, **24**, 400 (1952).
32. *ASTM 1966 Standards*, Part 15, Standard Method of Testing Cellulose Acetate, *ASTM Method D 871-63*, American Society for Testing and Materials, Philadelphia, p. 271.
33. C. R. Fordyce, L. B. Genung, and M. A. Pile, *Ind. Eng. Chem. Anal. Ed.*, **18**, 547 (1946).
34. J. A. Mitchell, C. D. Bockman, Jr., and A. V. Lee, *Anal. Chem.*, **29**, 499 (1957).
35. C. D. Bockman, Jr., *Appl. Spectry.*, **15**, 84 (1961).
36. C. J. Malm, L. J. Tanghe, B. C. Laird, and G. D. Smith, *Anal. Chem.*, **26**, 188 (1954).
37. L. J. Tanghe, L. B. Genung, and J. W. Mench, "Determination of Hydroxyl Content of Cellulose Acetate," in *Cellulose (Methods in Carbohydrate Chemistry*, R. L. Whistler, Ed., Vol. 3), Academic Press, New York, 1963, p. 203.
38. C. J. Malm, G. F. Nadeau, and L. B. Genung, *Ind. Eng. Chem. Anal. Ed.*, **14**, 292 (1942).
39. *ASTM 1966 Standards*, Part 15, Tentative Methods of Testing Cellulose Acetate Propionate and Cellulose Acetate Butyrate, *ASTM Method D 817-65*; American Society for Testing and Materials, Philadelphia, p. 229.
40. S. Zeisel, *Monatsh. Chem.*, **6**, 989 (1885).
41. F. Vieböck and C. Brecker, *Ber.*, **63B**, 3207 (1930).
42. E. P. Samsel and J. A. McHard, *Ind. Eng. Chem. Anal. Ed.*, **14**, 750 (1942).
43. *ASTM 1966 Standards*, Part 15, Standard Methods for Testing Ethyl Cellulose, *ASTM Method D 914-50*, American Society for Testing and Materials, Philadelphia, p. 300.
44. R. H. Cundiff and P. C. Markunas, *Anal. Chem.*, **33**, 1028 (1961).
45. J. G. Cobler, E. P. Samsel, and G. H. Beaver, *Talanta*, **9**, 473 (1962).
46. S. G. Cohen and H. C. Haas, *J. Am. Chem. Soc.*, **72**, 3954 (1950).
47. P. W. Morgan, *Anal. Chem.*, **18**, 500 (1946).
48. H. J. Lortz, *Anal. Chem.*, **28**, 892 (1956).
49. W. J. Alexander and T. E. Muller, Abstracts, Winter Meeting, American Chemical Society, Phoenix, Arizona, Jan. 1966, paper D-30.
50. E. D. Klug, "Hydroxyethyl Esters of Cellulose and Their Analytical Determination," in *(Cellulose, Methods of Carbohydrate Chemistry*, Vol. 3), R. L. Whistler, Ed., Academic Press, New York, 1963, p. 316.
51. R. Senju, *J. Agr. Chem. Soc. Japan*, **22**, 58 (1948).
52. H. Tai, R. M. Powers, and T. F. Pretzman, *Anal. Chem.*, **36**, 108 (1964).
53. F. J. Reidinger and N. Poginy, unpublished results.
54. L. Malaprade, *Bull. Soc. Chim. France*, [4], **43**, 683 (1928); [5], **1**, 833 (1934).
55. G. Mandric, *Rev. Chim., Acad. Rep. Populaire Roumaine*, **12**, 503 (1961).
56. C. E. Bricker and H. R. Johnson, *Ind. Eng. Chem. Anal. Ed.*, **17**, 400 (1945).
57. A. Davis, A. Roaldi, and L. E. Tufts, *J. Gas Chromatog.*, **2**, 306 (1964).
58. *Federal Register*, **29** F.R. **12871**, Sept. 12, 1964, § 121.2507.

59. *Federal Register*, **26** F.R. **938**, Jan. 31, 1961, § 121.101.
60. *Federal Register*, **25** F.R. **866**, Feb. 2, 1960, § 121.2001.
61. N. Allen, H. Y. Charbonnier, and R. M. Colman, *Ind. Eng. Chem. Anal. Ed.*, **12**, 384 (1940).
62. M. J. Cardone and J. Compton, *Anal. Chem.*, **24**, 1903 (1952).
63. M. J. Cardone and J. W. Compton, *Anal. Chem.*, **25**, 1869 (1953).
64. C. L. Whitman, G. W. Roecker, and C. F. McNerney, *Anal. Chem.*, **33**, 781 (1961).
65. G. W. Watt and J. D. Chrisp, *Anal. Chem.*, **26**, 452 (1954).
66. J. Mitchell, Jr., I. M. Kolthoff, E. S. Proskauer, and A. Weissberger, Eds., *Organic Analysis*, Vol. 1, Interscience, New York, 1953, p. 263.
67. R. C. Hirt, F. T. King, and R. G. Schmitt, *Anal. Chem.*, **26**, 1273 (1954).
68. R. Fosse, *Compt. Rend.*, **158**, 1076 (1914).
69. R. Fosse, *Compt. Rend.*, **159**, 253 (1915).
70. F. J. Paschmann, *Tappi*, **40**, 487 (1957).
71. D. Hummel, *Kunststoff- Lack- und Gummi-Analyse*, Carl Hanser Verlag, Munich, 1958.
72. J. Haslam and H. A. Willis, *Identification and Analysis of Plastics*, Van Nostrand, Princeton, N.J., 1965.
73. *The Sadtler Commercial Spectra of Polymers*, The Sadtler Research Laboratories, Philadelphia 2, Pa.
74. C. P. A. Kappelmeier, Ed., *Chemical Analysis of Resin Based Coating Materials*, Interscience, New York, 1959.
75. H. G. Nadeau and E. W. Neumann, *Mod. Packaging*, **37**, No. 6, 128 (Feb. 1964).
76. L. H. Phifer, *Mod. Packaging*, **38**, No. 3, 154 (Nov., 1964).
77. S. G. Gilbert, L. I. Oetzel, W. Asp, and I. L. Brazier, *Mod. Packaging*, **38**, No. 9, 167 (May 1965).
78. W. A. Moore and R. A. Kolbeson, *Anal. Chem.*, **28**, 161 (1956).
79. C. D. Cook, E. J. Elgood, G. C. Shaw, and D. H. Solomon, *Anal. Chem.*, **34**, 1177 (1962).
80. G. G. Esposito, *Anal. Chem.*, **35**, 1439 (1963).
81. J. Zulaica and G. Guiochon, *Anal. Chem.*, **35**, 1724 (1963).
82. F. J. Ludwig, *Anal. Chem.*, **37**, 1732 (1965).
83. F. J. Ludwig, *Soap Chem. Specialties*, March, 1966, p. 70.

CHAPTER 15

THE ANALYTICAL CHEMISTRY OF VINYL
FILM-FORMING POLYMERS

JOHN G. COBLER, MERTON W. LONG,
and E. GUY OWENS II

The Dow Chemical Co., Midland, Michigan

I. Introduction

The analytical chemistry of vinyl film-forming polymers as discussed in this chapter will be limited to those polymeric materials which contain, as the major component, vinyl chloride, vinylidene chloride, vinyl alcohol, or styrene. General analytical techniques are given which can be adapted by the analyst to his particular needs. The techniques are illustrated with specific examples, but it must be recognized that these examples are not all-inclusive and that the applications may not be limited to a particular polymer species.

II. Polymer Composition

A. Styrene Polymers

Polystyrene film is usually unplasticized. Although polystyrene is the principal film-forming polymer of the styrene polymer class, copolymers are commercially available. The principal styrene polymers are

$$\text{Styrene} \sim\!\!\!\left[-CH_2-\underset{\underset{C_6H_5}{|}}{CH}-\right]_n\!\!\!\sim$$

$$\text{Styrene}/\alpha\text{-methylstyrene} \sim\!\!\!\left[-CH_2-\underset{\underset{C_6H_5}{|}}{CH}-\right]_n\!\!\!\sim\!\!\!\left[-CH_2-\underset{\underset{C_6H_5}{|}}{\overset{\overset{CH_3}{|}}{C}}-\right]_n\!\!\!\sim$$

Styrene/acrylonitrile $\sim\left[\mathrm{CH_2-CH-}\atop\mathrm{C_6H_5}\right]_n\sim\left[\mathrm{-CH_2-CH-}\atop\mathrm{C\equiv N}\right]_n\sim$

Styrene/methyl methacrylate $\sim\left[\mathrm{-CH_2-CH-}\atop\mathrm{C_6H_5}\right]_n\sim\left[\begin{array}{c}\mathrm{CH_3}\\\mathrm{-CH_2-C-}\\\mathrm{C=O}\\\mathrm{O}\\\mathrm{CH_3}\end{array}\right]_n\sim$

B. VINYL CHLORIDE POLYMERS

Vinyl chloride $\sim\left[\mathrm{-CH_2-CH-}\atop\mathrm{Cl}\right]_n\sim$

Vinyl chloride/vinyl acetate $\sim\left[\mathrm{-CH_2-CH-}\atop\mathrm{Cl}\right]_n\sim\left[\begin{array}{c}\mathrm{-CH_2-CH-}\\\mathrm{O}\\\mathrm{C=O}\\\mathrm{CH_3}\end{array}\right]_n\sim$

Vinyl chloride/vinyl acetate/vinyl alcohol

$\sim\left[\mathrm{-CH_2-CH-}\atop\mathrm{Cl}\right]_n\sim\left[\begin{array}{c}\mathrm{-CH_2-CH-}\\\mathrm{O}\\\mathrm{C=O}\\\mathrm{CH_3}\end{array}\right]_n\sim\left[\begin{array}{c}\mathrm{-CH_2-CH-}\\\mathrm{O}\\\mathrm{H}\end{array}\right]_n\sim$

Vinyl chloride/vinyl acetate/maleic anhydride

$\sim\left[\mathrm{-CH_2-CH-}\atop\mathrm{Cl}\right]_n\sim\left[\begin{array}{c}\mathrm{-CH_2-CH-}\\\mathrm{O}\\\mathrm{C=O}\\\mathrm{CH_3}\end{array}\right]_n\sim\left[\begin{array}{c}\mathrm{-CH-\quad CH-}\\\mathrm{O=C\quad C=O}\\\mathrm{O}\end{array}\right]_n\sim$

Vinyl chloride/acrylic ester $\sim\left[\mathrm{-CH_2-CH-}\atop\mathrm{Cl}\right]_n\sim\left[\begin{array}{c}\mathrm{-CH_2-CH-}\\\mathrm{C=O}\\\mathrm{O}\\\mathrm{R}\end{array}\right]_n\sim$

Vinyl chloride/maleic acid ester $\sim\!\!\!\sim \left[-CH_2-\underset{\underset{Cl}{|}}{CH}-\right]_n \sim\!\!\!\sim \left[-\underset{\underset{\underset{\underset{R}{|}}{O}}{\underset{O}{||}}{C}}{CH}-\underset{\underset{\underset{\underset{R}{|}}{O}}{\underset{O}{||}}{C}}{CH}-\right]_n \sim\!\!\!\sim$

Vinyl chloride/vinylidene chloride $\sim\!\!\!\sim \left[-CH_2-\underset{\underset{Cl}{|}}{CH}-\right]_n \sim\!\!\!\sim \left[-CH_2-\underset{\underset{Cl}{|}}{\overset{\overset{Cl}{|}}{C}}-\right]_n \sim\!\!\!\sim$

These polymers are available in the form of rigid (unplasticized) or flexible films (containing up to 30% plasticizer). Typical plasticizers for flexible films are the various phthalate, phosphate, sebacate, adipate, azelate, and glycol esters and epoxidized oils. Polymeric modifiers such as butadiene–acrylonitrile, polyester resins and chlorinated polyethylene have also been employed to impart certain physical properties to the polymers.

The comonomeric esters (vinyl acetate, acrylate, fumarate, and maleate) actually act as internal plasticizers and impart greater flexibility, better low temperature properties, and increased solubility as compared with the uncopolymerized material.

Polyvinyl chloride loses hydrogen chloride when exposed to the action of heat and light, yielding an amber or dark colored polyene.

$$\sim\!\!\!\sim CH_2-CHCl\sim\!\!\!\sim \;\rightarrow\; \sim\!\!\!\sim CH=CH\sim\!\!\!\sim \;+\; HCl$$

Stabilization against the autocatalytic loss of hydrogen chloride is thus necessary. Alkali or alkaline earth oxides or hydroxides, fatty acid salts, organometallic compounds (lead or tin), and ethylene oxide compounds are used, in concentrations up to 5%, as hydrogen chloride acceptors.

C. Vinylidene Chloride Polymers

Polyvinylidene chloride is seldom used commercially. The principal polymers in this group are

Vinylidene chloride $\sim\!\!\!\sim \left[-CH_2-\underset{\underset{Cl}{|}}{\overset{\overset{Cl}{|}}{C}}-\right]_n \sim\!\!\!\sim$

Vinylidene chloride/acrylonitrile $\sim\left[-CH_2-\underset{\underset{Cl}{|}}{\overset{\overset{Cl}{|}}{C}}-\right]_n\sim\left[-CH_2-\underset{\underset{C\equiv N}{|}}{CH}-\right]_n\sim$

Vinylidene chloride/acrylic ester $\sim\left[-CH_2-\underset{\underset{Cl}{|}}{\overset{\overset{Cl}{|}}{C}}-\right]_n\sim\left[-CH_2-\underset{\underset{\underset{R}{|}}{\underset{O}{|}}}{\overset{}{CH}}-\right]_n$

with $C=O$

Vinylidene chloride/vinyl chloride $\sim\left[-CH_2-\underset{\underset{Cl}{|}}{\overset{\overset{Cl}{|}}{C}}-\right]_n\sim\left[-CH_2-\underset{\underset{Cl}{|}}{CH}-\right]_n\sim$

Plasticizers and stabilizers used for vinyl chloride polymers are normally satisfactory for use with vinylidene chloride polymers.

D. POLYVINYL ALCOHOL

Polyvinyl alcohol is prepared by the hydrolysis of polyvinyl acetate and may contain, depending upon the end use, a substantial concentration of vinyl acetate groups.

Polyvinyl alcohol $\sim\left[-CH_2-\underset{\underset{\underset{H}{|}}{\underset{O}{|}}}{CH}-\right]_n\sim$

Vinyl alcohol/vinyl acetate polymer

$$\sim\left[-CH_2-\underset{\underset{\underset{H}{|}}{\underset{O}{|}}}{CH}-\right]_n\sim\left[-CH_2-\underset{\underset{\underset{CH_3}{|}}{\underset{\underset{C=O}{|}}{O}}}{CH}-\right]_n\sim$$

The polymer may also be modified by reaction with formaldehyde or dibasic acids to decrease the sensitivity to water. Polyvinyl alcohol may be plasticized with 5–20 % of a polyhydroxy compound such as glycerol or triethylene glycol.

TABLE 1
Burning Characteristics of Vinyl Polymers

Polymer	Ease of ignition	Self-extinguishing	Character of flame	Fumes or odor	Behavior on heating	Residue
Polystyrene	Easy	No	Sooty, yellow	Monomer odor, sweet	Softens, becomes elastic, melts and distills	Small
Polyvinyl chloride	Difficult	Yes	Sooty, yellow	Acrid	Softens	Black residue
Vinyl chloride/vinyl acetate	Difficult	Yes	Smoky	Acrid fumes	Melts and chars	Black residue
Vinylidene chloride copolymers	Difficult	Yes	Yellow	White acrid fumes	Softens and melts	Black residue
Polyvinyl alcohol	Moderately difficult	Slowly	Slightly smoky	Sweetish aldehyde odor, brown fumes	Softens and chars	Black residue

III. Identification

Numerous qualitative schemes based on differences in solubility, burning characteristics, refractive index, and chemical classification tests have been published for the identification of polymeric materials (1). Such schemes, although largely supplanted by infrared spectrometric identification or pyrolysis–gas chromatographic procedures, are suitable as rapid confirmatory tests or when instrumentation is lacking. Physical property data are presented in Tables 1, 2, and 3. These data, together with simple chemical classification tests, will enable the analyst to develop his own identification scheme.

TABLE 2

Solubility of Vinyl Polymers[a]

Solvent	Poly-styrene	Poly-vinyl chloride	Vinyl chloride/ vinyl acetate	Vinylidene chloride copolymers	Poly-vinyl alcohol
Acetone	I	I	S–PS	I	I
Methyl ethyl ketone	S	S	S–PS	I	I
Cyclohexanone	PS	S	S	SΔ	I
Ether	PS	I	I	I	I
Dioxane	S	S	S	SΔ	I
Tetrahydrofuran	S	S	S	S	I
Ethanol, 95%	I	I	I	I	I
n-Butyl alcohol	I	I	I	I	I
Water	I	I	I	I	S
Ethyl acetate	PS	I	I	I	I
Butyl acetate	S	PS	PS	I	I
Ethylene dichloride	S	S	S	I	I
Chloroform	S	PS	S–PS	I	I
o-Dichlorobenzene	S	SΔ	S	SΔ	I
Benzene	S	I	I	I	I
Toluene	S	I	I	I	I
Nitroethane	PS	S	S–PS	I	I
Morpholine	PS	PS	S	I	I
Pyridine	S	PS	S–PS	I	I
Acetic acid	PS	I	I	I	I
Petroleum ether	I	I	I	I	I

[a] Key: S, soluble; PS, partially soluble; I, insoluble; Δ, hot.

TABLE 3
Density, Refractive Index, and Water Absorption of Vinyl Polymers

Polymer	Density[a] g/ml, 25°C	Refractive index,[b] n_D^{20}	Water absorption,[c] % in 24 hr, $\frac{1}{8}$ in. thickness
Polystyrene	1.04–1.065	1.59–1.60	0.03–0.05
Styrene/acrylonitrile	1.075–1.10	1.55–1.58	0.2–0.3
Styrene/methyl methacrylate	1.08–1.16	1.53–1.56	0.2
Polyvinyl chloride, unplasticized	1.35–1.45	1.52–1.55	0.07–0.4
Polyvinyl chloride, plasticized	1.25–1.35	up to 1.6	0.15–0.75
Vinyl chloride/vinyl acetate	1.16–1.45	1.4–1.5	0.15–1
Polyvinylidene chloride	1.65–1.72	1.60–1.63	0.1
Vinylidene chloride copolymers	1.6–1.7	1.5–1.6	0.1
Polyvinyl alcohol	1.26–1.29	1.49–1.53	

[a] *ASTM D 792.*
[b] *ASTM D 542.*
[c] *ASTM D 570.*

A. DETECTION OF POLYSTYRENE

1. Diazotization (2)

Reflux approximately 0.1 g of sample with 5 ml of concentrated nitric acid for approximately 1 hr or until a clear solution forms. Cool and pour into 20 ml of water. Extract the resulting suspension with two 10-ml portions of ethyl ether. Wash the combined ether extracts with water and then extract with two 5-ml portions of $1N$ sodium hydroxide. Make the solution acid with concentrated hydrochloric acid and add 1 g of granulated zinc. Warm on the steam bath for 30 min or until effervescence ceases. Cool and filter. Cool the filtrate in ice and add 1 ml of $0.5N$ sodium nitrite solution. Pour the diazotized solution into an excess of alkaline 2-naphthol solution. The presence of styrene polymers produces a rich red or scarlet color. Aniline–formaldehyde produces the same result without nitration and reduction.

2. Conversion to Phenol (3)

Add 4 or 5 drops of fuming nitric acid (sp gr 1.5) to 50–100 mg of sample in a test tube and evaporate to dryness.   Cover the mouth of the tube with a piece of filter paper previously dipped in an ether solution of 2,6-dichloro-N-chloro-p-benzoquinone imine and dried. Heat the tube with a bunsen flame for about 1 min starting at the middle and progressing downward.   Remove the paper and expose it to ammonia fumes.   The appearance of a blue stain is indicative of the presence of styrene polymers.   Phenol-formaldehyde and crosslinked epoxy resins yield negative results.

B. Detection of Polyvinyl Chloride
or of Vinyl Chloride Polymers

1. Pyridine–Sodium Hydroxide Test (4)

Dissolve approximately 50 mg of sample in 5 ml of pyridine. Heat the solution to a gentle boil for about 1 min.   To the hot solution add 0.5 ml of a 2% methanolic solution of sodium hydroxide.   A brown to black coloration gradually turning into a brown precipitate indicates the presence of polyvinyl chloride.   Under the same conditions neoprene and chlorinated rubber produce only a faint yellow or yellow-brown coloration.   Vinylidene chloride polymers produce a brown-black precipitate even in the cold. Thus, with this technique, it is impossible to identify vinyl chloride polymers in the presence of vinylidene chloride polymers.

2. Chloroacetic Acid Test (5)

Add 50–100 mg of the sample to 5 ml of molten monochloroacetic acid and to 5 ml of molten dichloroacetic acid contained in small test tubes.   Heat to boiling for 2–3 min with occasional swirling. Polyvinyl chloride produces a blue coloration in monochloroacetic acid and a reddish-purple coloration in dichloroacetic acid.   Vinyl chloride/vinyl acetate copolymers yield maroon to purple and blue to purple colorations, respectively.   Polyvinyl acetate and vinyl ether polymers react to give green or bluish-green colors in mono-chloroacetic acid and greenish-blue in dichloroacetic acid.

3. Reaction with Sodium Methoxide (6)

Dissolve the sample in tetrahydrofuran and transfer a few drops to a filter paper. Dry the paper in an oven and then place in a flask containing 50 ml of 0.1N sodium methoxide. Reflux the solution for 20 min. Remove the filter paper and wash it thoroughly with water. A yellow to red-brown spot changing to black or brown on standing shows the presence of vinyl chloride copolymers. Vinyl chloride homopolymer, chlorinated polyvinyl chloride, polyvinyl alcohol, and polyvinyl acetate yield negative results.

C. DETECTION OF VINYLIDENE CHLORIDE POLYMERS

1. Pyridine–Potassium Hydroxide Test (7)

Place 1 drop of pyridine on the surface of the film, followed by 1 drop of a saturated methanolic potassium hydroxide solution. In the presence of vinylidene chloride polymers, a brown-black coloration appears within 30 sec.

2. Morpholine Test (8)

Immerse the sample in morpholine. Vinylidene chloride polymers slowly darken and the liquid gradually turns black. Polyvinyl chloride does not discolor. Chlorinated polyvinyl chloride slowly dissolves, yielding a red-brown color.

D. DETECTION OF CHLORINE-CONTAINING POLYMERS

1. Beilstein Test

Beilstein's copper wire test is simple and convenient. Chlorine-containing polymers and copolymers give strong tests. Faint tests may be ignored and may be due to salts or volatile impurities.

Heat a copper wire in the blue flame of a gas burner until the flame is no longer green. Touch the hot wire to the plastic and reheat in the flame. A strong green coloration indicates the presence of chlorine.

2. Magnesium and Potassium Carbonate Fusion Procedure for Halogens (9)

a. Reagent. Fusion mixture. Grind together 2 parts of anhydrous potassium carbonate and 1 part of finely divided magnesium powder.

b. Procedure.   Heat a 10 × 100-mm test tube to redness and allow to cool.   Place approximately 0.1 g of polymer at the bottom of the test tube.   Hold the tube at an angle of 30° from the horizontal and allow 0.2 g of the fusion mixture to slide down the side of the test tube.   This mixture should extend about 3 cm up the tube and within 1 cm of the sample.   Allow several drops of ether (CAUTION) to run down on the mixture and then plug the mouth of the tube with glass wool.

Start the heating of the mixture near the mouth of the tube with a bunsen burner.   When the mixture begins to glow, bring the lower end of the tube into the flame so as to distil the pyrolyzate over the glowing mass.   The polymer in the test tube must not be pyrolyzed until the reaction mixture starts to glow.

Continue heating until finally the whole tube is brought to dull redness.   Drop the hot tube into 20 ml of distilled water in a beaker.   Break the tube up, stir the mixture thoroughly, and filter.   Acidify about 2 ml of the filtrate with dilute nitric acid and gently boil for several minutes to expel any hydrogen cyanide or hydrogen sulfide that may be present.   Add a few drops of silver nitrate solution.   A heavy precipitate indicates the presence of chlorine, bromine, or iodine.

3. Fusion with Manganese Dioxide (10)

Chlorine may be detected in nitrogen-free polymers, after ignition of the polymer in the presence of manganese dioxide, by reaction with thio-Michler's ketone.   Conduct the test in a small test tube which projects through a perforated asbestos sheet. Place a little of the polymer in the tube and mix intimately with manganese dioxide.   Cover the mouth of the tube with a disk of filter paper which has been moistened with a 0.1% solution of thio-Michler's ketone in benzene, and dried.   Heat the bottom of the tube strongly with a microflame for 1–2 min.   A positive response is indicated by the development of a blue color.

E. Detection of Nitrogen-Containing Polymers

1. Magnesium and Potassium Carbonate Fusion (11)

Nitrogen may be detected in a sample of the filtrate from Section III-D-2.   Adjust the pH of 1 ml of the filtrate to approximately 13 with Hydrion E paper.   Add 2 drops each of a saturated ferrous ammonium sulfate solution and of a 30% potassium fluoride solution. Boil the resulting solution gently for about 30 sec. Acidify the hot solution by careful addition of 30% sulfuric acid dropwise, until the iron hydroxide just dissolves. The appearance of the characteristic Prussian blue color indicates the presence of nitrogen. Care has to be exercised when making this test or false positive reactions may be obtained.

2. Fusion with Manganese Dioxide (12)

The test for nitrogen based on the formation of nitrous acid through pyrolytic oxidation in the presence of manganese dioxide is more sensitive and reliable than the Lassaigne test.

a. Reagents. *Solution 1.* Solution of sulfanilic acid 1% in 30% acetic acid.

Solution 2. Solution of 1-naphthylamine 0.1% in 30% acetic acid. Prepare the Griess reagent by mixing equal volumes of solutions 1 and 2 just prior to use.

Manganese Dioxide. Prepare by heating MnO_2 to 500–600°C to expel any nitrate nitrogen.

b. Procedure. Conduct the test in a small Pyrex test tube which projects through a perforated asbestos sheet. The test tube should be heated to a dull red heat and cooled prior to adding the sample; otherwise erratic and erroneous results may occur.

Place a little of the polymer in the test tube and intimately mix with manganese dioxide. Cover the mouth of the test tube with a disc of filter paper moistened with the Griess reagent, and heat the bottom of the tube strongly with a microflame. In general, it is sufficient to heat for 1–2 min. A positive response is indicated by the development of a pink or red circle on the colorless paper.

F. DETECTION OF CARBOXYL GROUPS IN POLYMERS (13)

a. Reagent. Dissolve 10–20 mg of the cationic dye pinacyanol in 1 l of a buffer solution of about pH 8 (0.2M disodium hydrogen phosphate and 0.1M citric acid in the proportion of 972.5 ml and 27.5 ml, respectively). This solution is fairly stable if kept shielded from light. A blank test with the reagent should not give any detectable coloration to benzene or the chosen solvent when the two are vigorously shaken in a test tube.

b. Procedure. Shake 5 ml of a 0.1% benzene solution of the polymer in a test tube with 5 ml of the test solution. Coloration of the benzene layer shows the presence of carboxyl groups in the polymer. A positive test is also given by strong acid groups such as SO_4^{2-}, SO_3^{2-}, etc. These strong acids, however, are easily distinguished by carrying out the dye partition test under acid conditions, whereby —COOH groups fail to respond to the test.

G. Detection of Vinyl Acetate in Polymers

When a copolymer containing vinyl acetate is pyrolyzed, acetic acid is formed. Place a small amount of shredded polymer in the bottom of a 10 × 100-mm test tube. Connect an inverted glass U delivery arm to the top of the test tube with the other end of the arm projecting into another small test tube acting as a condenser. Pyrolyze the polymer carefully with a microflame until a drop of pyrolyzate has been collected. Mix the pyrolyzate with 1 drop of 5% aqueous lanthanum nitrate. Add 1 drop of 0.1% iodine (in alcohol–water solution) to the above mixture, followed by a drop of 1N ammonium hydroxide. Acetic acid produces a blue to blue-brown color, which may slowly appear (14).

H. Detection of Polyvinyl Alcohol

1. Reaction with Iodine–Potassium Iodide (15)

Dissolve the sample in water. Add 2 drops of 0.1N iodine in 15% potassium iodide solution to 5 ml of the neutral solution. In the presence of polyvinyl alcohol, a blue-green color develops. Dilute the aqueous solution with water until the color almost disappears. Add 5 drops of concentrated hydrochloric acid. If polyvinyl alcohol is present, a green color develops. The sensitivity of the test may be increased by adding a drop of a saturated borax solution.

The blue color first formed in the aqueous solution is heat sensitive. When the solution is heated, the color changes through green, to greenish-yellow, to colorless. On cooling, the blue color reappears.

2. Conversion to Carboxylic Groups

Hydroxyl groups can be converted into strong acid groups by reaction with chlorosulfonic acid under mild conditions. The reaction is difficult to control and there is a tendency for some non-hydroxylic hydrogens also to react.   Hydroxyl groups can also react with the anhydride of a dibasic acid, e.g., phthalic anhydride or maleic anhydride, to build —COOH groups into the polymer which can then be tested for carboxyl groups (16).

a. Procedure.   Reflux a mixture of about 0.02–0.1 g of polymer and 0.5 g of phthalic anhydride in 25 ml of benzene for about 8 hr. Extract the solution with a 2% sodium hydroxide solution two or three times and then wash with water.   Test the benzene solution for —COOH groups by the dye partition test (Section III-F).   A blank test to check against any spurious impurity should also be run. Chloroform may be used since it frequently is a better solvent for polymers.   If both the polymer and anhydride are soluble in the solvent, about 3 hr of refluxing is sufficient.   The chloroform is evaporated after the reaction has taken place.   The excess phthalic anhydride is extracted from the residue with alcohol.   The polymer is then dissolved in benzene and tested by the usual procedure for —COOH group.

IV. Preparation of Sample

Most analytical procedures require a ground or otherwise finely subdivided sample.   Plastics may be reduced to the required size by comminution or by grinding in a mill such as a hammer mill, Wiley mill, or multicut mill.   Excessive heating in the mill results in the loss of volatile constituents and adherence of the softened plastic to the internal parts of the mill.   To prevent overheating, cool the mill and sample thoroughly with dry ice or liquid nitrogen. When both the mill and plastic have been thoroughly chilled, grind the plastic to pass an ASTM No. 35 (500-μ) sieve.

V. Isolation of Additives

A. SURFACE ADDITIVES

Surface additives on granules are normally lubricants used to prevent the polymer from sticking to the mold, antistatic agents, or antidusting agents.   Surface additives on films are primarily slip agents, antiblocking agents, or antifogging agents.   The concentration of these agents may vary from 0.05% to about 1%.   Surface additives are usually incompatible with the polymer, thus preventing their absorption into the body of the polymer. These agents may be mineral oil or waxes, ester waxes, fatty acids or their derivatives, aliphatic amides, and polyethers. Surface additives may be isolated by extraction procedures using a polymer nonsolvent, or from a film by surface abrasion with potassium bromide powder.

1. Extraction

a. Procedure. Wash 400 g of the granules or film with three 500-ml portions of a hot nonsolvent such as methanol or ethanol. Cool the combined extracts to room temperature. Some surface additives have a low solubility in alcohol and may separate on cooling. Filter off any insoluble material, dry, and weigh. Evaporate the filtrate to a volume of 50–60 ml and transfer to a tared 100-ml beaker. Dry the beaker and contents to constant weight at 70°C. The isolated surface additives may be characterized by infrared spectrometry. Dry the extracted polymer in a vacuum oven at 70°C, until all of the solvent has been removed. Retain the dried polymer for further characterization.

$$\% \text{ Total surface additives} = \frac{\text{Total weight of residues, g}}{\text{Sample weight, g}} \times 100$$

2. Surface Abrasion

One method for analyzing surface additives or coatings on films is to abrade lightly or polish the surface of the film with potassium bromide powder, collect the powder contaminated with the surface component, and press the mixture into a disk for infrared spectroscopic examination. This may be accomplished manually by sprinkling potassium bromide powder over the film and rubbing or polishing the surface by hand using fine steel wool.

Johnson (17) described a manually operated machine for abrasion which has been automated by Long and Cobler (18). The abrader is pictured in Figure 1. The film is taped to a pane of glass, supported on a carriage, and sprinkled with potassium bromide powder. A bar magnet wrapped with steel wool is positioned on the film over another bar magnet attached to a motor. The motor traverses the width of the film underneath the glass, while the carriage travels at right angles to the movement of the motor. The depth of abrasion

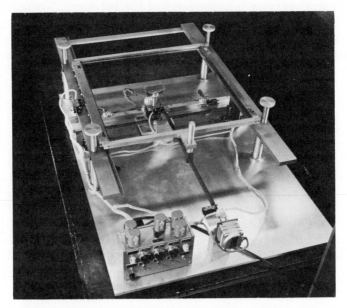

Fig. 1.   Automated abrader for abrading the surface of films.

is controlled by the weight of the bar magnet and the distance between the film and the motor-driven magnet.   After abrasion, the potassium bromide powder contaminated with the surface additive is collected in a cyclone separator, Figure 2, and pressed into a disk. If the disk is contaminated with some of the substrate in addition to the additive, it may be pulverized and extracted with solvents to effect a separation.   The solvents are then evaporated to dryness on potassium bromide powder for reexamination.

Figure 3 shows the infrared transmission spectrum of a vinylidene chloride/vinyl chloride film.   The spectrum has a strong carbonyl absorption band at 5.75 μ, resulting presumably from the plasticizer; however, the remainder of the spectrum is typical only of the copolymer.   A potassium bromide disk spectrum of the abrasion product is shown in Figure 4.   The spectrum has the characteristic absorption bands of a fatty acid glyceride essentially uncontaminated by the polymer substrate or the primary plasticizer.

Fig. 2.　Cyclone separator for collecting contaminated potassium bromide (left) and die for pressing potassium bromide disk (right).

B. Inorganic Additives

a. Procedure. Weigh 4 g of the washed and dried polymer into a 100-ml tared, conical centrifuge tube. Add 80 ml of a suitable solvent and warm on a water bath, stirring occasionally, until the polymer is in solution. Allow the solution to cool to room temperature. Centrifuge for 30 min or until the supernatant liquid is clear. Some insoluble additives may be removed by centrifuging at 2500 rpm while a finely ground additive may require a speed of 10–15,000 rpm. Decant the clear solution into a 150-ml beaker. Add 10 ml of solvent to the centrifuge tube, washing down the sides of the tube. Stir the residue with a stirring rod and repeat the centrifuging. Transfer the wash solution to the beaker. Repeat

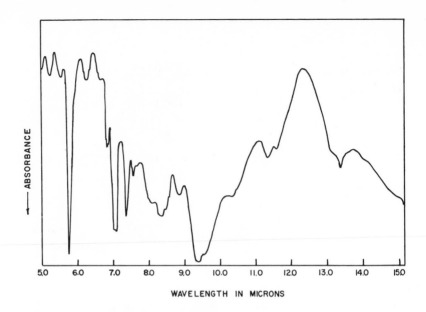

Fig. 3.   Infrared transmission spectrum of vinylidene chloride/vinyl chloride copolymer film.

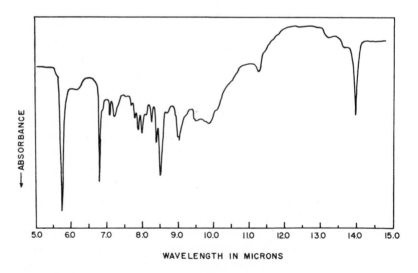

Fig. 4.   Infrared spectrum of potassium bromide disk prepared from material abraded from the surface of a vinylidene chloride/vinyl chloride copolymer film.

this washing with a second 10-ml portion of solvent. Dry the tube and contents on the water bath for 10–15 min and then for 1 hr at 105°C under vacuum. Cool the tube and contents in a desiccator and reweigh. The nature of the inorganic additives may be determined by suitable techniques such as emission spectrography or x-ray diffraction analysis.

$$\% \text{ Inorganic fillers } = \frac{\text{Weight of residue}}{\text{Weight of sample}} \times 100$$

C. Organic Additives

The organic additive fraction may be composed of products such as antioxidants, heat and light stabilizers, and low molecular weight polymer, but the primary component will be the plasticizer (or plasticizers), which may vary from 0.5% to several per cent. Plasticization is primarily a solvent action, the purpose being to aid in molding or compounding or to otherwise modify the properties of the finished product. Plasticizers include such materials as esters, waxes, or low molecular weight polymers. Plasticization obtained by building the plasticizer into the polymer chain (copolymer) is termed *internal plasticization*.

a. Procedure. Weigh 4 g of the washed and dried polymer into a 150-ml beaker. Add 100 ml of a suitable solvent and warm on a steam bath with stirring until dissolved. Transfer the solution rapidly, while stirring, to an 800-ml beaker containing 500 ml of a polymer nonsolvent. Rinse the 150-ml beaker with 25 ml of the nonsolvent and transfer all of the contents to the 800-ml beaker. Heat on the steam bath for 15–20 min until the precipitate has coagulated. Remove and allow to cool to room temperature. Filter with suction through a weighed medium-sintered glass crucible. Wash the beaker with two 50-ml portions of the nonsolvent, transferring the washings to the crucible. Dry the crucible and contents to constant weight in a vacuum oven at 70°C.

Combine the filtrate and washings and evaporate almost to dryness on the steam bath. Transfer the container and contents to a vacuum desiccator and dry to constant weight. If high-boiling solvents were used, the drying may be performed in a vacuum oven at 70°C. Monomers and certain low molecular weight additives may be volatilized during this operation, however. The additives may be characterized by suitable analytical procedures such as infrared spectrometry and functional group analysis.

% Organic additives
$$= \frac{\text{Weight of original sample} - \text{Weight of extracted sample}}{\text{Weight of the original sample}} \times 100$$

VI. Thermoanalytical Techniques

A. Differential Thermal Analysis (19)

1. *Glass Transitions*

The temperature of the glass transition (T_g) is usually taken as the temperature at which a maximum in mechanical loss occurs; however, differential thermal analysis provides another tool for studying this phenomenon.   Since the glass transition involves no latent heat, the thermograms do not show peaks typical of melting transitions.   Instead the T_g is accompanied by a change in specific heat which results only in a shift in the base line of the thermogram (Fig. 5).   This effect is small and high sensitivity and a stable base line are required in order to detect this shift.   To avoid an initial base line drift, Keavney and Eberlin (20) recommend the use of a

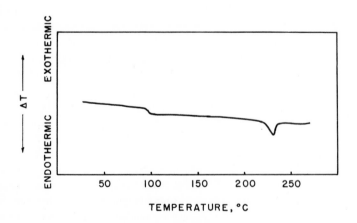

Fig. 5.   Differential thermogram of isotactic polystyrene showing T_g (100°C) and T_m (230°C).

stable organic material, such as isophthalic acid, for the reference substance. The thermal conductivity and heat capacity, and the rate of change of these quantities with temperature for an organic reference material, are more like those of a polymer than they are like those of aluminum oxide or glass. Polyethylene film has also been suggested as a reference material when studying other polymeric films. The transition temperature is essentially independent of heating rate between one and six centigrade degrees per minute, although more clear-cut transitions are observed at the higher rate of heating.

The glass transition temperatures for the vinyl polymers are given in Table 4. Literature values for polystyrene range from 82 to 100°C; however, the higher temperature appears to be correct for the completely devolatilized polymer. The T_g of a polymer is also decreased markedly by the addition of plasticizers. In copolymer systems, the comonomer whose homopolymer has the lower T_g tends to plasticize the copolymer internally (21). Thus, the copolymer will have a T_g lower than that of the higher softening homopolymer. For example, vinyl chloride/vinyl acetate systems exhibit lower T_g's than polyvinyl chloride whereas the T_g of a vinyl chloride/acrylonitrile copolymer is higher than that of polyvinyl chloride. The glass transition of many random copolymers is essentially a linear function of the composition along a line drawn

TABLE 4

Glass Transition Temperatures of Vinyl Polymers

Polymer	T_g, °C
Polystyrene	100
Polyvinyl alcohol	85
Polyvinylidene chloride	−17
Polyvinyl chloride	82
Vinylidene chloride/vinyl chloride (85:15)	65
Polyvinyl acetate	29
Vinyl chloride/vinyl acetate (88:12)	63
Polymethyl methacrylate	105
Polyacrylonitrile	107

between the T_g's of the respective homopolymers.   In no case do random copolymers have two glass transitions.

2. Crystallinity and Melting Temperatures

Although all polymers exhibit a T_g, only those polymers which have crystalline regions exhibit a first-order transition or melting temperature.   Crystallization is associated with the ability of polymer chains to pack in an orderly crystalline array.   For example, atactic polystyrene, with a completely ramdom configuration of phenyl groups, cannot be obtained in the crystalline state whereas isotactic polystyrene, with a regular repeating sequence, is partially crystalline.   Presumably, the bulkiness of the phenyl side groups prevents the formation of a crystal lattice unless the units are in a regular sequence.   Polyvinyl alcohol, on the other hand, with a smaller side group is highly crystalline even in the atactic form.   Polyvinyl chloride exhibits low crystallinity, with imperfect crystallites, except when highly oriented.   Symmetrical polymers such as polyvinylidene chloride are usually highly crystalline.

Introduction of comonomeric units into a polymer chain decreases both the degree of crystallinity and the first-order transition temperature.   External plasticization has the same effect.   The exception is the introduction of a comonomeric unit which is isomorphous with the primary monomer and which enters the chain without disrupting the crystal lattice.   Copolymers of this type have a melting point intermediate between that of the two homopolymers.

Differential thermal analysis is a suitable tool for determining the first-order transition temperature and for studying crystallization.   The first-order transition temperatures for vinyl polymers are given in Table 5.   Isotactic polystyrene has a low crystallization

TABLE 5
First-Order Transition Temperatures of Vinyl Polymers

Polymer	T_m, °C
Polystyrene, isotactic	230
Polyvinyl alcohol	215
Polyvinylidene chloride	205
Vinylidene chloride/vinyl chloride (85/15)	174

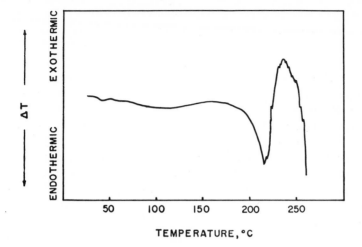

Fig. 6.   Differential thermogram of polyvinyl alcohol showing T_m with concomitant decomposition.

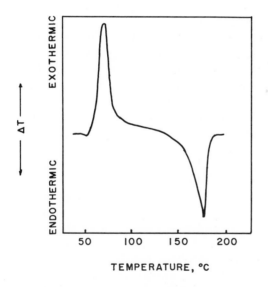

Fig. 7.   Differential thermogram of vinylidene chloride/vinyl chloride copolymer showing the cold crystallization exotherm and the T_m endotherm.

rate and must be annealed between 150 and 180°C for a period of time to obtain maximum crystallinity. The cooling curve for polystyrene shows no reproducible exotherm indicative of crystallization. Polyvinyl alcohol darkens before melting with decomposition at 215°C (Fig. 6). Polyvinyl chloride decomposes with loss of hydrogen chloride above 150°C or on prolonged exposure even at 100°C. Polyvinylidene chloride exhibits some decomposition at 140°C. Vinylidene chloride/vinyl chloride copolymers melt sharply at 174°C. When quenched from the melt, the copolymer is completely amorphous. When the temperature is raised past the second-order transition temperature, an exothermic reaction, indicative of cold crystallization, occurs, peaking at 67°C (Fig. 7).

B. Thermogravimetric Analysis

Thermogravimetric analysis (22) provides a rapid and simple method for assessing the thermal stability of polymeric materials (Fig. 8). It is also possible to estimate the activation energies of the various kinetic steps (23) and to characterize the degradation mechanism (24) from the weight loss curves. Most vinyl polymers are degraded by either a chain-scission or nonchain-scission mechanism. When a chain-scission reaction is initiated, it may proceed by either depropagation or transfer. The actual route taken by a polymer is dependent primarily upon its structure, although it will be influenced, to some extent, by factors such as temperature, presence of impurities, molecular weight, and crystallinity.

The different mechanisms may be illustrated by the thermal degradation of poly-α-methylstyrene, polystyrene, and polyvinyl chloride. Intramolecular disproportionation or depropagation degradation is characteristic of polymers, such as poly-α-methylstyrene, with a repeating unit of $-CH_2-CR_1R_2-$ where R_1 and R_2 are substituents other than hydrogen. Degradation is usually initiated at the chain ends which then split off monomer units, resulting in the complete depolymerization of the polymer chain. The rate of volatilization decreases linearly throughout the degradation (Fig. 9). The molecular weight of the residue at any particular time is representative of the remaining unaltered polymer chains. When one or both of the substituents is replaced by hydrogen, as in polystyrene, transfer reactions occur and the production of monomer decreases. The volatilization rate curve of polystyrene increases

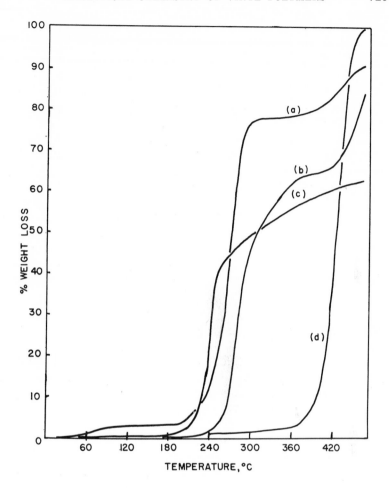

Fig. 8.   Thermogravimetric curves (*in vacuo*) for: (*a*) polyvinyl alcohol; (*b*) polyvinyl chloride; (*c*) polyvinylidene chloride; and (*d*) polystyrene.

initially and then decreases (Fig. 9).   The molecular weight of the residue decreases rapidly, owing to random chain scission which occurs as a result of intermolecular transfer.   Nonchain-scission reactions are common to short-chain esters and halogenated polymers which have a hydrogen atom attached to the carbon atom *beta* to the substituted group.   These molecules are dissociated into

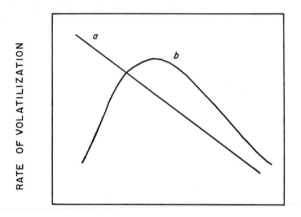

% VOLATILIZATION

Fig. 9.   Rates of volatilization as a function of per cent volatilization for:
(a) poly-α-methylstyrene; (b) polystyrene.

an acid and an olefin primarily by a free radical mechanism.   Thus, the initial step in the degradation of polyvinyl chloride involves dehydrochlorination without chain scission.   The resulting polyene chain is then degraded by chain scission.   Benzene and other aromatic compounds are formed during this latter process, presumably by cyclization of the unsaturated chain ends.

C. Pyrolysis and Gas Chromatography

When a polymer is pyrolyzed under controlled conditions, it decomposes into specific low molecular weight components.   This technique, when combined with gas chromatography for analysis of the pyrolyzate, provides a rapid method not only for studying degradation processes but for identification of polymers (25).   Absolute identification of complex polymers can be made only by comparing the chromatograms with those obtained from known polymers. Normally, however, sufficient information can be obtained if the chromatograms are examined for the main pyrolysis products of the various polymers, or if the pyrolysis products are condensed and examined by infrared spectrometry.   Figure 10 shows a syringe-type pyrolysis unit which can be attached to the injector port of a gas chromatograph with a Swagelok fitting.   The unit incorporates a cold and hot zone.   The sample is introduced into the cold zone which is cooled by two turns of $\frac{1}{8}$-in. copper tubing, soldered to the wall, through which cold water is allowed to flow.   The furnace zone contains a thermocouple for measuring the pyrolysis temperature.   The cool portion of the syringe can be maintained at less than 50°C when the furnace is at 750°C.   Pyrolysis temperatures have been varied from 150 to 950°C; however, a reactor temperature of 500–750°C appears to be satisfactory for most polymers.   The selection of a column packing is considerably more critical, two of the more useful being 20% di-2-ethylhexyl sebacate on 20–60-mesh firebrick and 25% LAC-2-R-446 (polydiethyleneglycolpentaerythritol adipate) plus 2% phosphoric acid on 80–100-mesh Chromosorb W.   The pyrolysis gas chromatograms of vinyl polymers, obtained according to the following procedure, are shown in Figures 11, 12, and 13.

a. Apparatus.   The gas chromatograph is equipped with a thermal conductivity detector and a $\frac{1}{4}$-in. × 8-ft column packed with 20% di-2-ethylhexyl sebacate on fire brick.

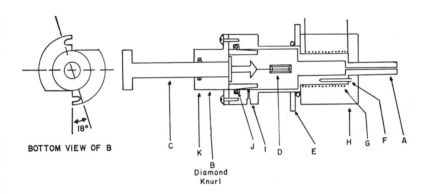

BOTTOM VIEW OF B

B
Diamond
Knurl

Fig. 10.   Pyrolysis unit for attachment to inlet port of gas chromatograph: A, $\frac{1}{4}$-in. tube to take compression tube fittings; B, quick disconnect sample loader sealed with O-ring; C, $\frac{3}{8}$-in. stainless steel push rod with stop and furnace seal; D, sample basket; E, two turns of copper tubing soldered to chamber; F, thermocouple well; G, 26-gauge insulated Nichrome wire; H, asbestos insulation; I, hole for carrier gas tube; J, O-ring and static seal; K, O-ring and dynamic seal.

The pyrolysis unit is shown in Figure 10.

b. Procedure. Weigh 5–10 mg of the sample into a micro-porcelain boat (Coors 00000) and place the boat in the basket of the syringe plunger. Assemble the pyrolysis unit and sweep with helium until all of the air has been removed. Pyrolyze the sample

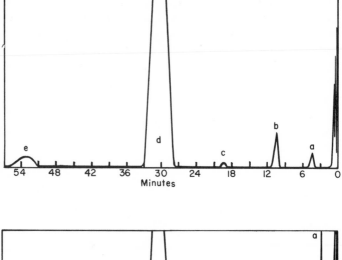

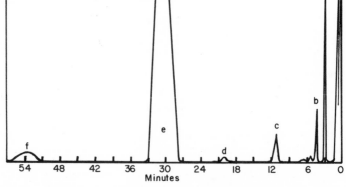

Fig. 11. Pyrolysis gas chromatograms. Top, polystyrene: *a*, benzene; *b*, toluene; *c*, ethylbenzene; *d*, styrene monomer; *e*, styrene dimer. Bottom, styrene/acrylonitrile: *a*, acrylonitrile; *b*, benzene; *c*, toluene; *d*, ethylbenzene; *e*, styrene monomer; *f*, styrene dimer.

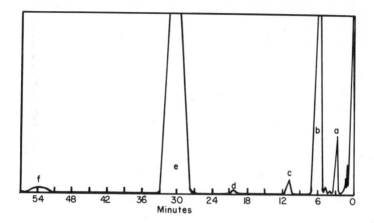

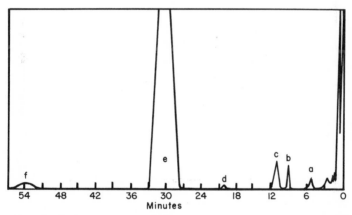

Fig. 12. Pyrolysis gas chromatograms. Top, styrene/methyl methacrylate: *a*, methanol; *b*, methyl methacrylate monomer; *c*, toluene; *d*, ethylbenzene; *e*, styrene monomer; *f*, styrene dimer. Bottom, styrene/butyl maleate: *a*, benzene; *b*, butyl alcohol; *c*, toluene; *d*, ethylbenzene; *e*, styrene monomer; *f*, styrene dimer.

at 750°C for 60 sec by inserting the boat into the furnace. Sweep the pyrolysis products into the column held at 110°C for polymers containing styrene or at 70°C for other vinyl polymers. Compare the retention times of the various peaks with the retention times of peaks obtained by the pyrolysis of known polymers. As an alter-

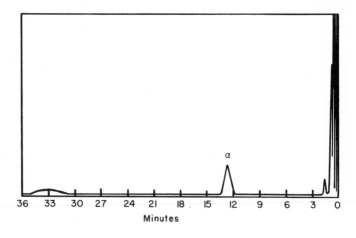

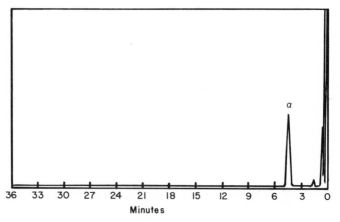

Fig. 13.   Pyrolysis gas chromatograms.   Top, polyvinyl chloride: *a*, benzene.
Bottom, polyvinylidene chloride: *a*, hydrocarbons.

native, the pyrolysis products may be trapped as they emerge from
the column and examined by infrared spectrometry.

VII. Infrared Spectrometry

Applications of infrared spectrometry discussed in this section
pertain primarily to the assignment of absorption band frequencies

to particular structural features of the polymers and to changes which occur on crystallization or orientation.

The absorption bands characteristic of the individual homopolymers and some typical copolymers are discussed below. The infrared spectrum of a copolymer is essentially the superposition of the component homopolymer spectra. Thus, interpretation of the infrared spectrum can easily lead to the identification of the components of a polymer. To determine whether the plastic is actually a copolymer or a blend of polymers, one must fractionate the plastic and examine the fractions. Although many functional groups possess a relatively high degree of invariance of frequency and intensity of absorption in different chemical environments, there are factors (change of phase, hydrogen bonding, molecular geometry, and mass of substituent groups) capable of causing frequency shifts. Examples of these effects are the shift in frequency and the appearance of new bands when comparing the copolymers of vinyl chloride with vinylidene chloride with the respective homopolymers, and the shifts resulting from stereoregularity changes in most polymers.

The detailed absorption band data of polymers under various conditions should be helpful for characterizing unknown materials, for studying changes which occur on processing, and for developing quantitative methods for analyzing complex systems. Variations of 5–10 cm^{-1} in the frequencies of specific absorption bands are observed frequently in the literature. In addition, some researchers have recorded bands not observed by other investigators. Undoubtedly, some of this variation results from differences in calibration. It appears reasonable, however, that differences in the state of the sample studied (crystallinity, orientation, or stereoregularity) are also responsible for discrepancies.

A. CALIBRATION OF INSTRUMENTS

Wavelength calibration is essential for precise work. Plyer and Peters (26) suggest the use of polystyrene for the calibration of spectrophotometers. Films are made by preparing a 10% solution of polystyrene in toluene. The solution is poured onto plate glass to form films 25–50 μ in thickness. The films are then dried at room temperature, protected from atmospheric dust. The dried films are floated from the glass with water and the infrared spectrum is

TABLE 6
Spectral Bands of Polystyrene Suitable for Wavelength Calibration

Wavelength, μ	Frequency, cm^{-1}
3.22	3104
3.24	3083
3.27	3061
3.30	3027
3.33	3003
3.42	2924
3.51	2850
5.14	1946
5.55	1802
6.24	1603
6.69	1495
9.72	1028
11.04	906

obtained from 2.5 to 16 μ.   Absorption bands suitable for calibration are given in Table 6.

B. CRYSTALLINITY, STEREOSPECIFICITY, AND ORIENTATION (27)

Infrared absorption bands result not only from vibrations of specific functional groups but also from vibrations of the entire polymer chain or chain segments.   These latter bands are therefore characteristic of the polymer as a whole and may reflect the geometrical shape of a polymer molecule.

The structure of most vinyl polymers may be approximated by a planar zigzag chain.   Accordingly, atactic, isotactic, and syndiotactic configurations are represented by the structures in Figure 14.

An individual polymer chain may traverse several distinct regions, some of which are amorphous and some of which are crystalline. The shape of the molecule in the amorphous region is probably similar to the random coil of a dilute solution, whereas the crystalline region is characterized by a highly ordered state.   Thus, spectral band differences between the amorphous and crystalline state are frequently obtained.   Transitions forbidden in the crystalline state may be allowed in the amorphous state, resulting in bands which are characteristic of the amorphous state only.   Interactions in the ran-

domly coiled chains are irregular and lead to a broadening of the
bands.   Because of the more regular intramolecular interactions of
the crystalline regions, the resulting absorption bands are usually
sharp and well defined.   Splitting of absorption bands is sometimes
observed as a result of interaction between adjacent chains of the
crystallites.   Since the area of the infrared beam or slit is much
larger than the individual amorphous or crystalline regions, absorption bands characteristic of each state are observed in the resulting
spectra.

The orientation of the crystalline and amorphous regions also
influences the intensity of the absorption bands.   Four types of
orientation are normally observed.

1. Random Orientation.   A randomly oriented sample is isotropic
and exhibits no dichroism.   Specimens for infrared spectrometric
examination prepared by mulling or by the potassium bromide pellet
technique show random orientations.   Films prepared by hot pressing, melt extrusion, or solution casting also exhibit little or no
orientation.

2. Uniaxial Orientation.   In uniaxial orientation, the long-chain
crystallographic axes (molecular chain axes) of the molecules become

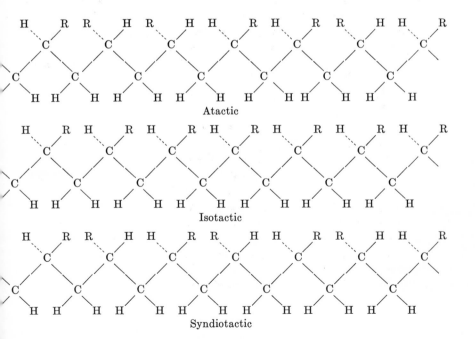

Fig. 14.   Optical stereoisomerism in vinyl polymers.

arranged parallel to each other and in one direction only.   However, all orientations of the crystallites about this axis will be probable. This type of orientation may be achieved in films by elongation either by cold drawing or hot stretching.

3. Uniplanar Orientation.   Uniplanar orientation occurs when the polymer chains are oriented parallel to the film surface.   When a specific plane of the polymer backbone or a particular crystallographic plane of a polymer crystallite is oriented parallel to the film surface, the term *selective uniplanar orientation* is sometimes used.

4. Double Orientation.   When a film has both uniaxial orientation and specific uniplanar orientation, the film is said to be doubly oriented.   Double orientation is obtained by rolling between heated rollers.

Orientation changes the microstructure of a polymer and therefore the physical properties of the film.   The effect is particularly noticeable with crystallizable polymers in which orientation of the chain molecules may be accompanied by crystallization.   Polarized infrared radiation may be used to study orientation and the accompanying changes in microstructure.   In oriented polymer films, changes in dipole moments of the molecular groups may be restricted to definite directions in relation to the chain axis.   When a beam of plane-polarized infrared radiation is passed through the film, an absorption band will be of maximum intensity, if the direction of the dipole moment change accompanying the vibration is parallel to the electric vector of the radiation.   If the electric vector of the polarized radiation is perpendicular to the vector of the vibration, the absorption band will be of minimum intensity.

Partial polarization of infrared radiation is obtained in the spectrometer itself; thus, the spectrum of an oriented sample will vary slightly, depending upon whether the draw direction of the film is parallel or perpendicular to the slit.   To obtain an unpolarized spectrum, the sample is placed so that the draw direction forms an angle of 45° with the slit.   Polarized infrared radiation is obtained by the reflection of a beam from a suitable surface.   The most commonly used polarizer consists of five or six silver chloride plates spaced about 5 mm apart and inclined at an angle of 26.5° to the beam direction.   The polarizer should be attached so that the plane of polarization makes a 45° angle with the spectrometer slit. The sample is also positioned so that the draw direction is at the

45° angle. It is preferable to rotate the draw direction of the sample from $+45°$ to $-45°$ and maintain the electric vector at $+45°$. Rotatable cells are available for this purpose. When possible, a screen should also be used in the reference beam of double-beam instruments.

The infrared spectra of samples are obtained with the vector parallel and then perpendicular to the draw direction of the sample. The absorbances of individual bands are measured for each spectrum and the dichroic ratio is calculated as follows:

$$R = A_\pi / A_\sigma$$

where A is absorbance or optical density, π is the parallel vector (sometimes denoted $\|$), and σ is the perpendicular vector (sometimes denoted $\perp$).

Errors in calculations arise from difficulty in measuring the absorbance of overlapping bands, imperfect polarization of the polarizer, and the fact that the beam is usually convergent rather than parallel when it passes through the sample.

C. General Methods of Sample Preparation

Polymer films approximately 1 mil in thickness may be examined directly. Thicker films and granules may be converted into films of suitable thickness by heat and pressure or by solution of the polymer in a suitable solvent and recasting a film from the solution. Many commercial plastics contain plasticizers or other additives which may contribute some infrared absorption characteristic of functional groups other than those of the polymer. Elimination of these interfering materials is sometimes necessary prior to characterization of the polymer.

1. Purification of Polymer

The purification of a polymer involves the dissolution of the sample in a suitable solvent and precipitation by the addition of a nonsolvent. This process may be repeated to ensure complete removal of any additives. The precipitated polymer is recovered by filtration or centrifugation.

2. Casting of Films from Solution

Add 2–2.5 g of the polymer to 50 ml of methylene chloride or other suitable low-boiling solvent contained in a 4-oz bottle. Stopper the bottle and shake the solvent–plastic mixture until the plastic is dissolved or dispersed into small gel particles. The mixture may be warmed on a steam bath to facilitate dissolution of the polymer. Spread a portion of the resultant solution uniformly over the surface of a rock-salt plate. Dry the solution on the plate slowly at room temperature and pressure to avoid the formation of bubbles in the film. After most of the solvent has been volatilized at room temperature, place the plate under an infrared heat lamp for 10–20 min to remove all traces of solvent. The resultant film should be 0.02–0.1 mm thick. The desired thickness may be obtained by building up layers successively. Polymers which are susceptible to oxidation should be dried under vacuum or in an atmosphere of nitrogen.

3. Potassium Bromide Disk Technique

The technique consists of grinding several milligrams of a solid sample with 1 g or less of finely powdered potassium bromide, placing this solid mixture in an evacuable die, and compressing the mixture under a pressure of 6,000–15,000 psig. Under this pressure, the potassium bromide fuses into a solid transparent matrix. Although potassium bromide is usually preferred, potassium iodide is used also since it flows more readily when pressed. The potassium bromide disk technique gives a representative spectrum of the sample without absorption bands contributed by the matrix, but moisture absorption bands may be present. Harvey, Stewart, and Achhammer (28) claim that exceptionally good spectra are obtained with polyvinyl chloride, presumably owing to the closeness of the respective refractive indices.

A few milligrams of sample are pulverized and intimately ground with approximately 1 g of potassium halide. Some samples which are difficult to grind may be ground with potassium bromide powder in the presence of a suitable solvent. The solvent should be completely removed before the pressing operation. Highly crosslinked resins may be powdered more efficiently by using either a mechanical vibrating ball mill or an abrasive mortar and pestle, such as agate, mullite, or Diamonite.

After the sample has been thoroughly mixed with the potassium bromide, it is pressed into a disk in an evacuable die. There are several different die designs available commercially in a variety of sizes.   One type has a split cylinder which can be taken apart for easy removal of the disk after the pressing operation.   Another presses the potassium bromide disk directly into an adapter for the spectrometer, and still another (Fig. 2) uses several types of paper rings to facilitate the removal of the disk.   The larger sized disks are much more difficult to prepare with clarity than the smaller disks, i.e., they require higher pressures.

After the disk has been pressed, it is placed in a spectrometer mounting adapter which properly positions the disk in the sample beam.   Very often reference beam attenuation will be necessary with the more opaque samples.

4. Preparation of Thin Films by Pressing

Although, the usual method for pressing thin films employs a Carver Laboratory Press or a similar press with heated platens, a very simple method has often been found to be satisfactory.

A plastic granule or section of film is placed on a microscope slide and covered with a cover glass.   On top of the cover glass is placed another microscope slide.   The assembly is heated on a hot plate which has been previously set to low, medium, or high, depending upon the melting point of the polymer.   Pressure is applied with a wooden dowel, which is moved with a circular motion until the film is thin enough for an infrared transmission spectrum.   In some cases the film may adhere to the glass; then it may be necessary to press between two pieces of Mylar.   The two pieces of Mylar, sandwiching the film, replace the cover glass.   Usually pressing in this manner yields a film which is suitable for infrared study.

D. Attenuated Total Reflectance

Attenuated total reflectance (ATR), introduced by Fahrenfort (29), is a technique whereby an infrared spectrum is obtained by reflectance of radiation from the surface of a sample instead of by transmission through the sample.   This technique will not only yield a reasonable spectrum of single films but is suitable for characterizing the exterior components of laminates.

The most useful ATR attachment is the microcell employing a multiple reflection crystal (Fig. 15).   In order to maintain internal reflection, the refractive index of the crystal must be higher than that of the sample.   The KRS-5 crystal (42% thallium bromide,

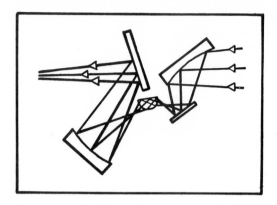

Fig. 15. Optical system of a typical attenuated total reflectance unit. The sample is in intimate contact with the base of the hemicylinder reflecting crystal.

58% thallium iodide) is recommended for most solid samples since it has a high refractive index (2.380 at 5 μ) and is tough and durable. To obtain a useful spectrum, the sample should be at least 10 μ thick and be in intimate contact with the face of the crystal. The angle of incidence of the radiation depends upon the desired depth of beam penetration into the sample. A steep angle produces a deeper pene-

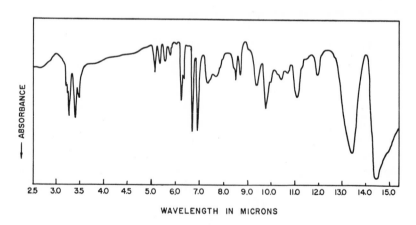

WAVELENGTH IN MICRONS

Fig. 16. Attenuated total reflectance spectrum of a polystyrene film obtained with a KRS-5 crystal.

tration than a shallow angle.   For most samples, when employing the KRS-5 crystal, the optimum angle is between 45 and 50°.   The ATR spectrum of a polystyrene film is shown in Figure 16.   The interpretation of ATR spectra is difficult since shifts in both the intensity and frequency of absorption bands, when compared with transmission spectra, frequently occur.

E. POLYVINYL CHLORIDE

1. Preparation of Sample

Polyvinyl chloride may be cast as a film on sodium chloride or potassium bromide plates from hot tetrahydrofuran (THF) or hot o-dichlorobenzene solution.   The tetrahydrofuran should be freshly distilled.   Tetrahydrofuran may be purified by refluxing over potassium hydroxide for 3 hr.   After refluxing, the solvent is distilled and stored in a brown glass bottle.   If unstabilized, the solvent should not be used after more than 2 weeks of storage. THF which has been allowed to age contains impurities (carbonyl compounds absorbing at about 5.8 μ) which are difficult to remove from films.   Residual THF in the film is shown by absorption bands at approximately 9.35 and 11 μ.   Grisenthwaite (30) suggests heating the films under high vacuum at 80°C for approximately 8 hr. An alternate procedure is to extract the films with a low-boiling solvent such as carbon disulfide.   The residual carbon disulfide may be removed by vacuum treatment at room temperature or under a heat lamp.   Films prepared from polymer solutions in unknown solvent systems should be carefully checked for traces of residual solvent.   Films are cast on salt plates and dried at both a low temperature (25°C) and at a moderately high temperature (80–90°C). If solvent retention is occurring, the sample dried at 80°C will exhibit weak absorption bands in some areas where the sample dried at 25°C shows medium absorption.   Drying is then continued until the weak bands completely disappear.   High-density polyvinyl chloride, which is only sparingly soluble in THF, may be dissolved in cyclohexanone at 100°C.   The mixture must be kept under a blanket of nitrogen during dissolution to avoid dehydrochlorination.

2. Frequency Assignment of Absorption Bands

The infrared spectrum of polyvinyl chloride in the 625–4000 cm⁻¹ region is shown in Figure 17a.   Assignments for the principal

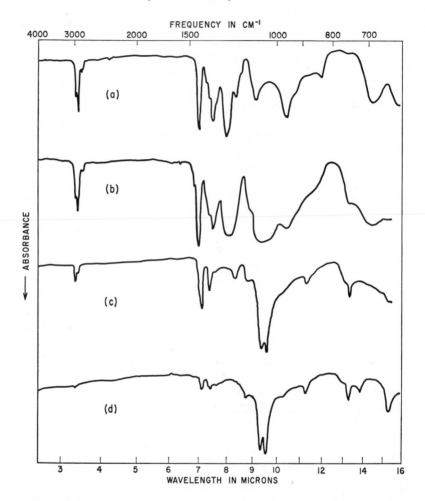

Fig. 17.   Infrared spectra of (a) polyvinyl chloride; (b) copolymer containing 60% vinyl chloride and 40% vinylidene chloride; (c) copolymer containing 10% vinyl chloride and 90% vinylidene chloride; and (d) polyvinylidene chloride.

TABLE 7
Principal Spectral Band Assignments of Polyvinyl Chloride[a]

Frequency, cm^{-1}	Wavelength, μ	Polarization	Intensity	Probable group assignment
2968	3.37	$\perp$	mw	CH
2930	3.41	$\perp$	w	CH$_2$
2914	3.43	$\perp$	ms	CH$_2$
2849	3.51	$\perp$	w	CH$_2$
2820	3.55	$\perp$	w	CH
1434	6.97	$\perp$	s	CH$_2$
1424	7.02	$\perp$	s	CH$_2$
1380	7.25	$\parallel$	vw	CH$_2$
1352	7.40	$\perp$	w	CH
1333	7.50	$\perp$	ms	CH
1254	7.97	$\perp$	s	CH
1230	8.13	$\parallel$	vw	CH
1197	8.35	$\parallel$	w	CH + CCl
1125	8.89	$\parallel$	vw	Skeletal
1095	9.13	$\perp$	m	Skeletal
960	10.42	$\perp$	ms	CH
833	12.00	$\parallel$	w	CH
693	14.43	$\perp$	m	CCl
642	15.58	$\perp$	m	CCl
615	16.26	$\perp$	s	CCl
604	16.58	$\perp$	s	CCl

[a] Key: s, strong; ms, medium strong; m, medium; mw, medium weak; w, weak; vw, very weak.

bands are given in Table 7. The assignments have been based primarily on the work of Krimm and Liang (31), which has been corroborated by more recent investigators (32,33).

3. Crystallinity Effects

Crystallinity in polyvinyl chloride is associated with a high syndiotactic structure while atacticity or low syndiotacticity is associated with amorphous polymers.

Kawasaki and associates (32) compared the intensity of various absorption bands of polyvinyl chloride with that of the >CH$_2$

absorbance at 2914 cm⁻¹. A linear increase in the intensity ratio
of several bands was correlated with an increase in specific gravity;
thus, these bands are assumed to be associated with crystallinity.
The most pronounced ratio changes were observed in absorption
bands at 1424, 1333, 1254, 1230, 1095, and 955 cm⁻¹. Apparently
there is a broad band or doublet in the region of 960 cm⁻¹. The
exact frequency of the maximum (or maxima) apparently shifts
with orientation, which may explain why Krimm failed to observe
the 955 cm⁻¹ absorption band. The 642 and 604 cm⁻¹ bands have
also been assigned to the crystalline region (34). Absorption bands
representative of the amorphous or disoriented regions are observed
at 1434 cm⁻¹ (shoulder on 1424 cm⁻¹ band), 1240 cm⁻¹ (shoulder on
1254 cm⁻¹ band), as a shoulder (960 cm⁻¹) on the 955 cm⁻¹ absorption band, and at 693 cm⁻¹. These bands decrease and eventually
disappear with increasing crystallinity.

4. Orientation and Polarization

The polarization of polyvinyl chloride (Fig. 18a) has been
reported by a number of investigators (31,34,35). Inversion of

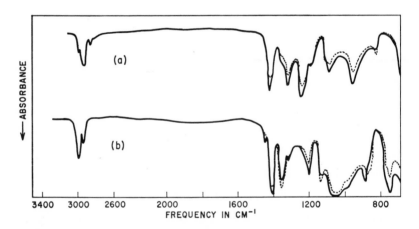

Fig. 18.   Infrared spectra of (a) oriented polyvinyl chloride and (b) oriented
copolymer containing 85% vinylidene chloride and 15% vinyl chloride. ———,
Radiation with electric vector perpendicular to direction of orientation; – – –,
radiation with electric vector parallel to direction of orientation.

dichroism of several absorption bands occurs with drawing and has been interpreted in terms of orientation (34). Films were extended to draw ratios up to 10 (length of drawn film to length of undrawn film) in an air bath at 180°C and in a glycerol bath at 110°C. The resulting observations of a frequency shift and an inversion of the dichroism of certain bands with extension should provide a valuable tool for the study of films and the correlation of physical properties with structure.

The 604 cm⁻¹ band which is representative of crystallinity (high syndiotacticity) exhibits parallel dichroism at low extension. The amount of parallel dichroism increases up to a draw ratio of 2.5. Above a draw ratio of 5, the band inverts and exhibits perpendicular dichroism. Similar inversion of dichroism is also observed for the 955 cm⁻¹ band and for the 1254 cm⁻¹ band. Concurrent with inversion of the 955 cm⁻¹ band is a shift in frequency of the perpendicular maximum towards a higher frequency at high draw ratios. A frequency shift is also observed with the amorphous-sensitive 693 cm⁻¹ band. At a low draw ratio, both the parallel and perpendicular band frequencies are essentially the same. On drawing, the parallel absorption maximum remains at 693 cm⁻¹ while the perpendicular absorption maximum shifts to about 680 cm⁻¹.

The amorphous bands generally show perpendicular dichroism and very weak parallel dichroism even at small draw ratios. This is consistent with the ease of orientation of the molecular chains in the amorphous regions along the draw direction. A change in the form and internal structure of the crystallites must occur to explain the difference of the peak positions for parallel and perpendicular absorption. A reasonable explanation is that the crystallites exist as laterally oriented domains along the length of the chain (molecular chain perpendicular to the long axis of the crystallite). At small draw ratios, the crystallites are aligned with their long axis parallel to the draw direction while the molecular chains in the crystallites are nearly perpendicular to this direction. With an increase in extension of the film, the packing of the chains in the crystallites becomes loose (alignment is destroyed) and the molecular chains become oriented parallel to the draw direction. After drawing, recrystallization can occur; however, the new crystallites may differ somewhat from the previous crystallites, resulting in frequency shifts in the absorption maxima.

5. *Infrared Spectrometric Analysis of Polyvinyl Chloride* (36)

This general method provides for the infrared identification and determination of resins, plasticizers, stabilizers, and fillers in polyvinyl chloride.

The sample is solvent-extracted to separate the plasticizer from the compound. The resin is dissolved from the remaining compound and the inorganic fillers and stabilizers are separated by centrifuging. Organometallic or organic stabilizers, if present, may be partially or wholly separated with either the plasticizer or resin components and should be considered when examining these components.

Polyvinyl chloride samples in the form of powder or sheets 0.002–0.005 cm thick may be extracted directly. Granules should be reduced in size by buffing on a coarse grinding wheel or grinding in a suitable mill (Section IV).

a. Identification and Determination of Plasticizers. Weigh approximately 1 g ± 0.2 mg of sample of fine particle size into a 27 × 100-mm paper extraction thimble. Place the thimble in a Soxhlet apparatus fitted with a tared 150-ml flask, and extract with 120 ml of anyhdrous ethyl ether for 6 hr. Remove the tared flask containing the solvent and the extracted plasticizers and heat gently on a steam bath to boil off the ethyl ether. Place the flask in an evacuated desiccator for a minimum of 1 hr to remove the last traces of ethyl ether. Weigh the flask containing the extracted plasticizers.

$$\% \text{ Total plasticizers} = \frac{\text{Weight of extracted plasticizers}}{\text{Weight of sample}} \times 100$$

The extracted plasticizers may be run on the infrared spectrometer as liquid films for identification or in carbon disulfide solution for quantitative determinations. A demountable cell with sodium chloride windows and a 0.025-mm spacer usually suffices to give a strong plasticizer spectrum. A capillary film is satisfactory for the ester plasticizers. Scan the spectrum from 2.5 to 15 μ. The plasticizers in the sample may be identified by reference to a collection of plasticizer spectra (37).

It is impossible to specify a single procedure that quantitatively determines all plasticizers with equal precision and accuracy. The following procedure is useful, however, for a number of plasticizers and their combinations, particularly if dioctyl phthalate or tricresyl phosphate is the primary plasticizer. The user should decide whether the efficiency, precision, and accuracy of the procedure are satisfactory for a specific combination of plasticizers to be analyzed.

Weigh 60 ± 0.2 mg of extracted plasticizer into a 25-ml erlenmeyer flask equipped with a glass stopper, add 20.0 ± 0.2 ml of carbon disulfide and dissolve the plasticizers. Run the resultant 3.0 mg/ml of plasticizer solution on the infrared spectrometer in a 1.0-mm liquid cell. Use a compensating 1.0-mm liquid cell, filled with carbon disulfide, in the reference beam. The dioctyl phthalate bands at 5.80 μ, 7.87 μ, and 8.92 μ, and the tricresyl phosphate band at 8.40 μ are satisfactory. The dioctyl phthalate band chosen will depend in part on possible secondary plasticizer interferences. Choice of bands for other plasticizers is left to the discretion of the user. At the analytical band wavelength chosen, measure the absorbances for the sample spectrum by the base line technique.

Prepare plasticizer standards in carbon disulfide covering the 3.0–0.5 mg/ml range for each plasticizer. These standard plasticizer solutions should be run under identical conditions as the samples to obtain the net absorbances of the components at a series of concentrations. Plot analytical curves of net absorbance versus concentration in milligrams per milliliter for each component.

Use the net absorbance of a specific plasticizer in conjunction with the appropriate analytical curve to determine the concentration in milligrams per milliliter.

$$\% \text{ Specific plasticizer} = (AB/3W) \times 100$$

where A is the concentration of plasticizer, in milligrams per milliliter, B is the total weight of extracted plasticizers, in milligrams, and W is the weight of sample, in milligrams.

In a two-plasticizer system, one plasticizer may be determined and the other calculated by difference.

The plasticizers may also be determined directly in a sample by extracting a known weight of the sample with carbon disulfide, diluting the extract to a known volume, and scanning as directed above. This technique is rapid and avoids possible losses during the evaporation of the solvent and the transfers required in the above method.

b. Identification and Determination of Stabilizers and Fillers.
Transfer the resin, stabilizers, and fillers from the extraction thimble
to a 50-ml beaker.   Add 20 ml of tetrachloroethane and heat the
sample gently until the resin has dissolved.   Wash the contents of
the beaker quantitatively into a tared 50-ml centrifuge tube with
20 ml of stabilized tetrahydrofuran (Section VII-E-1), swirl to mix,
and centrifuge for 30 min.   Decant the solution containing the
remaining resin.   Repeat the operation.   Dry the tared centrifuge
tube, containing the stabilizer and filler, at 110°C for 1 hr, cool, and
weigh.

% Inorganic stabilizer and filler =

$$\frac{\text{Weight of stabilizer and filler}}{\text{Weight of sample}} \times 100$$

The stabilizers and fillers may be identified by running them on the
infrared spectrometer as a Nujol mull or potassium bromide disk.
Prepare the Nujol mull by adding a few milligrams of powder to a
drop of Nujol in a small mortar and grind.   Run the resultant mull
as a film between two sodium chloride plates held in a demountable
cell mount.   Prepare the potassium bromide disk by adding approximately 1 mg of powder to 600 mg of dry potassium bromide powder
and mixing for 1 min in a vibrator mixer.   Place the mixture in a
0.5-in. diameter mold assembly and hold under vacuum for 3 min.
Press while still under vacuum for 3 min at a minimum pressure of
20,000 psig.   Higher pressures will produce more stable disks.
Place the resultant disk in a holder and run on the infrared spectrometer.   The stabilizer and filler components may be identified in
many cases by comparison with reference spectra (38).

Estimation of stabilizers and fillers may be accomplished by
the potassium bromide disk technique.   Weigh approximately
1 ± 0.2 mg of stabilizer and filler powder and add to approximately
600 ± 0.2 mg of dry potassium bromide powder, which has also
been weighed.   Prepare the potassium bromide disk and run on an
infrared spectrometer.   Use the base line techniques to determine
absorbances of the bands of interest.   The net absorbances are those
that would result if the stabilizers and fillers were exactly 1 mg in
600 mg of potassium bromide powder.   The percentage of component in the stabilizer fraction may be calculated from analytical
curves prepared from pure compounds.   The following bands are

usable in many cases: basic lead carbonate, 7.09 μ; calcined clay, 9.30 μ; calcium carbonate, 11.40 μ; antimony oxide, 13.50 μ; basic lead sulfate, 8.85 μ; dibasic lead phthalate, 6.51 μ.

c. Identification of Resin. Evaporate a few milliliters of the tetrachloroethane–tetrahydrofuran solution, a few drops at a time, on a microscope slide. Gentle heating will accelerate the drying. When the resultant film is dry, peel it from the microscope slide. Dry the film in a vacuum desiccator or vacuum oven to reduce solvent spectral interferences. Mount the film in the infrared spectrophotometer and record its spectrum from 2.5 to 15 μ. Alternately, a film may be cast directly on a sodium chloride plate.

The resin may be identified by its overall infrared spectrum. If the resin is a copolymer of vinyl chloride and vinyl acetate, a carbonyl band will be present at 5.74 μ; and, if the amount of acetate is greater than approximately 5%, a band attributed to the acetate group is present at 9.80 μ. Carbonyl bands in the spectrum of the resin may also arise from copolymers other than acetate (for example, acrylate), incomplete extraction of certain polymeric ester plasticizers, oxidation of the resin, esterification of the resin by certain compounding ingredients, or tetrahydrofuran oxidation products. Frequently, the carbonyl bands arising from oxidation are at lower frequencies than those of copolymers or caused by esterification of the resin.

F. Vinyl Chloride Copolymers

Copolymers of vinyl chloride with vinyl acetate have been manufactured commercially for many years. Recently, other comonomers such as acrylates, methacrylates, maleates, and acrylonitrile have been used for copolymerization with vinyl chloride. Most of the bands associated with polyvinyl chloride are present in the copolymer spectra with the exception of the weak CH band at 1197 cm^{-1} (8.35 μ).

1. Vinyl Chloride/Vinyl Acetate

Vinyl acetate copolymers (Fig. 19a) exhibit absorption bands contributed by the comonomeric vinyl acetate at 1742 cm^{-1} (5.74 μ), 1355 cm^{-1} (7.38 μ), and at 1025 cm^{-1} (9.76 μ). The usual mode characteristic of acetates at 1245 cm^{-1} is not observed as a distinct band, but probably accounts for the increased breadth of the

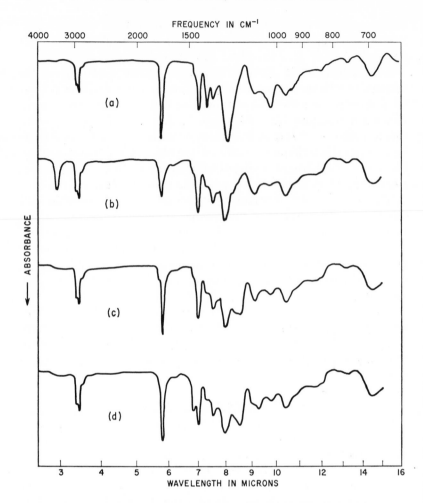

Fig. 19. Infrared spectra of (a) copolymer containing 87% vinyl chloride and 13% vinyl acetate; (b) copolymer containing 91% vinyl chloride, 3% vinyl acetate, and 6% vinyl alcohol; (c) copolymer containing 88% vinyl chloride and 12% methyl acrylate; and (d) copolymer of vinyl chloride and a maleate ester.

1250 cm^{-1} (8.00 μ) CH band of the vinyl chloride. Comonomeric vinyl acetate can be distinguished from methyl methacrylate and dioctyl maleates by the absence of an absorption band in the 1355 cm^{-1} (7.38 μ) region and from methyl acrylate, maleic acid, and dibutyl fumarate by the absence of absorption in the 1025 cm^{-1} (9.76 μ) region. A hydroxyl band at 3340 cm^{-1} (2.99 μ) is indicative of a vinyl-alcohol-modified or partially hydrolyzed vinyl chloride/vinyl acetate copolymer (Fig. 19b).

2. Vinyl Chloride Copolymers with Acrylates or Methacrylates

The acrylates (Fig. 19c) can be distinguished from the maleate and fumarate copolymers by the presence of a strong absorption band at 1199 cm^{-1} (8.34 μ). The methacrylate copolymers are distinguishable from the other ester copolymers (with the exception of vinyl acetate) by the absence of a band near 1170 cm^{-1} (8.55 μ) and from vinyl acetate by the presence of a band at 1193 cm^{-1} (8.38 μ).

3. Vinyl Chloride Copolymers with Maleic Acid and Maleates

The C=O stretching mode of the maleic acid comonomer occurs at a lower frequency, 1689 cm^{-1} (5.92 μ), than the corresponding mode of the esters (Fig. 19d). Ordinarily, fumarates and maleates can be identified from their infrared spectra. When present in a polymer as a minor comonomeric species, however, the normal variations are not detectable. Differences which do occur can be attributed to the chain length of the alcohol component. Fumarates and maleates may be distinguished from vinyl acetate copolymers by the presence of a band at 1170 cm^{-1} (8.55 μ) and from the acrylate copolymers by the absence of a band at 1199 cm^{-1} (8.34 μ).

4. Vinyl Chloride/Acrylonitrile

Acrylonitrile copolymers are easily characterized by a sharp band at 2257 cm^{-1} (4.43 μ) resulting from the nitrile group ($-$C$\equiv$N). Since the nitrile group absorbs only weakly, abnormally thick films are required to show concentrations below about 5%.

5. Analysis of Copolymers

Analytical curves for calculating the comonomer composition of the various copolymers of vinyl chloride and vinyl esters may be prepared from solutions of the individual vinyl ester homopolymers.

Alternately, the analytical curve for the acrylate, methacrylate, or maleate copolymers may be prepared from copolymers previously analyzed by the alkoxyl gas chromatographic technique (Section X-K-2). A solution of 4 g of copolymer in 20 ml of purified tetrahydrofuran is usually a satisfactory concentration. The solution in a 0.3-mm cell is scanned from 4 to 7 μ differentially against tetrahydrofuran in another 0.3-mm cell. The absorbance of the carbonyl band is measured by the base line technique using a line drawn between the shoulders at approximately 5.5 and 6.2 μ.

Standards for preparing analytical curves for the infrared spectrometric analysis of copolymers of vinyl chloride and acrylonitrile may be obtained from copolymers which have been analyzed for nitrogen (calculated as acrylonitrile) by the Kjeldahl procedure (Section X-D).

G. Polyvinylidene Chloride

1. Preparation of Sample

Although polyvinylidene chloride is soluble in hot o-dichlorobenzene, samples for infrared study are frequently prepared in the form of potassium bromide pellets. Correct grinding is the primary concern in preparation of satisfactory pellets. Too little grinding results in excessively large particle size, causing poorly defined spectra with inconsistent absorption intensities. Too much grinding yields an increased surface area of the potassium bromide, resulting in excessive moisture pickup. Thus, the sample should be ground to a small particle size prior to adding the potassium bromide powder. Clear pellets are then obtained by a short but thorough mixing of approximately 5 mg of the sample and 500 mg of potassium bromide. The entire mixture is ground thoroughly and pressed into a pellet at about 10 tons/cm² under reduced pressure.

2. Frequency Assignment of Absorption Bands

The infrared spectrum of polyvinylidene chloride in the 625–4000 cm^{-1} region is shown in Figure 17d. Assignments for the principal bands as made by Krimm and Liang (31) are given in Table 8. Narita, Ichinohe, and Enomoto (39) report that they obtained only two bands in the region of the CH$_2$ stretching modes (2966 and 3010 cm^{-1}). They frequently used a vinylidene chloride/

TABLE 8
Principal Spectral Band Assignments of Polyvinylidene Chloride[a]

Frequency, cm^{-1}	Wavelength, μ	Polari- zation	Intensity	Probable group assignment
2990	3.34	$\perp$	w	CH_2
2948	3.39	$\perp$	vw	CH_2
2930	3.41	$\perp$	mw	CH_2
2850	3.51	$\perp$	w	CH_2
1460	6.85	$\perp$	vw	CH_2
1407	7.11	$\perp$	m	CH_2
1360	7.35	$\parallel$	m	CH_2
1325	7.55	$\perp$	w	CH_2
1142	8.76	$\parallel$	m	Skeletal
1070	9.35	$\perp$	vs	Skeletal + CCl_2
1046	9.56	$\perp$	vs	Skeletal + CCl_2
980	10.20	$\perp$	w	CCl_2
887	11.27	$\perp$	m	Skeletal
778	12.85	$\perp$	vw	CH_2
754	13.26	$\perp$	m	CCl_2
657	15.22	$\perp$	m	CCl_2
603	16.58	$\perp$	s	CCl_2

[a] Key: vs, very strong; s, strong; m, medium; mw, medium weak; w, weak; vw, very weak.

vinyl chloride copolymer instead of polyvinylidene chloride, however, since it was difficult to prepare films of the latter. It seems possible that the CH_2 stretching mode assignments of Narita might have been made from the copolymer spectrum.

3. Crystallinity Effects

The intensities of the absorption bands at 887 and 754 cm^{-1} become very weak in molten samples. The same phenomenon occurs with vinylidene chloride/vinyl chloride copolymers. The intensities of the bands increase when the amorphous film is stretched. X-ray diffraction studies of polyvinylidene chloride show that the polymer becomes amorphous when it is molten and quenched but recrystallizes when kept at room temperature. Thus, the 887 and 754 cm^{-1} absorption bands are apparently crystallization-sensitive.

No orientation can be obtained with polyvinylidene chloride films because of the brittleness of the films.   Since most of the bands are also present in the copolymers of vinylidene chloride with vinyl chloride, the polarization of the polyvinylidene chloride bands is obtained from the copolymer spectrum (Fig. 18*b*).

H. Vinylidene Chloride Copolymers

Typical vinylidene chloride copolymers are those containing acrylonitrile, 2-ethylhexyl acrylate or ethyl acrylate.   Infrared spectra for these copolymers are shown in Figure 20*a*, *b*, and *c*, respectively.

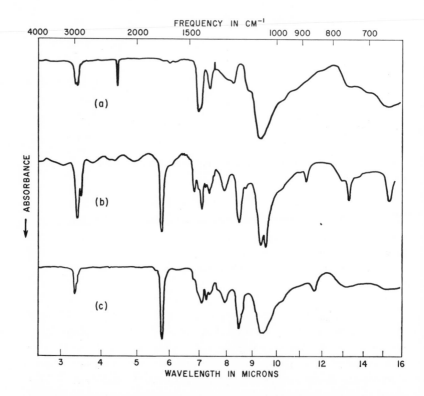

Fig. 20.   Infrared spectra of (*a*) copolymer containing 80% vinylidene chloride and 20% acrylonitrile; (*b*) copolymer containing 80% vinylidene chloride and 20% 2-ethylhexyl acrylate; and (*c*) copolymer containing 85% vinylidene chloride and 15% ethyl acrylate.

The comonomeric species can be identified from the absorption bands characteristic of the individual homopolymer. The techniques described for the analysis of vinyl chloride copolymers are applicable to vinylidene chloride copolymers, which are available in resin form or as a latex. Latexes may be cast directly on salt plates for infrared examination, even though sodium chloride is very susceptible to attack by moisture. Fogging of the plate will occur; however, it is usually not sufficient to interfere with the resulting spectrum. Cleaning and polishing of the plates is necessary after each use. Silver chloride plates are resistant to moisture and are recommended for the casting of latex films. Alternately, the latex may be cast on glass plates and the dried resin dissolved in a solvent such as tetrahydrofuran. Films are then cast on sodium chloride plates from solution.

I. Copolymers of Vinylidene Chloride and Vinyl Chloride

1. Preparation of Sample

Copolymers of vinylidene chloride and vinyl chloride may be prepared for infrared spectrometric examination according to the directions for treating polyvinyl chloride (Section VII-E-1).

2. Frequency Assignment of Absorption Bands

Infrared spectra for a 40/60 ratio of vinylidene chloride/vinyl chloride copolymer and a 90/10 ratio vinylidene chloride/vinyl chloride copolymer are shown in Figure 17b and c. The principal spectral band assignments for typical copolymers are given in Table 9 (31,40,41). Several significant differences between the spectra of the copolymer and of the homopolymer are apparent. Only two CH_2 stretching frequencies are observed in the spectrum of copolymers. Although these frequencies (2990 and 2930 cm^{-1}) are the same as the medium intensity bands in polyvinylidene chloride, the intensity ratio is reversed. This seems to suggest a different local environment for the CH_2 groups.

The CH_2 bending frequency maximum which appears at approximately 1424 cm^{-1} in polyvinyl chloride shifts continuously to a higher frequency with increasing vinylidene chloride content of the copolymers. In polyvinylidene chloride the maximum appears at 1407 cm^{-1}. A characteristic copolymer band at about 1200 cm^{-1}

TABLE 9

Spectral Band Shifts in Copolymers of Vinylidene Chloride
and Vinyl Chloride[a]

80% Vinylidene chloride/ 20% vinyl chloride			50% Vinylidene chloride/ 50% vinyl chloride			20% Vinylidene chloride/ 80% vinyl chloride		
Frequency, cm^{-1}	Wavelength, μ	Intensity	Frequency, cm^{-1}	Wavelength, μ	Intensity	Frequency, cm^{-1}	Wavelength, μ	Intensity
1409	7.10	s	1420	7.04	s	1428	7.00	s
1355	7.38	m	1347	7.42	m	1349	7.41	w
1328	7.53	w	1326	7.54	w	1328	7.53	m
			1237	8.08	w	1255	7.97	m
1205	8.30	m	1205	8.30	m	1205	8.30	m
1197	8.35	vw						
1140	8.77	w	1132	8.83	w	1124	8.90	sh
1070	9.35	vs	1070	9.35	vs	1089	9.18	m
1048	9.54	vs	1049	9.53	s			
987	10.13	sh	997	10.03	sh	960	10.42	w
884	11.31	m	885	11.30	sh			
864	11.57	sh	966	10.35	sh	941	10.63	sh
753	13.28	s	753	13.28	w	745	13.49	sh
						690	14.49	m
657	15.22	s	657	15.22	m			
			620	16.13	sh	635	15.75	s

[a] Key: vs, very strong; s, strong; m, medium; w, weak; vw, very weak; sh, shoulder.

is assigned to the CH bending mode of the vinyl chloride portion of the copolymer chain. The band appears initially at 1197 cm^{-1}, but as the vinylidene chloride component decreases, the frequency presumably shifts to about 1205 cm^{-1}. The intensity of this band increases up to a copolymer ratio of approximately 50:50. With increasing vinyl chloride content, additional bands appear successively at 1235 up to 1245 cm^{-1} and at 1247 up to 1255 cm^{-1}. This frequency shift occurs because the CH deformation mode of vinyl chloride is dependent upon its environment (effect of neighboring groups). Enomoto (40) has assigned the above frequencies to the following structural arrangements.

1197 cm^{-1}:

$$-CH_2-CCl_2-CH_2-\overset{*}{C}HCl-CH_2-CCl_2-$$

1235 cm^{-1}:

$$-CH_2-CCl_2-CH_2-\overset{*}{C}HCl-CH_2-CHCl-$$

1247 cm^{-1}:

$$-CH_2-CHCl-CH_2-\overset{*}{C}HCl-CH_2-CHCl-$$

The doublet at 1070 and 1048 cm^{-1} in the high vinylidene chloride copolymers loses its identity when the vinyl chloride concentration reaches approximately 50%. At this concentration a single peak with the maximum at about 1070 cm^{-1} and a shoulder at 1049 cm^{-1} replaces the above doublet. Additional changes are observed in the bands at approximately 885 and 753 cm^{-1}. The intensities of these bands become gradually weaker as the vinyl chloride content increases, almost disappearing at a vinyl chloride concentration of approximately 50%.

Several infrared procedures have been advanced for determining the vinyl chloride/vinylidene chloride ratio of copolymers. None of these yields satisfactory results, probably because of variations not only in intensity but also in the frequency of absorption bands with changes in copolymer composition. In addition, the sequence of monomers also affects the frequency of the absorption bands.

3. Crystallinity

Crystalline-sensitive bands in high vinylidene chloride content copolymers occur at approximately 1355, 1048, 884, and 753 cm^{-1}. The absorbance of these bands increases as the ratio of crystalline volume to amorphous volume (density) increases (39). Rector (42) has shown that the 1070 cm^{-1} band is stronger than that at 1048 cm^{-1} in the spectrum of unoriented samples, whereas the intensities are reversed in oriented samples. The ratio of these bands therefore yields a relative value of the degree of crystallinity.

J. POLYVINYL ALCOHOL

1. Preparation of Samples

Films may be cast from a 10% solution in water. The film must be dried at an elevated temperature (preferably in a vacuum oven at 70°C) to remove traces of water.

2. Frequency Assignment of Absorption Bands

The frequency assignments of the principal absorption bands (Fig. 21a) are given in Table 10 (43). The polarization spectrum of oriented polyvinyl alcohol film is shown in Figure 22 (43,44).

The absorption of the hydroxyl groups at 3340 cm⁻¹, together with the broad 640 cm⁻¹ band, indicates hydrogen bonding. No absorption is observed in the region where free or unbonded hydroxyl absorbs (3500–3700 cm⁻¹). Absorption bands are observed at 1710 and 1265 cm⁻¹ in the spectra of some polyvinyl alcohol films (Fig. 21b). Acid hydrolysis reduces or removes these absorption bands (45). They have, therefore, been attributed to residual vinyl ace-

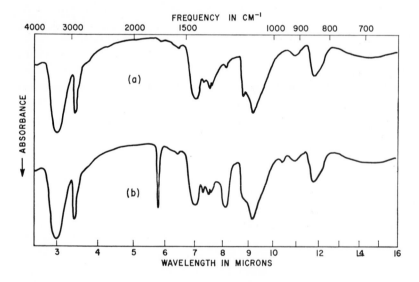

Fig. 21. Infrared spectra of (a) unoriented polyvinyl alcohol; (b) copolymer containing 87% vinyl alcohol and 13% vinyl acetate.

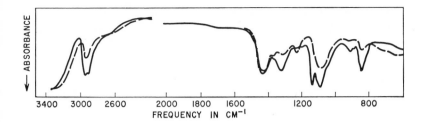

Fig. 22. Polarized infrared spectra of oriented polyvinyl alcohol: ———,
Radiation with electric vector perpendicular to direction of orientation; – – –,
radiation with electric vector parallel to direction of orientation.

TABLE 10
Principal Spectral Band Assignments of Polyvinyl Alcohol[a]

Frequency, cm^{-1}	Wavelength, μ	Polarization	Intensity	Probable group assignment
3340	2.99	⊥	vs	OH
2942	3.40	⊥	s	CH$_2$
2910	3.44	⊥	s	CH$_2$
2840	3.52	⊥	sh	CH
1446	6.92	‖	s	CH$_2$
1430	6.99	⊥	s	CH$_2$
1376	7.27		w	CH + OH
1320–1326	7.58–7.54	⊥	s	CH + OH
1235	8.10	‖	w	CH
1141	8.77	⊥	m	CO
1085–1096	9.22–9.12	⊥	s	CO
1040	9.62	⊥	sh	CC + CO
915	10.93	⊥	w	CC
850	11.77	⊥	m	CH$_2$
825	12.12	⊥	sh	CH$_2$
610–640	16.34–15.63	⊥, ‖	m, broad	OH

[a] Key: vs, very strong; s, strong; m, medium; w, weak; sh, shoulder.

tate groups.   A band is frequently observed at 1650 cm^{-1} which is attributed to the presence of water.   Although vacuum drying at room temperature does not reduce the intensity of this band, it does disappear when the film is dried at 65–100°C.   A relatively intense band is sometimes observed at 1595 cm^{-1}.   Krimm (43) attributes this absorption to a C=O group (possibly a β-diketone) within the chain.

3. Crystallinity Effects

A weak band is also apparent at 1141 cm^{-1}.   When the polymer film is heated, the intensity of the band increases.   When heated above 200°C, the band becomes weaker again.   Krimm (43) suggests that this band is indicative of a C—O— linkage and that it results from a thermal splitting off of water with resultant crosslinking.   The ether linkage could be an intermediate step in the formation of double bonds.   Tadokoro (44), however, shows a correlation between the intensity of the 1141 cm^{-1} band and density and concludes that the band is a crystalline-sensitive CO mode.   Nagai and associates and Nishino (47) consider it to be associated with the formation of a hydrogen bond.   Fujii (46) examined the infrared spectra of polyvinyl alcohol prepared under different conditions and concluded that the polymer exists in both syndiotactic and isotactic forms.   The isotactic polymer is rather exceptional in that it is less crystallizable than the syndiotactic polymer.   Murahoshi (48) found that the OH stretching band, normally at 3340 cm^{-1}, shifts towards higher frequencies in the isotactic spectrum and exhibits parallel dichroism.   This suggests the formation of intramolecular hydrogen bonding (instead of intermolecular bonding), which possibly explains the low degree of crystallizability.   The 915 cm^{-1} absorption band is not observed in the spectrum of the isotactic polymer.   Fujii (46) proposed that the degree of syndiotacticity can be calculated from the ratio of the absorbances at 915 and 850 cm^{-1}.

K. POLYSTYRENE

1. Preparation of Sample

Until recently polystyrenes have been prepared only in the amorphous (atactic) form.   Isotactic polystyrene can now be prepared by using a catalyst of the Ziegler type.   The atactic polymer may

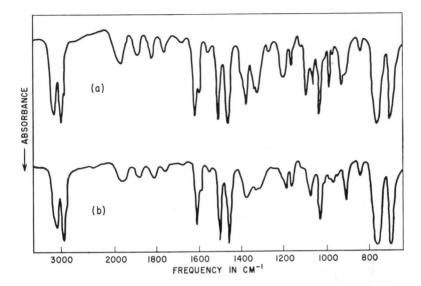

Fig. 23. Infrared spectra of (a) isotactic polystyrene and (b) atactic polystyrene.

be separated from the isotactic form by extraction with cold methyl
ethyl ketone.   Samples of either form may be prepared for infrared
examination by casting films from aromatic or chlorinated hydrocarbon solvents onto salt plates.

2. Frequency Assignment of Absorption Bands

The infrared spectra of atactic and isotactic polystyrenes are
shown in Figure 23.   The principal absorption bands and the most
probable group assignments for the atactic polymer are listed in
Table 11 (49).   Principal bands for the isotactic polymer are shown
in Table 12 (50).

3. Crystallinity Effects

The main differences between the crystalline and amorphous
forms are observed in absorption bands at 1080, 1048, and 920 cm⁻¹
in the spectrum of the crystalline form which are not present in
the spectrum of the amorphous polymer.   Other differences are
observed in minor frequency shifts of some of the bands.

TABLE 11
Principal Spectral Band Assignments of Atactic Polystyrene[a]

Frequency, cm^{-1}	Wavelength, μ	Intensity	Probable group assignment
3083	3.24	mw	
3061	3.27	s	Aromatic
3029	3.30	s	CH stretching
3003	3.33	w	
2923	3.42	vs	Assymmetric CH_2 stretching
2851	3.51	m	Symmetric CH_2 stretching
1945	5.14	w	
1875	5.33	w	
1800	5.56	w	Summation bands: typical of monosubstituted benzene
1745	5.73	w	
1675	5.97	w	
1602	6.24	m	Ring C=C
1585	6.31	w	Ring C=C
1543	6.48	w	
1493	6.70	s	Ring C=C
1450	6.90	vs	CH_2 deformation + ring C=C
1376	7.27	m	CH_2
1328	7.53	w	Ring
1310	7.63	w	Ring? CH_2 twist?
1182	8.46	m	Ring
1154	8.67	m	Ring
1070	9.35	m	Ring
1027	9.74	s	Ring
1005	9.95	w	Ring
982	10.18	w	Ring
965	10.36	w	Ring
945	10.58	w	Ring
906	11.04	m	Out-of-plane CH deformation
842	11.88	w	Ring
760	13.16	vs	In-phase, out-of-plane CH deformation
700	14.29	vs	Ring puckering

[a] Key: vs, very strong; s, strong; m, medium; w, weak; sh, shoulder.

TABLE 12
Principal Spectral Band Assignments of Isotactic Polystyrene[a]

Frequency, cm⁻¹	Wave length, μ	Polarization	Intensity	Probable group assignment
3083	3.24		mw ⎫	Aromatic
3061	3.27	⊥	s ⎬	CH stretching
3029	3.30	⊥	s ⎭	
2923	3.42	⊥	vs	Asymmetric CH_2 stretching
2851	3.51	‖ (?)	s	Symmetric CH_2 stretching
1602	6.24	⊥	s	Ring C=C
1585	6.31	⊥	ms	Ring C=C
1543	6.48	⊥	w	
1493	6.70	⊥	s	Ring C=C
1450	6.90	⊥	vs	CH_2 deformation + ring C=C
1440	6.94	⊥	sh	CH_2 deformation
1364	7.33	⊥	s	CH bending
1328	7.53	⊥	ms	Ring
1314	7.61	⊥	m ⎫	CH_2 twisting
1297	7.71	‖	m ⎭	
1182	8.46	‖	m	Ring
1154	8.67	⊥	m	Ring
1080	9.26	⊥	m	Ring
1048	9.54	⊥	ms	Ring
1027	9.74	⊥	ms	Ring
1005	9.95	‖ (?)	w	Ring
982	10.18	⊥	mw	Ring
965	10.36	⊥	w	Ring
920	10.87	⊥	m ⎫	Out-of-plane CH deformation
898	11.14	‖	mw ⎭	
842	11.88	⊥	mw	Ring
760	13.16	‖	vs	In-phase, out-of-plane CH-deformation
689	14.51	⊥	s ⎫	Ring puckering
680	14.71	‖	sh ⎭	

[a] Key: vs, very strong; s, strong; ms, medium strong; m, medium; mw, medium weak; w, weak; vw, very weak; sh, shoulder.

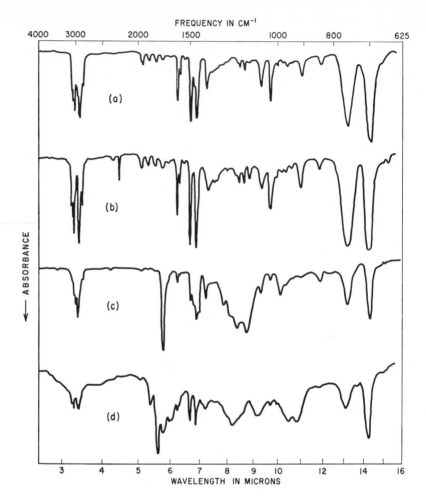

Fig. 24. Infrared spectra of (a) copolymer containing 60% styrene and 40% α-methylstyrene; (b) copolymer containing 92% styrene and 8% acrylonitrile; (c) copolymer of styrene and methyl methacrylate; and (d) copolymer of styrene and maleic anhydride.

L. STYRENE COPOLYMERS

Infrared spectra of typical styrene copolymers are shown in Figure 24. The comonomer can be identified readily by the absorption bands characteristic of the respective homopolymer species. Quantitative measurement of the composition of the copolymers may be made, employing a solution of the copolymer or cast film of indeterminate thickness. In the latter method a ratio of the concentrations of the two components is obtained. It is preferable to use absorption bands of comparable intensities for the ratio calculation. Polymeric styrene exhibits absorption bands at 6.2, 11.0, 13.2, and 14.3 μ which cover a broad intensity range and are suitable for use in the ratio calculation.

1. Determination of Monomers in Copolymers of Styrene and Acrylonitrile (51)

a. Calibration. Prepare individual solutions of styrene and acrylonitrile monomers in carbon disulfide covering the range from 0.1 to 0.5% monomer (10–50 mg in 25 ml of carbon disulfide based on a 10-g sample weight).

Scan the solutions from 2.5 to 16 μ in a 3.0-mm cell. Draw a base line from the adjacent shoulders of the 10.4 μ (acrylonitrile) and 12.9 μ (styrene) peaks. Measure the absorbance of the peaks and prepare analytical curves by plotting absorbance versus percentage of monomer.

b. Procedure. Weigh 10 g of sample into a 2-oz narrow mouth bottle and add 25 ml of carbon disulfide. Stopper the bottle securely and place on the shaker for 16 hr. Allow the polymer to settle and transfer a portion of the supernatant liquid to the 3.0-mm cell. Scan from 2.5 to 16 μ. Measure the absorbances of the characteristic absorption bands as indicated under Calibration. Read the concentrations of styrene monomer and acrylonitrile monomer, from the corresponding analytical curves.

2. Determination of Combined Acrylonitrile in Copolymers of Styrene and Acrylonitrile (51)

a. Calibration. Prepare additive-free samples of polystyrene and a styrene/acrylonitrile copolymer by dissolving the polymers separately in methylene chloride and reprecipitating by the addition of

methanol.   Recover the purified polymers by filtration and dry at
70°C *in vacuo*.   Calculate the acrylonitrile content of the styrene/
acrylonitrile copolymer from the nitrogen content determined by
the Kjeldahl procedure (Section X-D).   Prepare blends of varying
styrene/acrylonitrile content by dissolving known amounts of
polystyrene and analyzed styrene/acrylonitrile copolymer in
methylene chloride.

Cast films from the standard solutions directly on sodium chloride
plates and dry under an infrared heat lamp to remove all traces of
the solvent.   Scan the films from 2.5 to 16 μ.   Draw base lines
from 4.4 to 4.5 μ and from 6.1 to 6.4 μ and measure the absorbance
at 4.45 μ (acrylonitrile) and 6.2 μ (styrene).   Calculate the absorbance ratio ($A_{4.45}/A_{6.2}$) and plot an analytical curve of absorbance
ratio as a function of acrylonitrile concentration.

b. Procedure.   Weigh 2 g of reprecipitated polymer into a 2-oz
narrow mouth bottle and add 50 ml of methylene chloride.   Stopper
securely and place the bottle on the shaker until the sample is
dissolved.   Cast a polymer film approximately 1 mm thick directly
on a sodium chloride plate.   Dry the film under an infrared heat
lamp and scan from 2.5 to 16 μ.   Measure the absorbance of the
4.45 and 6.2 μ bands as indicated under calibration.   Calculate the
absorbance ratio and read the concentration of acrylonitrile from
the analytical curve.

VIII. Chromatographic Analysis of Nonpolymeric Components

Commercial films and resins contain various additives such as
plasticizer mixtures, stabilizers, lubricants, and brighteners.   The
number and kinds of these that are used tend to increase as new
developments are made.   It is not uncommon to separate three or
four additives from a polymer extract during analysis.   Thin layer
chromatography and gas chromatography are useful aids in such
separations.

A. Gas Chromatography

Gas chromatography is suitable for the separation and identification of substances which exhibit an appreciable vapor pressure
below 300–350°C.   Thus, many vinyl resin plasticizers, residual

monomers, residual solvents, and even some catalyst residues are amenable to gas chromatographic analysis.

Esposito (52) identified and determined various phthalate, sebacate, adipate, and phosphate esters using a 6-ft column packed with 20% silicone grease on washed Chromosorb W. The column temperature was programmed from 210 to 290°C at a rate of 4°C/min. The detector cell temperature was set at 300°C while the inlet port temperature was 330°C. Zulaica and Guiochon (53) used a mild pyrolysis (650°C for 10 sec) of the sample to liberate the plasticizers directly into the chromatographic column without prior extraction. Plasticizers boiling up to 400°C were eluted at 200–240°C from a 6-ft column packed with 0.5% Silicone Gum SE30 on 125–160-mesh glass beads or a 10-ft column packed with 0.5% poly(neopentyl glycol adipate) on glass beads.

1. Determination of High-Boiling Esters

This method is applicable to the determination of dibutyl sebacate, tributyl acetylcitrate, butyl benzyl phthalate, and diisooctyl adipate in polymer extracts. The additives are removed from the polymer by extraction with a suitable solvent, e.g., benzene or methylene chloride. The extract is then analyzed by gas chromatography. If alcohols are used as extractants, erratic data may be obtained as a result of ester interchange that occurs in the chromatographic column.

a. Apparatus. Gas chromatograph equipped with a flame ionization detector, or detector of similar sensitivity, and a $\frac{1}{4}$ in. $\times$ 4 ft stainless steel column packed with 5% by weight Silicone 410 on 80–100-mesh Chromosorb W.

b. Standardization. Prepare a standard solution containing from 0.2 to 0.5 g each of dibutyl sebacate, tributyl acetylcitrate, butyl benzyl phthalate, and diisooctyl adipate in 100 ml of benzene. Inject a 10-μl aliquot of the standard solution into the gas chromatograph at a column temperature of 240°C. A typical chromatogram is shown in Figure 25. Measure the peak area of each component.

c. Procedure. Cut a 10-g sample of the plastic film, weighed to the nearest 0.1 g, into small pieces. Place the sample in a 16-oz bottle with a foil-lined cap. Add 100 ml of benzene and place the sample on a shaker for 1 hr. Inject a 10-μl aliquot of the benzene extract into the instrument and record the chromatogram. Measure the peak area of each component and compare the peak areas with those of the standards.

% Component

$$= \frac{\text{Peak height component} \times \text{Grams standard}/100 \text{ ml}}{\text{Peak height of standard} \times \text{Weight of sample, g}}$$

2. Determination of Residual Styrene Monomer in Styrene Polymers

a. Apparatus. Gas chromatograph with hydrogen flame detector or apparatus of equivalent sensitivity.

Chromatograph column; $\frac{1}{4}$-in. outer-diameter metal tubing (0.028 in. wall thickness), 4 ft in length, packed with 20% Carbowax 20 M-alkaline on 60–80-mesh firebrick (the packing is a product of Wilkens Instrument and Research, Inc., Walnut Creek, California, Catalog No. 0033).

b. Standardization. Prepare a standard solution by weighing accurately 15–20 mg of styrene monomer into a 2-oz bottle containing 25.0 ml of methylene chloride. Cap the bottle tightly and shake to mix the solution thoroughly.

Inject 1 μl of the standard solution into the gas chromatograph

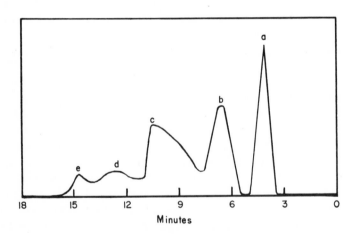

Fig. 25. Gas chromatogram of mixed ester plasticizers: (a) dibutyl sebacate; (b) tributyl acetylcitrate; (c) butyl benzyl phthalate; and (d) and (e) commercial diisooctyl adipate.

at a column temperature of 100°C.   Measure the area of the styrene monomer peak which emerges after approximately 12 min.

c. Procedure.   Transfer approximately 1 g of sample (accurately weighed to the nearest 0.001 g) to a 2-oz bottle and add several glass beads.   Pipet 20.0 ml of methylene chloride into the bottle and cap tightly.   Place the bottle on the mechanical shaker and shake it until the polymer is completely dissolved.   Transfer the solution to a 25-ml volumetric flask and dilute to volume.   If any insoluble residue remains, allow the flask to stand until a clear supernatant layer appears.   Inject 3 μl of the clear supernatant liquid into the gas chromatograph.   Measure the area of the resulting styrene monomer peak.   Compare the sample peak area with the area produced by the standard styrene monomer solution.

% Styrene monomer

$$= \frac{\text{Monomer in standard, mg} \times \text{Peak area of sample}}{\text{Peak area of monomer standard} \times \text{Sample weight, g} \times 30}$$

B. Thin Layer Chromatography

Thin layer chromatography has the advantage of being faster than conventional column chromatography and having greater capacity than paper chromatography.   In many cases, thin layer chromatography is superior to gas chromatography, for less drastic conditions are required to separate the components.   Also, high-boiling components that are very difficult or impossible to separate by gas chromatography may be separated by thin layer chromatography.

Commonly, as with other chromatographic methods, the elution of the components is not specific.   In particular, the fraction that moves with the solvent front and the fraction that does not migrate should be examined closely for purity.   Generally, the more polar compounds are strongly retained.   The following series illustrates the gradation from high to low mobility: alkanes, ethers, ketones, amines, alcohols, and acids.   The more polar solvents have a stronger eluting action.   The following series illustrates a gradation from weak to strong eluting power: hexane, carbon tetrachloride, methylene chloride, methanol, and water.

The most common and most versatile coating for a chromatographic plate is silica gel with a binder such as calcium sulfate.

Other coatings are diatomaceous earth, cellulose powders, and alumina. The coating is applied either from a water slurry or from an organic solvent slurry.

Water Slurry. A typical water slurry is a mixture of 1 part of adsorbent (containing 5–10% calcium sulfate) and 1.5–2 parts of water. Water slurries set up rapidly; therefore, the plates should be coated quickly with either a drawbar or a stationary doctor blade. Thicker and more uniform layers, with less tendency to flake, can be prepared from water slurries.

Organic Solvent Slurries. Typical slurries are mixtures containing 20–30% silica gel, with 10% binder, in chloroform. Such slurries are stable and may be stored for several months. The slurry must be mixed thoroughly just prior to dip coating. After air drying, the coated plates are steamed in order to set the binder. If the coating is too thick, the adsorbent may flake off. Plates made from frosted glass sometimes yield better separation of the components than clear glass.

The adsorbent, after air drying, is activated by heating in an oven for 30 min at 110°C, or other temperatures and times as required. Longer times and higher temperatures yield plates with more retentive power. Air drying without heat may be desirable for the separation of hydrophilic mixtures. A 5–10% solution of the sample in a volatile solvent may be applied to the plate with a syringe or eyedropper with a constricted end. Poor separation of components will result from overloading the plates. Such overloading may be detected, if the developing solvent does not wet the sample spot, or if it takes unusually long to pass the application point. Single solvents or binary mixtures are usually employed for the development of the components, although more complex solvent systems may be used occasionally. The simple solvent systems are used for components that are relatively easy to separate, e.g., three or four components from different generic classes. The more elaborate solvent mixtures are used for the separation of homologs or isomers. A convenient practice is to make preliminary separations using small, coated microscope slides. Several solvents representing a range from polar to nonpolar may be tested simultaneously in 10–15 min. The optimum conditions can then be applied to separations on a larger scale.

Ascending development is usual. The plates are placed in an airtight enclosure containing about 1 cm depth of solvent. The use of filter paper strips to prevent surface evaporation is effective. The paper is placed so that it does not touch the plate. Solvent, brought up by a wicking action, helps saturate the air space around the plates. Ultraviolet light may allow location of the spots by contrast, even though the components themselves do not fluoresce. Under ordinary light, the component areas are sometimes better seen from the back of the plate. A general method for locating components is by exposure to iodine vapor.   The iodine is adsorbed by the component which is stained brown-yellow to pink in color, depending upon the polarity of the compound, i.e., polar compounds are brown-yellow and nonpolar compounds tend to be pink or violet.   The coloration does not persist but disappears in a very short tıme.   The treatment is nondestructive, except to highly sensitive materials such as organic sulfur compounds.

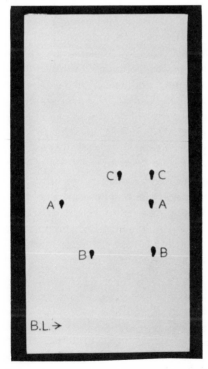

Fig. 26.   Thin-layer chromatogram of ester plasticizers: (A) dibutyl sebacate; (B) tributyl acetylcitrate; and (C) dioctyl phthalate.

The fractions are removed by scraping the adsorbent from the plates. The adsorbed component is dissolved in a solvent and separated from the adsorbent by filtration. If the quantity of the fraction is suspected to be small, the sample should be filtered directly into a small aluminum evaporating dish or into a small mortar. After the solvent is evaporated, potassium bromide (for the disk technique) or Nujol (for mulling) may be added directly to the sample. Care should be taken that the eluting solvent is strong enough to desorb the sample. If additional sample is needed for further identification, fractions from five or six plates may be combined. Figure 26 shows the separation of dibutyl sebacate, tributyl acetylcitrate, and dioctyl phthalate on silica gel G, using methylene chloride containing 1% methanol as the developer. The spots were developed by exposure to iodine vapor.

IX. Polymer Molecular Weights

Many of the physical properties of a plastic such as heat distortion, elongation, tensile strength, and impact strength are dependent on or affected by the average molecular weight and the molecular weight distribution of the polymer. Vinyl polymers can be made by the bulk, solution, emulsion, or suspension polymerization process. As would be expected, the average molecular weight and molecular weight distribution vary, depending upon the process used. More important, however, are the conditions under which a particular polymerization is carried out. Factors such as temperature, presence of chain terminator, type and amount of catalyst, solvent, or emulsifier exert considerable influence on the composition of the finished polymer. The experimental determination of molecular weights and molecular weight distributions is necessary not only to provide suitable control of the polymerization, but also to understand variations in physical properties. Methods commonly used for these determinations are discussed in this section.

A. Average Molecular Weight

Most polymers are mixtures of different molecular weights. These differences are so pronounced that a single property measurement is usually not adequate for accurate description of the mixture. The common practice is to measure a property proportional to the number of molecules present and, in addition, a property proportional to the square of the mass of the molecules. In the former method all molecules contribute equally, even though they are of widely different molecular weight.   In the latter, the higher molecular weight species contribute relatively more than the lower. These measurements result in an average biased by the lower species and an average biased by the higher species.

1. Methods which Measure Properties Proportional to the Number of Molecules Present (Number Average, $\bar{M}_n$)

Most conventional methods dependent on molal response are applicable only to the low molecular weight range.   As might be expected, the accuracy decreases with increasing molecular weight. Cryoscopic and ebulliometric methods based upon temperature detection fail, because of decreasing temperature differences, at a molecular weight of about 2.5×10^3.   Osmotic pressure measurements, however, are more sensitive on a molal basis and are accurate over a larger range, to about one million.   They depend upon use

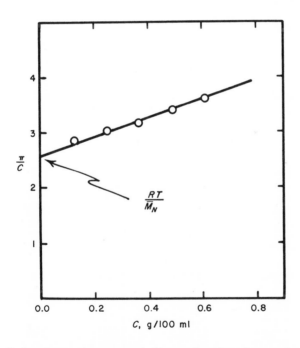

Fig. 27.   Reduced osmotic pressure (π/c) as a function of concentration c for a solution of polystyrene in toluene (77).

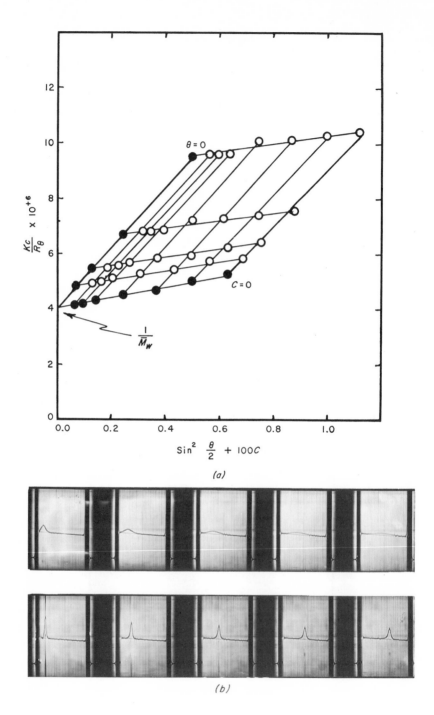

(a)

(b)

of a semipermable membrane which may allow diffusion of molecular weights below about 2×10^4. This is a serious source of error in static osmometers but less important in dynamic (high-speed) systems. Figure 27 shows a typical plot of π/c as a function of c for polystyrene in toluene obtained with a high speed osmometer.

Methods based on end-group determination are not normally applicable to vinyl polymers because of the lack of distinctive terminal groups.

If the distribution of molecular weights in a polymer is broad, the number-average molecular weight may not be representative of its physical properties. In these situations, an average more indicative of the higher molecular species is needed to help describe the mixture.

2. Methods which Measure Properties Proportional to the Square of the Particle Mass (Weight Average, $\bar{M}_w$)

The light scattered by polymers in solution is proportional to the square of the mass of the molecules (or aggregates) after subtracting solvent scattering and extrapolating to infinite dilution and zero angle. The light scattering method is versatile and capable of spanning a very wide range, actually becoming more accurate for higher molecular weights. A common method of data treatment involves extrapolation to zero angle and concentration using a grid type plot (Zimm method), Figure 28a. The weight average can be obtained directly in favorable cases by ultracentrifugation. Sedimentation velocity treatment to obtain molecular weight distributions from distributions of sedimentation coefficients involves calibration with standards of known molecular weight (Fig. 28b). It is necessary to have a condition of "little interaction" between solvent and polymer. Interpolymer entanglements which occur when molecules are extended (solution in a good solvent) can prevent separations according to molecular size (54). In a poor solvent at a selected temperature where the molecules are tightly coiled, interpolymer entanglements cause less difficulty.

Fig. 28. (a) Zimm plot for polystyrene in benzene at 25°C. (b) Ultracentrifuge schlieren patterns illustrating the sedimentation velocity behavior of broad (upper frames) and narrow (lower frames) molecular weight distribution polystyrenes in cyclohexane at 35°C. A double sector cell was employed with a Beckman Model E ultracentrifuge operating at 59,780 rpm. Exposures were made at 16-min intervals.

3. Solution Viscosity

Intrinsic viscosity obtained from dilute polymer solutions is proportional to molecular weight as expressed by the Mark-Houwink equation, $[\eta] = KM^a$, where K and a are constants, depending on the system. Calibration against known standards is necessary. This relation yields a straight line for log $[\eta]$ vs. log M, if the standards are narrow fractions, or if they have a similar molecular weight distribution.

The relationship between viscosity average, number average, and weight average is illustrated for a series of polyvinyl chloride

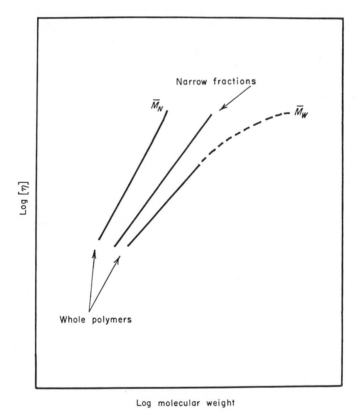

Fig. 29. Qualitative relationship of $\overline{M}_v$, $\overline{M}_n$, and $\overline{M}_w$ for a series of polyvinyl chlorides as found by Freeman and Manning (55).

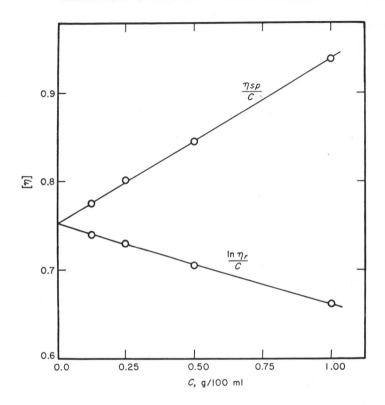

Fig. 30. Reduced specific viscosity and reduced inherent viscosity of toluene solutions of polystyrene as a function of concentration.

polymers in Figure 29 (55). Relationships for whole polymers with similar broad distributions result in displacement from the narrow fraction (unbranched) line as shown. This figure, however, shows the combined displacing effects of branching (more noticeable at the high end, as indicated by the dashed line) and polydispersity.

Expected behavior in the absence of branching would be qualitatively similar to the solid lines except that the divergence between $\bar{M}_w$ of whole vs. narrow polymers as indicated by $[\eta]$ would not be so large. If the intrinsic viscosity is used to determine the molecular weight of a polymer with a broad distribution and the narrow-fraction relationship is used to calculate the molecular weight average, it is seen that the result will be divergent to an extent depending on the polydispersity of the sample. In this case, if $\bar{M}_n$ and $\bar{M}_w$ have been determined independently, $\bar{M}_w > \bar{M}_v > \bar{M}_n$. Since the solution viscosity is influenced to a great extent by the longer chains, $\bar{M}_v$ is normally closer to $\bar{M}_w$ than to $\bar{M}_n$. Branching effects, shown in Figure 29, would tend to make $\bar{M}_n$ closer to the narrow fraction line at the high end rather than $\bar{M}_w$. A similar situation might exist if gel particles caused very high $\bar{M}_w$ results by light scattering. This is not the normal situation, however. Determination of $[\eta]$ for polystyrene in toluene is illustrated in Figure 30 (56).

It is evident from the foregoing that some idea of the molecular weight distribution can be obtained by comparing the different types of average molecular weights. This is usually done by comparing, for example, the number and weight averages by osmotic pressure and light scattering expressed as a ratio $\bar{M}_w/\bar{M}_n$. The values converge for narrow distributions and diverge as the distributions broaden. Fractionation may be necessary if a more descriptive definition is required.

B. Polymer Fractionation

The most common type of fractionation depends on solubility differences of the various molecular weights in the mixture. This usually involves a liquid/liquid solution boundary or a solid polymer/liquid interface and may employ a temperature or solvent gradient to dissolve, precipitate, or extract successive increments of the polymer. Other less well known techniques such as zone refining and turbidimetric titration also depend in part on solubility differences. For the purposes of this discussion, fractionation techniques are divided into two groups: (1) techniques in which the fractions are not recovered, and (2) preparative techniques.

Some fractionations require solutions so dilute that fraction collection and characterization is not normally possible. The nature of the fractionation may also make collection of fractions difficult. These methods include gel permeation, turbidimetric titration, ultracentrifugation, and variations of some column fractionation techniques. These latter methods may employ very dilute solutions and a continuous detector based on ultraviolet absorbance or refractive index rather than fraction collection.

Since the fractions are not available, these methods (except perhaps equilibrium ultracentrifugation) require calibration against known standards. As a consequence, reproducibility must be well established.

The efficiency of fractionation is best when dilute solutions are employed and a large number of fractions separated. Preparative techniques including even those designed for analytical use (with microviscosity etc.) present a conflicting choice between resolution and the practical need to recover enough material for analysis. A compromise is clearly necessary. For this reason, preparative fractionations are not ideally suited for measuring small differences such as might be expected from a comparison between batches of the same polymer or other closely related samples. Analysis time for preparative methods is usually measured in days or weeks but some of the nonpreparative methods require a few hours or less. Moreover, a nonpreparative method may actually be more definitive due to a combination of better reproducibility and perhaps better resolution. On the other hand, some nonpreparative methods may be lacking in ability to define differences in absolute terms. A good example of this is the turbidimetric titration method.

The most common method used for fraction characterization is dilute solution viscosity. From strictly analytical and control standpoints, concentrated solution viscosity (in the vicinity of 10% concentrations, for example) may have more differentiating power, but the usual small fractions recovered favor the use of dilute solutions (intrinsic viscosity). Even in a relatively simple intrinsic viscosity determination there is a chance for weight error.

Some polymers have a unique ability to retain solvents, even though dried under vacuum at elevated temperatures. For this reason, it is desirable to determine the viscosity of the whole polymer before and after precipitation from the solvents to be used in fractionation. Normally, the result after precipitation would be expected to be slightly higher, assuming that some of the lower molecular weights are not recovered. If it is lower, it could be the result of retained solvent or degradation. In the latter respect, time of contact is important and it may be desirable to analyze a sample reconstituted from the fractions, if possible, to check for degradation effects. Some polymers such as polybutadiene may pose a problem in the reverse direction because of crosslinking or gel formation and this should also be considered if the nature of the polymer warrants it.

Solvent contact can drastically alter the degree of crystallinity in partly crystalline or potentially crystalline polymers. This is important in fractionations involving a solid polymer/solvent interface. Solvent-induced or otherwise, the crystallinity can have a larger effect on solubility than differences in molecular weight. One alternative might be to carry out fractionation at an elevated temperature where the polymer is uniformly noncrystalline. Another possibility, though drastic, may be to convert the polymer to a noncrystalline form as was done in the fractionation of polyvinyl alcohol after acetylation (57). Fractionation based on a liquid/liquid boundary should be considered, if crystallinity is a problem.

Possible bias of the methods used to characterize the fractions should be considered in addition to precision. It has already been shown that intrinsic viscosity is subject to bias, depending on the sharpness of the fractions. When establishing the narrow-fraction relationship between intrinsic viscosity and molecular weight, it has been pointed out (54) that it is better to use $[\eta]$ vs. $\bar{M}_w$ relationships than $[\eta]$ vs. $\bar{M}_n$ because the former should be less subject to error from polydispersity of the fractions. Light scattering results may be high because of dust or small amounts of gel. High results because of the gel may be legitimate but not if the gel was formed by or during the fractionation treatment. Solutions to be used in light scattering work should be checked visually immediately before measurement (some instruments have a window for this purpose) and the dust, if present, is readily apparent at a low angle using the arc with no filter in place. Osmotic pressure results may be biased because of diffusion of some of the lower molecular weights through the membrane. Comparison of preparative results with one of the nonpreparative techniques may help bring both techniques into better perspective (58).

Figure 31 shows three superimposed gel permeation curves for polystyrene samples with different molecular weight distributions. The samples with $\bar{M}_w/\bar{M}_n = 2.2$ and 1.08 have the same weight-average molecular weight (within 10%) and the sample with $\bar{M}_w/\bar{M}_n = 2.6$ has the same number-average molecular weight (within 10%) as the narrow polystyrene. All three were analyzed

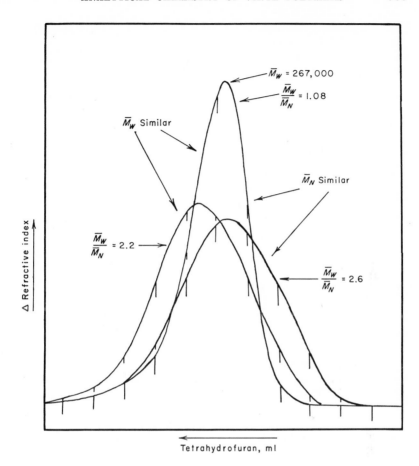

Fig. 31.   Gel permeation chromatograms for three polystyrenes with different
molecular weight distributions.

under the same conditions: flow rate, 1 ml/min; sample size, 1 ml
of a 1% solution; columns, three 4-ft columns in series of 10³,
1.5 × 10⁴, and 10⁶ Å permeability limits.   Molecular weight, if
plotted on the elution axis is logarithmic, the lower species eluting
later and with better resolution than the higher.   Because of this,
the right side of the curves is condensed relative to the left in terms

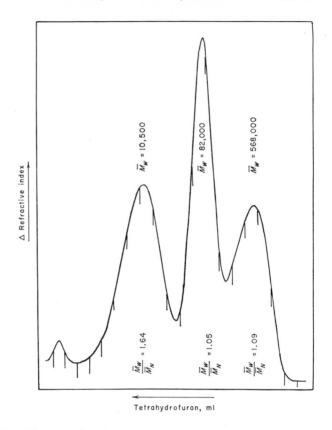

Fig. 32.   Gel permeation chromatograms for three polystyrenes with relatively narrow molecular weight distributions with different $\bar{M}_w$.

of a linear molecular weight scale.   Figure 32 illustrates this and the resolution of three narrow polystyrenes used for calibration purposes.   The short vertical lines mark 5-ml elution intervals. The gel columns are styrene–divinylbenzene beads crosslinked in the presence of an appropriate diluting solvent to give pore sizes with varying permeability to the constituents of the sample (59).

Tables 13 and 14 summarize some typical fractionation data but are not intended to be inclusive.   Polystyrene is by far the best characterized polymer serving in some cases as a molecular weight calibration standard as well as an infrared calibration standard. The literature prior to 1952 is well summarized in reference 60 for polystyrene. More specific details concerning the other polymers can be obtained by consulting the literature cited. A recent review of the analytical problems of molecular weight determination can be found in reference 61.

C. COPOLYMERS

Some errors that occur in molecular weight determinations are the fault of the detector system or failure of the polymer to behave in the predicted way. It has already been shown that homopolymers do not follow the expected viscosity–molecular weight relationship, if branched. Copolymers are even more difficult to characterize. Errors arising from nonuniformity become important if the distribution of the respective monomer units is not uniform throughout the molecular weights present.

For example, light scattering, ultracentrifugation, and gel permeation depend on refractive index for detection, although ultraviolet absorption can be used, if applicable in the latter two methods. A problem may arise if the segments comprising the copolymer have a different refractive index and their distribution is not uniform. In this case, the whole-polymer refractive index increment will not be the same as the actual increment for some of the fractions. This can result in a large error, amplified in light scattering by use of a solvent which gives a small average increment, such as toluene for styrene/butadiene copolymers of the GR-S type (62,63). The dependence of apparent molecular weight by light scattering in a graded series of solvents with varying refractive index increments has been used as a measure of nonhomogeneity of this type (64,65). Actually, this is only one aspect of the unpredictability problem. For example, solution properties on which viscosity determinations are based could also be affected if viscosity increments are different. Fractionations of nonuniform copolymers may be more affected by chemical affinities than by solubility differences owing to differences in molecular weights. Thus, the composition distribution of copolymers must be dealt with, in addition to the distribution of molecular weights.

TABLE 13
Preparative Fractionation

Method	Polymer	Solvents	Method of analysis	Molecular weight, range	Viscosity equation
Precipitation from solution by addition of a nonsolvent	Polyvinyl chloride (55)	Tetrahydrofuran/water	Light scattering, osmometry, viscosity	2.5×10^4–3.0×10^5	$[\eta]_{THF}^{25°} = 1.63 \times 10^{-4} \bar{M}_w^{0.766}$
	Polystyrene (66)	Dioxane/methanol	Light scattering, viscosity	2.0×10^3–2.0×10^6	$[\eta]_{toluene}^{25°} = 1.75 \times 10^{-4} M^{0.69}$ $[\eta]_{butanone}^{25°} = 3.9 \times 10^{-4} M^{0.58}$ $[\eta]_{dichloroethane}^{25°} = 2.1 \times 10^{-4} M^{0.66}$
Precipitation from solution by evaporation	Polystyrene (66)	Butanone/1-butanol			
Extraction with temperature and solvent gradients	Polystyrene (67)	Methyl ethyl ketone/ethanol	Light scattering, viscosity	2.2×10^4–5.4×10^5	$[\eta]_{toluene}^{25°} = 1.28 \times 10^{-4} M^{0.709}$
Coacervation, liquid–liquid boundary	Polystyrene (68)				
	Polystyrene (69)	Benzene/methanol Tetrachloroethane/n-heptane			

Extraction from coated screen with solvent gradient	Polyvinyl acetate (57)	Ethyl acetate/ petroleum ether	Light scattering, viscosity	2.5×10^4–4.0×10^5	$[\eta]_{\text{methanol}}^{25°} = 1.71 \times 10^{-4} \bar{M}_w^{0.65}$
	Polyvinyl alcohol (57)[a]	Water/n-propyl alcohol	Viscosity	1.1×10^4–3.0×10^5	$[\eta]_{\text{water}}^{25°} = 5.38 \times 10^{-4} \bar{M}_w^{0.63}$
Zone refining, freezing front traps larger molecules first	Polystyrene (70)	Benzene	Viscosity		
General	Polystyrene (60,71)				

[a] Better results when converted to polyvinyl acetate prior to fractionation.

TABLE 14
Nonpreparative Fractionation

Method	Application	Reference
Ultracentrifugation		
(a) Sedimentation velocity	General review	72
	Polystyrene in cyclohexane, 11,000–1,000,000 mol. wt.	54
(b) Direct method for $\bar{M}_w$	Polyvinyl chloride in tetrahydrofuran	73
(c) Density gradient	Separation of atactic and isotactic polystyrenes	74
Gel permeation chromatography	Nature of separation	59
	Polyvinyl chloride	75
Turbidimetric titration	Polystyrene	76

D. Definition of Terms and Calculations

1. Definition of Molecular Weight

a. Number Average Molecular Weight

$$\bar{M}_n = \frac{\Sigma n_i M_i}{\Sigma n_i}$$

b. Weight Average Molecular Weight

$$\bar{M}_w = \frac{\Sigma n_i M_i^2}{\Sigma n_i M_i}$$

where n_i is the molar concentration of ith component and M_i is the molecular weight of ith component.

2. Osmometry

a. General

$$\pi/c = (RT/M) + RTA_2c + \cdots$$

where π = osmotic pressure in cm of toluene,

A_2 = second virial coefficient from slope π/c vs. c/RT,

c = concentration in g/100 ml,

T = temperature, absolute,

R = gas constant, 848/density of solvent, and

M = molecular weight ($\bar{M}_n$).

b. Calculations for Figure 27 (76).

$$\pi/c = RT/M, \qquad c \to 0$$

$$\bar{M}_n = \frac{RT}{\text{intercept } [\pi/c]_{c=0}}$$

$$= \frac{(848)(310)}{(2.56)(0.8632)} = 119,000$$

where 0.8632 is the density of toluene at 37°C.

3. Light Scattering

a. General

$$Kc/R_\theta = (1/M) + 2A_2c + \cdots$$

where R_θ = scattering intensity ratio due to dissolved polymer at
   angle θ,
M = molecular weight ($\bar{M}_w$, if polydisperse),
A_2 = second virial coefficient,
c = concentration of polymer (g/ml),
K = $(2\pi^2 n^2/N\lambda_0^4)(dn/dc)^2$,
θ = angle of observation relative to incident beam,
N = Avagadro's number,
λ_0 = wave length of incident light,
n = refractive index of solvent, and
dn/dc = specific refractive index increment (ml/g).

b. Calculations for Figure 28.   The data shown in Figure 28
were obtained with an instrument manufactured by Société
Française d'Instruments de Contrôle et d'Analyses (SOFICA).
Benzene was used in the bath surrounding the cell.   Incident light
was unpolarized.   Angles were 30, 37.5, 45, 60, 75, 90, and 105°.
Concentrations were 0.5, 0.25, 0.125, and 0.0625 g/100 ml.

$$1/\bar{M}_w = (Kc/R_\theta)_{c\to 0,\theta\to 0}$$

$$1/\bar{M}_w = 4 \times 10^{-6}$$

$$\bar{M}_w = 2.5 \times 10^5$$

The values of Kc/R_θ were obtained individually for each point
by computer calculations.   Without computer aid, it may be more
convenient to correct only the intercept to absolute terms.   In this

case, values of c/I are used where I is the solution scattering minus the solvent scattering multiplied by $(\sin \theta)/(1 + \cos^2 \theta)$. A grid-type graph similar to that shown can be obtained by plotting these c/I values as a function of $\sin^2 (\theta/2) + kc$. The intercept $(c/I)_{c \to 0, \theta \to 0}$ then is converted to absolute terms by use of instrument calibration values.

$$\lambda_0 = 546 \ m\mu$$

$$dn/dc = 0.106 \text{ for polystyrene in benzene at } 546 \ m\mu$$

$$R_b = 16.3 \times 10^{-6} \text{ for benzene as a standard}$$

$$\frac{2\pi^2 n^2}{N\lambda_0^4 R_b} = 0.506$$

$$K = 0.506(dn/dc)^2(\text{calibration reading for benzene})$$

$$1/\bar{M}_w = (0.506)(0.106)^2(\text{calibration reading})(c/I)_{c \to 0, \theta \to 0}$$

4. Viscosity

$$\eta_{rel} = \frac{\text{Viscosity of solution}}{\text{Viscosity of solvent}}$$

$$\eta_{rel} - 1 = \text{Specific viscosity, } \eta_{sp}$$

$$[\eta] = \text{Intrinsic viscosity} = \lim_{c \to 0} \frac{\ln \eta_{rel}}{c}$$

$$= \lim_{c \to 0} \frac{\eta_{sp}}{c}$$

X. General Methods

A. Volatile Loss and Solvent-Solubility

The general procedures for determining volatile loss and solvent-solubility are empirical. Reproducible results can be obtained only if the test conditions are strictly followed.

1. Volatile Loss from Polystyrene

Grind the sample to pass a U.S. No. 12 (1680 μ) sieve. Weigh exactly 1 g of the ground sample into a tared aluminum dish (0.5 in. deep and 2 in. in diameter). Place the dish in the tray (Fig. 33).

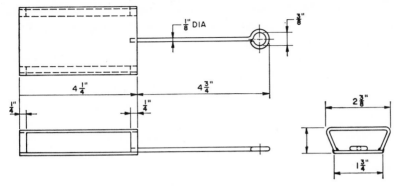

No. 16 GA. TYPE 18-8 STAINLESS (Weld)

Fig. 33. Sample tray for Abderhalden oven.

Four samples should be run at one time. Transfer the tray to a preheated Abderhalden oven employing 1,2,4-trichlorobenzene as the heat transfer liquid. Evacuate the oven to 1–5 mm Hg pressure. Withdraw the samples after exactly 25 min and place them in a desiccator. Weigh the samples when cool.

$$\% \text{ Volatile loss} = \frac{\text{Loss in weight, g}}{\text{Sample weight, g}} \times 100$$

2. Volatile Loss from Vinyl Chloride Polymers

Accurately weigh about 5 g of the ground sample into a tared weighing bottle. Place the dish in an Aberhalden oven preheated to 113.5°C (employing 1,1,2-trichloroethane as the heat transfer agent) and evacuate to 10–20 mm Hg pressure. After 4 hr remove the samples from the oven, cool in a desiccator, and reweigh.

$$\% \text{ Volatile loss} = \frac{\text{Loss in weight, g}}{\text{Sample weight, g}} \times 100$$

3. Methanol Solubility of Polystyrene (78)

a. Reagents. *Methanol*, C.P., redistilled.
Dioxane, stored over sodium hydroxide and redistilled prior to use.

b. Procedure. Grind the sample to pass an ASTM No. 12 (1680 μ) sieve. Weigh about 0.3 g of the ground polymer into a tared 50-ml beaker, weighing to the nearest 0.0001 g. Add 15 ml of dioxane, cover the beaker with a watch glass, and let stand until the sample is dissolved. Stir the solution thoroughly and transfer rapidly to a 400-ml beaker containing 200–250 ml of methanol. Stir the methanol vigorously during the addition. Rinse the 50-ml beaker with 25–50 ml of methanol, transferring all of the sample to the 400-ml beaker. Heat to 65°C on a steam bath, while stirring, until the precipitate coagulates; then remove and allow to settle. Filter with suction through a tared fine-porosity, sintered-glass crucible, first decanting as much liquid as possible and then transferring the solids. Wash the beaker and crucible with about 125 ml of methanol. Continue suction until the precipitate is nearly dry. Dry to constant weight at 65–70°C. Cool in a desiccator and weigh.

c. Calculation.

$$\% \text{ Methanol-soluble content} = \frac{A - B}{A} \times 100$$

where A is the weight of sample in g and B is the weight of precipitate in g.

B. Solution Viscosity

1. Viscosity of 10% Polystyrene Solution

a. Apparatus. *Viscometer.* Ostwald viscometer (Cannon–Fenske modification) having a capillary diameter of 1.4–1.5 mm. Calibrate at 25°C against National Bureau of Standards standard viscosity oils.

Electric Timer reading to 0.1 sec.

Constant Temperature Bath regulated to 25.00° ± 0.01°C.

b. Procedure. Grind the sample to pass an ASTM No. 12 (1680 μ) screen. Weigh 0.9593 g of the sample into a 2-oz wide-mouth bottle and add exactly 10 ml of redistilled toluene to make a 10% solution of the polymer.

Cap the bottle securely with a foil-lined screw cap and shake to dissolve the sample. Affix a small piece of cotton to the tip of a 5-ml pipet as a filter and transfer exactly 5 ml of solution into the viscometer.

Place the viscometer in the constant temperature bath. When the solution attains the temperature of the bath, force the liquid 5–8 mm above the upper calibration mark by pressure applied with a rubber bulb.

Remove the pressure and measure the time required for the liquid level to pass between the two calibration marks. Make triplicate flow-time measurements which should check to 0.1 sec.

c. Calculation.

$$\text{Absolute viscosity, cps} = \frac{T_1 \times D_1 \times V_2}{T_2 \times D_2}$$

where T_1 = time of flow of unknown solution,
T_2 = time of flow of standard viscosity oil,
D_1 = density of unknown solution,
D_2 = density of standard viscosity oil, and
V_2 = viscosity of standard viscosity oil in cps.

The density of a 10% solution of polystyrene in toluene at 25°C is 0.8810 g/ml.

2. Viscosity of 2% Solutions of Vinyl Chloride Polymers

a. Apparatus. *Viscometer.* Ostwald viscometer (Cannon-Fenske modification) having a capillary diameter of about 0.40 mm. Calibrate against National Bureau of Standards standard viscosity oils.

Constant temperature oil bath maintained at 120°C.
Constant temperature bath maintained at 165°C.
Electric timer reading to 0.1 sec.

b. Procedure. Weigh exactly 0.2663 g of the sample into a 20 × 150 mm test tube. Add exactly 10 ml of redistilled o-dichlorobenzene and place the tube in the 165°C constant temperature bath for 5 min. Use a stirring rod to hasten solution. If the sample turns dark, indicating decomposition, it should be discarded.

Suspend a viscometer in the 120°C constant temperature bath for a sufficient time to reach the bath temperature. Affix a small piece of cotton to the tip of a preheated 5-ml pipet as a filter and transfer 5 ml of the solution into the viscometer. When the solution attains the temperature of the bath, force the liquid 5–8 mm above the upper calibration mark by pressure applied with a rubber bulb. Release the pressure and measure the time required for the

liquid level to pass between the two calibration marks. Make triplicate flow-time measurements which should agree within 0.3 sec.

Drain the viscometer tube and rinse it immediately with boiling
hot o-dichlorobenzene. Cool, rinse with acetone, and dry.

c. Calculation.

$$\text{Absolute viscosity, cps} = \frac{T_1 \times D_1 \times V_2}{T_2 \times D_2}$$

where $T_1 =$ time of flow of unknown solution,

$T_2 =$ time of flow of standard viscosity oil,

$D_1 =$ density of unknown solution,

$D_2 =$ density of standard viscosity oil, and

$V_2 =$ viscosity of standard viscosity oil in cps.

3. Viscosity of 2% Solution of Vinylidene Chloride Polymers

The test is carried out using the procedure for vinyl chloride
polymers (Section X-B-2).

4. Viscosity of 4% Solution of Polyvinyl Alcohol

a. Apparatus. *Hoeppler Viscometer* or other suitable ball–drop
viscometer such as described in *ASTM-D1343* (79).

Electric timer reading to 0.1 sec.

Constant temperature bath regulated to $20.00° \pm 0.01°C$.

b. Procedure. Weigh 192.0 ± 0.1 g of water into a 350-ml pressure bottle. Chill to 0°C and add exactly 8 g of polymer. If the
polymer is already in solution, calculate, from the total solids content, the amount of sample required to yield 8 g of solids. Weigh
sufficient water into the bottle to give a total of 192 g and then add
the calculated amount of aqueous sample. Stopper the bottle
securely and shake vigorously to disperse the polymer. Place the
bottle in a water bath at $80 \pm 5°C$. Shake the bottle frequently
until the polymer is dissolved. Cool the solution to 20°C and determine the viscosity by the ball–drop method.

c. Calculation.

$$\text{Absolute viscosity, cps} = K \times D \times T$$

where K is a factor dependent upon the ball and temperature used,
D is the density of the sample solution, and T is time of fall.

C. Combined Styrene in Styrene/Acrylonitrile Polymers

a. Apparatus.   Ultraviolet recording spectrophotometer.

b. Procedure.   Weigh 100 mg of the purified polymer into a 100-ml volumetric flask.   Dissolve the sample in redistilled tetrahydrofuran and dilute to volume with the same solvent.   Transfer a portion of the solution to a 1.0-cm absorption cell.   Record the absorption curve from 300 to 240 mμ, using the pure solvent as the reference solution.   Draw a base line across the absorption minima located at 250 and at 267 mμ.   Measure the net absorbance of the maximum at approximately 259 mμ.   Calculate the number of mg of styrene per 100 ml from an analytical curve prepared by treating purified polystyrene in the same manner.

$$\% \text{ Combined styrene} = \frac{\text{Styrene, mg}}{\text{Sample, mg}} \times 100$$

D. Determination of Acrylonitrile in Plastics

a. Apparatus.   *Kjeldahl Digestion and Distillation Equipment* with 800-ml flasks.

Sodium Sulfide Solution.   Dissolve 40 g of sodium sulfide (Na_2S) in distilled water and dilute to 1 liter.

Sodium Hydroxide Solution.   Dissolve 1000 g of technical grade sodium hydroxide in 1 liter of distilled water.

b. Procedure.   Transfer approximately 1 g of purified polymer accurately weighed, to a kjeldahl flask.   Add 0.5 g of mercuric oxide, 10 g of potassium sulfate, and 25 ml of concentrated sulfuric acid, washing down any sample adhering to the neck of the flask.   Mix the contents of the flask and place it on the digestion rack.   Heat the contents gently at first and then gradually bring the temperature up to boiling.   Rotate the flask occasionally to bring the acid into contact with any undigested material.   Continue the digestion for 2 hr after the solution becomes colorless or nearly so.

Pipet 50 ml of 0.2N hydrochloric acid into a 500-ml erlenmeyer flask and place it so that the tip of the condenser dips just below the surface of the solution.

Allow the kjeldahl flask to cool and add cautiously 500 ml of water.   Mix the solutions thoroughly and cool again.   Add 1–2 g of mossy zinc and 25–30 ml of the sodium sulfide solution.   Add 80–90 ml of the sodium hydroxide solution, pouring it slowly down the side of the flask to prevent mixing of the solutions. Immediately connect the flask to the connecting bulb and condenser. Swirl the contents of the flask carefully to mix the solutions. Start to heat immediately and distill over about 300 ml of solution. Add several drops of methyl red indicator and titrate the excess acid with 0.2N sodium hydroxide.

Make a blank determination using the same amounts of all reagents.

c. Calculation.

$$\% \text{ Acrylonitrile} = \frac{(A - B)N \times 0.053 \times 100}{\text{Sample weight in g}}$$

where A = ml of sodium hydroxide solution required for titration of the blank, B = ml of sodium hydroxide solution required for titration of the sample, and N = normality of the sodium hydroxide solution.

E. TOTAL CHLORINE IN CHLORINE-CONTAINING POLYMERS

Various methods are available for determining chlorine in chlorine-containing polymers. One of the most widely used is based on the calcination of the polymer with calcium oxide, dissolution of the calcined residue in nitric acid, and argentimetric determination of the chloride. Parlashkevich (80) proposed a procedure based on pyrolytic liberation of hydrogen chloride, absorption of the hydrogen chloride in water, and determination of the acid concentration by measuring the electrical conductance of the solution. The Parr peroxide bomb technique is still perhaps the most satisfactory.

a. Apparatus. *Parr peroxide bomb* (Series 2100, 22 ml, and accessories).

Blast burner.

Safety shield.

b. Reagents. *Ferric nitrate solution.* Dissolve 50 g of ferric nitrate in distilled water and dilute to 1 liter.

c. Procedure. Examine each bomb prior to use. Discard any which appears to be corroded. Place 6–7 g of sodium peroxide in the bomb. Add about 0.2 g of the sample, weighed accurately, and 0.50 g of powdered sugar onto the peroxide. Cover the sample with 9–10 g of sodium peroxide. Assemble the bomb, making sure the head gasket is in place, and tighten with a wrench. Shake the bomb vigorously to mix the contents. Tap the bomb lightly on the bench to dislodge any particles sticking to the upper portions of the bomb.

Set the bomb in the ignition housing stand and place a safety shield in front of the assembly. Apply a sharp, intense flame to the bottom of the bomb, and heat until a dark red spot appears on the side of the bomb. Underheating will result in incomplete oxidation of the organic matter and low chlorine results. Overheating may cause possible damage to the bomb and even an explosion. Allow the bomb to cool in air for several minutes and then to cool to room temperature under tap water. Remove the screw caps taking care not to disturb the bomb lid. Wash the outside of the bomb with distilled water. Remove the lid and wash any adhering particles into a 600-ml beaker. Place the bomb on its side in the beaker, add sufficient water to cover about one-third of the bomb, and immediately cover the beaker with a watch glass. When the reaction has completely subsided, remove the bomb, and wash thoroughly with a stream of water. Neutralize the solution by adding 50 ml of 1:1 nitric acid slowly with constant stirring.

Add an additional 10 ml of 1:1 nitric acid, 2 ml of nitrobenzene and 50.0 ml of $0.1N$ silver nitrate. Stir vigorously until the precipitated silver chloride becomes a spongy mass. Add 10 ml of ferric nitrate indicator and titrate the excess silver nitrate with $0.1N$ ammonium thiocyanate to the first pink end point. Run a blank determination on the reagents.

d. Calculation.

$$\% \text{ Vinyl chloride} = \frac{(A \times N_A) - (B \times N_B) \times 0.06246}{\text{Weight of sample}} \times 100$$

where A = net ml of silver nitrate solution,
   N_A = normality of silver nitrate solution,
   B = net ml of ammonium thiocyanate solution, and
   N_B = normality of ammonium thiocyanate solution.

F. VINYLIDENE CHLORIDE IN COPOLYMERS OF VINYLIDENE CHLORIDE AND VINYL CHLORIDE

Determine the total chloride content of the dry additive-free polymer as prescribed in Section X-E.

$$\% \text{ Vinylidene chloride} = \frac{\% \text{ Chloride} - 56.73}{0.1641}$$

G. Determination of Vinyl Acetate in its Copolymers with Vinyl Chloride

There is, at present, no highly accurate method for the determination of vinyl acetate in its copolymers with vinyl chloride. Usually vinyl acetate is determined indirectly, the vinyl chloride being calculated from the total chlorine content, and the vinyl acetate content then being calculated as the difference. For the direct determination of vinyl acetate, the polymer is dissolved in dioxane in a stream of nitrogen. Sodium ethoxide is added and the solution refluxed for 1 hr to saponify the vinyl acetate. The solution is acidified with sulfuric acid and silver sulfate is added to precipitate silver chloride. The acetic acid is then steam distilled into a standard sodium hydroxide solution. At least 2 liters of distillate must be collected to assure complete recovery of acetic acid. Traces of sulfuric acid may also distil over. Lardera (81) suggests passing the saponified mixture through an ion-exchange resin in the hydrogen form. The total released acids are determined acidimetrically and then hydrochloric acid is determined argentimetrically. The acetic acid content, determined by difference, is calculated as vinyl acetate.

A rapid method for the estimation of vinyl acetate in its copolymers utilizes a pyrolysis and gas chromatographic technique (82).

a. Apparatus. *Pyrolysis tube*, Vycor, Figure 34.

Gas chromatograph with hydrogen flame detector and a $\frac{1}{4}$-in. × 7-ft column packed with 25% LAC-2-R-446 (polydiethyleneglycolpentaerythritol adipate) plus 2% phosphoric acid on 80–100-mesh Chromosorb W.

b. Standardization. Place about 20 ml of chloroform in a 50-ml volumetric flask and weigh. Add about 400 mg of glacial acetic acid to the flask, stopper, and reweigh. Dilute to volume with chloroform. Inject 3 μl of the standard solution into the gas chromatograph operated at a column temperature of 132°C. Measure the area of the resulting acetic acid peak.

c. Procedure. Weigh 25–50 mg of the copolymer into the pyrolysis tube and evacuate to a pressure of 1 mm. Close the stopcock and place the tube in a heating block or furnace which has previously been heated to 350°C. Allow the tube to remain at 350°C for 10 min. Remove the tube and, after cooling to room temperature, place it in a Dry Ice–acetone bath at −70°C for 15 min.

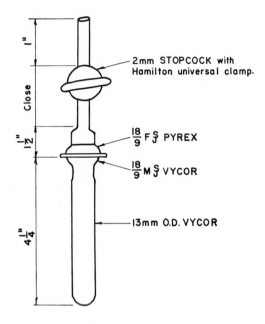

Fig. 34.   Pyrolysis tube.

Pipet 5 ml of chloroform into the tube.   Close the stopcock and
allow the tube and contents to warm to room temperature.   Shake
the tube vigorously to dissolve the acetic acid.   Inject 3 µl of the
solution into the gas chromatograph.

d. Calculation.   Measure the area of the resulting acetic acid
peak.   Compare the sample peak area with the area produced by
the standard acetic acid solution.

$$\% \text{ Vinyl acetate} = \frac{A \times B}{C \times D} \times 14.3$$

where A = mg of acetic acid in standard, B = peak area of sample,
C = peak area of standard, and D = sample weight in mg.

H. SEPARATION OF POLYVINYL CHLORIDE AND POLYVINYL ACETATE

Two techniques available for analyzing polymer blends are fractional precipitation of an individual polymer species from solution

by the addition of a specific nonsolvent and extraction of a polymeric material with a series of liquids of increasing solvent action (83).

a. Procedure.   Dissolve 1 g of the plastic in 15 ml of redistilled o-dichlorobenzene, warming slightly, if necessary.   Add slowly with stirring 100 ml of warm 96% ethanol.   Reflux the mixture for 2 hr on a steam bath.   Allow the mixture to cool to room temperature. Separate the precipitated polymer from the solution by filtration on a tared medium-porosity sintered-glass crucible or by centrifugation. Rinse the precipitate with three 20-ml portions of a 1:3 solution of ethyl ether and ethanol.   Dry the crucible and contents to constant weight in a vacuum oven at 70°C.

$$\% \text{ Polyvinyl chloride} = \frac{\text{Weight of residue}}{\text{Weight of sample}} \times 100$$

Combine the filtrate and the washings and evaporate almost to dryness on a steam bath.   Dissolve the residue in 30 ml of acetone and add 10 ml of water dropwise with stirring.   Filter the mixture through a tared medium-porosity sintered-glass crucible.   Wash the precipitate with three 5-ml portions of a 4:1 solution of acetone and water.   Dry the crucible and contents in a vacuum oven at 70°C.

$$\% \text{ Polyvinyl acetate} = \frac{\text{Weight of residue}}{\text{Weight of sample}} \times 100$$

Combine the filtrate and washings in a beaker and evaporate to dryness on a steam bath.   Transfer the beaker to a vacuum desiccator and dry to constant weight.

$$\% \text{ Plasticizer} = \frac{\text{Weight of residue}}{\text{Weight of sample}} \times 100$$

I. VINYL ALCOHOL IN COPOLYMERS

Acetylation with acetic anhydride–pyridine is a satisfactory procedure for determining comonomeric vinyl alcohol.   The excess acetic anhydride is decomposed with water and the liberated acetic acid titrated with potassium hydroxide.   Kennett (84) recommends addition of a surfactant to the potassium hydroxide solution.

a. Reagents. *Acetic anhydride–pyridine reagent.* Add 60 g of c.p. acetic anhydride to 440 g of freshly distilled pyridine. Store in a brown glass-stoppered bottle. Prepare fresh weekly. Slight discoloration does not indicate deterioration; however, the blank value should not change significantly.

Pyridine–water solution. Add 8 ml of distilled water to 92 ml of freshly distilled pyridine.

Potassium hydroxide, 0.5N. Dissolve 33 g of potassium hydroxide in 200 ml of anhydrous methanol. Add 20 ml of Tergitol 4. Dilute the solution to a volume of 1 liter with freshly distilled pyridine and store in a brown glass-stoppered bottle. Standardize against benzoic acid.

b. Procedure. Weigh 2–3 g of sample into a 200-ml pressure bottle containing 20.0 ml of acetic anhydride–pyridine reagent. Wash down the neck of the bottle with 35 ml of pyridine. Stopper the bottle and place on a steam bath behind a safety shield. Swirl the bottle frequently until the polymer has dissolved. Allow the bottle to remain on the steam bath for 20 min after complete solution. Remove from the steam bath and allow to cool to room temperature. Add 25 ml of the pyridine–water solution. Stopper the bottle and swirl gently to mix the contents. Replace the bottle on the steam bath for 30 min to hydrolyze the excess acetic anhydride. Remove the bottle and allow it to cool to room temperature. Wash down the sides of the bottle with 25 ml of pyridine. Cool the bottle in an ice bath at 0°C for 15 min. Titrate the liberated acetic acid to the phenolphthalein end point with 0.5N hydrochloric acid. Run a blank on the reagents.

c. Calculation.

$$\% \text{ Vinyl alcohol} = \frac{\text{Net ml KOH} \times N \times 0.0440}{\text{Sample weight, g}} \times 100$$

J. VINYL ACIDS

a. Reagents. *Potassium hydroxide, 0.1N,* in methanolic solution.

Ethylene dichloride. Neutralize to phenolphthalein end point with 0.1N methanolic potassium hydroxide.

b. Procedure. Weigh 5 g of sample accurately into a 250-ml erlenmeyer flask and dissolve in 150 ml of ethylene dichloride. Solution may be hastened by warming to 60–70°C with gentle swirling. Cool the flask and contents to room temperature and titrate to the phenolphthalein end point with 0.1N potassium hydroxide.

c. Calculation.

$$\% \text{ Vinyl acid as maleic acid} = \frac{\text{ml KOH} \times N \times 0.085}{\text{Sample, g}} \times 100$$

K. DETERMINATION OF ACRYLATES, METHACRYLATES, OR MALEATES IN COPOLYMERS

Pyrolysis–gas chromatography and infrared spectrometry are suitable for the identification of comonomeric esters such as the acrylates, methacrylates, and maleates. Frequently, however, quantitative measurement by these techniques is difficult. A total alkoxyl determination is suitable for determining the methyl, ethyl, or butyl esters (including isomers) when the alcohol component is known and when only one species is present. An alkoxyl–gas chromatographic procedure may be used for the determination of mixtures of methyl, ethyl, and butyl esters (85). Comonomeric esters containing higher alcohol constituents, 2-ethylhexanol, etc., may be determined by a saponification–gas chromatographic procedure (85). The procedures are also applicable to the analysis of ester plasticizers.

1. Total Alkoxyl Determination

a. Apparatus. *Distillation unit*, Figure 35.

b. Reagents. *Bromine–acetic acid solution.* Dissolve 100 g of potassium acetate in a solution composed of 900 ml of glacial acetic acid and 100 ml of acetic anhydride. Dissolve 5 ml of bromine in 145 ml of the acetic acid solution. Prepare the bromine solution fresh daily.

Sodium acetate solution. Dissolve 220 g of sodium acetate in water and dilute to 1 liter.

Dilute sulfuric acid. Carefully mix one volume of reagent grade sulfuric acid with nine volumes of water.

Hydriodic acid, 57%. Hydriodic acid forms with water a constant-boiling mixture (boiling point 126–127°C) which contains 57% hydriodic acid. The concentration of hydriodic acid in the reagent should be not less than 56.5%.

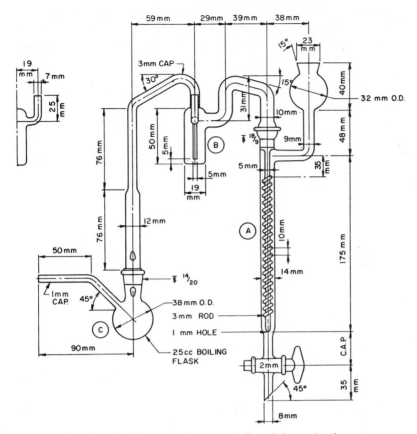

Fig. 35. Distillation unit for total alkoxyl determination.

c. Procedure. Add 3 ml of water to the scrubbing trap and 10 ml of bromine–acetic acid solution to the receiver and attach the receiver to the distillation apparatus (Fig. 35). Weigh a 50-mg sample into the side arm flask, add 3 ml of melted phenol, and warm on a steam bath until the sample has dissolved. Occasionally it is necessary to dissolve the sample in o-dichlorobenzene prior to the addition of the phenol. Add 6 ml of constant boiling hydriodic acid to the flask and attach it at once to the apparatus. If the addition of hydriodic acid precipitates the sample, gently shake the reaction flask from time to time so that the precipitate floats freely in the boiling mixture. Connect the side arm to a source of nitrogen and pass a current of the gas into the apparatus at a rate of 2 bubbles/sec. Immerse the flask in an oil bath at 150°C and heat for 3 hr.

Wash the contents of the receiver into a flask containing 15 ml of sodium acetate solution. Add formic acid dropwise to reduce the bromine, and then add about 6 drops more. A total of 10–12 drops is usually required. Add 3 g of potassium iodide and 15 ml of dilute sulfuric acid. Titrate the liberated iodine with $0.1N$ sodium thiosulfate. Make a blank determination.

d. Calculation.

$$\% \text{ Ester} = \frac{A \times B}{\text{Sample wt.} \times 6} \times 100$$

where A is ml of $0.1N$ sodium thiosulfate and B is the ester equivalent weight of monomer.

2. Alkoxyl Determination by Gas Chromatography

a. Apparatus. *Distillation unit*, Figure 36.

Gas chromatograph with thermal conductivity detector system and a 0.25-in. × 9-ft stainless steel column packed with di-2-ethylhexyl sebacate on fire brick.

b. Reagents. *o-Methoxybenzoic acid* (standard for methyl esters).

p-Ethoxybenzoic acid (standard for ethyl esters).

Dibutyl phthalate (standard for butyl esters).

The purity of the standards may be determined by the total alkoxyl procedure.

c. Standardization. Hydrolyze the standards as described for the total alkoxyl (Section X-K-1), substituting the absorber tube (Fig. 36) for the bromine receiver. Place a small amount of magnesium perchlorate in the inlet section of the absorber and add 0.25 ml of n-heptane to the outlet section. Assemble the apparatus and place the absorber in a Dry Ice bath at −80°C. After hydrolysis, disconnect the trap and allow it to warm to room temperature. Inject 2 μl of the heptane solution into the gas chromatograph at a

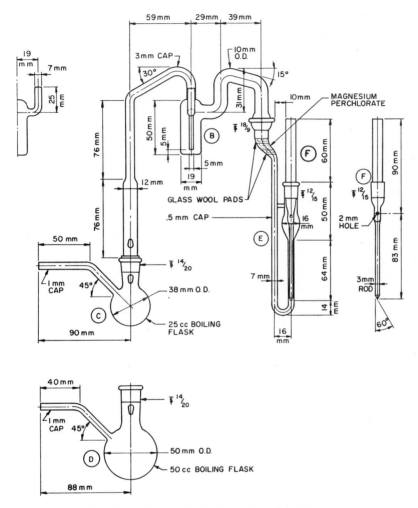

Fig. 36.   Alkoxyl distillation unit with cold trap.

column temperature of 70°C.   Measure the peak area of the individual alkyl iodides.

  d. **Procedure.**   Weigh 50 mg of sample into the side-arm flask
and proceed as directed under standardization.   Measure the peak
area of the individual iodides.

e. Calculations.

$$\% \text{ Methoxyl (CH}_3\text{O)} = \frac{\text{Peak area sample (methyl iodide)}}{\text{Peak area of standard (methyl iodide)}}$$
$$\times \frac{\text{Methoxyl standard mg} \times 0.2533}{\text{Sample, mg}}$$

$\% \text{ Ethoxyl (C}_2\text{H}_5\text{O)}$

$$= \frac{\text{Peak area sample (ethyl iodide)} \times \text{Ethoxyl standard mg} \times 0.2767}{\text{Peak area of standard (ethyl iodide)} \times \text{Sample mg}}$$

$\% \text{ Butoxyl (C}_4\text{H}_9\text{O)}$

$$= \frac{\text{Peak area sample (butyl iodide)} \times \text{Butoxyl standard, mg} \times 0.2316}{\text{Peak area standard (butyl iodide)} \times \text{Sample, mg}}$$

Calculate the concentration of the desired ester from the alkoxyl concentration.

3. Saponification and Gas Chromatography

a. Apparatus. *Gas chromatograph* with hydrogen flame detector and a $\frac{1}{4}$-in. $\times$ 4-ft stainless steel column packed with 23% Oronite N1-W Dispersant on 42–60-mesh firebrick.

Parr bomb, 22-ml, stainless steel, with lead gasket.

Oven operating at 160°C and equipped with a device for slowly rotating the bombs.

b. Reagent. Alcoholic potassium hydroxide. Dissolve 15 g of reagent grade potassium hydroxide in 50 ml of Formula 2B absolute ethanol.

c. Standardization. Accurately weigh 25–50 mg of the desired monomeric ester into the Parr bomb. Add 13 ml of tetrahydrofuran and 3 ml of alcoholic potassium hydroxide solution. Assemble the bomb, making sure the head gasket is in place, and tighten with a wrench. Place the bomb in the rotating device in the oven and rotate slowly at 160°C for 4 hr. Remove the bomb from the oven and allow it to cool to room temperature. Transfer the contents of the bomb to a 25-ml volumetric flask. Rinse the bomb and bomb lid with Formula 2B absolute ethanol, and add the rinsings to the flask. Dilute to volume with the ethanol.

Transfer a 10-ml aliquot of the solution to a 1-oz French square bottle, and add 0.50 ml of concentrated hydrochloric acid. Cap

the bottle and shake for several seconds.   Allow the precipitate to
settle.   Inject 3 μl of the supernatant liquid into the gas chromatographic column at a temperature of 160°C.   Measure the area of
the resulting alcohol peak.

d. Procedure.   Accurately weigh 400–500 mg of the copolymer
into the Parr bomb and dissolve in 13 ml of tetrahydrofuran.   Add
3 ml of the alcoholic potassium hydroxide solution and proceed as
directed under standardization.

e. Calculation.   Measure the area of the resulting alcohol peak.

$$\% \text{ Ester} = \frac{A \times B}{C \times D} \times 100$$

where A is mg of monomer in the standard, B is the peak area of
the sample, C is the peak area of the standard, and D is the sample
weight in mg.

L. Vinyl Acetate in Polymers of Vinyl Alcohol and Vinyl Acetate

Weigh 1 g of polymer (or equivalent of polymer solution) into a
125-ml flask.   Dissolve the sample in 10 ml of water, warming if
necessary.   Add 25 ml of methanol and 25 ml of 0.5N sodium
hydroxide.   Reflux for 1 hr and then cool to room temperature.
Titrate the excess sodium hydroxide with 0.5N hydrochloric acid
using phenolphthalein as the indicator.   Carry a reagent blank
through the same procedure.

$$\% \text{ Vinyl acetate} = \frac{(A - B) \times 0.086}{\text{Weight of dry sample}} \times 100$$

where A is ml of hydrochloric acid required for blank, B is ml of
hydrochloric acid required for sample, and N is the normality of
hydrochloric acid.

M. Specific Gravity

1. Pycnometric Method

The pycnometric procedure (86) is perhaps the most versatile
procedure for determining the specific gravity of plastics, being
applicable to powdered or ground plastics as well as to fabricated

articles.   The pycnometer is filled with water at 23.0° ± 0.1°C, weighed, emptied, dried, and reweighed.   Approximately 1–5 g of specimen is added to the dry pycnometer, the exact weight of the specimen being determined by difference.   The pycnometer containing the specimen is filled with water at 23.0° ± 0.1°C (air bubbles are removed by placing the pycnometer in a vacuum desiccator for several minutes and reweighing).   Kerosene or other suitable liquid may be used if the specimen is soluble in or otherwise affected by water.   Methanol may be used if the specific gravity of the specimen is less than 1.0.   When nonaqueous liquids are used, the calculations are corrected by multiplying by the specific gravity of the liquid at 23°C.

$$\text{sp gr, } 23/23°C = S/(A - B + S)$$

where S is the sample weight, A is the weight of pycnometer filled with water, and B is the weight of pycnometer containing specimen filled with water.

2. Specific Gravity of Films

A rapid method for determining the specific gravity of polyvinyl chloride films has been described by Khoroshoya (87).   The method appears to be applicable to most plastic films that are not affected by water.

a. Reagents.   Prepare the following aqueous potassium iodide solutions.

Solution	Potassium iodide, g/liter	Sp gr of solution, 20/4°C
A	772.9	1.5458
B	465.8	1.3308
C	155.5	1.1104

b. Procedure.   Adjust the temperature of the solutions to approximately 20°C and pipet 10.0 ml of each into separate 50-ml beakers.   Place a 4 × 4-mm piece of the film on the surface of each liquid.   Select the solution of lowest specific gravity on which the film floats and titrate with water at 20°C.   Add the water slowly, mixing the solution thoroughly after each addition.   If bubbles appear on the surface of the film, remove them with a fine wire.   Record the amount of water added when the film just falls to the bottom of the beaker.

c. Calculation.

$$\text{sp gr, } 20/4°C = \frac{\text{ml water} + 10S}{\text{ml water} + 10}$$

where S is the specific gravity of the potassium iodide solution.

Acknowledgment

The authors wish to thank Dr. W. J. Potts and Dr. V. A. Stenger for their helpful criticisms and suggestions.

References

1. T. P. G. Shaw, *Ind. Eng. Chem. Anal. Ed.*, **16**, 541 (1944); H. Nechamkin, *Ind. Eng. Chem. Anal. Ed.*, **15**, 40 (1943).
2. J. E. Ford and W. J. Roff, *J. Textile Inst.*, **45**, T580 (1954).
3. F. Feigl and V. Anger, *Mod. Plastics*, **37**, No. 9, 151 (1960).
4. H. Wechsler, *J. Polymer Sci.*, **11**, 233 (1953).
5. H. Winterscheidt, *Seifen-Oele-Fette-Wachse*, **80**, 239 (1954).
6. H. Rath and L. Heiss, *Kunststoffe*, **44**, 341 (1954).
7. J. H. Player, *Analyst*, **80**, 633 (1955).
8. B. Verdcourt, *Patra. J.*, **10**, 135 (1947).
9. R. L. Shriner, R. C. Fuson, and D. Y. Curtin, *Systematic Identification of Organic Compounds*, 4th ed., Wiley, New York, 1960, p. 58.
10. F. Feigl, *Spot Tests in Organic Analysis*, 6th ed., Elsevier, New York, 1960, p. 84.
11. Ref. 10, p. 98.
12. F. Feigl and J. R. Amaral, *Anal. Chem.*, **30**, 1148 (1958).
13. S. R. Palit and P. Ghosh, *J. Polymer Sci.*, **58**, 1225 (1962).
14. T. P. G. Shaw, *Ind. Eng. Chem. Anal. Ed.*, **16**, 541 (1944).
15. W. Gallay, *Can. J. Res.*, **14B**, 105 (1936).
16. S. R. Palit and A. R. Mukherjee, *J. Polymer Sci.*, **58**, 1243 (1962).
17. W. T. M. Johnson, *Offic. Dig. Federation Soc. Paint Technol.*, **32**, 427, 1067 (1960).
18. M. W. Long and J. G. Cobler, *Conf. Anal. Chem. Appl. Spectry.*, Pittsburgh, March 1964.
19. H. E. Kissinger and S. B. Newman, in *Analytical Chemistry of Polymers*, Part II, (*High Polymers* Vol. 12), G. M. Kline, Ed., Interscience, New York, 1962, p. 159; B. Ke, in *Newer Methods of Polymer Characterization* (*Polymer Reviews*, Vol. 6), B. Ke, Ed., Interscience, New York, 1964, p. 347; J. G. Cobler, in *Standard Methods of Chemical Analysis* (*Instrumental Methods of Analysis*), F. J. Welcher, Ed., Vol. 3, Part B, Van Nostrand, Princeton, 1966, p. 1637.
20. J. J. Keavney and E. C. Eberlin, *J. Appl. Polymer Sci.*, **3**, 47 (1960).

21. F. P. Reding, J. A. Faucher, and R. D. Whitman, *J. Polymer Sci.*, **57**, 483 (1962).
22. L. A. Wall, in *Analytical Chemistry of Polymers*, G. M. Kline, Ed., Vol. XII, Part II, Interscience, New York, 1962, p. 181; L. A. Wall and J. H. Flynn, *Rubber Chem. Technol.* **35**, 1157 (1962); J. G. Cobler, in *Standard Methods of Chemical Analysis* (*Instrumental Methods of Analysis*), F. J. Welcher, Ed., Vol. 3, Part B, Van Nostrand, Princeton, 1966, p. 1630.
23. H. H. G. Jellinek, *J. Polymer Sci.*, **4**, 13 (1949); S. L. Madorsky, *J. Polymer Sci.*, **9**, 133 (1952); E. S. Freeman and B. Carroll, *J. Phys. Chem.*, **62**, 394 (1958); D. A. Anderson and E. S. Freeman, *J. Polymer Sci.*, **54**, 253 (1961).
24. H. Melville, *Sci. Progr.* (*London*), **38**, 1 (1950).
25. J. G. Cobler and E. P. Samsel, *Soc. Plastics Engrs. Trans.*, **2**, 145 (1962); J. E. Guillet, W. C. Wooten, and R. L. Combs, *J. Appl. Polymer Sci.*, **3**, 61 (1960); E. A. Radell and H. C. Strutz, *Anal. Chem.*, **31**, 1890 (1959).
26. E. K. Plyer and C. W. Peters, *J. Res. Natl. Bur. Std.*, **45**, 462 (1950).
27. A. Elliott, E. J. Ambrose, and R. B. Temple, *J. Chem. Phys.*, **16**, 877 (1948); R. Zbinden, *Infrared Spectroscopy of High Polymers*, Academic Press, New York, 1964; S. Krimm, in *Infrared Spectroscopy and Molecular Structure*, M. Davies, Ed., Elsevier, New York, 1963, p. 270; C. Y. Liang, in *Newer Methods of Polymer Characterization* (*Polymer Reviews*, Vol. 6), B. Ke, Ed., Interscience, New York, 1964, p. 33.
28. M. R. Harvey, J. E. Stewart, and B. G. Achhammer, *J. Res. Natl. Bur. Std.*, **56**, 225 (1956).
29. J. Fahrenfort, *Spectrochim. Acta*, **17**, 698 (1961); J. Fahrenfort and W. M. Visser, *Spectrochim. Acta*, **18**, 1103 (1962).
30. R. J. Grisenthwaite, *Plastics* (*London*), **27**, 119 (1962).
31. S. Krimm and C. Y. Liang, *J. Polymer Sci.*, **22**, 95 (1956).
32. A. Kawasaki, J. Furukawa, T. Tsuruta, and S. Shiotani, *Polymer*, **2**, 143 (1961).
33. J. Schurz, H. Bayzer, and H. Stubchen, *Makromol. Chem.*, **23**, 152 (1957); T. Shimanouchi and M. Tasumi, *Bull. Chem. Soc. Japan*, **34**, 359 (1961); A. Narita, S. Ichinohe, and S. Enomoto, *J. Chem. Phys.*, **31**, 1151 (1959).
34. M. Tasumi and T. Shimanouchi, *Spectrochim. Acta*, **17**, 731 (1961).
35. A. Elliott, E. J. Ambrose, and R. B. Temple, *J. Chem. Phys.*, **16**, 877 (1948); T. Shimanouchi, S. Tsuchiya, and S. Mizushima, *J. Chem. Phys.*, **30**, 1365 (1959).
36. *ASTM Method D 2124-62T*. American Society for Testing and Materials, Philadelphia, Pa. *ASTM, Standards on Plastics*, Part 9, 1962, p. 1295.
37. D. N. Kendall, R. R. Hampton, H. Hausdorff, and F. Pristera, *Appl. Spectry.*, **7**, 179 (1953); R. B. DuVall, *Infrared Spectra of Plasticizers and Other Additives*, Dow Chemical Co., Midland, Mich., 1962.
38. J. M. Hunt, M. P. Wisherd, and L. C. Bonham, *Anal. Chem.*, **22**, 1478 (1950); F. A. Miller and C. H. Wilkins, *Anal. Chem.*, **24**, 1253 (1952).
39. S. Narita, S. Ichinohe, and S. Enomoto, *J. Polymer Sci.*, **37**, 251 (1959).
40. S. Enomoto, *J. Polymer Sci.*, **55**, 95 (1961).
41. S. Narita, S. Ichinohe, and S. Enomoto, *J. Polymer Sci.*, **36**, 389 (1959).

42. M. E. Rector, Dow Chemical Co., Midland, Mich., private communication, 1963.
43. S. Krimm, C. Y. Liang, and G. B. B. M. Sutherland, *J. Polymer Sci.*, **22**, 227 (1956).
44. H. Tadokoro, *Bull. Chem. Soc. Japan*, **32**, 1334 (1959).
45. E. R. Blout and R. Karplus, *J. Am. Chem. Soc.*, **70**, 862 (1948); E. J. Ambrose, A. Elliott, and R. B. Temple, *Proc. Roy. Soc. London, Ser. A*, **199**, 183 (1949).
46. K. Fujii, T. Mochizuki, S. Imoto, S. Ukida, and M. Matsumoto, *J. Polymer Sci.*, Part A, **2**, 2327 (1964).
47. Y. Nishino, J. Ukita, and T. Kominami, *J. Chem. Soc. Japan*, **58**, 159 (1955); E. Nagai and S. Kuribayashi, *Kobunshi Kagaku*, **12**, 322 (1955).
48. S. Murahashi, H. Yuki, H. Tadokoro, Y. Chatani, U. Yonemura, K. Fukuda, and M. Tsutsui, Paper Presented at Annual Meeting Soc. Polymer Sci. Tokyo, Japan, May, 1961.
49. C. Y. Liang and S. Krimm, *J. Polymer Sci.*, **27**, 241 (1958); A. Palm, *J. Phys. Chem.*, **55**, 1320 (1951); G. W. King, R. M. Hainer, and H. O. McMahon, *J. Appl. Phys.*, **20**, 559 (1949).
50. M. Kobayashi, *Bull. Chem. Soc. Japan*, **33**, 1416 (1960); H. Tadokoro, N. Nishiyama, S. Nozakura, and S. Murahashi, *J. Polymer Sci.*, **36**, 553 (1959); T. Onishi and S. Krimm, *J. Appl. Phys.*, **32**, 2320 (1961).
51. R. T. Scheddel, Dow Chemical Co., Midland, Mich., private communication, 1964.
52. G. G. Esposito, *Anal. Chem.*, **35**, 1439 (1963).
53. J. Zulaica and G. Guiochon, *Anal. Chem.*, **35**, 1725 (1963).
54. H. W. McCormick, *J. Polymer Sci.*, **36**, 341 (1959).
55. M. Freeman and P. P. Manning, *J. Polymer Sci. A*, **2**, 2017 (1964).
56. D. J. Streeter, Dow Chemical Co., Midland, Mich., private communication, 1957.
57. A. Beresniewicz, *J. Polymer Sci.*, **35**, 321 (1959).
58. D. J. Harmon, Meeting Am. Chem. Soc., Polymer Div., Chicago, *Polymer Preprints*, **5**, No. 2, 712 (Sept. 1964).
59. J. C. Moore and J. G. Hendrickson, Meeting Am. Chem. Soc., Polymer Div., Chicago, *Polymer Preprints*, **5**, No. 2, p. 706 (Sept. 1964).
60. R. F. Boyer and R. Simha, in *Styrene*, R. H. Boundy and R. F. Boyer, Eds., Am. Chem. Soc. Monograph No. 115, Reinhold, New York, 1952, p. 314.
61. F. W. Billmeyer, Jr., Meeting Am. Chem. Soc., Polymer Div., Chicago, *Polymer Preprints*, **5**, No. 2, 627 (Sept. 1964).
62. R. Tremblay, M. Rinfret, and R. Rivest, *J. Chem. Phys.*, **20**, 523 (1952).
63. W. S. Bahary and L. Bsharah, Meeting Am. Chem. Soc., Polymer Div., Philadelphia, *Polymer Preprints*, **5**, No. 1, 1 (1964).
64. W. Bushuk and H. Benoit, *Can. J. Chem.*, **36**, 1616 (1958).
65. M. Leng and H. Benoit, *J. Polymer Sci.*, **57**, 263 (1962).
66. P. Outer, C. Carr, and B. Zimm, *J. Chem. Phys.*, **18**, 830 (1950).
67. J. L. Jungnickel and F. T. Weiss, *J. Polymer Sci.*, **49**, 437 (1961); N. S.

812   J. G. COBLER, M. W. LONG, AND E. G. OWENS II

Schneider, J. D. Loconti, and L. G. Holmes, *J. Appl. Polymer Sci.*, **5**, 354 (1961); C. A. Baker and R. J. P. Williams, *J. Chem. Soc.*, **1956**, 2352.
68. J. H. S. Green and M. F. Vaughan, *Chem. Ind. (London)*, **1958**, 829.
69. E. Turska and L. Utracki, *J. Appl. Polymer Sci.*, **2**, 46 (1959).
70. J. D. Loconti and J. W. Cahill, *J. Polymer Sci.*, **63**, 3163 (1963).
71. F. C. Goodrich and M. J. R. Cantow, Meeting Am. Chem. Soc., Polymer Div., Chicago, *Polymer Preprints*, **5**, No. 2, 728 (Sept. 1964).
72. J. E. Blair, Meeting Am. Chem. Soc., Polymer Div., Chicago, *Polymer Preprints*, **5**, No. 2, 740 (Sept. 1964).
73. G. Kegeles, S. M. Klainer, and W. J. Salem, *J. Phys. Chem.*, **61**, 1286 (1957).
74. R. Buchdahl, H. A. Ende, and L. H. Peebles, *J. Phys. Chem.*, **65**, 1468 (1961).
75. L. E. Maley, Meeting Am. Chem. Soc., Polymer Div., Chicago, *Polymer Preprints*, **5**, No. 2, 720 (Sept. 1964).
76. F. W. Peaker, *Analyst*, **85**, 235 (1960).
77. D. A. Mapes, Dow Chemical Co., Midland, Mich., private communication, 1962.
78. American Society for Testing and Materials, Philadelphia, Pa., *ASTM Standards*, Part 6, 1955, p. 17.
79. *ASTM Method D1343-56*. American Society for Testing and Materials, Philadelphia, Pa., *ASTM Standards*, Part 8, 1958, p. 486.
80. N. Y. Parlashkevich, R. N. Gribkova, V. P. Bildina, and T. S. Ivanovskaya, (*Plasticheskie Massy* **1961**, No. 6, 55 [*Soviet Plastics Engl. Transl.*, **9**, 46 (1961)]; *Chem. Abstr.*, **57**, 6112g (1962).
81. M. R. Lardera, E. Cernia, and A. Mori, *Ann. Chim. Rome*, **46**, 194 (1956).
82. J. G. Cobler and E P. Samsal, *Soc. Plastics Engrs. Trans.*, **2**, 145 (1962).
83. W. J. Taat and W. vander Heul, *Chem. Weekblad*, **44**, 393 (1948).
84. C. J. Kennett, in *Analytical Chemistry of Polymers*, G. M. Kline, Ed., Part I, (*High Polymer*, Vol. 12), Interscience, New York, 1959, p. 481.
85. J. G. Cobler, in *Standard Methods of Chemical Analysis*, Vol. 2, Part B, F. J. Welcher, Ed., 6th ed., Van Nostrand, Princeton, 1963, pp. 2093, 2097.
86. *ASTM Method D792-60T*. American Society for Testing and Materials, Philadelphia, Pa., *ASTM Standards*, Part 9, 1961, p. 734.
87. E. S. Khoroshaya, G. I. Kovrigina, and V. I. Alekseenko, *Zavodsk. Lab.*, **28**, 205 (1962) [*Soviet Plastics Eng. Transl.*, **10**, 52 (1961)]; *Chem. Abstr.* **57**, 1062b (1962).

CHAPTER 16

POLYMER SAFETY FOR USES IN CONTACT WITH FOODS

RICHARD HENDERSON

Olin Mathieson Chemical Corp., New Haven, Conn.

Introduction

People in a primitive society obtain their food supplies from their immediate environment.   They learn by tribal custom and folklore what foods are safe and what animal and vegetable materials may be unsafe.   Much of the people's time is spent in obtaining food.

As a society develops advances in agriculture that increase productivity per man hour, fewer people are needed to supply the nutritional requirements. Methods of processing, packaging, and distributing agricultural products become increasingly necessary as the areas of agricultural production move away from the population centers of consumption. At the present time in the United States, only approximately 10% of the population is engaged in agriculture. In such a society, a large segment of the population can devote a major portion of its time to intellectual, technological, cultural, and social activities unrelated to supplying its nutritional needs. But tribal custom and folklore no longer serve as a means of evaluating the safety of different foodstuffs. The majority of the population depends on others for the wholesomeness of their daily bread.

An article in *The Wall Street Journal* of November 25, 1964 (1) mentions French bread flown from Paris to New York, St. Peter's fish caught in the Sea of Galilee and flown to the U.S., and papaya flown from Hawaii to a store in Hollywood. Foodstuffs handled by air freight in 1964 were double the amount handled in 1963. It seems obvious that the average individual who consumes foods produced, processed, and shipped to him from remote sources is unable to determine their safety. The packaging material in which a food is distributed from the producer or processor to the consumer is one factor that must be evaluated in determining the safety of the food.

The chemist who makes a new polymer that may be useful as a food-packaging film is frequently impatient with the time required to establish the safety of the new materials. He should be thankful that present analytical and toxicological procedures make it possible to establish such safety in a few years, and therefore he is not dependent on the generations of human experience that established the safety of primitive man's food supply. This does not mean that we should accept present methods as the ultimate and not strive for better analytical and toxicological procedures that may improve our ability to predict the safety of a new material.

Although the Food Additives Amendment of 1958 (2) and the general regulations pertaining to it focussed attention on procedures for establishing the safety of food-packaging films, a review of the history of the present food, drug, and cosmetic laws shows that

Congress recognized the importance of packaging as a factor in providing a safe food supply by drafting the Pure Food Act of 1906 (3).

II. Historical Development of the U.S. Food, Drug, and Cosmetic Act

The change from a rural agricultural economy to an urban industrial one in the United States started about the middle of the 19th century.   The building of railroads followed by the development of the automobile provided means of food transportation over great distances.   The reaper and the tractor reduced manpower needs on the farm while the need for manpower in the new industries was increasing rapidly.   The investigations of Professor H. L. Russell in Wisconsin in 1898, and those of Professors S. C. Prescott and W. Lyman Underwood in Massachusetts in 1896, demonstrating the role of bacteria in the spoilage of canned foods, furnished a scientific basis for the canning industry.   By the beginning of the 20th century, there were increasing numbers of persons who depended on others to produce their foods.   There also was growing concern regarding the quality and safety of the food supply.

There were food laws in the United States as early as 1848, but these early laws were not very effective.   The first effective law was the Pure Food Act passed by Congress on June 30, 1906 (3). This act has been more commonly known as the Wiley Act because it was largely through the efforts of Harvey W. Wiley that it was enacted and became effective.

The following section of this act can be construed to include adulteration of food from the components of packaging (3).

Section 7.   That for the purposes of this Act an article shall be deemed to be adulterated:

In the case of food:

First.   If any substance has been mixed and packed with it so as to reduce or lower or injuriously affect its quality or strength.

At that time, the primary concern was with the grosser forms of adulteration.   Wooden boxes and crates, paper bags, metal cans, and glass bottles were the major food-packaging materials.   Recent developments in the field of bacteriology had focussed attention on bacterial spoilage.   The packaging materials had a history of safe use and there was little thought given to testing the migration of packaging components into food or the toxicity of trace components that might migrate from the packaging into the food.

It is interesting to note the following statement in the Pure Food Act of 1906 in the light of present efforts to have complete factory inspection and complete disclosure of all formulations.

> That nothing in this Act shall be construed as requiring or compelling proprietors or manufacturers of proprietary foods which contain no unwholesome added ingredient to disclose their trade formulas, except insofar as the provisions of this Act may require to secure freedom from adulteration or misbranding.

This act was amended six times in the first 30 years of its existence. It was repealed and replaced by the Food, Drug, and Cosmetics Act of 1938. The 1938 Act contains the provision (4):

> *Section 402.* A food shall be deemed to be adulterated—(a) (6) if its container is composed, in whole or in part, of any poisonous or deleterious substance which may render the contents injurious to health; . . .

The burden of proving adulteration rested with the government. The supplier of a new packaging material had no legal procedure for establishing whether food packed in his material would be considered adulterated by the Food and Drug Administration. Similarly, the 1938 act did not provide any means other than in Standards of Identity for establishing the F.D.A. acceptability of new compounds added directly to foods for the purpose of improving their texture, flavor, or other qualities. George P. Larrick, then Commissioner of Food and Drugs, U.S. Department of Health, Education, and Welfare, summarized, in 1963, the opinion of Food and Drug officials (5):

> We thought in 1938 that the problem of poisonous additives could best be handled by a rule banning all of them except when required in production or unavoidable in good manufacturing practice. But this left wholly unregulated the additives of unknown or uncertain toxicity. And enforcement against cumulative poisons was difficult indeed. The recent amendments have faced the problem by requiring advance proof that all proposed additives will be safe for their intended uses, even though in larger amounts and in other circumstances it may be clear that the additives cannot be classed as harmless or even nontoxic.

Although the 1938 Act did not provide any legal means for obtaining Food and Drug Administration approval of food additives, including indirect additives that might migrate from a packaging material, F.D.A. personnel were consulted by members of the food industry and the packaging industry regarding the acceptability of new materials. F.D.A. personnel provided informal opinions relative to the acceptability of new materials. Some of the data on packaging materials furnished for these informal opinions have been published (6–8). Other data are available only in the files of the company providing such data and in F.D.A. files. Such data are important because they furnished the basis for "prior sanctions" accepted under F.D.A. regulations for enforcement of the Food Additives Amendment of 1958 discussed below.

Prior to September 6, 1958, food manufacturers and suppliers of food-packaging materials could establish legal acceptability of a new food additive other than pesticides only by:

1. Having the additive listed in a standard of identity for a particular food.

2. Obtaining a favorable court decision that a food containing the new additive was not adulterated within the meaning of the Food, Drug, and Cosmetics Act of 1938.

Experience in establishing standards of identity led the food industry to the opinion that this method was not a suitable one for determining F.D.A. acceptability of a new food additive. Disadvantages of cost and adverse publicity inherent in obtaining a court decision, regardless of whether the decision might be favorable, made this route of establishing acceptability of a new food additive undesirable. The Food Additives Amendment of 1958 resulted from the efforts of Congress and the industries involved in providing a safe food supply to establish a legal procedure for acceptance by the Food and Drug Administration of new food additives, either incorporated in foods by direct addition or by migration from a food-packaging material.

The Food, Drug, and Cosmetics Act as amended and the general regulation for its enforcement (2) formally recognized the fact that technological progress makes available new chemicals that can be safely added to foods to improve them and to benefit consumers. This law provides a means for industry to obtain from the Food and Drug Administration acceptance of new food additives. The basic philosophy underlying the law, to establish the safety of a food additive before its widespread use, is sound even though the mechanics of administering it are frequently extremely frustrating to industries trying to obtain F.D.A. acceptance of a new product.

The sections of the Food Additives Amendment of 1958 (Public Law 85-929, 72 Stat. 1784) pertinent to this discussion are:

SEC. 2. Section 201, as amended, of the Federal Food, Drug, and Cosmetic Act is further amended by adding at the end of such section the following new paragraphs:

(*s*) The term "food additive" means any substance the intended use of which results or may reasonably be expected to result, directly or indirectly, in its becoming a component or otherwise affecting the characteristics of any food (including any substance intended for use in producing, manufacturing, packing, processing, preparing, treating, packaging, transporting, or holding food; and including any source of radiation intended for any such use), if such substance is not generally recognized, among experts qualified by scientific training and experience to evaluate its safety, as having been adequately shown through scientific procedures (or, in the case of a substance used in food prior to January 1, 1958, through either scientific procedures or experience based on common use in food) to be safe under the conditions of its intended use; except that such term does not include—

(*3*) any substance used in accordance with a sanction or approval granted prior to the enactment of this paragraph pursuant to this Act, the Poultry Products Inspection Act (*21 U.S.C. 451* and the following) or the Meat Inspection Act of March 4, 1907 (*34 Stat. 1260*), as amended and extended (*21 U.S.C. 71* and the following).

(*t*) The term "safe," as used in paragraph (*s*) of this section and in *section 409*, has reference to the health of man or animal.

SEC. 3. (*a*) Clause (*2*) of *section 402* (*a*), as amended, of such Act is amended to read as follows: "(*2*) (*A*) if it bears or contains any added poisonous or added deleterious substance (except a pesticide chemical in or on a raw agricultural commodity and except a food additive) which is unsafe within the meaning of *section 406*, or (*B*) if it is a raw agricultural commodity and it bears or contains a pesticide chemical which is unsafe within the meaning of *section 408* (*a*), or (*C*) if it is, or it bears or contains, any food additive which is unsafe within the meaning of *section 409*."

FOOD ADDITIVES

Unsafe Food Additives

SEC. 409. (*a*) A food additive shall, with respect to any particular use or intended use of such additives, be deemed to be unsafe for the purposes of the application of clause (*2*) (*C*) of *section 402* (*a*), unless—

(*1*) it and its use or intended use conform to the terms of an exemption which is in effect pursuant to subsection (*i*) of this section; or

(*2*) there is in effect, and it and its use or intended use are in conformity with, a regulation issued under this section prescribing the conditions under which such additive may be safely used.

While such a regulation relating to a food additive is in effect, a food shall not, by reason of bearing or containing such an additive in accordance with the regulation, be considered adulterated within the meaning of clause (*1*) of *section 402* (*a*).

Petition To Establish Safety

(*b*) (*1*) Any person may, with respect to any intended use of a food additive, file with the Secretary a petition proposing the issuance of a regulation prescribing the conditions under which such additive may be safely used.

(*2*) Such petition shall, in addition to any explanatory or supporting data, contain—

(*A*) the name and all pertinent information concerning such food additive, including, where available, its chemical identity and composition;

(*B*) a statement of the conditions of the proposed use of such additive, including all directions, recommendations, and suggestions proposed for the use of such additive, and including specimens of its proposed labeling;

(*C*) all relevant data bearing on the physical or other technical effect such additive is intended to produce, and the quantity of such additive required to produce such effect;

(*D*) a description of practicable methods for determining the quantity of such additive in or on food, and any substance formed in or on food, because of its use; and

(*E*) full reports of investigations made with respect to the safety for use of such additive, including full information as to the methods and controls used in conducting such investigations.

(*3*) Upon request of the Secretary, the petitioner shall furnish (or, if the petitioner is not the manufacturer of such additive, the petitioner shall have the manufacturer of such additive furnish, without disclosure to the petitioner) a full description of the methods used in, and the facilities and controls used for, the production of such additive.

(*4*) Upon request of the Secretary, the petitioner shall furnish samples of the food additive involved, or articles used as components thereof, and of the food in or on which the additive is proposed to be used.

(*5*) Notice of the regulation proposed by the petitioner shall be published in general terms by the Secretary within thirty days after filing.

Action on the Petition

(*c*) (*1*) The Secretary shall—

(*A*) by order establish a regulation (whether or not in accord with that proposed by the petitioner) prescribing, with respect to one or more proposed uses of the food additive involved, the conditions under which such additive may be safely used (including, but not limited to, specifications as to the particular food or classes of food in or on which such additive may be used, the maximum quantity which may be used or permitted to remain in or on such food, the manner in which such additive may be added to or used in or on such food, and any directions or other labeling or packaging requirements for such additive deemed necessary by him to assure the safety of such use), and shall notify the petitioner of such order and the reasons for such action; or

(B) by order deny the petition, and shall notify the petitioner of such order and of the reasons for such action.

(2) The order required by paragraph (1) (A) or (B) of this subsection shall be issued within ninety days after the date of filing of the petition, except that the Secretary may (prior to such ninetieth day), by written notice to the petitioner, extend such ninety-day period to such time (not more than one hundred and eighty days after the date of filing of the petition) as the Secretary deems necessary to enable him to study and investigate the petition.

(3) No such regulation shall issue if a fair evaluation of the data before the Secretary—

(A) fails to establish that the proposed use of the food additive, under the conditions of use to be specified in the regulation, will be safe: Provided, That no additive shall be deemed to be safe if it is found to induce cancer when ingested by man or animal, or if it is found, after tests which are appropriate for the evaluation of the safety of food additives, to induce cancer in man or animal; or

(B) shows that the proposed use of the additive would promote deception of the consumer in violation of this Act or would otherwise result in adulteration or in misbranding of food within the meaning of this Act.

(4) If, in the judgment of the Secretary, based upon a fair evaluation of the data before him, a tolerance limitation is required in order to assure that the proposed use of an additive will be safe, the Secretary—

(A) shall not fix such tolerance limitation at a level higher than he finds to be reasonably required to accomplish the physical or other technical effect for which such additive is intended; and

(B) shall not establish a regulation for such proposed use if he finds upon a fair evaluation of the data before him that such data do not establish that such use would accomplish the intended physical or other technical effect.

(5) In determining, for the purposes of this section, whether a proposed use of a food additive is safe, the Secretary shall consider among other relevant factors—

(A) the probable consumption of the additive and of any substance formed in or on food because of the use of the additive;

(B) the cumulative effect of such additive in the diet of man or animals, taking into account any chemically or pharmacologically related substance or substances in such diet; and

(C) safety factors which in the opinion of experts qualified by scientific training and experience to evaluate the safety of food additives are generally recognized as appropriate for the use of animal experimentation data.

Regulation Issued on Secretary's Initiative

(*d*) The Secretary may at any time, upon his own initiative, propose the issuance of a regulation prescribing, with respect to any particular use of a food additive, the conditions under which such additive may be safely used, and the reasons therefor. After the thirtieth day following publication of such a proposal, the Secretary may by order establish a regulation based upon the proposal.

Publication and Effective Date of Orders

(*e*) Any order, including any regulation established by such order, issued under subsection (*c*) or (*d*) of this section, shall be published and shall be effective upon publication, but the Secretary may stay such effectiveness if, after issuance of such order, a hearing is sought with respect to such order pursuant to subsection (*f*).

Objections and Public Hearing

(*f*) (*1*) Within thirty days after publication of an order made pursuant to subsection (*c*) or (*d*) of this section, any person adversely affected by such an order may file objections thereto with the Secretary, specifying with particularity the provisions of the order deemed objectionable, stating reasonable grounds therefor, and requesting a public hearing upon such objections. The Secretary shall, after due notice, as promptly as possible hold such public hearing for the purpose of receiving evidence relevant and material to the issues raised by such objections. As soon as practicable after completion of the hearing, the Secretary shall by order act upon such objections and make such order public.

(*2*) Such order shall be based upon a fair evaluation of the entire record at such hearing, and shall include a statement setting forth in detail the findings and conclusions upon which the order is based.

(*3*) The Secretary shall specify in the order the date on which it shall take effect, except that it shall not be made to take effect prior to the ninetieth day after its publication, unless the Secretary finds that emergency conditions exist necessitating an earlier effective date, in which event the Secretary shall specify in the order his findings as to such conditions.

Judicial Review

(*g*) (*1*) In a case of actual controversy as to the validity of any order issued under subsection (*f*), including any order thereunder with respect to amendment or repeal of a regulation issued under this section, any person who will be adversely affected by such order may obtain judicial review by filing in the United States Court of Appeals for the circuit wherein such person resides or has his principal place of business, or in the United States Court of Appeals for the District of Columbia Circuit, within sixty days after the entry of such order, a petition praying that the order be set aside in whole or in part.

(*2*) A copy of such petition shall be forthwith served upon the Secretary, or upon any officer designated by him for that purpose, and thereupon the Secretary shall certify and file in the court a transcript of the proceedings and the record on which he based his order. Upon such filing, the court shall have exclusive jurisdiction to affirm or set aside the order complained of in whole or in part. The findings of the Secretary with respect to questions of fact shall be sustained if based upon a fair evaluation of the entire record at such hearing. The court shall advance on the docket and expedite the disposition of all cases filed therein pursuant to this section.

(3) The court, on such judicial review, shall not sustain the order of the Secretary if he failed to comply with any requirement imposed on him by subsection (f) (2) of this section.

(4) If application is made to the court for leave to adduce additional evidence, the court may order such additional evidence to be taken before the Secretary and to be adduced upon the hearing in such manner and upon such terms and conditions as to the court may seem proper, if such evidence is material and there were reasonable grounds for failure to adduce such evidence in the proceedings below. The Secretary may modify his findings as to the facts and order by reason of the additional evidence so taken, and shall file with the court such modified findings and order.

(5) The judgment of the court affirming or setting aside, in whole or in part, any order under this section shall be final, subject to review by the Supreme Court of the United States upon certiorari or certification as provided in *section 1254* of *title 28* of the United States Code. The commencement of proceedings under this section shall not, unless specifically ordered by the court to the contrary, operate as a stay of an order.

In summary, prior to 1906 it was a case of "let the buyer beware" relative to the safety of foods. The Pure Food Act of 1906 defined in general terms foods that would be considered adulterated; by inference, food could also be considered adulterated by components migrating from the packaging. The Food, Drug, and Cosmetics Act of 1938 included a specific statement regarding safety of containers but did not require proof of safety before a packaging material was used for food packaging. The Food Additives Amendment of 1958 and the Definitions and Procedural Regulations under this Amendment require proof of safety and approval by the Food and Drug Administration before a packaging material is used for food packaging. The supplier of a polymer film for food packaging can market a new film for food use without first establishing its safety but few food packers will be likely to use such a film. Therefore, it has become necessary for suppliers of polymers that may come in contact with food to establish the safety of the polymers in contact with food if they wish to participate in the food industry market for their products.

III. Procedures for Establishing Safety of
Polymers for Contact with Foods

A. DEFINITION OF A FOOD ADDITIVE

The Food Additives Amendment of 1958 (2) defines a food additive as "any substance, the intended use of which results or may reasonably be expected to result, directly or indirectly, in its becoming a component or otherwise affecting the characteristics of any food . . . " unless the substance is generally recognized as safe for its intended use or is used in accordance with a sanction or approval granted prior to enactment of the 1958 Amendment. Substances generally recognized as safe or having prior sanctions as defined under *Section 201* of the Food Additives Amendment of 1958 are not considered to be food additives.   A food additive will be considered unsafe unless it and its use or intended use are in conformity with an effective regulation issued in accordance with the requirements of the Food Additives Amendment of 1958.

B. FOOD ADDITIVE STATUS OF POLYMER COMPONENTS

Polymer components that may migrate into food may be generally recognized as safe or included in prior sanctions relative to their intended use.   If this is the case, such components are not food additives within the meaning of the Food Additives Amendment of 1958.   Subpart *B* of the Food and Drug Administration Food Additives Regulations lists chemicals that are generally recognized as safe for specific uses.   Pertinent sections of this Subpart are included in Appendix I.   Subpart *E* of the Regulations lists chemicals that have prior sanction for use in food packaging materials; this list is given in Appendix II.   New information may necessitate changes in these Subparts; up-to-date copies can be obtained from the Division of Public Information, Food and Drug Administration, Department of Health, Education, and Welfare, Washington, D.C.

Polymer additives—antioxidants, stabilizers, release agents, or other additives—that are generally recognized as safe or have prior sanctions for a particular use may interact with the polymer to form new products.   If such reaction products are formed and may migrate into food, their safety as food additives must be established.

Components of a new polymer that may migrate into food may be included as safe food additives in Food and Drug Administration regulations. Subpart F of the regulations lists those components of packaging materials that are regarded as safe food additives. The listing of a chemical as a safe food additive when used in one type of packaging material does not imply that the chemical will be regarded as safe when used in another one. A current copy of Subpart F should be obtained from the Division of Public Information, Food and Drug Aministration, Department of Health, Education, and Welfare, Washington, D.C., for review relative to the status of components of a new polymer that may become a food additive. It is advisable to consult with Food and Drug Administration personnel, if there is any question regarding the applicability of an existing regulation to a new polymer composition.

The safety of polymer components that may become food additives as defined above, should be established before marketing the polymer for food contact use. If the safety is not established and a Food and Drug Administration regulation is not obtained, there is little possibility of establishing a market for use of the polymer in the food industry.

C. Procedures for Establishing Safety of Polymer Components that May Become Food Additives

The Food and Drug Administration has prepared outlines of recommended procedures to be followed and information to be included in a petition requesting issuance of a regulation relative to safe use of food additives, including those that may migrate from polymers in contact with food. These outlines are revised from time to time. It is advisable to obtain the latest revision and to discuss proposed experiments with Food and Drug Administration personnel before undertaking work to establish the safety of a new food additive.

The general areas of information that should be included in a petition are as follows:

1. Proposed Use

Types of foods that may be in contact with the polymer, temperature and time of contact, and area of contact per unit weight of food.

2. Composition of Polymer

Composition of polymer, including major and minor components, basic polymer, monomers, plasticizers, stabilizers, catalyst residues, fillers, colors, and antioxidants. A description of the manufacturing process which includes side-reaction products and impurities.

3. Extractability of Components

Extractability of components into simulated food solvents. The simulated food solvents should be representative of the types of foods the polymer will touch. Examples are distilled water, 3% sodium chloride solution, 3% acetic acid solution, 3% sodium bicarbonate solution, 20% sucrose solution containing 1% citric acid, liquid food fat such as vegetable oil, and aqueous ethyl alcohol.

The American Society for Testing and Materials, *Tentative Method F34–63T* (9) describes an acceptable procedure for determining components that may be extracted from polymer films.

4. Abrasion

Abrasion data, if the polymer may be abraded by contact with dry foods. *Method 6191, Federal Test Method Standard No. 141* (10) describes a procedure for determining abrasion. Details of abrasion tests relative to contact of a polymer with a particular food should be agreed upon in consultation with Food and Drug Administration personnel.

5. Analytical Methods

Analytical methods of suitable sensitivity and reliability to permit quantitative determination in or on foods of polymer components that may become food additives. This area can be particularly difficult and expensive. Analytical methods in the parts-per-million range may be required, if the polymer component that may become a food additive has adverse physiological effects.

6. Toxicology and Safety Data

Adequate data on toxicity and physiological effects of components that may become food additives should be furnished. Publication 750, National Academy of Sciences—National Research Council,

Principles and Procedures for Evaluating the Safety of Food Additives (11) furnishes a brief outline of the work necessary in this area. The publication of the Association of Food and Drug Officials of the United States, *Appraisal of the Safety of Chemicals in Foods, Drugs, and Cosmetics* (12), provides more detailed information.

Details of acute, sub-acute, and chronic toxicity experiments, establishment of metabolic pathways and physiological effects of chemicals are outside the scope of this chapter. Those who are unfamiliar with the science of toxicology should consult toxicologists within their own organization or in independent biological research laboratories for assistance in designing experiments in this area.

The necessary toxicological experiments to establish the safety of a new food additive may take several years and can cost over $100,000. The time and money involved should not be committed without close coordination of the program with Food and Drug Administration personnel. There must be strong evidence that the new polymer has good potential markets and that components of the polymer that may become food additives are essential to the required properties of the polymer to justify the cost of establishing its safety.

7. Preparation of Petition

Good data can be spoiled by a poor presentation. The petition to the Food and Drug Administration should be carefully prepared and the data clearly presented. Petitions are reviewed by many specialists in the Food and Drug Administration. Counterpart specialists in the organization preparing the petition should check the final draft of the petition for accuracy. Marketing personnel should review proposed uses; production should be sure that process descriptions are accurate; the analytical group should be sure that descriptions of analytical procedures are given in sufficient detail to allow Food and Drug Administration analysts, both in Washington and in field laboratories, to replicate reported results. Toxicologists must include protocols of experiments, adverse observations as well as favorable ones, and explanations of any unusual findings. It is advisable to review the various sections of the petition with Food and Drug Administration personnel on an informal basis before formal submission of a petition. Even with close coordination of the work with the Food and Drug Administration at all stages, experience has shown that additional data may be required after formal submission of a petition.   The job cannot be considered completed until the Food and Drug Administration issues the regulation requested in the petition.   Once a regulation has been issued there is always the possibility that new information may necessitate a revision of the regulation.

IV. Summary

Individual consumers cannot determine economically the safety of foods produced, processed, and packaged for them.   Data to establish the safety of many food-packaging components that might migrate into foods had been presented to the Food and Drug Administration prior to enactment of the Food Additives Amendment of 1958.   This amendment made it mandatory to establish the safety of food additives *prior to their use.*   The required procedures to establish safety may take 3 to 5 years and several hundred thousand dollars.   The developer of a new polymer having potential uses in contact with foods should be prepared to commit the time and money if he expects to profit from the food industry market.   The potential share of the food industry market that a new polymer may be able to obtain may not be sufficient to warrant the expense involved in the demonstration of safety.

Food and Drug Administration regulations providing for the safe use of many polymeric materials in contact with food have been issued since 1958.   This indicates that food-industry uses of polymeric materials can be sufficiently profitable to warrant the expense of establishing the safety of components that may migrate into foods.   Examples of polymeric materials for which Food and Drug Administration regulations have been issued include:

Cellulose-based materials (cellophane and paper)
Olefin polymers
Copolymers of ethylene and methyl acrylate
Polyurethane resins
Polyethylene terephthalate
Polyamide–epichlorohydrin resins
Perfluorocarbon resins
Melamine–formaldehyde resins

References

1. R. Ricklefs, *The Wall Street Journal*, Nov. 25, 1964, p. 1.
2. Federal Food, Drug, and Cosmetic Act, June 25, 1938, Ch. *675, Sec. 1, 52 Stat. 1040; Title 21*, Code of Federal Regulations, *Sec. 301* et seq.
3. Fifty-ninth Congress, Session I, *Ch. 3915* (1906), pp. 768–72.
4. Food Additives Amendment of 1958, Sept. 6, 1958, Public Law *85–929, 72 Stat. 1784; Title 21*, Code of Federal Regulations, *Sec. 409.*
5. G. P. Larrick, Statement Before the Subcommittee on Public Health and Safety of the House Committee on Interstate and Foreign Commerce, April 25, 1963.
6. A. J. Lehman, Assoc. Food & Drug Officials, U.S., *Quart. Bull.*, **15**, 82(1951).
7. A. J. Lehman, Assoc. Food & Drug Officials, U.S., *Quart. Bull.*, **18**, 129 (1954).
8. A. J. Lehman, Assoc. Food & Drug Officials, U.S., *Quart. Bull.*, **20**, 159 (1956).
9. *1964 Book of Standards, Part 15*, "Tentative Method for Exposing Flexible Barrier Materials to Liquids For Extraction, ASTM Designation *F34–63T*," pp. 691–5. American Society for Testing and Materials, Philadelphia, 1964.
10. *Federal Test Method Standard Number 141*, "Method 6191, Abrasion Resistance (Falling Sand)." General Services Administration, Business Services Center, Washington, D.C., 1958.
11. *"Principles and Procedures for Evaluating the Safety of Food Additives," Publ. 750*, Natl. Acad. Sci.—Natl. Res. Council, Washington, D.C., 1960.
12. *Appraisal of the Safety of Chemicals in Foods, Drugs, and Cosmetics,"* Assoc. of Food & Drug Officials, U.S., Texas State Department of Health, Austin, 1959.

Appendix I

FOOD AND DRUG ADMINISTRATION
FOOD ADDITIVE REGULATIONS†

SUBPART B—EXEMPTION OF CERTAIN FOOD ADDITIVES
FROM THE REQUIREMENTS OF TOLERANCES

§121.101. *Substances that are Generally Recognized as Safe*

(a) It is impracticable to list all substances that are generally recognized as safe for their intended use. However, by way of illustration, the Commissioner regards such common food ingredients as salt, pepper, sugar, vinegar, baking powder, and monosodium glutamate as safe for their intended use. The lists in paragraph (d) of this section include additional substances that, when used for the purposes indicated, in accordance with good manufacturing practice, are regarded by the Commissioner as generally recognized as safe for such uses.

(b) For the purposes of this section, good manufacturing practice shall be defined to include the following restrictions:

(1) The quantity of a substance added to food does not exceed the amount reasonably required to accomplish its intended physical, nutritional, or other technical effect in food; and

(2) The quantity of a substance that becomes a component of food as a result of its use in the manufacturing, processing, or packaging of food, and which is not intended to accomplish any physical or other technical effect in the food itself, shall be reduced to the extent reasonably possible.

(3) The substance is of appropriate food grade and is prepared and handled as a food ingredient. Upon request the Commissioner will offer an opinion, based on specifications and intended use, as to whether or not a particular grade or lot of the substance is of suitable purity for use in food and would generally be regarded as safe for the purpose intended, by experts qualified to evaluate its safety.

(c) The inclusion of substances in the list of nutrients does not constitute a finding on the part of the Department that the substance is useful as a supplement to the diet for humans.

(d) Substances that are generally recognized as safe for their intended use within the meaning of *section 409* of the act are as follows:

† Amendment published in the Federal Register Jan. 31, 1961; **26** F.R. 938.

Product	Tolerance	Limitations or restrictions
(1) Anticaking agents		
Aluminum calcium silicate	2%	In table salt.
Calcium silicate	5%	In baking powder.
Calcium silicate	2%	In table salt.
Magnesium silicate	2%	In table salt.
*Sodium aluminosilicate (sodium silicoaluminate)	2%	
*Sodium calcium alumino-silicate, hydrated (sodium calcium silicoaluminate)	2%	
Tricalcium silicate	2%	In table salt.
(2) Chemical preservatives		
Ascorbic acid		
Ascorbyl palmitate		
Benzoic acid	0.1%	
Butylated hydroxyanisole	Total content of antioxidants not over 0.02% of fat or oil content, including essential (volatile) oil content of food.	
Butylated hydroxytoluene	Total content of antioxidants not over 0.02% of fat or oil content, including essential (volatile) oil content of food.	
Calcium ascorbate		
Calcium propionate		
*Calcium sorbate		
Caprylic acid		In cheese wraps.
Dilauryl thiodipropionate	Total content of antioxidants not over 0.02% of fat or oil content, including essential (volatile) oil content of the food.	
Erythorbic acid		
Gum guaiac	0.1% (equivalent antioxidant activity 0.01%).	In edible fats or oils.

* Substances added from February 2 and August 4, 1960, proposed lists.

(continued)

Product	Tolerance	Limitations or restrictions
*Methylparaben (methyl-*p*-hydroxybenzoate)	0.1%	
Nordihydroguaiaretic acid	Total content of antioxidants not over 0.02% of fat or oil content including essential (volatile) oil content of the food.	
Potassium bisulfite		Not in meats or in food recognized as source of vitamin B_1.
Potassium metabisulfite		Not in meats or in food recognized as source of vitamin B_1.
Potassium sorbate		
Propionic acid		
Propyl gallate	Total content of antioxidants not over 0.02% of fat or oil content, including essential (volatile) oil content of the food.	
*Propylparaben (propyl-*p*-hydroxybenzoate)	0.1%	
Sodium ascorbate		
Sodium benzoate	0.1%	
Sodium bisulfite		Not in meats or in foods recognized as a source of vitamin B_1.
Sodium metabisulfite		Not in meats or in foods recognized as a source of vitamin B_1.
Sodium propionate		
Sodium sorbate		
Sodium sulfite		Not in meats or in food recognized as source of vitamin B_1.
Sorbic acid		
*Stannous chloride	0.0015% calculated as tin	
Sulfur dioxide		Not in meats or in food recognized as source of vitamin B_1.

* Substances added from February 2 and August 4, 1960, proposed lists.

(*continued*)

Product	Tolerance	Limitations or restrictions
Thiodipropionic acid	Total content of anti-oxidants not over 0.02% of fat or oil content, including essential (volatile) oil content of the food.	
Tocopherols		
(3) Emulsifying agents		
Cholic acid	0.1%	Dried egg whites.
Desoxycholic acid	0.1%	Dried egg whites.
†Diacetyl tartaric acid esters of mono- and diglycerides of edible fats or oils, or edible fat-forming fatty acids		
Glycocholic acid	0.1 %	Dried egg whites.
†Mono- and diglycerides of edible fats or oils, or edible fat-forming fatty acids		
†Monosodium phosphate derivatives of mono- and diglycerides of edible fats or oils, or edible fat-forming fatty acids		
Propylene glycol		
Ox bile extract	0.1%	Dried egg whites.
Taurocholic acid (or its sodium salt)	0.1%	Dried egg whites.
(4) Nonnutritive sweeteners		
*Ammonium saccharin		
Calcium cyclamate (calcium cyclohexyl sulfamate)		
Calcium saccharin		
*Magnesium cyclamate (magnesium cyclohexyl sulfamate)		
*Potassium cyclamate (potassium cyclohexyl sulfamate)		
Saccharin		
Sodium cyclamate (sodium cyclohexyl sulfamate)		
Sodium saccharin		

* Substances added from February 2 and August 4, 1960, proposed lists.
† Amendment published in the Federal Register Dec. 2, 1964; **29**F.R. 16079.

(continued)

Product	Tolerance	Limitations or restrictions
**(5) *Nutrients and/or dietary supplements*		
*Alanine (L and DL forms)		
*Arginine (L and DL forms)		
Ascorbic acid		
*Aspartic acid (L and DL forms)		
*Biotin		
Calcium carbonate		
*Calcium citrate		
*Calcium glycerophosphate		
Calcium oxide		
Calcium pantothenate		
Calcium phosphate (mono-, di-, tribasic)		
*Calcium pyrophosphate		
Calcium sulfate		
Carotene		
*Choline bitartrate		
*Choline chloride		
Copper gluconate	0.005%	
Cuprous iodide	0.01%	In table salt as a source of dietary iodine.
*Cysteine (L form)		
*Cystine (L and DL forms)		
Ferric phosphate		
Ferric pyrophosphate		
Ferric sodium pyrophosphate		
*Ferrous gluconate		
*Ferrous lactate		
Ferrous sulfate		
*Glycine (aminoacetic acid)		In animal feeds
*Histidine (L and DL forms)		
*Inositol		
Iron, reduced		
*Isoleucine (L and DL forms)		
*Leucine (L and DL forms)		
*Linoleic acid (prepared from edible fats and oils and free from chick edema factor).		
Lysine (L and DL forms)		

* Substances added from February 2 and August 4, 1960, proposed lists.
‡ Amino acids listed may be free, hydrochloride salt, hydrated, or anhydrous form, where applicable.

(continued)

Product	Tolerance	Limitations or restrictions
*Magnesium oxide		
*Magnesium phosphate (di-, tribasic)		
*Magnesium sulfate		
*Manganese chloride		
*Manganese citrate		
*Manganese gluconate		
*Manganese glycerophosphate		
*Manganese hypophosphite		
*Manganese sulfate		
*Manganous oxide		
*Mannitol	5%	In special dietary foods.
*Methionine		Animal feeds.
*Methionine hydroxy analog and its calcium salts		Animal feeds.
Niacin		
Niacinamide		
D-Pantothenyl alcohol		
*Phenylalanine (L and DL forms)		
Potassium chloride		
*Potassium glycerophosphate		
Potassium iodide	0.01%	In table salt as a source of dietary iodine.
*Proline (L and DL forms)		
Pyridoxine hydrochloride		
Riboflavin		
Riboflavin-5-phosphate		
*Serine (L and DL forms)		
Sodium pentothenate		
Sodium phosphate (mono-, di-, tribasic)		
Sorbitol	7%	In foods for special dietary use.
Thiamine hydrochloride		
Thiamine mononitrate		
*Threonine (L and DL forms)		
Tocopherols		
α-Tocopherol acetate		

* Substances added from February 2 and August 4, 1960, proposed lists.

(continued)

Product	Tolerance	Limitations or restrictions
*Tryptophane (L and DL forms		
*Tyrosine (L and DL forms)		
*Valine (L and DL forms)		
Vitamin A		
Vitamin A acetate		
Vitamin A palmitate		
Vitamin B_{12}		
Vitamin D_2		
Vitamin D_3		
*Zinc sulfate		
*Zinc gluconate		
*Zinc chloride		
*Zinc oxide		
*Zinc stearate (prepared from stearic acid free from chick edema factor)		

†(6) Sequestrants

Product	Tolerance	Limitations or restrictions
Calcium acetate		
Calcium chloride		
Calcium citrate		
Calcium diacetate		
Calcium gluconate		
Calcium hexametaphosphate		
Calcium phosphate, monobasic		
Calcium phytate		
Citric acid		
Dipotassium phosphate		
Disodium phosphate		
Isopropyl citrate	0.02%	
Monoisopropyl citrate		
Potassium citrate		
Sodium acid phosphate		
Sodium citrate		
Sodium diacetate		
Sodium gluconate		
Sodium hexametaphosphate		
Sodium metaphosphate		

* Substances added from February 2 and August 4, 1960, proposed lists.

† For the purpose of this list, no attempt has been made to designate those sequestrants that may also function as chemical preservatives.

(continued)

Product	Tolerance	Limitations or restrictions
Sodium phosphate (mono-, di-, tribasic)		
Sodium potassium tartrate		
Sodium pyrophosphate		
Sodium pyrophosphate, tetra		
Sodium tartrate		
Sodium thiosulfate	0.1%	In salt.
Sodium tripolyphosphate		
Stearyl citrate	0.15%	
Tartaric acid		

(7) Stabilizers

*Acacia (gum arabic)
Agar-agar
*Ammonium alginate
*Calcium alginate
Carob bean gum (locust bean gum)
Chondrus extract (carrageenin)
*Ghatti gum
Guar gum
*Potassium alginate
*Sodium alginate
*Sterculia gum (karaya gum)
*Tragacanth (gum tragacanth)

(8) Miscellaneous and/or general-purpose food additives

Acetic acid		
*Adipic acid		Buffer and neutralizing agent.
Aluminum ammonium sulfate		
Aluminum potassium sulfate		
Aluminum sodium sulfate		
Aluminum sulfate		
Ammonium bicarbonate		

* Substance added from February 2 and August 4, 1960, proposed lists.

(continued)

Product	Tolerance	Limitations or restrictions
Ammonium carbonate		
Ammonium hydroxide		
Ammonium phosphate (mono- and dibasic)		
*Ammonium sulfate		
*Beeswax (yellow wax)		
*Beeswax, bleached (white wax)		
*Bentonite		
Butane		
Caffeine	0.02%	In cola-type beverages.
Calcium carbonate		
Calcium chloride		
Calcium citrate		
Calcium gluconate		
Calcium hydroxide		
Calcium lactate		
Calcium oxide		
Calcium phosphate (mono-, di-, tribasic)		
Caramel		
Carbon dioxide		
Carnauba wax		
Citric acid		
*Dextrans (of average molecular weight below 100,000)		
Ethyl formate	0.0015%	As fumigant for cashew nuts.
*Glutamic acid		Salt substitute.
*Glutamic acid hydrochloride		Salt substitute.
Glycerol		
Glyceryl monostearate		
Helium		
*Hydrochloric acid		Buffer and neutralizing agent.
*Hydrogen peroxide		Bleaching agent.
Lactic acid		
*Lecithin		
Magnesium carbonate		
Magnesium hydroxide		

* Substances added from February 2 and August 4, 1960, proposed lists.

(continued)

Product	Tolerance	Limitations or restrictions
Magnesium oxide		
Magnesium stearate		As migratory substance from packaging materials when used as a stabilizer.
*Malic acid		
*Methylcellulose (USP methylcellulose, except that the methoxy content shall not be less than 27.5% and not more than 31.5% on a dry-weight basis).		
Monoammonium glutamate		
*Monopotassium glutamate		
Nitrogen		
*Nitrous oxide		Propellant for certain dairy and vegetable-fat toppings in pressurized containers.
Papain		
Phosphoric acid		
Potassium acid tartrate		
Potassium bicarbonate		
Potassium carbonate		
Potassium citrate		
Potassium hydroxide		
*Potassium sulfate		
Propane		
Propylene glycol		
*Rennet (rennin)		
*Silica aerogel (finely powdered microcellular silica foam having a minimum silica content of 89.5%		Component of antifoaming agent.
Sodium acetate		
Sodium acid pyrophosphate		
Sodium aluminum phosphate		
Sodium bicarbonate		
Sodium carbonate		
Sodium citrate		

* Substances added from February 2 and August 4, 1960, proposed lists.

(continued)

Product	Tolerance	Limitations or restrictions
*Sodium carboxymethylcellulose (the sodium salt of carboxymethylcellulose not less than 99.5% on a dry-weight basis, with maximum substitution of 0.95 carboxymethyl groups per anhydroglucose unit, and with a minimum viscosity of 25 cp for 2% by weight aqueous solution at 25°C)		
*Sodium caseinate		
Sodium citrate		
Sodium hydroxide		
*Sodium pectinate		
Sodium phosphate (mono-, di-, tribasic)		
Sodium potassium tartrate		
Sodium sesquicarbonate		
Sodium tripolyphosphate		
*Succinic acid		
Sulfuric acid		
Tartaric acid		
Triacetin (glyceryl triacetate)		
Triethyl citrate	0.25%	Dried egg whites.

* Substances added from February 2 and August 4, 1960, proposed lists.

(e) Spices, seasonings, essential oils, oleoresins, and natural extractives that are generally recognized as safe for their intended use, within the meaning of *section 409* of the act,† are as follows:

(1) Spices and Other Natural Seasonings and Flavorings (Leaves, Roots, Barks, Berries, etc.)

Common name	Botanical name of plant source
Alfalfa herb and seed	*Medicago sativa* L.
Allspice	*Pimenta officinalis* Lindl.
Ambrette seed	*Hibiscus abelmoschus* L.
Angelica	*Angelica archangelica* L. or other spp. of *Angelica*.
Angelica root	*Angelica archangelica* L. or other spp. of *Angelica*.
Angelica seed	*Angelica archangelica* L. or other spp. of *Angelica*.
Angostura (cusparia bark)	*Galipea officinalis* Hancock.
Anise	*Pimpinella anisum* L.
Anise, star	*Illicium verum* Hook. f.
Balm (lemon balm)	*Melissa officinalis* L.
Basil, bush	*Ocimum minimum* L.
Basil, sweet	*Ocimum basilicum* L.
Bay	*Laurus nobilis* L.
Calendula	*Calendula officinalis* L.
Camomile (chamomile), English or Roman	*Anthemis nobilis* L.
Camomile (chamomile), German or Hungarian	*Matricaria chamomilla* L.
Capers	*Capparis spinosa* L.
Capsicum	*Capsicum frutescens* L. or *Capsicum annuum* L.
Caraway	*Carum carvi* L.
Caraway, black (black cumin)	*Nigella sativa* L.
Cardamom (cardamon)	*Elettaria cardamomum* Maton.
Cassia, Chinese	*Cinnamomum cassia* Blume.
Cassia, Padang or Batavia	*Cinnamomum burmanni* Blume.
Cassia, Saigon	*Cinnamomum loureirii* Nees.
Cayenne pepper	*Capsicum frutescens* L. or *Capsicum annuum* L.
Celery seed	*Apium graveolens* L.

† Added by Amendment of June 10, 1961; **26** F.R. 5221.

(*f,g*) [These sections are omitted as not pertinent.]

(*h*) Substances migrating to food from paper and paperboard products used in food packaging that are generally recognized as safe for their intended use, within the meaning of *section 409†* of the act, are as follows:

Acetic acid
Alum (double sulfate of aluminum and ammonium potassium, or sodium)
Aluminum hydroxide
Aluminum oleate
Aluminum palmitate
Ammonium chloride
Ammonium hydroxide
Calcium chloride
Calcium hydroxide
Calcium sulfate
Casein
Cellulose acetate
Clay (kaolin)
Copper sulfate
Corn starch
Corn sugar (sirup)
Dextrin
Diatomaceous earth filler.
Ethylcellulose
Ethylvanillin
Ferric sulfate
Ferrous sulfate
Formic acid or sodium salt
Glycerol
Guar gum
Invert sugar
Iron, reduced
Locust bean gum (carob bean gum)
Magnesium carbonate
Magnesium chloride
Magnesium hydroxide
Magnesium sulfate
Methyl and ethyl acrylate
Mono- and diglycerides from glycerolysis of edible fats and oils

Oleic acid
Oxides of iron
Potassium sorbate
Propionic acid
Propylene glycol
Silicon dioxides
Pulps from wood, straw, bagasse, or other natural sources
Soap (sodium oleate, sodium palmitate)
Sodium aluminate
Sodium carbonate
Sodium chloride
Sodium hexametaphosphate
Sodium hydrosulfite
Sodium hydroxide
Sodium phosphoaluminate
Sodium silicate
Sodium sorbate
Sodium sulfate
Sodium thiosulfate (additive in salt)
Sodium tripolyphosphate
Sorbitol
Soy protein, isolated
Sulfamic acid
Sulfuric acid
Starch, acid-modified
Starch, pregelatinized
Starch, unmodified
Sucrose
Talc
Urea
Vanillin
Zinc hydrosulfite
Zinc sulfate

† Added by Amendment June 10, 1961; **26** F.R. 5225.

(*i*) Substances migrating to food from cotton and cotton fabrics used in dry food packaging that are generally recognized as safe for their intended use, within the meaning of *section 409* of the act,† are as follows:

Acacia (gum arabic)
Acetic acid
Beef tallow
Calcium chloride
Carboxymethylcellulose
Coconut oil, refined
Corn dextrin
Corn starch
Fish oil (hydrogenated)
Gelatin
Guar gum
Hydrogen peroxide
Japan wax
Lard
Lard oil
Lecithin (vegetable)
Locust bean gum (carob bean gum)
Oleic acid
Peanut oil
Potato starch
Sodium acetate

Sodium bicarbonate
Sodium carbonate
Sodium chloride
Sodium hydroxide
Sodium sulfate
Sodium silicate
Sodium tripolyphosphate
Sorbose
Soybean oil (hydrogenated)
Stearic acid
Talc
Tall oil
Tallow (hydrogenated)
Tallow flakes
Tapioca starch
Tartaric acid
Tetrasodium pyrophosphate
Urea
Wheat starch
Zinc chloride

† Added by Amendment of June 10, 1961; **26** F.R. 5221.

Appendix II

Food and Drug Administration
Food Additive Regulations*

Subpart E—Substances for which Prior Sanctions have been Granted

§121.2001 Substances Employed in the Manufacture of Food-Packaging Materials

Prior to the enactment of the food additives amendment to the Federal Food, Drug, and Cosmetic Act, sanctions were granted for the usage of the following substances in the manufacture of packaging materials. So used, these substances are not considered "food additives" within the meaning of *section 201(s)* of the act, provided that they are of good commercial grade, are suitable for association with food, and are used in accordance with good manufacturing practice. For the purpose of this section, good manufacturing practice for food-packaging materials includes the restriction that the quantity of any of these substances which becomes a component of food as a result of use in food-packaging materials shall not be intended to accomplish any physical or technical effect in the food itself, shall be reduced to the least amount reasonably possible, and shall not exceed any limit specified in this section:

(*a*) *Antioxidants (limit of addition to food, 0.005%)*
Butylated hydroxyanisole
Butylated hydroxytoluene
Dilauryl thiodipropionate
Distearyl thiodipropionate
Gum guaiac
Nordihydroguairetic acid
Propyl gallate
Thiodipropionic acid
2,4,5-Trihydroxybutyrophenone

(*b*) *Antimycotics*
Calcium propionate
Methylparaben (methyl-*p*-hydroxybenzoate)
Propylparaben (propyl-*p*-hydroxybenzoate)
Sodium benzoate
Sodium propionate
Sorbic acid

* Amendment published in Federal Register Feb. 18, 1966; **31** F.R. 2897.

(c) *Driers*
Cobalt caprylate
Cobalt linoleate
Cobalt naphthenate
Cobalt tallate [from tall oil]
Iron caprylate
Iron linoleate
Iron naphthenate
Iron tallate [from tall oil]
Manganese caprylate
Manganese linoleate
Manganese naphthenate
Manganese tallate [from tall oil]

(d) *Drying oils (as components of finished resins)*
Chinawood oil (tung oil)
Dehydrated castor oil
Linseed oil
Tall oil

(e) *Plasticizers*
Acetyl tributyl citrate
Acetyl triethyl citrate
p-tert-Butylphenyl salicylate
Butyl stearate
Butylphthalyl butyl glycolate
Dibutyl sebacate
Di-(2-ethylhexyl) phthalate (for foods of high water content only)
Diethyl phthalate
Diisobutyl adipate
Diisooctyl phthalate (for foods of high water content only)
Diphenyl-2-ethylhexyl phosphate
Epoxidized soybean oil (iodine number maximum 6; and oxirane oxygen mini-
    mum, 6.0%)
Ethylphthalyl ethyl glycolate
Glycerol monooleate
Monoisopropyl citrate
Mono, di-, and tristearyl citrate
Triacetin (glycerol triacetate)
Triethyl citrate
3-(2-Xenoyl)-1,2-epoxypropane

(f) *Release agents*
Dimethylpolysiloxane (substantially free from hydrolyzable chloride and alkoxy
    groups, no more than 18% loss in weight after heating 4 hr at 200°C; viscosity
    300 centistokes, 600 centistokes at 25°C, sp gr 0.96–0.97 at 25°C, refractive
    index 1.400–1.404 at 25°C)
Linoleamide (linoleic acid amide)

Oleamide (oleic acid amide)
Palmitamide (palmitic acid amide)
Polyethylene glycol 400
Polyethylene glycol 1500
Polyethylene glycol 4000
Stearamide (stearic acid amide)

(g) *Stabilizers*
Aluminum mono-, di-, and tristearate
Ammonium citrate
Ammonium potassium hydrogen phosphate
Calcium acetate
Calcium carbonate
Calcium glycerophosphate
Calcium phosphate
Calcium hydrogen phosphate
Calcium oleate
Calcium ricinoleate
Calcium stearate
Disodium hydrogen phosphate
Magnesium glycerophosphate
Magnesium stearate
Magnesium phosphate
Magnesium hydrogen phosphate
Mono-, di-, and trisodium citrate
Mono-, di-, and tripotassium citrate
Potassium oleate
Potassium stearate
Sodium pyrophosphate
Sodium stearate
Sodium tetrapyrophosphate
Stannous stearate (not to exceed 50 ppm tin as a migrant in finished food)
Zinc orthophosphate (not to exceed 50 ppm zinc as a migrant in finished food)
Zinc resinate (not to exceed 50 ppm zinc as a migrant in finished food)

†(h) *Substances used in the manufacture of paper and paperboard products used in food packaging*
Aliphatic polyoxyethylene ethers*
1-Alkyl (C_6-C_{18})-amino-3-aminopropane monoacetate*
Borax or boric acid for use in adhesives, sizes, and coatings*
Butadiene–styrene copolymer
Chromium complex of perfluorooctane sulfonyl glycine for use on paper and paperboard which is waxed*

* Under the conditions of normal use, these substances would not reasonably be expected to migrate to food, based on available scientific information and data.
† Amendment published in the *Federal Register*, March 1, 1960; **25** F.R. 1773.

Disodium cyanodithioimidocarbamate with ethylenediamine and potassium N-methyl dithiocarbamate and/or sodium 2-mercaptobenzothiazole (slimicides)*

Ethyl acrylate and methyl methacrylate copolymers of itaconic acid or methacrylic acid for use only on paper and paperboard which is waxed*

Hexamethylenetetramine as a setting agent for protein, including casein*

1-(2-Hydroxyethyl)-1-(4-chlorobutyl)-2-alkyl (C_6-C_{17}) imidazolinium chloride*

Itaconic acid (polymerized)

Melamine formaldehyde polymer

Methyl acrylate (polymerized)

Methyl ethers of mono-, di-, and tripropylene glycol*

Myristo chromic chloride complex

Nitrocellulose

Polyethylene glycol 400

Polyvinyl acetate

Potassium pentachlorophenate as a slime control agent*

Potassium trichlorophenate as a slime control agent*

Pyrethrins in combination with piperonyl butoxide in outside plies of multiwall bags*

Resins from high and low viscosity polyvinyl alcohol for fatty foods only

Rubber hydrochloride

Sodium pentachlorophenate as a slime control agent*

Sodium trichlorophenate as a slime control agent*

Stearatochromic chloride complex

Titanium dioxide*

Urea formaldehyde polymer

Vinylidine chlorides (polymerized)

* Under the conditions of normal use, these substances would not reasonably be expected to migrate to food, based on available scientific information and data.

AUTHOR INDEX

Numbers in parentheses are reference numbers and show that an author's work is referred to although his name is not mentioned in the text. Numbers in *italics* indicate the pages on which the full references appear.

A

Aartsen, J. J. van, 348(106), *364*
Abbott, A. D., 58(30), *81*
Abitz, W., 35(11), *43*
Abramo, S. V., 56(26a), *81*
Achhammer, B. G., 740, *810*
Adams, C. H., 575(28), *648*
Adams, N., 437, 438, *456*
Addor, R. W., 69(56), *82*
Adelman, R. L., 64(46), *81*
Adkins, J. E., 238(20), 247(20), *251*
Aelion, R., 51(14), *80*
Aggarwal, S. L., 40(15), *43*, 62(35), *81*, 243(61), 250(61), *252*, 461(4), *479*
Albohn, A. R., 178(9), 179(9), *194*
Alderson, T., 54(23), *81*
Aldrich, P. H., 76(84), 77(84), *83*
Alekssenko, V. I., 808(87), *812*
Alexander, F. M., 387(23,24), *421*, *422*
Alexander, L. E., 268(22), 269(22), *362*
Alexander, P., 393(34), *422*
Alexander, W. J., 95(14,16), 104(37), 105(40,41), *113*, *114*, 677, *701*
Alfrey, T., Jr., 162(87), *171*, 228–232(1), 234(1), 238(1), 242(1), 247(1), 249(85,87–89), *250*, *253*
Allan, A. J. G., 445, 448, 453(60a), *456*
Allen, N., 682(61), 687(61), *702*
Alles, F. P., 466, 477(26), *480*
Allison, J. B., 62(37), *81*
Amaral, J. R., 716(12), *809*
Ambrose, E. J., 736(27), 746(35), 760(45), *810*, *811*
Amdur, E. J., 636(59), *650*

American Cyanamid Co., 67(51), *82*
American Society for Testing and Materials, 241(39), 242(59), 243(59), *251*, *252*, 428(5a), 447, *455*, 538(7), 539, 542(13), *543*, 549, 555(10,11), 558(11), 571(23), 575(27), 585(31), 586(32), 592(35), 593(36), 595(37), 599(40), 604(43), 624(49), 629(51), 633(54), 636(55), *647–649*, 653(1), 654(3), 655(4), 656(9), 659(32), 661(32), 663(32), 665(39), 671(43), *700*, *701*, 748(36), 791(78), 794(79), 807(86), *810*, *812*, 825(9), *828*
American Viscose, 186(13), *194*, 468
Amontons, G., 426, *455*
Anderson, A. W., 103(34), *114*
Anderson, D. A., 728(23), *810*
Andreeva, M. A., 222(50), *226*
Andrew, F. L., 454(108), *458*
Andrews, R. D., 242(60), *252*
Angelo, R. J., 73(71), *82*
Anger, V., 713(3), *809*
Arai, T., 247(84), *253*
Arbran, K. G., 241(42), *251*
Archard, J. F., 438, *456*
Ardis, A. E., 68(53,54), *82*
Arnold, H. R., 451(73), *457*
Arnold, K. A., 583, *648*
Arnoult, J. E., 448(62), *456*
Aronsen, A. N., 408(44), *422*
Asp, W., 695(77), 696(77), *702*
Assaf, A. G., 156(80), 157(80), *171*
Audrieth, L. F., 222(53), *226*
Austin, P. R., 451(73), *457*, 465
Avrami, M., 239(29), 240(29), *251*
Ayer, E., 158(84), *171*

847

SUBJECT INDEX

A

Acetylacetone, 204, 207, 209
Acrylamide, 50
Acrylates, determination of, in
  copolymers, 802
Acrylonitrile, determination of, in
  plastics, 753, 767–768, 795–796
Addition polymers, 26
Additives generally recognized as
  safe, 829–842
Additives, inorganic, in vinyl poly-
  mers, 721
  organic, in vinyl polymers, 723
  surface, 721
Adhesion, 538
Adhesion of metals, 428
Adipic acid, 47
Aging of films, 538
Air-conditioned cabinets, 643
Air-conditioned rooms, 637–643
Air conditioning equipment, 642
Air doctor coating, 506, 518
Alkali cellulose, 97
Alkoxyl, total, in polymers, 802
Alkyd resins, 30
Alkyl sulfites, 59
Aluminum phosphinate polymers,
  212
Allyl chloride, 66, 67
Allyldimethyl silane, 79
Allyl mercaptan, 79
Allyl polymers, 67
Alpha-olefins, analytical chemistry
  of polymers of, 703–812
  polymerization of, 62–74
Amides as polymer lubricants, 444
p-Aminobenzoic acid, 51
ε-Aminocaproic acid, 55
4-Aminocyclohexanecarboxylic acid,
  52

9-Aminononanoic acid, 51
9-(or 10-)Aminostearic acid, 52
12-Aminostearic acid, 52
Aminotriazoles, 57
11-Aminoundecanoic acid, 51
Amontons' laws of friction, 426, 620
Amorphous orientation, 287–299
  by sonic modulus, 292
  system model, 288
Amorphous orientation function of
  polycrystalline, polymers, by
  sonic modulus, 287
Amorphous polymers, 35, 39, 40, 42,
  64, 69
Amphiboles, 217
Amylopectin, 175, 177
Amylose, 173, 175, 177
Amylose films, 176
  plasticizers, 176
  properties of, from corn, 176
Anchoring agents for cellophane
  coatings, 151
  detection of formaldehyde in, 690
  determination of melamine in,
    690–693
  mechanism of action, 151
  self-anchorage, 152
  use of, 151
Anisotropy of films, cellophane,
  133, 141
  cellulose, 94
  general, 238, 242, 247
  polycrystalline polymers, 264
Annealing of films, 393
Anticaking agents in food, list, 830
Antimycotics permitted in food
  packing materials, list, 843
Antioxidants permitted in food
  packaging materials, list, 843
Ardil, 174
Atactic polymers, 63, 65

867

N

Natta-Ziegler polyolefins, 38
Neck down of film, 370
Necking of polymers, 463–464
Newtonian fluid, 230, 245
  stress exponent, 372
Newton's law for fluids, 229
Ninnemann pendulum-impact tester,
  580
Nitrocellulose, cellophane coatings,
  12–14, 694
  (See also *Cellulose nitrate.*)
Nonheat-sealable films, 609–610
Non-Newtonian fluid, 230
Nonnutritive sweeteners in food,
  list, 832
Nutrient supplements in food, list,
  833
6-Nylon, 28, 31, 41, 42, 47
6,6-Nylon, 26, 47, 437
6,10-Nylon, 47
  sonic modulus, 290
7-Nylon, 41
11-Nylon, 42
Nylon film, 13, 541
Nylon spherulites, 334

O

Octadecanedioic acid, 57
Odor and flavor transmission, 534,
  607
Odor testing, 607
Oleic acid, 52
Optical activity of polymers, 40
Optical retardation, 303–306
Organoleptic tests, 607
Orientation, 37, 65
Orientation effects in testing, 528
Orientation of film, 256–364, 459–480
  achievement of, 249
  amorphous orientation, 287–299
  biaxial orientation, 462, 471
  birefringence, 242, 247, 299
  characterization of, 264
  definitions, 460

equipment patents for, 468
factors involved in, 460
heat setting after, 462
measurement, 242
patents claiming improvement by,
  464–467
process steps, 478
shrinkage, 243, 248
stress-release, 243, 248
uniaxial orientation, 462, 467
variables that affect, 477
Orientation function for isotactic
  polypropylene, 284
Orientation of melts in extrusion, 232
  uniaxial and biaxial, 232
Osmometry, 788
Owens film friction tester, 449
Oxadiazoles, 56
Oxalic acid, 47, 53
β,β'-Oxydiethyl sulfide, 77
Oxygen-bridged complexes, 210
Oxygen transmission, 533, 603

P

Paraffin wax in cellophane coatings,
  695
Peanut protein, 174
Pendulum impact test for measuring
  impact strength, 578
Pentaerythritol tetrastearate in
  cellophane coatings, 695
Penton, 76–77
Permeability, 532–534, 596–607 (See
  also *Gas transmission.*)
Petition to establish safety of food
  additives, 824–827
Phenyldichlorophosphine, 58, 59
4-Phenylenediamine, 49
Phenylsilsesquioxane, 219
Phosphonitrilic chloride, 221
  compared with silicone resonance,
  223
  conversion to chlorofluoride, 223
  polymers related to, 221, 222
Phosphonoanhydrides, 222
Phosphorus polymers, 220